7-16-73

Organic Compounds in Aquatic Environments

Organic Compounds in Aquatic Environments

Edited by

SAMUEL D. FAUST

and

JOSEPH V. HUNTER

College of Agriculture and Environmental Science
Rutgers—The State University
New Brunswick, New Jersey

MARCEL DEKKER, INC. New York 1971

First Printing: October, 1971
Second Printing: August, 1972

MARCEL DEKKER, INC.
95 Madison Avenue, New York, New York 10016

LIBRARY OF CONGRESS CATALOG CARD NUMBER 72-172938

ISBN 0-8247-1188-2

PRINTED IN THE UNITED STATES OF AMERICA

DEDICATION

To our wives,

with affection,

Anne and Ann.

PREFACE

This volume, "Organic Compounds in Aquatic Environments", found its origin in the Fifth Rudolfs Research Conference held at Rutgers, The State University in July, 1969. These conferences, held periodically, stress an interdisciplinary approach to an important problem of man's enivironment. Also, these conferences are based on the recognition of the need for a greater understanding of the basic sciences upon which applications and solutions to problems in water and wastewater technology depend. In order to achieve this objective an entire conference is arranged so as to stress a particular theme in depth. At the same time it is the purpose to bring together the pure scientist and the applied scientist so that an interdisciplinary cross-fertilization could evolve whereby, on the one hand, the basic concepts are set forth and, on the other, the problem of the environment as they relate to these basic concepts are discussed and brought into sharp focus.

For many years, most recommendations for solutions to problems of environmental pollution were not only narrow in scope but were approached from the art to invention and then to science. As the problems were increased in number and complexity with the vast outpouring of new chemical compounds man finds himself confronted with a dire need to consider total solutions to these problems. However, and most significant, is the fact that man is now confronted in many areas with complex organic contaminants in low concentrations which will become even more complex and numerous in the not-too-distant future. The concern about organic compounds as they relate to tastes and odors, toxicity, carcinogenic effects, etc. - must be carefully weighed with regard to their total impact upon man's environment. In addition, compound persistence, interaction, or, more specifically, synergistic effects must be established. Problems must be approached from the top downward instead of from the bottom upward, simply because man cannot rely on the old safety factor in that there exists in all areas or will exist in all areas sufficiently clean water, clean air, and abundant land to compensate for a lack of adequate control or consideration. We must now reverse our direction and resort initially to science, then to invention and art.

There is little doubt that one of the most important needs in the field of water pollution is additional knowledge as to the nature, distribution, and fate of organic contaminants in man's environment. This position is reinforced by the fact that once complete treatment of all pollution is achieved man will be confronted in large measure with the complex and residual organics associated with increased population and industrial growth. Total solutions to problems are nigh impossible unless the correct questions are asked and the proper solutions evolved.

A. Joel Kaplovsky
Chairman, Department of
Environmental Sciences
Rutgers, The State University,
New Brunswick, New Jersey.

CONTRIBUTORS

OSMAN M. ALY, Campbell Soup Co., Napoleon, Ohio

JULIAN B. ANDELMAN, Graduate School of Public Health, University of Pittsburgh, Pittsburgh, Pennsylvania

*ROBERT A. BAKER, Carnegie - Mellon Institute, Pittsburgh, Pennsylvania

*ARTHUR W. BUSCH, Department of Chemical Engineering, Rice University, Houston, Texas

M. ARIF CAGLAR, U.S. Army Natick Laboratories, Natick, Massachusetts

H. A. G. CHERMIN, Central Laboratory, Staatsmijnen/DSM, Geleen, The Netherlands

RUSSELL F. CHRISTMAN, Department of Civil Engineering, University of Washington, Seattle, Washington

MARGARET O. DAYHOFF, National Biomedical Research Foundation, Silver Spring, Maryland

EGON T. DEGENS, Woods Hole Oceanographic Institution, Woods Hole, Massachusetts

DONALD B. DENNEY, Department of Chemistry, Rutgers, The State University, New Brunswick, New Jersey

M. A. EL-DIB, National Research Center, Dokki, Cario, U.A.R.

*MORRIS B. ETTINGER, The Dow Chemical Co., Midland, Michigan

SAMUEL D. FAUST, Department of Environmental Sciences, Rutgers, The State University, New Brunswick, New Jersey

* Session Chairman of the Fifth Rudolfs Research Conference.

HASSAN M. GOMAA, Kerachemie, Corp., Donmills, Ontario

SOTIRIOS G. GRIGOROPOULOS, Environmental Research Center, University of Missouri, Rolla, Missouri

RICHARD L. GUSTAFSON, Rohm and Haas, Philadelphia, Pennsylvania

ROBERT H. HARRIS, Harvard University, Cambridge, Massachusetts

THOMAS W. HEALY, Department of Physical Chemistry, University of Melbourne, Victoria, Australia

C. W. HOUSTON, University of Rhode Island, Kingston, Rhode Island

JOSEPH V. HUNTER, Department of Environmental Sciences, Rutgers, The State University, New Brunswick, New Jersey

A. J. KAPLOVSKY, Department of Environmental Sciences, Rutgers, The State University, New Brunswick, New Jersey

ARTHUR E. MARTELL, Department of Chemistry, Texas A & M University, College Station, Texas

JOHANN MATHEJA, Massachusetts Institute of Technology, Cambridge, Massachusetts

PERRY L. McCARTY, Department of Civil Engineering, Stanford University, Stanford, California

W. G. MEINSCHEIN, Department of Geology, Indiana University, Bloomington, Indiana

F. J. MICALE, Center for Surface and Coatings Research, Lehigh University, Bethlehem, Pennsylvania

ROGER A. MINEAR, Illinois Institute of Technology, Chicago, Illinois

*JAMES J. MORGAN, California Institute of Technology, Pasadena, California

WALTER J. NICKERSON, Institute of Microbiology, Rutgers, The State University, New Brunswick, New Jersey

E. D. OWEN, Department of Chemistry, University College of South Wales and Monmouthshire, Cardiff, Wales

JOHN PALEOS, Rohm and Haas, Philadelphia, Pennsylvania

V. C. ROSE, University of Rhode Island, Kingston, Rhode Island

JOSEPH ROSEN, Department of Agricultural Chemistry, Rutgers, The State University, New Brunswick, New Jersey

*R. R. SAUERS, Department of Chemistry, Rutgers, The State University, New Brunswick, New Jersey

M. SCHNITZER, Soil Research Institute, Canada Department of Agriculture, Ottawa, Ontario

ALVIN SIEGEL, Southampton College of Long Island University, Southampton, New York

JOHN W. SMITH, Esso Research and Engineering Co., Florham Park, New Jersey

ELISABETH STUMM - ZOLLINGER, Harvard University, Cambridge, Massachusetts

M. J. SUESS, Graduate School of Public Health, University of Pittsburgh, Pittsburgh, Pennsylvania

A. R. THOMPSON, University of Rhode Island, Kingston, Rhode Island

*COOPER H. WAYMAN, Department of Chemistry, Colorado School of Mines, Golden, Colorado

*WALTER J. WEBER, JR., University of Michigan, Ann Arbor, Michigan

P. M. WILLIAMS, Institute of Marine Resources, University of California, San Diego, California

A. C. ZETTLEMOYER, Center for Surface and Coatings Research, Lehigh University, Bethlehem, Pennsylvania

ACKNOWLEDGMENTS

Many people contributed to the success of the Fifth Rudolfs Research Conference. It would only prolong the length of this book to name everyone. The editors are indebted, however, for the cooperation and services provided by Dr. Richard H. Merritt, Director, Resident Instruction, College of Agriculture and Environmental Science and Roger R. Locandro, Assistant Director, Resident Instruction and their secretarial staff. All arrangements for the conference were provided by this office. Many graduate students (they had no choice), technicians, and secretaries of the Department of Environmental Sciences are acknowledged for their excellent cooperation. Mrs. Carl J. Schaefer, Mrs. Samuel D. Faust, and Miss Donna Novak provided excellent typing services. As no financial support for this conference was provided by any Federal agency, the editors are especially appreciative of the administrative agility by which Dr. A. Joel Kaplovsky, Chairman, Department of Environmental Sciences, provided the money. We are also appreciative of the splendid cooperation of Marcel Dekker, Inc., N.Y.

Any successful conference evolves from excellent speakers. Their chapters in this book provide the evidence. Spirited discussions followed many of the papers. The editors are appreciative of the following session chairmen who refereed the arguments: Mr. Morris B. Ettinger, The Dow Chemical Co. (whose opening monologue was, as usual, hilarious); Dr. Walter J. Weber, Jr., University of Michigan; Professor A.W. Busch, Rice University; Dr. James J. Morgan, California Institute of Technology; Dr. Robert A. Baker, Senior Fellow, Carnegie-Mellon Institute; Dr. Cooper H. Wayman, Colorado School of Mines; and Dr. R. R. Sauers, Rutgers, The State University.

New Brunswick, N. J.
August, 1970

Samuel D. Faust
Joseph V. Hunter

CONTENTS

Organic Compounds in Aquatic Environments

CHAPTER 1

ORIGIN OF ORGANICS FROM INORGANICS

M. O. Dayhoff

National Biomedical Research Foundation

I. INTRODUCTION

Prerequisite to an understanding of pollution is an understanding of the world of thermodynamic equilibrium; a world very different from the unstable one we know, in which efficient biocatalysts selectively channel the energy from sunlight for the purposes of living things.

What is the contribution of equilibrium processes to pollution? Thermodynamic or chemical equilibrium gives the most stable state of a system. There are assumed to be no practical barriers to the formation of any compounds; the distribution of molecular species is independent of the rate or nature of the specific reactions by which equilibrium was reached. All real systems have a tendency to move in the direction of equilibrium when activation energy barriers are surmounted by the thermal motion of the molecules or by catalysts.

There are a few systems which can be closely approximated by equilibrium, particularly where long periods of time, high temperatures, or suitable catalysts are involved. These include: combustion product mixes, volcanic gases, the hot atmosphere of Venus and other astronomical systems, and possibly the evolving solar system and the prebiological earth. Other systems, including the present day earth, tend to change in the direction of equilibrium when possible.

Our laboratory became interested in the subject of equilibrium systems of organic compounds through our interest in the origin of life. What were the compounds available in quantity during the early evolution of life in the strange and unfamiliar reducing atmosphere? E.R. Lippincott, R.V. Eck, G. Nagarajan, and the author made a survey of the equilibrium concentrations of a great number of organic compounds in gaseous systems of varying carbon, hydrogen, oxygen, nitrogen, phosphorous, sulfur, and chlorine compositions over a wide range of temperatures and pressures (1). The calculations were extended to aqueous solutions by C. M. Park, G. Atkinson, D. W. Ebdon, D. Wallace, and the author (2). The results from these surveys are applicable to the prediction of the organic compounds which are currently originating in our environment because they are stable at equilibrium. The nature of these products suggests measures which can help to solve the pollution problem.

II. COMPUTER METHOD

A very general computer method (1,2), originally suggested by White, Johnson, and Dantzig (3) and extended to multiphase systems by Boynton (4), was used in our calculations. For this approach one needs only the chemical formula of each compound and its standard free energy of formation. Each system is defined by its elemental composition as well as by the compounds which can occur in it. The question is asked, "How do the elements distribute themselves among these compounds?" If, in the laboratory, one started with a system of pure benzene, or one of pure acetylene or one containing hydrogen and graphite in an atomic ratio of 1:1, and let each come to thermodynamic equilibrium at the same temperature and pressure, the final system would be the same in each case.

The mathematical procedure for finding the equilibrium distribution of compounds is based upon the fact that, at equilibrium, the total free energy of the system is a minimum. This total free energy, G, is the sum of a contribution from each of the constituent chemical species in the mixture, $g_{i,\alpha}$; the contribution of each species depends on its standard free energy of formation, its concentration, and the temperature and pressure of the system. Thus:

$$G = \sum_i \sum_\alpha g_{i,\alpha} \qquad (1\text{-}1)$$

where i ranges over all compounds and α ranges over all phases.

$$g_{i,\alpha} = X_{i,\alpha} \left\{ G^o_{i,\alpha} + \delta_\alpha RT\ln P + RT\ln\left(\frac{X_{i,\alpha}}{\bar{X}_\alpha}\right)\right\} \qquad (1\text{-}2)$$

where $G^o_{i,\alpha}$ is the standard free energy of formation of compound i in phase α, $X_{i,\alpha}$ is the number of moles of compound i in phase α, $\bar{X}_\alpha$ is the total number of moles in phase α, δ_α is 1 for gas and 0 for liquids and solids, R is the gas constant, 1.986 cal/mole-deg, T is the temperature in degrees Kelvin, and P is the pressure of the system in atmospheres.

The total free energy of the system is minimized subject to the constraints that the quantity of each element remains fixed and the total number of moles in each phase is the sum of the mole fractions of the compounds in that phase.

The mathematical problem reduces to the solution of a system of linear simultaneous equations involving the standard free energies of formation of the major compounds, their elemental compositions, the temperature, and the pressure. The number of equations is equal to the number of elements plus the number of phases. Any number of chemical compounds may be considered, so long as all of the most stable species are included. The stable species occur in nature and are familiar to chemists, so that it is no problem to think of and include all of them.

In obtaining the equilibrium balance, an iterative procedure is followed by the computer. Initially, the concentrations of some of the more important compounds are chosen to comprise the desired elemental composition of the system, while the concentrations of the other compounds are assumed to be negligible. This system constitutes a first approximation to the solution. At each iteration thereafter, a new distribution of compounds is calculated, closer to the equilibrium than the previous one. Eventually, the correct solution is obtained.

Table 1 shows a typical result for a system containing 40% hydrogen, 5% oxygen, 5% carbon, and 50% nitrogen at one atmosphere pressure and 298.15°K. Ideal gaseous and aqueous phases are present.

The programs (1,2) will consider system after system, producing pages and pages of output. The process of understanding and drawing conclusions from results in this form is very tedious. Graphical methods, therefore, have been developed for the reorganization and display of the results. One interesting sort of question that arises is, "Does methanol ever become an important chemical species at any elemental composition?" In order to conveniently consider questions like this, ternary plots are frequently employed. Figure 1 shows such a coordinate system for the elements carbon, hydrogen, and oxygen. Every system composed of these three elements can be plotted at one point on this diagram. The plot point for a system containing a single pure element is at the corresponding corner of the diagram. The percent of a given element in a system is given by the altitude from the system plot point to the side opposite the corresponding corner. Any possible composition can be

TABLE 1

Thermodynamic Equilibrium System Computed

	Pressure	Temperature
	1.0 Atm.	298.15^{o}K.

Elemental Composition

Element or Charge	Mole Fraction	Partial Molal Free Energy of Element or Charge
N	.50	- .16
H	.40	- 2.97
C	.05	-10.46
O	.05	-89.87
+	$.02 \times 10^{-4}$	-24.68
-	$.02 \times 10^{-4}$	24.68

Phase Composition

	Mole Fraction
Gas (G)	.877
Aqueous (L)	.123

Major Constituents

		Mole Fraction of Total System
N_2	G	.64
CH_4	G	.14
H_2O	L	.11
NH_3	G	$.78 \times 10^{-1}$
H_2O	G	$.25 \times 10^{-1}$
NH_3	L	$.11 \times 10^{-1}$
H_2	G	$.23 \times 10^{-2}$
OH^-	L	$.20 \times 10^{-4}$
NH_4^+	L	$.20 \times 10^{-4}$

TABLE 1 (continued)

All Constituents	Phase	Standard Free Energy of Formation*	Mole Fraction Composition: Each Phase	Mole Fraction Composition: Total System	Aqueous Phase Molality
N_2	G	0.000	.73	.64	
CH_4	G	-12.130 a	.16	.14	
H_2O	L	-56.688 a	.91	.11	
NH_3	G	-3.940 a	$.89 \times 10^{-1}$	$.78 \times 10^{-1}$	
H_2O	G	-54.635 a	$.28 \times 10^{-1}$	$.25 \times 10^{-1}$	
NH_3	L	-3.971 a	$.93 \times 10^{-1}$	$.11 \times 10^{-1}$	5.72
H_2	G	0.000	$.26 \times 10^{-2}$	$.23 \times 10^{-2}$	
$\bar{O}H$	L	-35.215 a	$.17 \times 10^{-3}$	$.20 \times 10^{-4}$	$.10 \times 10^{-1}$
NH_4^+	L	-16.591 a	$.16 \times 10^{-3}$	$.20 \times 10^{-4}$	$.10 \times 10^{-1}$
N_2	L	6.730 e	$.84 \times 10^{-5}$	$.10 \times 10^{-5}$	$.52 \times 10^{-3}$
CH_4	L	-5.840 a	$.38 \times 10^{-5}$	$.47 \times 10^{-6}$	$.23 \times 10^{-3}$
H_2	L	6.613 e	$.37 \times 10^{-7}$	$.46 \times 10^{-8}$	$.23 \times 10^{-5}$
CO_3^{2-}	L	-123.810 a, c	$.39 \times 10^{-9}$	$.48 \times 10^{-10}$	$.24 \times 10^{-7}$
HCO_3^-	L	-137.900 a, c	$.82 \times 10^{-11}$	$.10 \times 10^{-11}$	$.50 \times 10^{-9}$
CO_2	G	-94.258 a	$.33 \times 10^{-13}$	$.29 \times 10^{-13}$	
$HCOO^-$	L	-81.490 a	$.38 \times 10^{-13}$	$.47 \times 10^{-14}$	$.24 \times 10^{-11}$
H^+	L	+2.379	$.18 \times 10^{-13}$	$.22 \times 10^{-14}$	$.11 \times 10^{-11}$
$CO(NH_2)_2$	L	-46.340 c	$.13 \times 10^{-14}$	$.16 \times 10^{-15}$	$.78 \times 10^{-13}$
CH_3NH_2	G	7.670 a	$.20 \times 10^{-16}$	$.18 \times 10^{-16}$	

TABLE 1 (continued)

All Constituents	Phase	Standard Free Energy of Formation*	Mole Fraction Composition: Each Phase	Mole Fraction Composition: Total System	Aqueous Phase Molality
CH_3NH_2	L	7.320 a	$.37 \times 10^{-16}$	$.45 \times 10^{-17}$	$.23 \times 10^{-14}$
CO_2	L	-89.883 a	$.20 \times 10^{-16}$	$.25 \times 10^{-17}$	$.12 \times 10^{-14}$
CH_3COO^-	L	-86.640 c	$.17 \times 10^{-16}$	$.21 \times 10^{-17}$	$.10 \times 10^{-14}$
$CH_3NH_3^+$	L	-7.170 a	$.15 \times 10^{-17}$	$.19 \times 10^{-18}$	$.94 \times 10^{-16}$
H_2CO_3	L	-143.040 d	$.47 \times 10^{-19}$	$.58 \times 10^{-20}$	$.29 \times 10^{-17}$
CH_3OH	G	-38.720 a	$.46 \times 10^{-20}$	$.40 \times 10^{-20}$	
CH_3OH	L	-39.320 c	$.13 \times 10^{-19}$	$.16 \times 10^{-20}$	$.80 \times 10^{-18}$
HCOOH	L	-86.600 a	$.21 \times 10^{-21}$	$.26 \times 10^{-22}$	$.13 \times 10^{-19}$
HCOOH	G	-83.850 f	$.20 \times 10^{-23}$	$.18 \times 10^{-23}$	
CH_3COOH	L	-93.130 c	$.97 \times 10^{-24}$	$.12 \times 10^{-24}$	$.59 \times 10^{-22}$
HCHO	L	-28.600 c	$.65 \times 10^{-24}$	$.85 \times 10^{-26}$	$.41 \times 10^{-23}$
C_2H_5OH	G	-40.300 c	$.50 \times 10^{-26}$	$.43 \times 10^{-26}$	
C_2H_5OH	L	-41.000 c,g	$.16 \times 10^{-25}$	$.20 \times 10^{-26}$	$.99 \times 10^{-24}$
HCHO	G	-26.265 b	$.13 \times 10^{-26}$	$.11 \times 10^{-26}$	
O_2	G	0.000	.0	.0	
NO_3^-	L	-24.230 a	.0	.0	.0
HNO_3	G	-17.870 a	.0	.0	
O_2	L	+6.330 e	.0	.0	.0

* See Ref. 5a, b, c, d, e, f, and g.

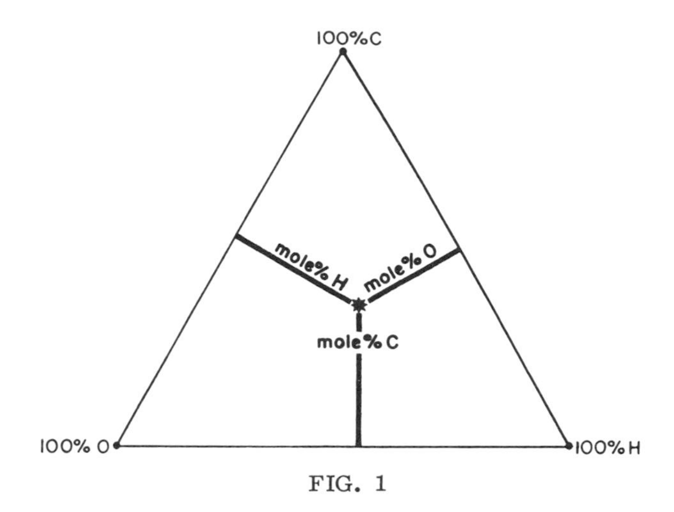

FIG. 1

The ternary plot. The plotted point shows a system with 35% carbon, 30% oxygen, and 35% hydrogen.

represented on the diagram. A very interesting and convenient property of this representation makes it possible to predict the effect on a system of adding or removing a compound. The resulting system will lie somewhere along a line drawn through the plot point of the initial system and that of the element or compound removed. Figure 2 shows the direction of change of a system, P, upon the loss of oxygen. The length of the vector to the new system plot point depends on the relative amount of oxygen lost.

III. GENERAL FEATURES OF THE C:H:O SYSTEM

To explore the general characteristics of the C:H:O system in the range of interest, consider a gaseous system at 500°K. The details of such a system are intermediate between two circumstances of concern in pollution studies, combustion systems, and ecological systems. The main features of gaseous phases of a given elemental composition are similar over a broad range of temperatures and pressures. However, a given system may not always be stable as a single gas phase. At room temperature, many systems which are gases at high temperature, exist as two or three phases: an aqueous phase, a gas phase containing only a small amount of water vapor, and sometimes a graphite phase. The constituents of the gas phase are very similar to those in gas phases of the same total composition at other temperatures.

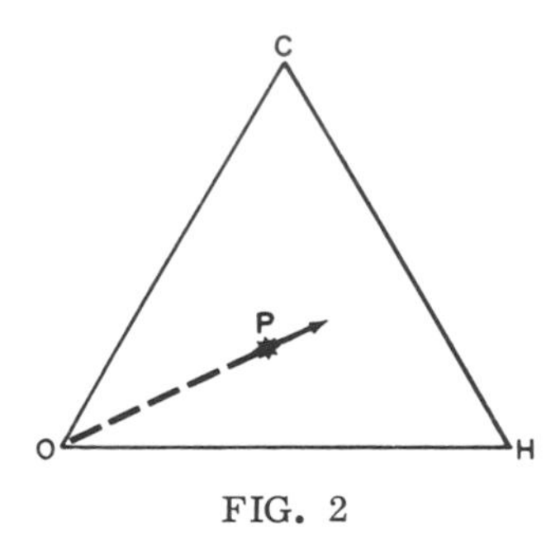

FIG. 2

Vector addition of systems. When oxygen is lost from a system, P, the new system will lie on a line drawn through the oxygen point and the system point, as indicated by the end of the vector. In general, all combinations of two systems in any proportions will lie on the line through both of them.

Figure 3 summarizes the results at 500°K, and .2 to 50 atm. The only gaseous compounds ever present in significant concentrations are CO_2, CH_4, H_2O, H_2, and O_2. At higher temperatures, CO would also become important. When the plot points of neighboring major compounds are connected, the ternary diagram is divided into triangular regions. In each region there are three major constituents, the compounds plotted at the corners of the triangles. All other organic compounds are minor constituents at most. In the oxygen corner, organic compounds are so dilute that no molecules at all should be present in a typical laboratory size system. In the earth's atmosphere, then, all organics would be oxidized. Over other regions, where the system is reducing, organic compounds are present in small concentrations.

At complete equilibrium, graphite forms in systems of high carbon content. The curved line shows the phase boundary. Systems above this line are composed of two phases, a gas phase of some composition along the phase boundary line and graphite. These results lead to the conclusion that, except for CH_4, CO_2, and a little CO, no organic chemical is stable. The whole field of organic chemistry, therefore, deals with the transitions of metastable states of the atoms. It is little wonder that yields of reactions are usually low, and never perfect. At every opportunity the compounds degrade to more stable configurations.

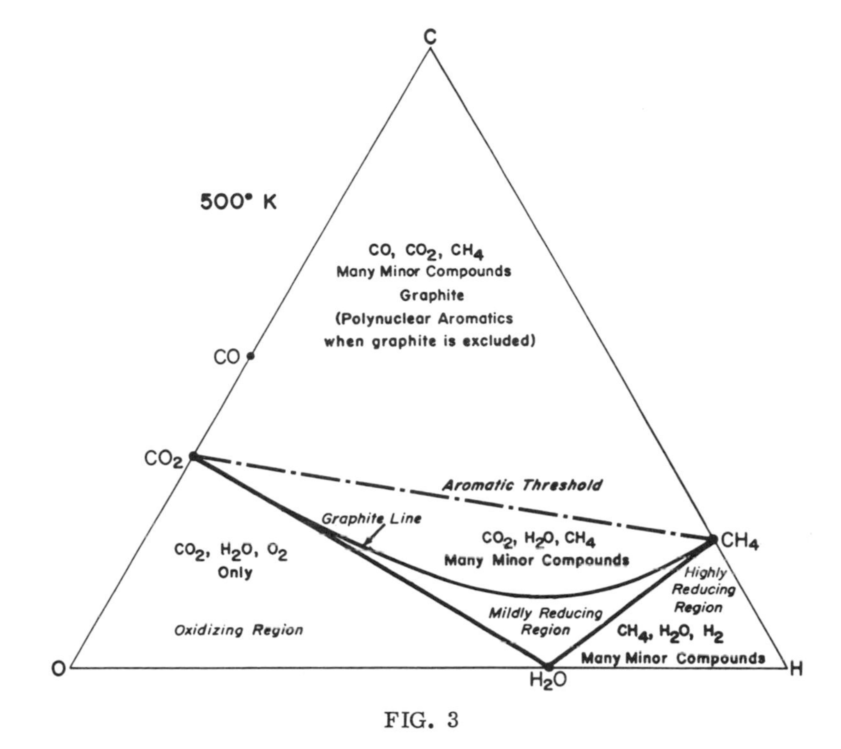

FIG. 3

Summary of compounds present at equilibrium in atmospheres of varying elemental proportions at 500°K. and pressures ranging from .2 to 50 Atm. The points corresponding to pure atmospheres of the major compounds are indicated and the important compounds in different regions are shown. The solid curve indicates the phase boundary above which graphite is stable. The activation energy for the formation of graphite is so high that under many practical conditions it does not form in significant quantities. The gaseous systems above the curve may still approximate equilibrium mixtures. Polynuclear aromatics are formed in high concentration in these systems.

IV. C:H:O SYSTEMS, LIMITED EQUILIBRIUM

Under many experimental conditions, graphite does not form when expected because of its high activation energy of formation. For example, whenever one processes organic chemicals, a dark, tarry

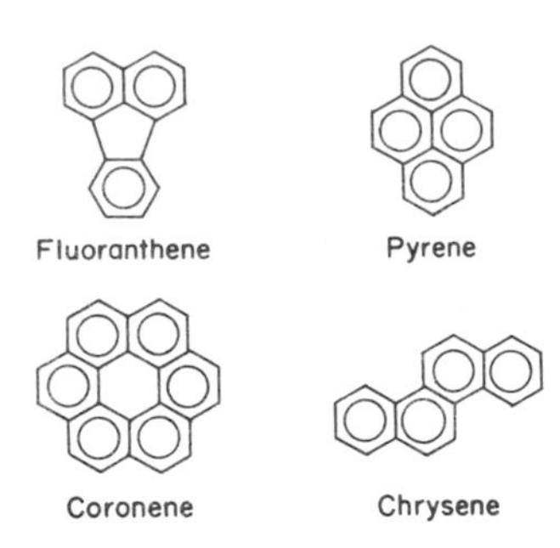

FIG. 4

Aromatic compounds detected experimentally in products of gaseous plasma discharge reactions. Some of these, and related compounds which would be expected to form also, are carcinogenic.

mass is prone to accumulate. This mass is not graphite but is a mixture of organic compounds soluble in such liquids as benzene. Even when graphite can not form, other compounds may still be approximately in equilibrium with each other. To explore such systems theoretically, a limited equilibrium has been used in which it is presumed that the rates of all reactions leading to graphite are zero, while all the compounds included in the computation equilibrate relatively rapidly. This gross approximation is useful and relevent. In such systems, above the CH_4-CO_2 line, it is possible to calculate that major quantities of one class of organic compounds, the aromatics, should be stable. Higher homologues are favored over benzene and naphthalene. As an approximation to the behavior of this class, a composite called "asphalt" is included whose free energy is derived from a polynuclear aromatic of formula $C_{22}H_{12}$. This free energy has been modified so that the concentration of polynuclear aromatics calculated will be the sum of concentrations for 100 isomers of this composition. In the "aromatic" region, other organics are still of minor importance, although their concentrations usually increase gradually as systems become richer in carbon.

Dr. Lippincott and colleagues have subjected organic gases of varying elemental compositions to a radio frequency plasma discharge (1). This energetic field permits the elements to dissociate and recombine into more stable configurations. Results have qualitatively confirmed our computations. Copious amounts of a gray precipitate were obtained in the regions where our computations predict aromatics. On analysis, this precipitate yielded quantities of polynuclear aromatics. Those shown in Figure 4 were separated and positively identified. Many other similar structures must have been present also.

Figures 5a and 5b display the concentrations of two particular compounds, methane and methanol, in systems of many possible compositions. Methane is a major constituent and methanol is typical of the minor organic constituents. In the oxygen region, organic compounds are not present (concentration $< 10^{-38}$ mole fraction). Methane has a significant concentration in the three triangular regions which include its plot point. Methanol has a very low concentration even in the reducing regions. It is most important in systems whose total composition is close to the molecular composition of methanol itself. The diagram shows this point with a star. Even here, methanol has a concentration of only 0.4×10^{-11} mole fraction. Higher homologues of methanol, such as ethanol or propanol, are even less stable than methanol; the higher the homologue, the less stable is the compound.

Other organic compounds behave in a manner similar to methanol. The maximum mole fraction concentrations of the more important compounds of carbon, hydrogen, and oxygen in the set of systems plotted in Figure 5 are shown in Table 2.

V. C:H:O:N:S:Cl SYSTEMS

Many organic compounds of importance contain nitrogen. In C:H:O:N systems, N_2 is usually a major constituent, while ammonia can also be important in the highly reducing region and the oxides of nitrogen in the oxidizing region. Organic compounds containing nitrogen are very minor constituents even in the reducing regions. Table 3 shows the maximum concentrations of nitrogen-containing compounds in systems with the same proportions of C:H:O as those shown in Figure 5. Nitrogen constituted 80% of each system, the total pressure was 1 atm., and the temperature was 500°K.

When small amounts of sulfur are included, the main constituents containing sulfur are SO_3 in the oxidizing corner, SO_2 along the CO_2 - H_2O line, and H_2S in the reducing regions. COS becomes important in the aromatic regions while CS_2 is a minor constituent there when graphite is forbidden to form. The organics containing sulfur occur in low concentrations. However, they are slightly more abundant than their oxygen analogues. Chlorinated organics are less stable by several orders of magnitude than the corresponding hydrogen analogues. The major constituents containing chlorine are HCl in all systems and Cl_2 in the oxidizing region. Figure 6 shows the regions where the various major constituents containing nitrogen, sulfur, and chlorine are important.

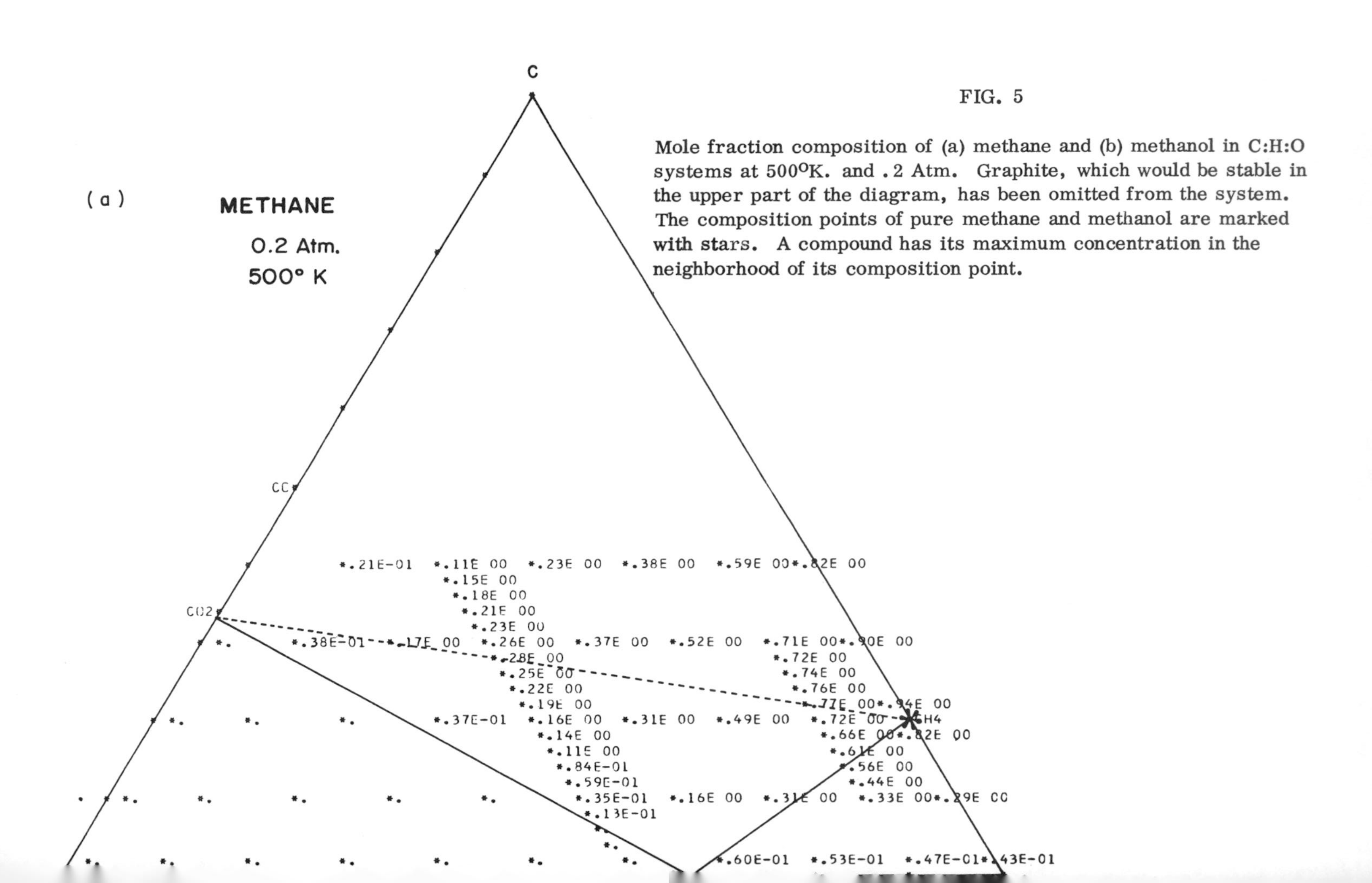

FIG. 5

Mole fraction composition of (a) methane and (b) methanol in C:H:O systems at 500°K. and .2 Atm. Graphite, which would be stable in the upper part of the diagram, has been omitted from the system. The composition points of pure methane and methanol are marked with stars. A compound has its maximum concentration in the neighborhood of its composition point.

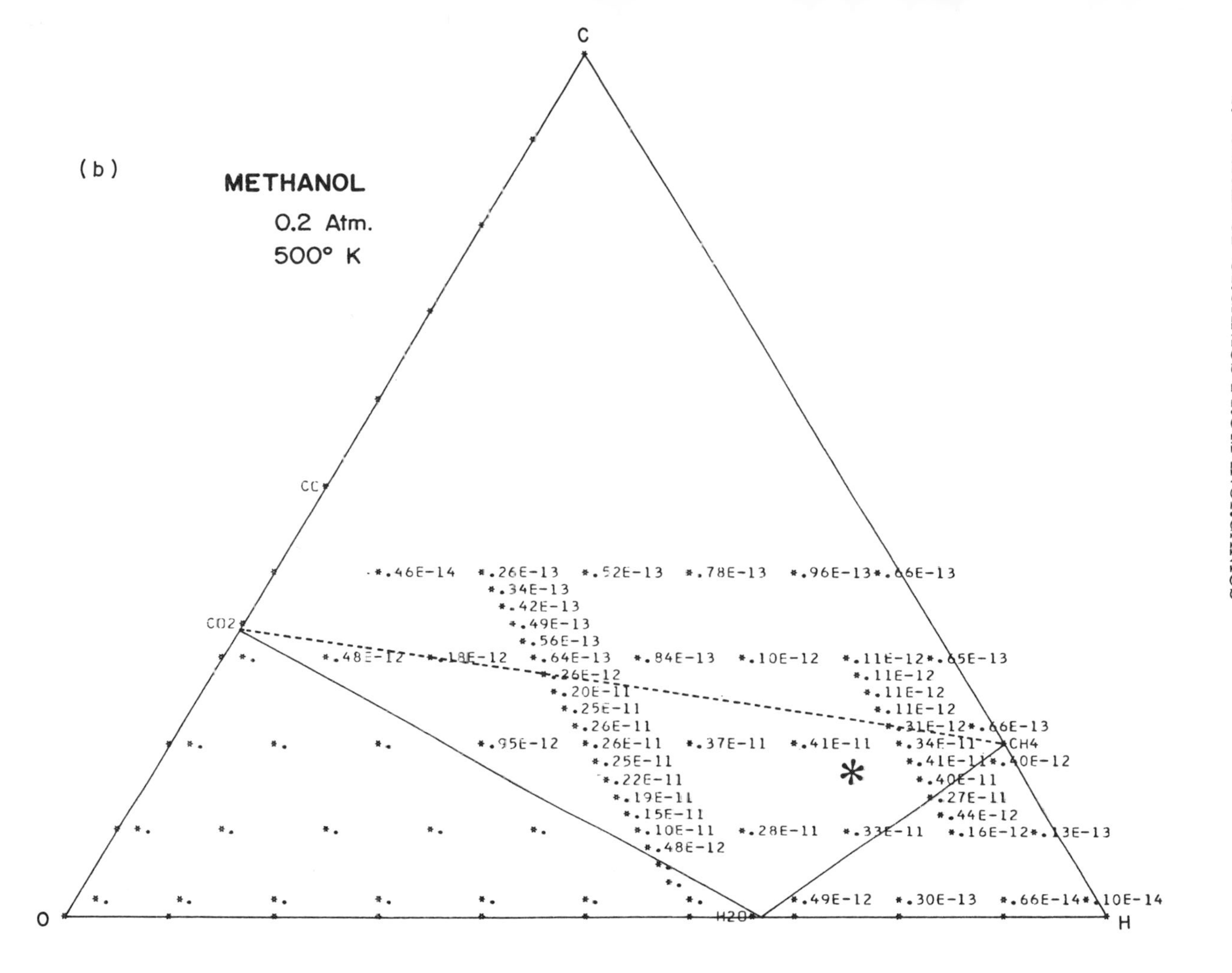
(b)
METHANOL
0.2 Atm.
500° K
C
O
H
CC
CO2
CH4
H2O
.46E-14 .26E-13 .52E-13 .78E-13 .96E-13 .66E-13
.34E-13
.42E-13
.49E-13
.56E-13
.48E-12 .18E-12 .64E-13 .84E-13 .10E-12 .11E-12 .65E-13
.26E-12
.11E-12
.20E-11
.11E-12
.25E-11
.11E-12
.26E-11
.31E-12 .66E-13
.95E-12 .26E-11 .37E-11 .41E-11 .34E-11
.25E-11
.41E-11 .40E-12
.22E-11
.40E-11
.19E-11
.27E-11
.15E-11
.44E-12
.10E-11 .28E-11 .33E-11 .16E-12 .13E-13
.48E-12
.49E-12 .30E-13 .66E-14 .10E-14

TABLE 2

Maximum Mole Fraction of Selected Organic Compounds in Systems with the C:H:O Ratios Plotted in Figure 5
$500^{\circ}K.$, .2 Atm.

	Graphite Excluded	Graphite Included
Methane	.9	.8
Asphalt	.1	.0
Ethane	$.1 \times 10^{-3}$	$.3 \times 10^{-6}$
Hexane	$.2 \times 10^{-15}$	$.4 \times 10^{-27}$
Cyclohexane	$.1 \times 10^{-13}$	$.1 \times 10^{-27}$
Ethylene	$.2 \times 10^{-7}$	$.5 \times 10^{-12}$
Acetylene	$.2 \times 10^{-15}$	$.4 \times 10^{-22}$
Methanol	$.4 \times 10^{-11}$	$.3 \times 10^{-11}$
Ethanol	$.3 \times 10^{-14}$	$.4 \times 10^{-15}$
Acetaldehyde	$.4 \times 10^{-11}$	$.4 \times 10^{-14}$
Formic Acid	$.2 \times 10^{-9}$	$.2 \times 10^{-9}$
Acetic Acid	$.4 \times 10^{-9}$	$.3 \times 10^{-10}$
Oxalic Acid	$.2 \times 10^{-12}$	$.9 \times 10^{-13}$
Benzene	$.3 \times 10^{-3}$	$.2 \times 10^{-23}$
Naphthalene	$.3 \times 10^{-3}$	.0
Xylene	$.2 \times 10^{-6}$	$.2 \times 10^{-31}$
Furan	$.1 \times 10^{-16}$	$.1 \times 10^{-28}$

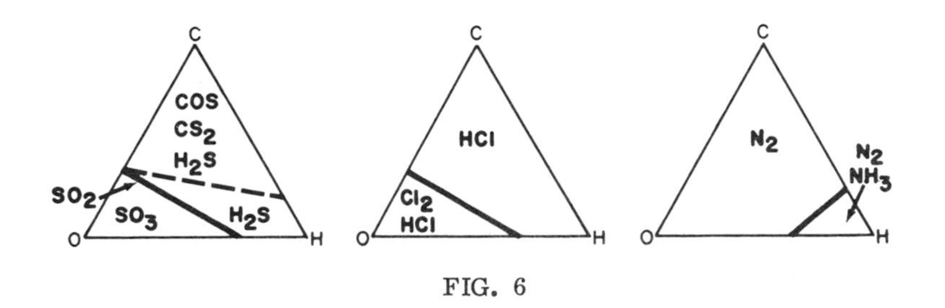

FIG. 6

Main compounds containing S, Cl, and N when these elements are trace constituents of C:H:O systems.

TABLE 3

Maximum Mole Fraction of Nitrogenous Organic Compounds in Systems with the C:H:O Ratios of Figure 5
500°K., 1 Atm., 80% N Atoms

	Graphite Excluded	Graphite Included
Nitrogen	.9	.9
Ammonia	$.2 \times 10^{-1}$	$.2 \times 10^{-1}$
NO_2	$.3 \times 10^{-7}$	$.3 \times 10^{-7}$
Nitric Acid	$.1 \times 10^{-9}$	$.1 \times 10^{-9}$
Hydrogen Cyanide	$.2 \times 10^{-9}$	$.8 \times 10^{-13}$
Methylamine	$.2 \times 10^{-12}$	$.2 \times 10^{-12}$
Pyrrole	$.6 \times 10^{-17}$	$.2 \times 10^{-29}$
Pyridine	$.9 \times 10^{-13}$	$.1 \times 10^{-29}$
Formamide	$.5 \times 10^{-21}$	$.2 \times 10^{-21}$
Ethanolamine	$.3 \times 10^{-27}$	$.8 \times 10^{-28}$
Glycine	$.5 \times 10^{-22}$	$.2 \times 10^{-22}$

VI. ELIMINATION OF CARCINOGENIC POLYNUCLEAR AROMATICS

The most important case in which equilibrium organic products contribute to pollution arises from the processing of carbon rich gases, for example, the exhaust mixtures from internal combustion engines or industrial flues. Polynuclear aromatics, including carcinogenic compounds, are stable products under certain conditions. They will form whenever suitable catalysts or sufficient heat provide the proper environment. Elimination of these products requires a change in combustion conditions or in the elemental composition of the combustion mixture.

Consider a system of octane, whose composition point is shown in Figure 7. A combustion mixture of octane and air would lie somewhere on the line from the octane point to the oxygen corner. If the mixture is overly rich in octane, as at point A, the system will lie in the aromatic region and carcinogenic polynuclear aromatics would be expected products of combustion. In order to prevent the formation of these carcinogens, the elemental composition of such a mixture must be removed from the aromatic region. This may be effected for practical purposes by the addition to the combusion mixture of more oxygen leading to a mixture of composition B, or by the addition of

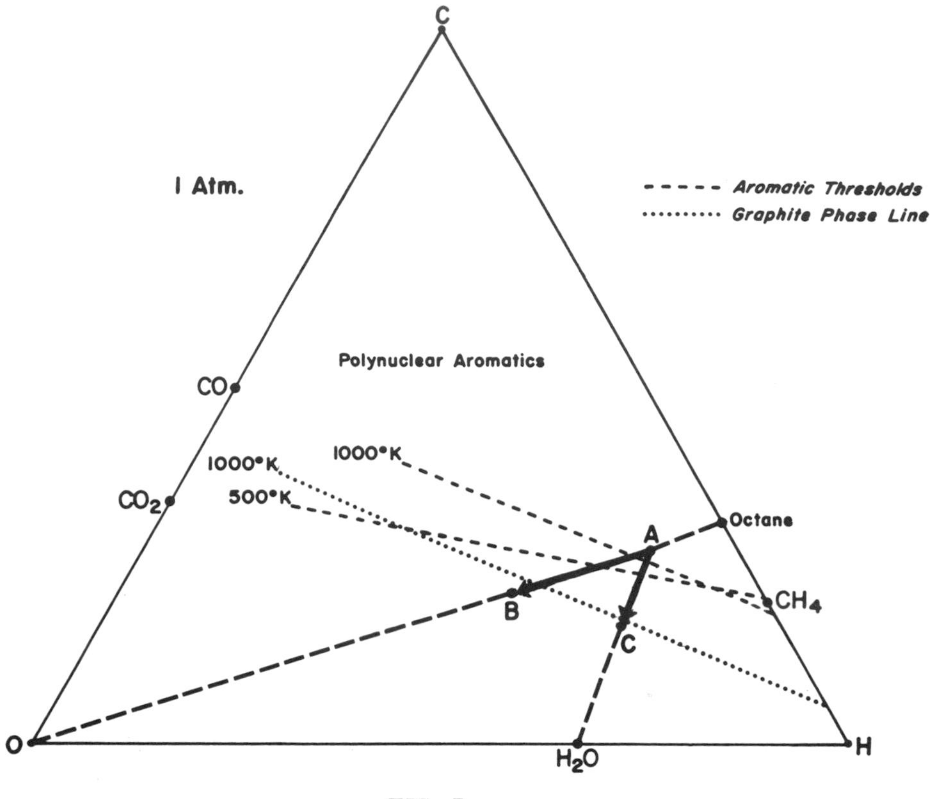

FIG. 7

Elimination of carcinogenic hydrocarbons from combustion products. A mixture shown at point A, for example, a rich mixture of octane and oxygen, would be expected to form polynuclear aromatics as stable combustion products. If more oxygen or water was added to the system, it would then be represented by composition points B or C. Both of these are outside the region where asphalt forms. The dashed lines show the position of the aromatic threshold at 500° and 1000°K. and 1 Atm. The many mixtures between the two lines would form polynuclear aromatics at the lower temperatures but not at the higher one. The elevation of the effective combustion temperature in such systems, as for example by the introduction of a platinum catalyst, would also eliminate aromatic formation. The dotted line shows the graphite phase boundary. Graphite could form above this if catalysts were present.

water leading to a mixture of composition C, or by the addition of both oxygen and water. Figure 7 shows the effect of the addition of these compounds. The vectors to the new systems now end outside the aromatic region. Carcinogenic hydrocarbons are no longer stable products of these systems nor is graphite a stable product at high temperatures.

The position of the aromatic line is temperature dependent. Some systems which form carcinogens at low effective combustion temperatures would not form them at higher temperatures. The position of the aromatic line is almost independent of pressure in the range 1 to 100 atmospheres.

VII. EQUILIBRIUM IN AQUEOUS SOLUTIONS

It is conceivable that an aqueous environment might further stabilize certain hydrophilic organics or ionic species, leading to a significant increase in the abundance of organics in equilibrium systems. The order of magnitude of such effects turns out to be very small in all cases which have been examined.

Figure 8 shows the main characteristics of the phase diagram for the C:H:O system. In the upper part of the diagram, above the aromatic threshold, no liquid water is present, nor is the aqueous phase present in systems near the sides of the triangle below the aromatic line. Because water and polynuclear aromatics are never simultaneous major constituents in the same system, aqueous phases do not contain appreciable amounts of polynuclear aromatics. A system such as that indicated by point P, with overall composition in the lower central region of the diagram, is really composed of two phases, one an aqueous phase with composition close to that of pure water, and one a gaseous phase whose composition is given by the intersection, Q, with the phase line, of the line through water and the system point.

Table 4 shows the equilibrium balance in typical liquid systems of carbon, hydrogen, oxygen, and nitrogen. Nitrogen has an abundance equal to the sum of the abundances of C, H, and O. In systems containing free oxygen gas, the major carbonaceous constituent is CO_2. Formic acid, one of the most stable organics, occurs with a molal concentration of only 10^{-46}. Nitric acid is stable at about .1 molal, a concentration sufficiently high so that, at equilibrium, most of the oxygen in the earth's atmosphere would be found as HNO_3 dissolved in the ocean. In the hydrogen rich region, ammonia is important in both aqueous and gas phases. All organics are present in low concentration, significantly below the detection level of the human nose. In the central reducing region, the organics reach their highest concentrations, the formate

TABLE 4

Equilibria of Organic Compounds in Aqueous Solutions
298.15°K., 1 Atm., 50% N

	A Reducing	B Weakly Reducing	C Weakly Reducing	D Oxidizing
Elemental Composition; Mole Fraction				
C	.050	.050	.050	.050
H	.375	.300	.300	.050
O	.075	.150	.150	.400
N	.500	.500	.500	.500
Phase; Mole Fraction				
Graphite	.0	.108	----	.0
Gas	.823	.579	.774	.976
Aqueous Solution	.177	.313	.226	.024
Compound Concentrations				
Solid Phase				
Graphite	not stable	present	forbidden to form	not stable
Gas Phase; Mole Fraction				
H_2	$.98 \times 10^{-3}$	$.19 \times 10^{-5}$	$.19 \times 10^{-5}$	$.49 \times 10^{-41}$
O_2	.0	.0	.0	.30
CH_4	.16	$.27 \times 10^{-2}$	$.81 \times 10^{-1}$	.0

TABLE 4 (continued)

	A Reducing	B Weakly Reducing	C Weakly Reducing	D Oxidizing
CO_2	$.20 \times 10^{-11}$	$.27 \times 10^{-2}$	$.81 \times 10^{-1}$	.11
H_2O	.03	$.31 \times 10^{-1}$	$.31 \times 10^{-1}$	.03
N_2	.79	.96	.81	.55
NH_3	.02	$.19 \times 10^{-5}$	$.18 \times 10^{-5}$	.0
HNO_3	.0	.0	.0	$.35 \times 10^{-8}$
CH_3OH	$.14 \times 10^{-19}$	$.12 \times 10^{-18}$	$.37 \times 10^{-17}$	.0
C_2H_5OH	$.40 \times 10^{-25}$	$.34 \times 10^{-23}$	$.29 \times 10^{-20}$	.0
HCHO	$.10 \times 10^{-25}$	$.49 \times 10^{-22}$	$.14 \times 10^{-20}$	.0
CH_3CHO	$.31 \times 10^{-28}$	$.14 \times 10^{-23}$	$.12 \times 10^{-20}$	.0
HCOOH	$.46 \times 10^{-22}$	$.12 \times 10^{-15}$	$.35 \times 10^{-14}$	$.13 \times 10^{-49}$
CH_3NH_2	$.13 \times 10^{-16}$	$.11 \times 10^{-19}$	$.30 \times 10^{-18}$	.0
Aqueous Solution; Molality				
H^+	$.21 \times 10^{-11}$	$.14 \times 10^{-7}$	$.81 \times 10^{-7}$	$.95 \times 10^{-1}$
OH^-	$.48 \times 10^{-2}$	$.71 \times 10^{-6}$	$.12 \times 10^{-6}$	$.11 \times 10^{-12}$
CO_2	$.70 \times 10^{-13}$	$.94 \times 10^{-4}$	$.28 \times 10^{-2}$	$.38 \times 10^{-2}$
H_2CO_3	$.18 \times 10^{-15}$	$.24 \times 10^{-6}$	$.71 \times 10^{-5}$	$.98 \times 10^{-5}$
HCO_3^-	$.14 \times 10^{-7}$	$.29 \times 10^{-2}$	$.15 \times 10^{-1}$	$.18 \times 10^{-7}$
CO_3^{2-}	$.32 \times 10^{-6}$	$.96 \times 10^{-5}$	$.87 \times 10^{-5}$	$.86 \times 10^{-17}$
NH_3	1.27	$.11 \times 10^{-3}$	$.10 \times 10^{-3}$	.0
NH_4^+	$.48 \times 10^{-2}$	$.29 \times 10^{-2}$	$.15 \times 10^{-1}$	.0

TABLE 4 (continued)

	A Reducing	B Weakly Reducing	C Weakly Reducing	D Oxidizing
NO_3^-	.0	.0	.0	$.95 \times 10^{-1}$
CH_4	$.22 \times 10^{-3}$	$.37 \times 10^{-5}$	$.11 \times 10^{-3}$	.0
CH_3OH	$.21 \times 10^{-17}$	$.19 \times 10^{-16}$	$.57 \times 10^{-15}$	.0
C_2H_5OH	$.75 \times 10^{-23}$	$.61 \times 10^{-21}$	$.53 \times 10^{-18}$	.0
$HCHO$	$.30 \times 10^{-22}$	$.14 \times 10^{-18}$	$.42 \times 10^{-17}$	.0
$HCOOH$	$.27 \times 10^{-18}$	$.69 \times 10^{-12}$	$.20 \times 10^{-10}$	$.72 \times 10^{-46}$
$HCOO^-$	$.24 \times 10^{-10}$	$.86 \times 10^{-8}$	$.45 \times 10^{-7}$	.0
CH_3COOH	$.34 \times 10^{-20}$	$.79 \times 10^{-13}$	$.69 \times 10^{-10}$	.0
CH_3COO^-	$.29 \times 10^{-13}$	$.97 \times 10^{-10}$	$.15 \times 10^{-7}$	.0
C_6H_5COOH	.0	$.42 \times 10^{-46}$	$.82 \times 10^{-36}$	.0
$C_6H_5COO^-$	.0	$.20 \times 10^{-42}$	$.67 \times 10^{-33}$	.0
$(COO)_2^{2-}$	.0	$.30 \times 10^{-16}$	$.80 \times 10^{-15}$	.0
$(COOH)_2$	.0	$.16 \times 10^{-26}$	$.14 \times 10^{-23}$	.0
CH_3NH_2	$.14 \times 10^{-14}$	$.11 \times 10^{-17}$	$.30 \times 10^{-16}$	.0
$CH_3NH_3^+$	$.12 \times 10^{-15}$	$.67 \times 10^{-15}$	$.10 \times 10^{-12}$	.0
$CO(NH_2)_2$	$.23 \times 10^{-12}$	$.26 \times 10^{-11}$	$.64 \times 10^{-10}$	.0
$C_6H_{12}O_6$	.0	.0	.0	.0
O_2	.0	.0	.0	$.39 \times 10^{-3}$
N_2	$.52 \times 10^{-3}$	$.62 \times 10^{-3}$	$.52 \times 10^{-3}$	$.36 \times 10^{-3}$
H_2	$.79 \times 10^{-6}$	$.15 \times 10^{-8}$	$.15 \times 10^{-8}$	$.38 \times 10^{-44}$

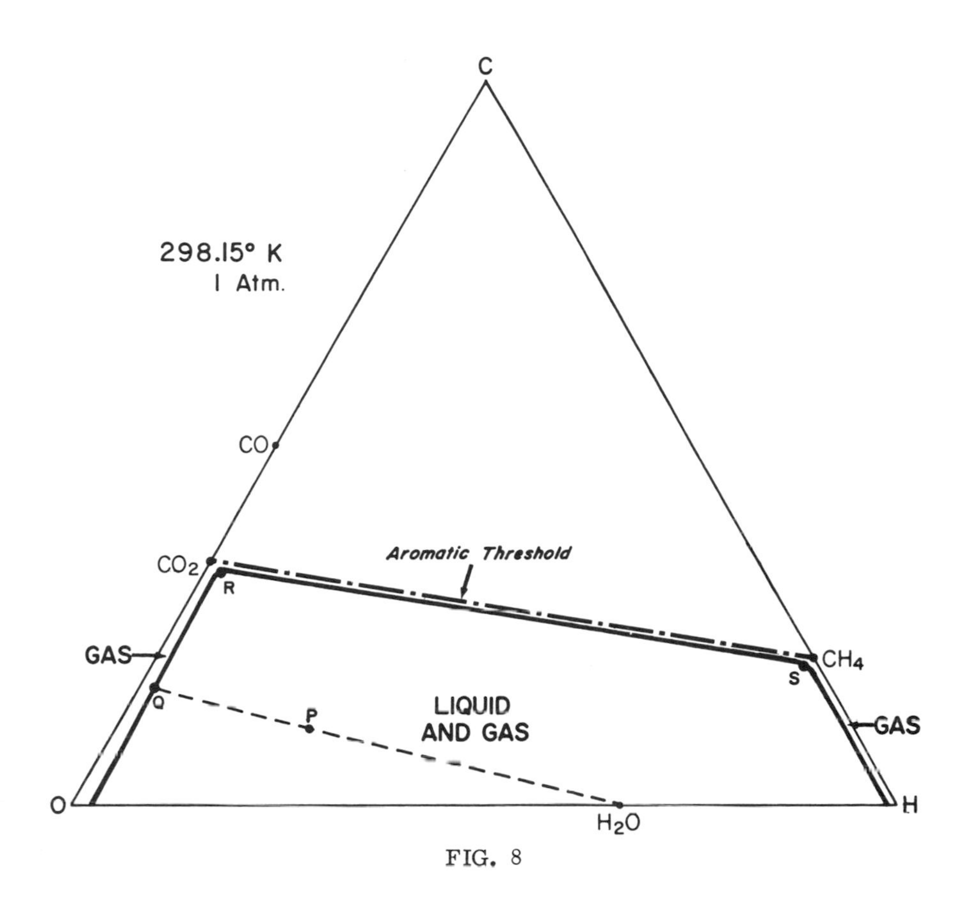

FIG. 8

Characteristics of limited equilibrium systems of C, H, and O at 1 Atm. and 300°K. Graphite has been excluded. A system of composition P would be composed of two phases, a gas of composition Q and an aqueous phase of almost pure water. If graphite has been included, gas phases between R and S would not be stable, but would further decompose into graphite and gas systems of composition R or S.

ion having a molal concentration of 10^{-8}, while glycine, the simplest amino acid, would have a molal concentration of 10^{-19}. This simple building block of proteins is very dilute; there would be only a few molecules in a drop of water.

VIII. CATALYSTS FOR THE REMOVAL OF ORGANIC POLLUTANTS

The results of our search for significant quantities of organic compounds in aqueous solutions at equilibrium showed that none existed in significant concentration. From the point of view of pollution control, this negative result has favorable implications. Any organic compound, under the influence of a suitable catalyst, breaks down into CO_2, CH_4, H_2O, and H_2 or O_2. When nitrogen is included, N_2 and HNO_3 or NH_3 may also form. The main problem in the control of organic pollutants is then to find a suitable catalyst for their decomposition.

Simple inorganic or organic compounds may act as catalysts even though their efficiency and versatility is quite limited; it may well be impossible to find a suitable form for a particular purpose.

Whole organisms, such as bacteria, have been used, in effect, as catalysts. For example, one sewage disposal process employs organisms which view the sewage as food and, using oxygen, metabolize it for energy, giving off such simple products as CO_2 and H_2O. Where the nature and concentration of the food is sufficient to maintain a colony of acceptable and effective organisms, this method works very well. The method is limited by the restriction that a viable organism, which will effectively remove the pollutant, be found. There is a limited number of viable species, containing a limited number of proteins which function as catalysts within the constraints in the living system.

A third type of catalyst is a complex biopolymer such as a protein. In contrast to small compounds and whole organisms (with their exacting requirements for survival), the catalytic potential of the protein enzyme is enormous. In aqueous solutions at room temperature, enzymes found in nature are outstanding catalysts, highly efficient, and highly specific. Figure 9 illustrates how complex the active center of a protein molecule is. The figure is taken from a drawing by Irving Geis in the "Atlas of Protein Sequence and Structure 1969" (6), based on the x-ray crystallographic analysis of D. C. Phillips and co-workers (7). The atoms are C, N, O, and S; the hydrogens have been omitted for simplicity. This protein, lysozyme, roughly globular in shape, and 129 amino acids in length, hydrolyses certain specific sugars and their derivatives. It presents a groove down one side into which the substrate will fit. Hydrogen bonding regions and van der Waals surfaces occur in just the right places to hold the

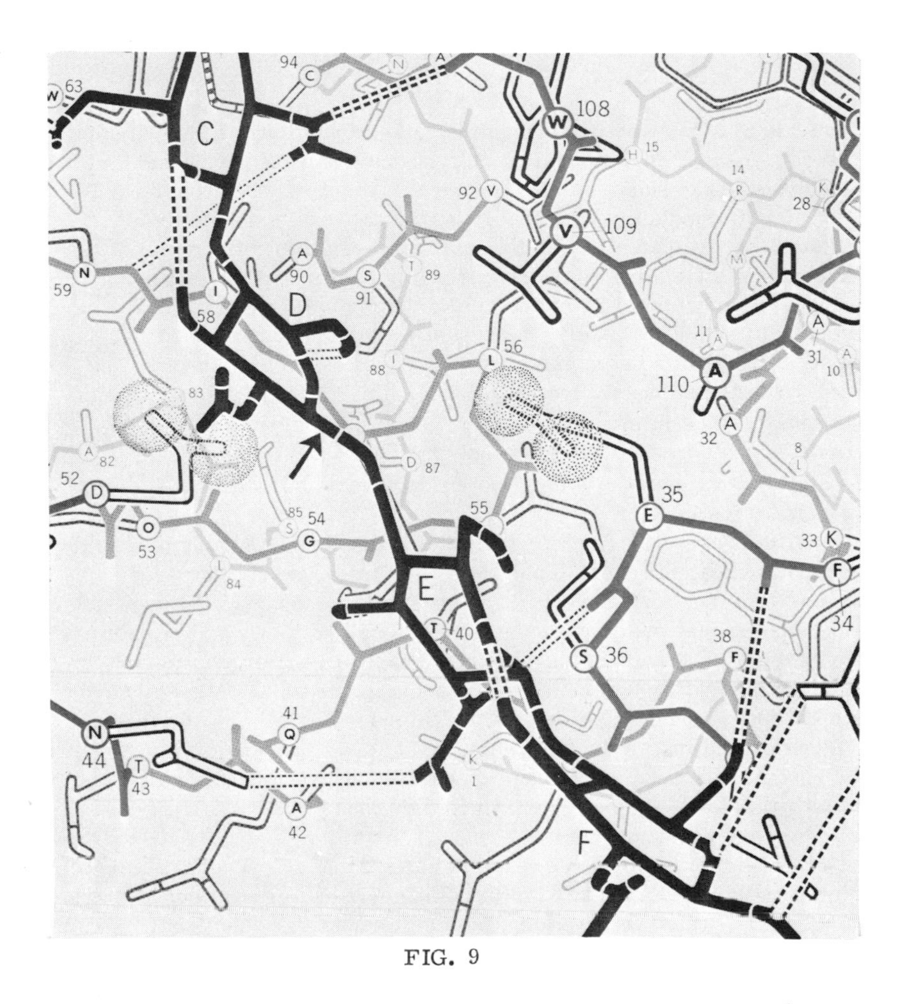

FIG. 9

Catalytic portion of the enzyme lysozyme with the substrate in place. Phillips and co-workers (7), who performed the X-ray crystallographic analysis of the protein, suggest that the substrate, a hexamer of n-acetyl glucosamine, shown in solid black, fits into a cleft in the enzyme very well, forming many hydrogen bonds. However, a slight distortion must be made in ring D to achieve the simultaneous fit of both ends. An electronic rearrangement, involving glutamic and aspartic acid residues at positions 35 and 52 of the protein and the substrate bond indicated by the arrow, becomes possible. Hydrolysis of the substrate results. The drawing made by Irving Geis is reprinted from Ref. (6), p. D-209, by courtesy of the National Biomedical Research Foundation. The skeletal bonds between carbon, nitrogen, oxygen, and sulfur are shown in black. Short lines across the bonds to nitrogen, oxygen, and sulfur atoms distinguish them from carbon. Hydrogen atoms have been omitted.

substrate in a position contiguous to catalytic portions of the enzyme, in such a manner that a covalent change occurs resulting in the hydrolysis of the substrate.

It is immediately apparent from the mode of construction of proteins that the potential for the formation of different complex three-dimensional structures is enormous. Any position in the chain can be occupied by any one of 20 amino acid side chains. A change in a single amino acid residue not only produces a new chemical reactivity at the site of introduction but it may also influence the overall shape of the whole protein and consequently the accessibility and reactivity of side chains at distant sites. In view of the tremendous number of such minor changes available, it is clear that there are many activities possible. The specificity of a protein is a very delicate function of covalent structure.

The practical utilization of proteins depends on having a feasible method for designing functional molecules and a method for producing these structures. Chemists do not presently have sufficient knowledge or computational power to design a functional protein a priori, nor are their methods of synthesizing proteins feasible on a production basis, although laboratory syntheses have been reported (8). Presently, the proper design must be selected from things which occur in living organisms and the production is dependent upon farming of the appropriate species. For the long range, there is the possibility of "evolving" better organisms and better enzymes by a process of artifically induced mutation and selection.

Among the procedures feasible now, there are several levels of sophistication in the search for living organisms, or particular enzymes inside them, which will perform particular catalytic functions. First, one can search in the vast storehouse of different well-adapted wild-type organisms for those which can metabolize a given contaminant. The enzymes responsible may even function alone in unfavorable environments where the whole organisms could not propagate. The soap industry has shown that enzymes can be procured in massive quantities and incorporated in detergent products where they are effective in breaking down organic compounds.

The protein found in the wild species is not necessarily optimal in the performance of a function such as the rapid degradation of a pollutant. Each functional protein in a species must coordinate with many other structures and activities in the cell. It has evolved to a structure which is optimal for the survival of the total organism and the species. For example, in the organism there is a point where too great a chemical efficiency becomes deleterious; the enzyme breaks down and destroys useful structures in the cell too rapidly.

Sub-optimal organisms, containing super-efficient enzymes, can occur through mutations in genes. A small number of the mutations occurring lead to proteins more efficient than the wild type under particular circumstances. In order to locate such mutants, an artificial environment can sometimes be found which will enhance the rate of reproduction of a desirable mutant over that of the wild type. One would expect that an organism with an increased ability to metabolize a particular pollutant would often be selectively encouraged to reproduce in a nutrient solution rich in this contaminant. Once such an organism has demonstrated its superior enzymatic ability, it can be cultivated in pure form and the enzyme harvested.

The rate of occurrence of mutants may be enhanced by irradiation or chemical treatment and desirable forms may be selected and cultivated, just as in the case of the natural mutants. By these, and other methods of genetic engineering, useful and efficient organisms and enzymes may be produced.

IX. CONCLUSION

We have seen that the world of equilibrium is very strange indeed. Not only are biopolymers and other large biochemical structures unstable, but all of the compounds commonly dealt with in organic chemistry are unstable. It is little wonder that the yields of organic reactions are fractional and dependent on conditions. At every opportunity a system tends toward a more stable state. At thermodynamic equilibrium, if oxygen is present in excess, the concentrations of organic compounds are negligible. Carbon is found as CO_2 or CO. Even in reducing atmospheres such as one might find on lake bottoms or in combustion chambers, only methane is stable in addition to the oxides of carbon.

In systems of high carbon content (in a limited equilibrium which excludes the formation of graphite) benzene and polynuclear aromatics may form in quantity. These routinely appear as tars in the processing of complex organic compounds. However, they can also form from certain mixtures of CO and H_2 under selected changes of pressure and temperature. Polynuclear aromatics, including carcinogens, are sometimes found as products of combustion, for example, in automobile exhaust. The production of these carcinogens may be eliminated through careful control of the elemental composition and the effective temperature of the combustion mixtures. Addition of oxygen or water to the system can change the elemental composition to a region where polynuclear aromatics are not stable products.

In the aqueous phase, organic compounds are found only in very low concentrations; even polynuclear aromatics are not stable in equilibrium with water.

The negative result of our search for significant quantities of organics at equilibrium in aqueous solutions is in fact fortunate for the solution of the problem of the removal of organic pollutants. One need only provide suitable catalysts and the organic compounds will remove themselves. Catalytic methods presently employed involve either inorganic and simple organic compounds, or whole organisms. In addition to these methods, protein enzymes, whose discovery or design is made feasible by genetic engineering, may prove to be very powerful antipollutants.

ACKNOWLEDGMENTS

This work was supported by NASA Contract 21-003-002 and NIH Grant GM-08710 to the National Biomedical Research Foundation.

REFERENCES

1. a. M. O. Dayhoff, E. R. Lippincott, R. V. Eck, and G. Nagarajan, "Thermodynamic Equilibrium in Prebiological Atmospheres of C, H, O, N, P, S, and Cl," NASA SP-3040, National Aeronautics and Space Administration, Washington, D. C., 1967. (This report contains a listing of the computer program to calculate equilibria in the gas phase as well as results for many systems.) Figures 1-6 are adapted from this report.

b. M. O. Dayhoff, E. R. Lippincott, and R. V. Eck, Science, 146, 1461 (1964).

c. R. V. Eck, E. R. Lippincott, M. O. Dayhoff, and Y. T. Pratt, Science, 153, 628 (1966). (Computations and results of plasma discharge experiments.)

d. E. R. Lippincott, R. V. Eck, M. O. Dayhoff, and C. Sagan, Astrophys. J., 147, 753 (1967).

2. a. M. O. Dayhoff and C. M. Park, NASA Special Report, (In preparation). "Computer Program for Calculating Thermodynamic Equilibria in Multiphase Systems."

b. G. Atkinson, M. O. Dayhoff, D. Wallace, and D. W. Ebdon, "A Thermodynamic Model of Seawater," Submitted for publication.

3. W. B. White, S. M. Johnson, and G. B. Dantzig, J. Chem. Phys., 28, 751 (1958).

4. F. P. Boynton, J. Chem. Phys., 32, 1880 (1960).

5. The standard free energy of formation data has come from the sources following. The computer program deals with concentrations in mole fractions. In order to convert the free energy values ordinarily given for aqueous solutes from a molality standard state to the mole fraction standard state, it was necessary to add RTln 55.51 or 2.379 to each value as customarily listed. Thus, the standard free energy of H^+ is given here as +2.379 instead of the usual 0.0.

a. D.D. Wagman, W.H. Evans, I. Halow, V.B. Parker, S. M. Bailey, and R.H. Schumm, Selected Values of Chemical Thermodynamic Properties, Part I. Technical Note 270-1., National Bureau of Standards, Washington, D.C., 1965.

b. G. T. Furukawa, M. L. Reilly, G.D. Mitchell, E.S. Domalski, I. Halow, and G. T. Armstrong, Ninth Preliminary Report on "A Survey of Thermodynamic Properties of the Compounds of the Elements CHNOPS," NBS Report 9449, National Bureau of Standards, Washington, D.C., 1966.

c. W. M. Latimer, Oxidation Potentials, Prentice-Hall, Inc., Englewood Cliffs, N.J., 1952.

d. See Reference 2b above.

e. Estimated from the table of solubility of gases in water. Handbook of Chemistry and Physics, p. 1706, Chemical Rubber Publishing Company, Cleveland, Ohio, 1953.

f. J.H.S. Green, J. Chem. Soc. London, Part 2, 2241 (1961).

g. H.W. Foote and S. R. Scholes, J. Amer. Chem. Soc., 33, 1309 (1911).

6. M.O. Dayhoff, Atlas of Protein Sequence and Structure, National Biomedical Research Foundation, Silver Spring, Md., 1969.

7. a. C.C.F. Blake, L.N. Johnson, G.A. Mair, A.C.T. North, D.C. Phillips, and V. R. Sarma, Proc. Royal Soc. London, Series B, 167, 378 (1967).

   b. C.C.F. Blake, G.A. Mair, A.C.T. North, D.C. Phillips, and V.R. Sarma, Proc. Royal Soc. London, Series B, 167, 365 (1967).

8. a. B. Gutte, and R.B. Merrifield, J. Amer. Chem. Soc., 91, 501 (1969).

   b. R. Hirschmann, R.F. Nutt, D.F. Veber, R.A. Vitali, S.L. Varga, T.A. Jacob, F.W. Holly, and R.G. Denkewalter, J. Amer. Chem. Soc., 91, 507 (1969).

CHAPTER 2

FORMATION OF ORGANIC POLYMERS ON MINERALS AND VICE VERSA

Egon T. Degens and Johann Matheja

Woods Hole Oceanographic Institution
and
Massachusetts Institute of Technology

I. INTRODUCTION

We are living in a world of coordination: socially, politically, and economically. Let us, for example, consider the present situation here in this auditorium; you are coordinated to me individually and as a group, and I am coordinated to you in a specific manner. In the course of my talk this coordination may prove to be perfect at one time and less perfect at another time; some of you may even leave this room and thus lower the coordination number. As a speaker I should be aware that there is always the probability of coordination change, and adjust accordingly.

This paper will be concerned with coordination phenomona in systems where minerals and organic compounds interact. The principles can be discussed conveniently under the heading epitaxis.

II. EPITAXIS

The term epitaxis is derived from the Greek tassein meaning to arrange or to order and in this context simply implies oriented growth of some crystalline matter upon some other crystalline compound. Four types of systems come to mind in which one partner represents the template and the other partner the epitaxial product:

	Template	Product
a.	mineral	mineral
b.	protein	mineral
c.	mineral	protein
d.	phospholipid	protein

A. System Mineral -- Mineral.

Much work has been done on the oriented growth of minerals on minerals (1 a, 1 b). The reaction principles are well understood. For illustration, however, a simple experiment is outlined:

The orthorhombic aragonite with the crystallochemical formula, $CaCO_3$, should serve as a template for strontianite, $SrCO_3$. One will notice that strontianite will readily nucleate on aragonite surfaces. This is so, because calcium and strontium are chemically similar and, in addition, form $Ca^{+2}O_9$ and $Sr^{+2}O_9$ coordination polyhedra in aragonite and strontianite. Both minerals are isostructural and follow the space group Pmcn.

In the next experiment, the orthorhombic aragonite should serve as a template for the rhombohedral calcite, $CaCO_3$. Calcite growth on aragonite surfaces will commonly not proceed, because calcium is six-fold coordinated in calcite $Ca^{+2}O_6$ and the space group is $R\bar{3}c$.

In conclusion, unless a chemical and structural fit is established between two minerals, epitaxial growth will be restricted.

B. System Protein -- Mineral.

Mineralizations in biological systems are the consequence of interactions between metal ions and distinct organic polymers. A specific organic matrix serves as a template in the nucleation of minerals that occur in teeth, bones, or shell structures. A common template is protein, but other organic tissues such as glycoproteins, polysaccharides, or phospholipid membranes are employed as mineralization agents.

It has been suggested (2) that the acidic and basic side chains of amino acids are involved in the fixation of cations and anions, respectively. Although there is certain truth in this suggestion, the molecular mechanism leading to mineralization is fundamentally different from that proposed earlier (2). For mollusc shells, the essentials of calcification are presented briefly:

Proteins and glycoproteins are secreted from cells located in the outer rim of the first mantle fold of the organism. Metal ions, in particular calcium, become coordinated to oxygen-containing functions causing the formation of metal ion coordination polyhedra (Fig. 1 a). This, in combination with the simultaneous development of crosslinkages, will stabilize the protein structure and the molecular order will increase. The carbonate ion, on the other hand, is attached via hydrogen bridges to the peptide chain (Figs. 1 b, 1 c). Interaction of metal ion polyhedra with carbonate ions results in an exchange of oxygen at the polyhedra, whereby calcium becomes more stably coordinated to oxygen. This step is favored thermodynamically and represents the moment of crystal seed formation. In case calcium is coordinated to six oxygens, $Ca^{+2}O_6$, a calcite nucleus will form; in contrast, a ninefold coordination will cause the development of aragonite. These nucleation processes are controlled by the availability of oxygen functions; they will find their reflection in the texture of the crystal phases and are thus within the realm of knowledge of crystal physics.

The protein matrix in the shell structures of molluscs is species specific (3a,b). A phylogenetic tree constructed on the basis of this specificity conforms with a tree constructed on morphological evidence. In conclusion, the availability of oxygen in the shell organic matrix determines the number of nucleation sites and the type of minerals. Evolutionary changes in shell morphology are a macroscopic expression of the molecular changes of the mineralized tissues.

C. System Mineral -- Protein.

The chemical synthesis of a number of biologically interesting polymers has been successful; this particularly concerns the formation of peptides (4). Present techniques involve carboxyl activation and the inactivation of functional groups not participating in the formation of the amide bond by so-called protective groups.

Mineral matrices can also introduce carboxyl activation and will also function as solid phase protective groups. For demonstration, experiments are referred to in which kaolinite serves as a template for the adsorption and polymerization of amino acids (5):

In the presence of kaolinite, $Al_4(OH)_8Si_4O_{10}$ (space group: Cc or $P\bar{1}$), amino acids will be picked up from an aqueous solvent and will be brought into solid solution. Amino groups become hydrogen bonded to the structural oxygen, or in the case of the basic amino acids, occur as positive ions. They are tightly fixed to the silicate surface and thus are rendered inactive. The carboxyl groups are fixed to the positively charged Al-oxy-hydroxy groups by means of ionic bridges, or become directly attached to the aluminum. Due to its high oxygen coordination, aluminum may pick up protons or discharge hydroxyls:

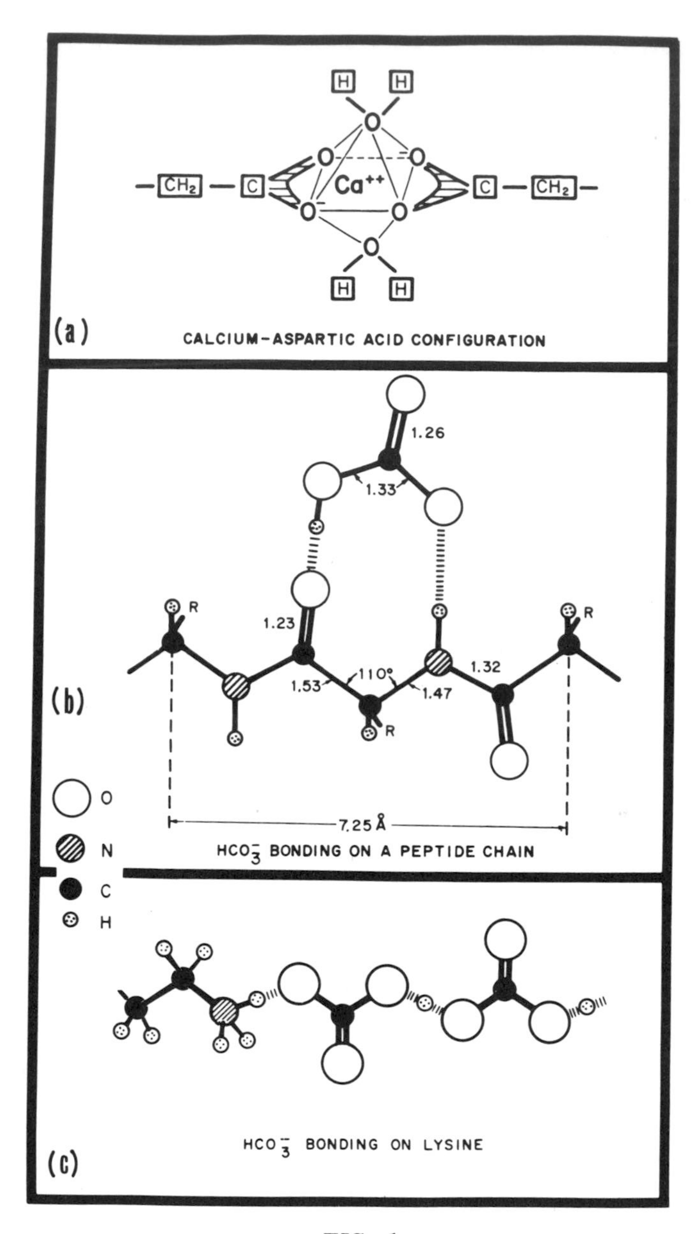

FIG. 1

Metal ion coordination polyhedron $Ca^{+2}O_6$ involving the carboxyl groups of aspartic acid (a); bonding of bicarbonate to a peptide chain (b); and to lysine (c) by means of hydrogen bonds.

$$\begin{matrix} Al \\ Al \end{matrix} > O \cdots + H^{\oplus} \longrightarrow \begin{matrix} Al^{\oplus} \\ Al \end{matrix} > OH^{\ominus}$$

or

$$Al - OH \longrightarrow Al^{\oplus} + OH^{\ominus}$$

In contrast, the silica in the clay structure is charged negatively since the hydroxyls at the surface will release protons:

$$Si - OH \longrightarrow Si - O^{\ominus} + H^{\oplus}.$$

Due to the tight fixation of oxygen to silicon, a release of OH^- is unlikely.

In water, amino acids cannot polymerize as a result of dipole-dipole interactions. In solid solution, however, amino acids will polymerize, because the solvent medium does not interfere, and because this reaction step is favored energetically (6). Inasmuch as in the kaolinite experiment, 1000 times more amino acids were polymerized to peptides than could conceivably become adsorbed to the clay surface, a flow of freshly polymerized molecules across the catalytically effective mineral surface has to be postulated. The preferential polymerization of aspartic and glutamic acids can be explained on the grounds that electrostatic interactions cause the attachment of amino acids to the silicate surface. In experiments we could also show that in presence of kaolinite the L-isomers of aspartic acid exhibit a higher polymerization efficiency than the D-isomers of aspartic acid. This feature is a consequence of the polar nature of the kaolinite lattice. The structural relationships in a kaolinite-amino acid-peptide system are shown in Figure 2.

In conclusion, minerals can serve as asymmetric polymerization templates by virtue of electrostatic interaction and carboxyl activation. Metal ion-oxygen surfaces are catalytically effective and permit a continuous flow of organic molecules.

D. System Phospholipids -- Protein.

To the extent as nitrogen increases the molecular order of organic molecules via hydrogen bridges, so do metal ions by forming oxygen coordination polyhedra. For instance, in biophosphates the molecular structures created by such a coordination pattern are identical to those observed in ordinary minerals (Fig. 3).

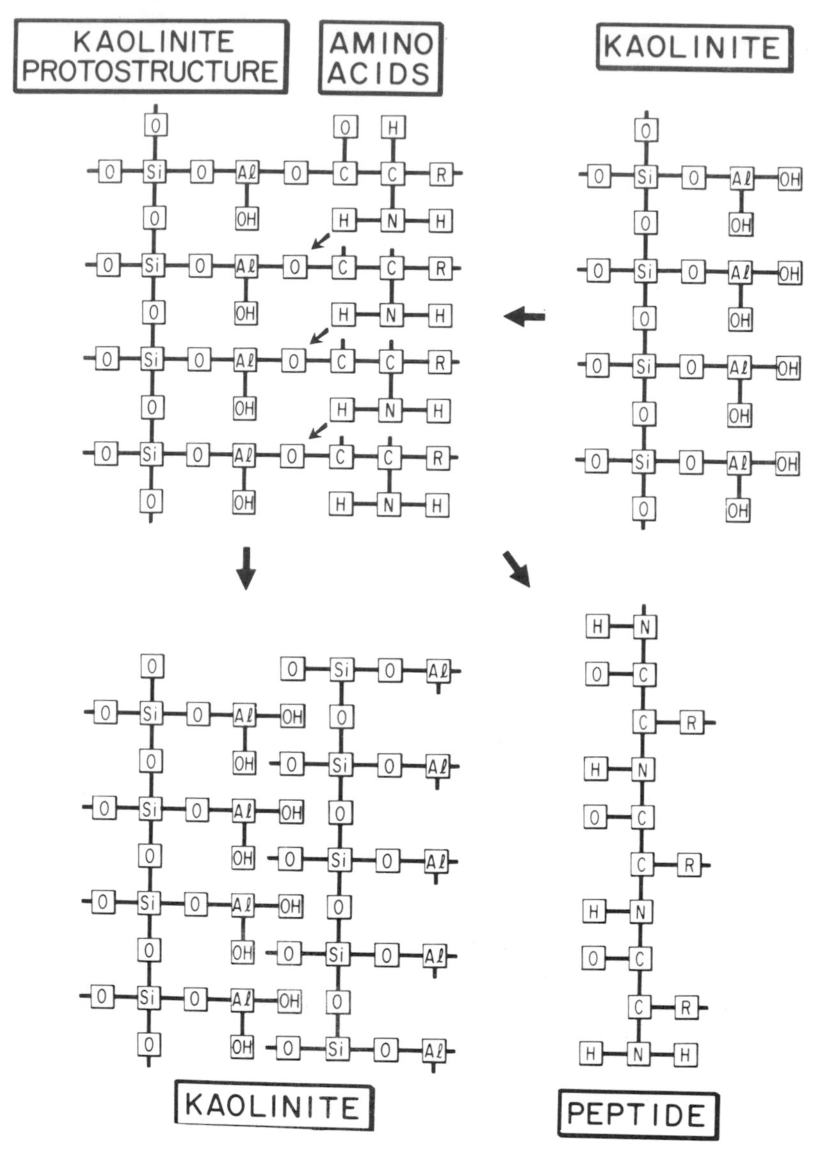

FIG. 2

Structural relationships in an amino acid-peptide-kaolinite system (schematic). The significance of the aluminum-oxy-hydroxy surface for carboxyl activation and solid phase protection is apparent. Such type of interaction may not only accomplish polmerization of amino acids and selection of specific isomers but will also lead to epitaxial growth of kaolinite.

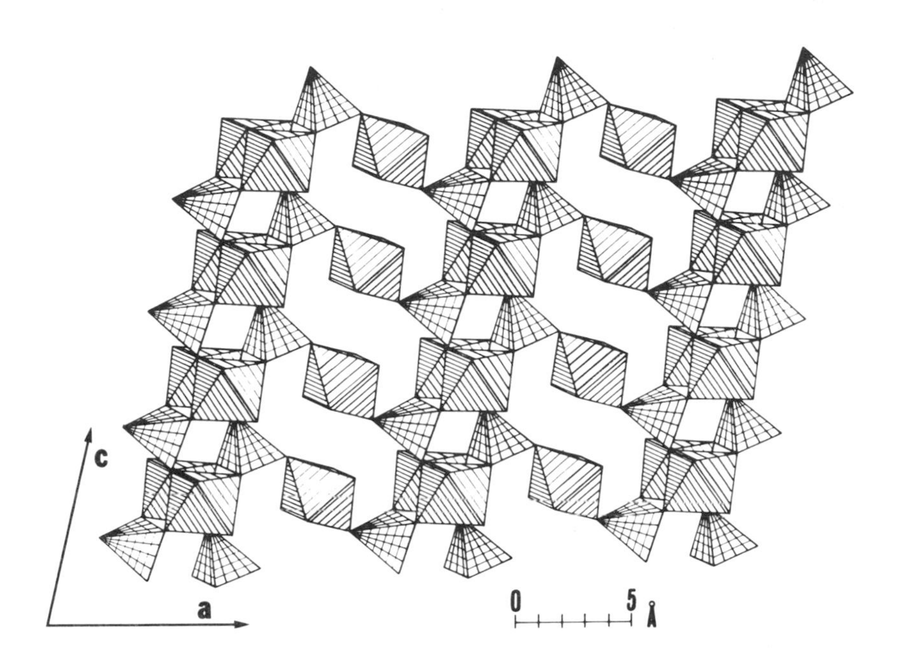

FIG. 3

Molecular structure of vivianite, $Fe_3(PO_4)_2 \cdot 8H_2O$. The structure is built up of single and double octahedral groups of oxygen and H_2O around Fe. The double group, $Fe_2O_6(H_2O)_4$, is linked to two neighboring similar groups and four other single groups, $FeO_2(H_2O)_4$ by means of phosphate tetrahedra. This will result in a complex band parallel to (010); parallel bands are linked to each other by H_2O molecules. The cell dimensions are: a = 10.08, b = 13.43, and c = 4.70 Å.

A structural model of DNA in Na^+ coordination is shown (Fig. 4) to emphasize the significance of metal ions in the structural conformation of nucleic acids. Since phosphate backbones can act as proton conduction bands (7) some of the functional aspects of metal ion coordination polyhedra become apparent. Also, in phospholipid membranes, metal ions serve as a structural element due to their ability to coordinate to the oxygen in PO_4 and glycerol (Fig. 5). As a result, a fabric comes into existence comparable to an ordinary phosphate mineral (Fig. 3). This structurally well-ordered ionic surface lattice will not only give membranes a structural rigidity, but, at the same time, will function as a dynamic molecular sieve and promote charge transfer (7).

The well developed molecular surface structure of phospholipid bileaflets will introduce a specific spacial geometry (e.g., 40 Å periodicity) for the structural protein layer. It is proposed that these proteins come into presence not via the conventional route of protein synthesis involving ribosomes and the genetic code; instead, they are formed by epitaxial growth on the crystalline surface structure of phospholipids. On the other hand, the crystalline order of the structural protein will introduce a specific spacing for the adjacent protein which participates in metabolic processes. These structural and functional relationships suggest that membranes are the site of a metabolic code.

In conclusion, metal ion coordination polyhedra and hydrogen bonds are the two most fundamental structural elements in bio-polymers. The formation of oxide chains and layers in phospholipid membranes allows the establishment of a topochemical coding and information readout device. The formation of structural proteins in membranes is a consequence of this molecular order.

III. CONCLUSION

Reactions between organic molecules and minerals follow identical coordination principles independent of whether we are dealing with inorganic or biogenic systems. As a rule, metal ions will form oxygen coordinated metal ion polyhedra. These polyhedra have three major functions:

1. ordering of randomly distributed compounds into a specific three-dimensional structure, thus increasing the molecular order,
2. control of interaction mechanism and kinetics by introducing a specific reaction pattern, and
3. incorporation of oxygens of different molecules into the polyhedra which will introduce intramolecular linkage elements.

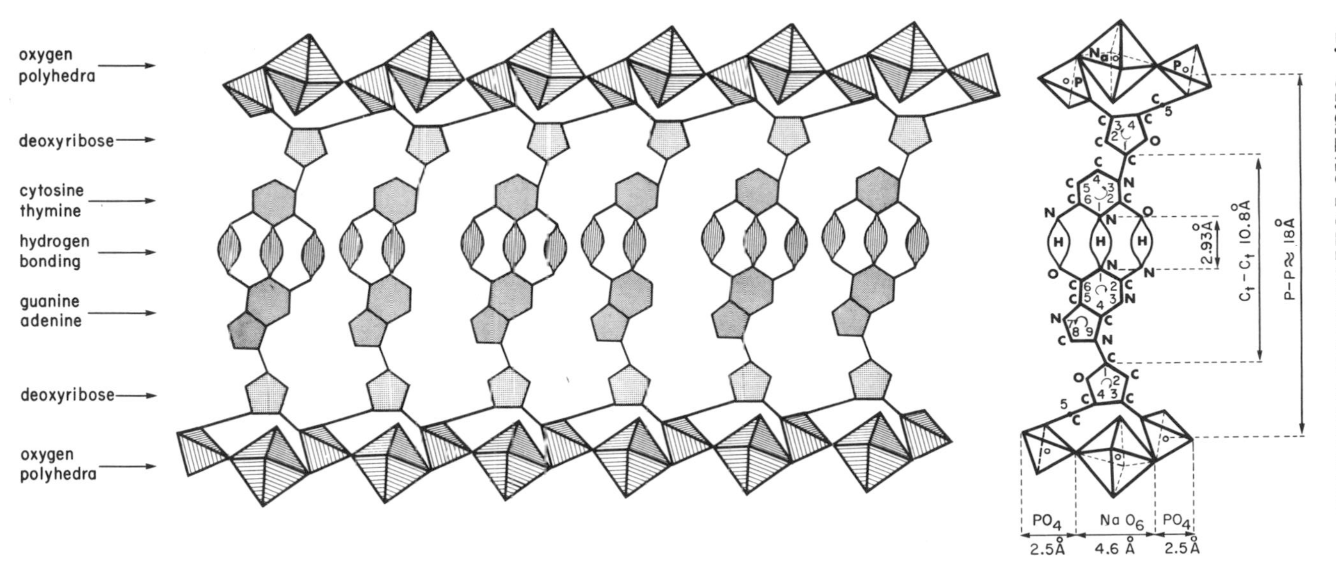

FIG. 4

Model of DNA (schematic) showing metal co-ordinated phosphate groups. A sodium ion serves as linkage element between two phosphate tetrahedra by establishing an oxygen polyhedron. The intermolecular distance between (a) the hydrogen linked bases of the purines and pyrimidines, and (b) the glycosidic carbons are the most probable ones. In addition, a weak C-H · · ·O bonding could exist between thymine and adenine; its consequence for the electron bonding confirguration in the aromatics would be considerable. It is also interesting to note that extremely short C -O distances (C_3 - O = 2.88 Å; C_2H - O_4 = 2.5 Å) are reported between deoxyribose and phosphate. The ultimate reason for the incorporation of sugars into nucleic acids is related to the fact that they will supply the oxygen necessary for the construction of the metal ion oxygen polyhedron (C_3 atom).

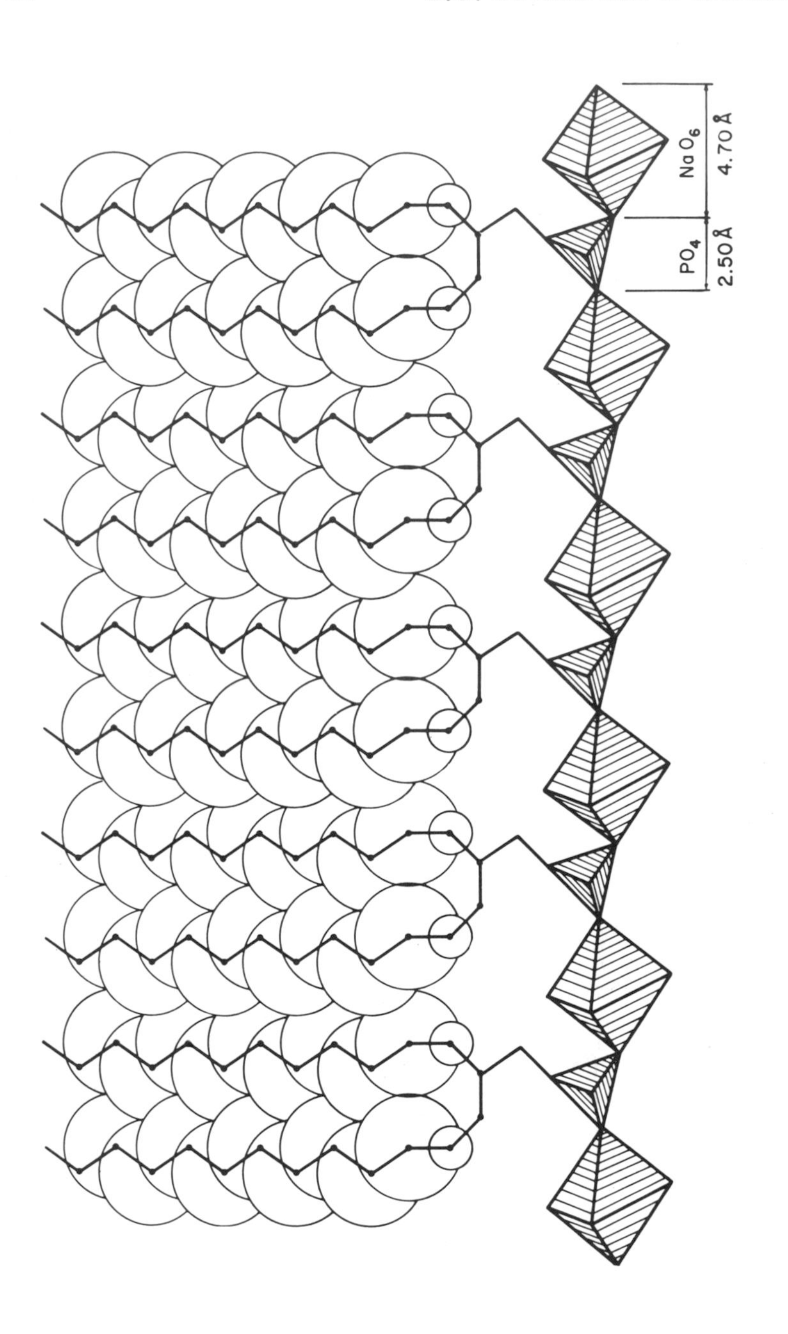
Na O6
4.70 Å
PO4
2.50 Å

ACKNOWLEDGMENTS

This work was sponsored by the National Aeronautics and Space Administration (NASA - 22-014-001) and by a grant from the Petroleum Research Fund administered by the American Chemical Society (PRF - 1943-A2). Grateful acknowledgment is hereby made to NASA and the donors of said fund. Contribution No. 2350 from the Woods Hole Oceanographic Institution.

REFERENCES

1. a. A. Neuhaus and H. Beckmann, Kolloid-Z. (Z. Polymere), 182, 121 (1962).
   b. A. Neuhaus, Fortschr. Miner., 29/30, 136 (1950/51).

2. M. J. Glimcher, in: Calcification in Biological Systems, (R. F. Sognnaes, ed.), Amer. Assoc. Advan. Sci., Publ. No. 64, 1960, p. 421.

3. a. E. T. Degens, D. W. Spencer, and R. H. Parker, Comp. Biochem. Phys., 20, 553 (1967).
   b. J. Matheja and E. T. Degens, N. Jb. Geol. Paläont. Mh., p. 215 (1968).

4. M. Bodanszky and A. A. Bodanszky, Amer. Sci., 55, 2 (1967).

FIG. 5

Membrane cross-section composed of double chain lipid-oxygen polyhedra. Glycerol is only schematically represented; the carbonyl groups and the PO_4 residues have been omitted. The molecular dimensions are based on the NaO_6 octahedron. The molecular aggregation, as presented here, results in a corresponding interfacial area fit for both chains concerned. Assuming the surface area of a hydrocarbon chain is identical to the surface area empirically determined for fatty acids (20 $Å^2$ per molecule), we obtain the same values for the surface area of a tetrahedron-octahedron unit. The cross-section for the minimum six oxygens falls in the range of 33-37.5 $Å^2$. For metal ions, the mean surface area amounts to 4$Å^2$. This would add up to a total surface area of 39.5$Å^2$ for a polyhedron unit, a value which would conform with the surface area of the associated lipids.

5. E. T. Degens and J. Matheja, in: Structure, Function, and Origin of Nucleic Acids and Proteins, (J. Oró and A. P. Kimball, eds.), Academic Press, New York, 1969.

6. E. T. Degens and J. Matheja, J. Brit. Interplan. Soc., 21, 52 (1968).

7. J. Matheja and E. T. Degens, Structural Molecular Biology of Phosphates, in Fortschritte der Evolutionsforschung, (G. Heberer and F. Schwanitz, eds.), Fischer Verlag, Frankfurt, 1970.

CHAPTER 3

THE OCCURRENCE OF ORGANICS IN ROCKS

W. G. Meinschein

Indiana University

I. INTRODUCTION

Carbon compounds play an important but inadequately understood role in many geochemical processes that occur within the earth's crust. The chemical versatility of carbon is unmatched by any other element. Carbonaceous substances are encountered in nature as solids and fluids either concentrated in deposits or widely dispersed throughout sedimentary rocks. Organic materials, such as coal, natural gas, and petroleum serve as indispensable sources of energy and of chemical intermediates, but the scientific potential of organic materials in rocks may not be fully appreciated.

Because organic compounds may have innumerable and complex structures, these compounds may retain vast amounts of information. Almost any chemical or physical difference required to define processes within the accessible regions of earth may be found between organic compounds of various functionalities or sizes. Thus, systematic changes in the structures and distributions of organic molecules in associated rocks may indicate the direction and magnitude of redistributions and the types and conditions of reactions that have been operative in these rocks. Also, sedimentary organic materials contain remnants of preexistent organisms. These remnants constitute our only direct ties to the biological past. The properties of carbon compounds and the essential involvement of carbon in metabolic reactions suggest that organic substances in ancient rocks may provide the most effective means of determining paleoenvironments and biological evolution.

The carbonaceous materials in sedimentary deposits may be viewed as products of the processes that made, altered, and redistributed these materials. Major emphasis has been paid the chemical conversion of biological remnants into fossil fuels. In certain cases this emphasis appears unwarranted, and, in other cases, the roles of organic materials in natural processes are not understood. Whereas coal is composed principally of altered biological materials, the chief components of which were probably cellulose and lignin, physical rather than chemical processes may control the rates of formation, quantities, and location of gas and oil accumulations (1,2). Organic-inorganic interactions are clearly indicated also in the concentrations of many ores and the formation of soils.

II. SEDIMENTARY ORGANIC MATERIALS

More than a century of study has not resolved the issue of whether the organic substances in rocks are almost totally or only partially derived from biological remnants. This issue is of greater import than may be generally recognized because the information that organic materials may provide is dependent on their origins. Plant and animals produce select arrays of organic compounds, thus a biological origin defines within limits the initial states of organic systems which have been involved in natural processes. Abiotic products are geochemically less useful than certain biological compounds, for the composition of a randomly synthesized carbon compound defines neither the reactants nor reactions that formed it. Likewise, the geochemical usefulness of simple or extensively altered biological remnants is limited, since these remnants may be indistinguishable from abiotic products.

The disparity in the scientific potential of abiotic and biological materials may be deduced indirectly from the use that has been made of these materials. An abiotic origin is seldom proposed for any organic substances that may not be degradation and/or condensation products of biological molecules. This ambiguity concerning an abiotic origin renders such a proposal valueless, but the value of some biological products is amply demonstrated. Morphologically or molecularly preserved remnants of pre-existent organisms have been widely employed in the development of important theories. Fossils are a significant source of pre-historical information. The shells of marine organisms beneath continental surfaces, the imprints of tropical ferns in Antarctican coals, and algal remains in Precambrian rocks attest to the continuity and antiquity of life on earth and to marked changes that have altered the features, climates, and inhabitants of ancient worlds.

Although an apparent disparity in scientific value may not constitute a valid reason for accepting a biological origin and ignoring an abiotic origin of the carbonaceous materials in sedimentary rocks, it does suggest that care should be exerted not to diminish the effectiveness of organic geochemical research by unfounded claims of the equivalences of abiotic and biological materials. As suggested already, many labile biological products can not be recognizably preserved for geologic time. Admittedly a major portion of the carbonaceous substances in rocks may not be convincingly related to remnants of former life, but fossil records, observations of plants and animals, and measurements of exchange rates of CO_2 between the atmosphere and biosphere as well as considerations of the conditional dependencies of abiotic and metabolic reactions and of the properties and distributions of organic molecules make it evident that life had probably dominated the production of effectively all the carbon compounds that were initially incorporated into the sediments which have formed extant sedimentary rocks (3).

Acceptance of the concept that pre-existent organisms were the only significant sources of sedimentary organic materials makes possible certain hypotheses as to the processes that control the compositions of these materials. If, as proposed above, living things dominate the production of carbon compounds, it is reasonable to expect that, as life has evolved, the molecular constituents of organisms have also changed or evolved. Such evolutionary changes, however, may not be reliably detected until a better understanding of metabolic evolution and of other processes affecting the compositions of biological remnants is obtained. These other processes include the carbon cycle, abiotic alterations, complex formations, and physical redistributions.

Most students of organic geochemistry or biogeochemistry, who are generally cognizant of the compositions of sedimentary organic matter, commonly relate structurally similar compounds in rocks and organisms. Various investigators, for example, have considered the chemical transitions required to convert cholorophyll into porphyrins that are found in shales and petroleum. This practice has merit in that it seemingly provides an understanding of geochemical processes.

There are reasons, however, for suspecting that the selectivity of the carbon cycle may destroy many direct proportionalities between the abundances of biological compounds and their structural relatives in ancient rocks. For it is not in the best interest of organisms to squander energy and carbon in the production of compounds that are incorporated in rocks for geologic time. Based on estimates of biological productivity and of fossil fuels, organisms have been remarkably efficient in retaining carbon. Less than one part in 10^4 to 10^8 parts of the organic materials made by former life has been preserved apparently in ancient rocks (4). This efficiency probably results from rate differentials in metabolic and abiotic reactions, for if metabolic reaction rates did not greatly exceed abiotic reaction rates, the compositions of biological molecules could be significantly controlled by abiotic reactions. A rapid turnover of the reactive and major molecular constituents within organisms appears a prerequisite to maintaining the high efficiency of the carbon cycle and the persistence of life. Thus, it may be assumed that only minor anabolic products may have low catabolic activities, which could permit their escape from the biosphere, and many of the relatively stable fossil organic molecules that closely resemble biological compounds may be concentrates or derivates of minor and catabolically inactive rather than major constituents of pre-existent organisms. Furthermore, because metabolically inert materials are eliminated preferentially by organisms and with anaerobes occupying a terminal position in the food chain, the contributions of elimination products and anaerobic remains to sedimentary organic materials may proportionally exceed the contributions from the cells of aerobes.

Solid carbonaceous substances encountered in rocks are grossly defined as humus, kerogen, bitumen, chitin, etc. A voluminous scientific literature has been devoted to descriptions and characterizations of these materials and their properties. These organophilic substances are mineralogically active. Direct correlations between the concentrations of organics and minerals in many rocks effectively establish that organisms and their remnants are involved essentially in the accumulation of various commercially important metals. Humic acid and fulvic acids dissolve rocks. Humates of calcium, iron, aluminum, manganese, and of other elements participate in the redistribution of these elements in soils. Humus is an important natural adsorbent, and its role in soil formation and agriculture is apparent. A comprehensive review of the significance of complex and economically important but incompletely defined organic materials has been written by Manskaya and Drozdova (5).

Numerous biological sources and various reactions have been proposed to explain compositional differences in the many classes of solid carbonaceous materials found in rocks. Tremendous efforts have been expended in the isolation and classification of these materials. Their nomenclature has been repeatedly revised and endlessly debated. Although the importance of the economic roles that solid organic substances play may not be doubted, it is questionable that scientists may ever obtain a fundamental understanding of the compositions and origins of a major portion of these substances.

In recent years, the contribution of microorganisms to the formation of mineralized organic matter has received much attention. Enormous numbers of microorganisms take part not only in the decompositions of proteins, carbohydrates, and fats but of relatively stable compounds as well. High molecular weight organic substances such as lignins, tannins, and wax are attacked by microorganisms, who add their remains, active enzymes, and other metabolites to sedimentary organic materials, but the fates of these materials are not controlled solely by biological processes.

Sedimentation subjects the remnants of pre-existent organisms to progressively changing conditions. Organisms frequently deplete the supply of oxygen at relatively shallow depths beneath sediment surfaces, and anaerobes replace aerobes as the principal kinds of life under reducing conditions. Oxygen becomes a vital commodity to be extracted from organic remnants. With increasing depth in sediments, the availabilities of certain essential biological substances generally decrease and consequently biological activity decreases, while temperatures and pressures increase. Overall the sedimentation process removes organic materials from oxidizing environments at moderate temperatures that are biologically active into reducing environments with high temperatures and pressures that favor abiotic alterations of these materials.

Changes in the biological and abiotic activity during sedimentation decrease the oxygen and increase the carbon and hydrogen contents of fossil organic substances. Nitrogen compounds are always present in humic acids. It was previously assumed that these nitrogen compounds were proteins, but many investigators now believe that melanoidins are the principal source of humic nitrogen. It is postulated that the melanoidins are products of reactions between amino acids and the aromatic portion of humic acids (5), but considerations based on the catabolic activities of biological molecules, as discussed above, suggest plant alkaloids as a likely source of the nitrogen compounds in humic acids. Identifications and definitive characterizations of nitrogen compounds in petroleum indicate that these compounds were derived from alkaloids (6).

Irrespective of the economic importance of humic acids, peats, coals, and related materials, the potentials of these substances in paleoenvironmental and paleobiological investigations will not be great until a more fundamental knowledge is obtained of their molecular structures. The task of determining the structures of complex mixtures of high molecular weight compounds with low solubilities and volatilities is a formidable one. No established means exist by which most individual constituents of these mixtures can be isolated and analyzed by modern instruments of high resolving powers. Abiotic alterations probably have contributed appreciably to the structural complexity and variety of the solid carbonaceous materials in sedimentary deposits. This complexity and variety hamper the scientific exploitation of these materials.

III. MOLECULAR INDICATORS

Abelson has stressed that geochemically useful organic compounds must be stable enough to partially retain their structures for geologic time (7). As noted above, much of the complexity of the solid carbonaceous materials in rocks may be traced to random abiotic alterations of labile biological molecules. Most molecular remnants of preexistent organisms are enriched in carbon and hydrogen and depleted in oxygen as compared to the major compound types in plants and animals. For reasons already cited, the selectivity of the carbon cycle and the conditions within sedimentary columns serve to increase the carbon and hydrogen contents of fossil organic matter. Only the organic soluble or lipid fractions of organisms, which are composed of fats, waxes, and hydrocarbons, have elemental compositions that are equivalent to the mean elemental compositions of sedimentary organic matter. It is not surprising, therefore, that the greatest resemblances between fossil and biological organic compounds are observed between certain benzene soluble constituents of rocks and organisms. Many of the saturated hydrocarbons in sediments and sedimentary rocks are structurally similar to the fatty acids, aliphatic alcohols, isoprenoids, and steroids found in biological lipids (8-11). These resemblances have led to the widespread use of alkanes as molecular fossils (8-10, 12-18). The potential of molecular fossils is indicated by the calculations of Morrison who estimates that the information gathered by observations at the molecular level can be 10^8 times greater than the information gathered by observations through a microscope.

Extensive analytical data have been gathered on biological and sedimental lipids for the purpose of determining the controls that different natural processes may exert on hydrocarbon compositions. Arbitrarily, we have concentrated our attention on C_{15} and larger ($> C_{15}$) lipids. This restriction simplifies sample recovery and reduces ambiguities about the origins of compounds, because: (a) $> C_{15}$ lipids can be partially or totally recovered by low temperature evaporative processes from solutions in volatile solvents used in extracting and chromatographically analyzing these lipids, (b) the probability that abiotic reactions may yield a specific compound is many times less likely for a $> C_{15}$ compound than for a compound in the C_1 to C_{15} range, (c) the biological compounds that contain 15 or more directly bonded carbon atoms, which may be readily converted into $> C_{15}$ hydrocarbons, are limited in number and are restricted principally to compounds appearing in biological lipids.

Comparisons of the structural types and distributions of $> C_{15}$ hydrocarbons, acids, alcohols, and terpenes from organisms, fecal matter, and sedimentary deposits indicate that: (a) the concentrations of olefinic hydrocarbons, chiefly terpenes, in organisms greatly exceed the concentrations of these compounds in fossil organic matter, (b) the alkanes in organisms and sediments or sedimentary rocks are largely n-paraffins, isoprenoids, and steranes that structurally resemble biological acids, alcohols, and isoprenoids, (c) the relative abundances of fossil n-paraffins and certain polycyclic alkanes differ significantly from the relative abundances of their structurally related acids, alcohols, and isoprenoids in most biological and sedimental lipids, (d) the concentrations of alkanes in extracts of fecal matter and Recent sediments are comparable, and are an order of magnitude greater than the concentrations of alkanes in the average biological lipid, (e) aromatic hydrocarbons appear in negligible or trace quantities in biological lipids, and the aromatic hydrocarbons in Recent marine sediments are most commonly detected in sediments which are buried at depths in excess of a few feet, (f) non-alkyl substituted phenanthrene, pyrene, chrysene, fluoranthene, triphenylene, 1,2-benzanthracene, perylene, 1,12-benzperylene, and coronene are prominent polycylic aromatic constituents of Recent marine sediments, whereas alkyl substituted polycyclic aromatic hydrocarbons predominate in the aromatic fractions from ancient rocks and crude oils (8,19).

Permitted interpretations of the above analytical observations are: (a) the low concentrations of biologically abundant terpenes in sedimentary deposits suggest that these compounds are chemically and/or catabolically too active to accumulate unaltered in ancient rocks, (b) the low catabolic and chemical activities of alkanes serve to concentrate these compounds preferentially in the lipid fractions of fecal matter and rocks, (c) most aromatic hydrocarbons are not formed in living cells, (d) the simplicities and locations of aromatic fractions within Recent marine sediments suggest that anaerobes participate in the production of these fractions, (e) the complexity of the aromatic fractions and the positions of the alkyl substituents on aromatic hydrocarbons indicate that abiotic alterations of steroids and isoprenoids as well as physical redistributions exert compositional controls on aromatics in ancient rocks and crude oils (8, 19, 20).

Johns et al. (21) and McCarthy and Calvin (22) have discussed the distributions of C_{16} to C_{22} isoprenoids. The C_{17} isoprenoid is far less abundant than its C_{16}, C_{18}, C_{19}, C_{20}, and C_{21} homologs in alkanes from 3×10^7 year-old to 2×10^9 year-old rocks. Other fossil alkanes, such as n-paraffins, steranes, and certain triterpenoids also display a general uniformity in their distributional and structural patterns. The compositional resemblances between alkanes from different geological environments are in sharp contrast to the compositional variations observed for the principal organic constituents. If the latter variations are attributable to conditionally dependent abiotic alterations of labile biological remnants, the compositions of fossil alkanes attest to the stability and biological origin of these compounds. It appears unlikely that abiotic reactions have been involved extensively either in the formation or in the degradation of the $> C_{15}$ alkanes in most sedimentary rocks (23). Paradoxically alkanes, which may play a relatively insignificant metabolic role, may retain the most legible record of metabolic evolution. Because of their low catabolic activities and their great abilities to retain and display structural order, the $> C_{15}$ alkanes will probably remain the most useful paleobiological indicators. Furthermore, the negligible polarities, abundances, chemical equivalences, and physical differences of n-paraffins and certain other homologous alkanes make these compounds outstanding indicators of the physical redistribution processes that yield natural gas and petroleum (1, 2).

As the studies of Blumer indicate, aromatic hydrocarbons may serve as the best indicators of the chemical changes that alter fossil organic materials (24). Presently our knowledge of the organic compounds that interact with minerals and of the nature of these interactions is so inadequate that it is not possible to make predictions of what course research in this field may follow.

REFERENCES

1. E. G. Baker, Fundamental Aspects of Petroleum Geochemistry, (B. Nagy and U. Colombo, eds.), Elsevier, New York, 1967, p. 299.

2. W. G. Meinschein, Y. M. Sternberg, and R. W. Klusman, Nature, 220, 1185 (1968).

3. W. G. Meinschein, Accounts Chem. Res., in press.

4. J. R. Vallentyne, Arch. Hydrobiol., 58, 423 (1962).

5. S. M. Manskaya and T. V. Drozdova, Geochemistry of Organic Substances, (Leonard Shapiro and I. A. Breger, eds.), Pergamon Press, New York, 1968.

6. L. R. Snyder, Nature, 205, 277 (1965).

7. P. H. Abelson, in Researches in Geochemistry, (P. H. Abelson, ed.), John Wiley & Sons, New York, 1959, p. 79.

8. W. G. Meinschein, Bull. Amer. Assoc. Petrol. Geologists, 43, 925 (1959); Geochim. Cosmochim. Acta, 22, 58 (1961); Space Sci. Rev., 2, 653 (1963).

9. A. L. Burlingame, P. Haug, T. Belsky, and M. Calvin, Proc. Nat. Acad. Sci., 54, 405 (1965).

10. G. Eglinton and M. Calvin, Sci. Amer., 216, 32 (1967).

11. I. R. Hills, G. W. Smith, and E. V. Whitehead, Nature, 219, 243 (1968).

12. B. Nagy, W. G. Meinschein, and D. J. Hennessy, Ann. N.Y. Acad. Sci., 93, 25 (1961).

13. W. G. Meinschein, B. Nagy, and D. J. Hennessy, ibid, 108, 553 (1963).

14. W. G. Meinschein, E. S. Barghoorn, and J. W. Schopf, Science, 145, 262 (1964).

15. G. Eglinton, P. M. Scott, T. Belsky, A. L. Burlingame, and M. Calvin, Science, 145, 263 (1964).

16. E. S. Barghoorn, W. G. Meinschein, and J. W. Schopf, Science, 148, 461 (1965).

17. J. Oro, D. W. Nooner, A. Zlatkis, S. A. Wikstrom, and E. S. Barghoorn, Science, 148, 77 (1965).

18. W. G. Meinschein, Science, 150, 601 (1965).

19. W. G. Meinschein, Life Sci. Space Res., 3, 165 (1965).

20. B. J. Mair and J. L. Martinez-Pico, Chem. Eng. News, 40, 54 (1962).

21. R. B. Johns, T. Belsky, E. D. McCarthy, A. L. Burlingame, P. Haug, H. K. Schnoes, W. Richter, and M. Calvin, Geochim. Cosmochim. Acta, 30, 1191 (1966).

22. E. D. McCarthy and M. Calvin, Abstract for A.C.S. Meeting, Petroleum Chemistry Division, New York, September 1966.

23. W. G. Meinschein, in Organic Geochemistry: Methods and Results, (G. Eglinton and M.T.J. Murphy, eds.), Springer-Verlag, Chapter 13, in press.

24. M. Blumer, Science, 149, 722 (1965).

CHAPTER 4

ORIGIN OF ORGANICS FROM ARTIFICIAL CONTAMINATION

J. V. Hunter

Rutgers, The State University

I. INTRODUCTION

Since the discovery by man that water could be used to dilute and to transport unwanted materials from the location at which they were generated to some distant place where their adverse effects would not be so unpleasantly evident, the aquatic environment has been the unfortunate recipient of much of his waste products. Intensified by man's development of water carriage systems to facilitate waste transferral to the aquatic environment and his ability to alter or to create new organic species, such materials, both liquid and solid, soluble and insoluble, animate and unanimate have been discharged in considerable quantities into rivers, lakes, estuaries, and oceans.

Much of the organic wastes produced by man originate from the simple processes of living, arising from cooking, washing, urine, feces, etc. Such water-carried wastes are usually designated domestic waste water or sewage. Those arising from other activities are more definitively described but frequently are referred to as industrial waste water or sewage. The relative volumes and organic matter contributions of these wastes are presented in Table 1. However, indicative of total or relative organic contributions, general organic pollution parameters such as the BOD give no inkling as to the nature of the organic materials present in waste waters. As such, detailed knowledge would aid in assessing the environmental consequences of organic waste discharge, in evaluating the potentialities of recovery of useable materials from such discharges, and in better predicting the behavior of waste water treatment facilities (especially those based on physio-chemical removal principles), many studies have been made of waste water organics. These have varied from analysis

TABLE 1

Estimated Domestic and Industrial Sources of Organic Pollutants for 1963[a]

Waste Water Source	Waste Water Characteristics	
	Volume Billion Gallons/yr.	Strength (BOD) Million Lbs./yr.
Chemical	3,700	9,700
Pulp and Paper	1,900	5,900
Food Processing	690	4,300
Textile Mills	140	890
Petroleum and Coal	1,300	500
Primary Metals	4,300	480
Transportation Equipment	240	120
Electrical Machinery	91	70
Machinery	150	60
Rubber and Plastics	160	40
Domestic Sewage	5,300[b]	7,300[c]

[a] From Ref. (1) for the year 1963.
[b] Flow estimated by assuming that 120 million people are sewered and have a contribution of 120 gallons/capita/day.
[c] BOD estimated by assuming that 120 million people have a BOD contribution of 0.16 lbs/capita/day.

for a single compound of interest to attempts to characterize most of the waste water organics. It is the purpose of this chapter to review the problems encountered and the results obtained from such studies, and briefly note certain of the environmental consequences of these organic contaminants.

II. ANALYTICAL PROBLEMS AND REQUIRED TECHNIQUES

Domestic and industrial waste waters are usually physically inhomogeneous mixtures of many simple and complex organic constituents. The analytical problems due to these considerations are intensified frequently by low organic matter concentrations, its biological availability, and temporal variations in both the nature and concentration of the organic contaminants. As interpretation of the results of waste water analyses requires an understanding of these factors, they will be discussed in some detail.

A. Temporal Variation.

Considerable variations in both waste water flow and the concentration and nature of its organic constituents occur as a function of time. An example of the flow variations of a domestic sewage is shown in Figure 1. The indicated variations of MBAS (methylene blue active substance, more or less equivalent to anionic surfactants) may be considered typical of domestic waste water organics. Corresponding variations that occur in industrial waste waters are even more extreme.

Although much information concerning waste water organics can be and has been obtained by simple grab sampling, only an extended series of such analyses as a function of time will reveal the average composition of organics as well as infrequent or minor constituents. Although less information is achieved, it is usually much easier and simpler to composite the waste waters according to flow and run exhaustive analyses on a single sample. It should be noted that both of these approaches are used, depending on whether average pictures or the actual variations of the waste water organic constituents are desired. In addition, the actual time over which composites are prepared is limited by such considerations as sample deterioration, a factor that does not have to play a role in analysis of individual grab samples, even when these are part of an extended study.

B. Biological Availability.

Many of the organic constituents are available to microorganisms as a source of energy, cell carbon, or cell nitrogen. Such degradability is intensified by the presence of large numbers of microorganisms in many waste waters, the presence of suitable nutrients, and a favorable pH value. This is especially true of domestic waste water. If knowledge is required only of a single constituent or a group of

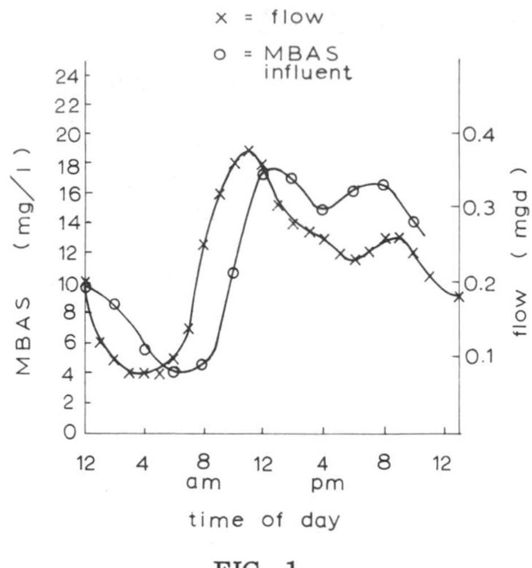

FIG. 1

Temporal variations in waste water flow and influent MBAS levels (Bernardsville, N. J., 9/19/66).

constituents, suitable preservatives can be added to prevent their deterioration. However, such additives may effect the other organics present or even interfere in their determination and, therefore, are not too well suited for extensive analyses. In such cases, cooling waste waters to $4^{o}C$. or below will inhibit significant biological attack for about one day and, thus, it seems to be the only truly general approach available.

C. Physical Inhomogeneity.

Not all of the organics present in waste waters are soluble, many are insoluble and exist in various states of subdivision. This is an important factor in chemical analysis, as the nature of soluble organics is likely to differ from insoluble organics and, therefore, require different analytic approaches. When information concerning the concentration of a specific organic is required, knowledge of the solubility of the constituent should enable the investigator to select the appropriate physical fraction or fractions for analysis. Even in this case, care must be exercised in interpretation, as adsorption or partition may transfer soluble organics to the particulate or insoluble

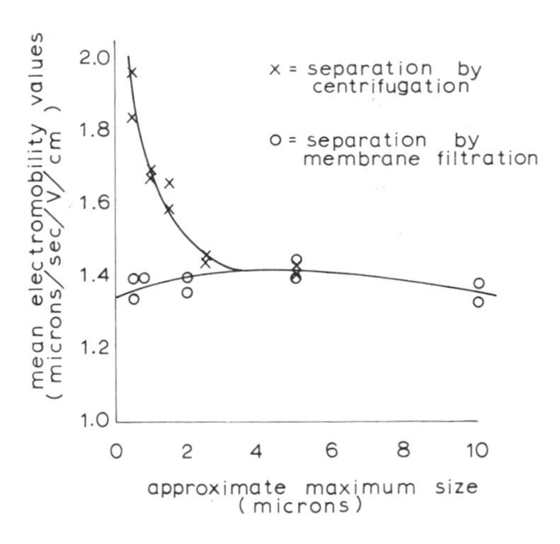

FIG. 2

Effect of particulate matter separation procedure on mean electromobility values. Reproduced from Ref. (10), p. 565 by courtesy of Int. J. Air - Water Pollut.

phases. An example of this for the removal of soluble alkyl benzene sulfonates by domestic waste water particulates is presented in Table 2.

When attempts at a total analysis are made, some type of physical fractionation of waste waters would be desirable in simplfying the analytic problems. These have included preliminary settling and removal of gross particulates, removal of finer particulates by centrifugation, and colloidal particulates by candle filtration (4), high pressure cellulose membrane filtration (6, 8), and high speed centrifugation (9). These techniques of separation presumably can divide waste water particulates into size range classifications, but it should be noted that they are based on different mechanisms and, thus, should be considered equivalent only with considerable caution. Figure 2 indicates the difference in behavior (electromobility) of particulates separated by centrifugation and membrane filtration techniques (10).

In all probability, such physical fractionation techniques do not markedly allow the organic matter distribution, but they are laborious and slow. A study was made by the author to evaluate a rapid procedure to separate waste water into a single particulate and a single

TABLE 2

ABS Removed from Sewage During Primary Sedimentation

ABS Removed %	ABS Concentration mg/l	Reference
25.0	16	(2)
13.5	10.1	(3)
12.0	25.8	(4)
2-3	5 - 10	(5)
1.8	4.3	(6)
0.3	6.4	(7)

soluble fraction using chemical coagulation. As would be expected, the distribution of organics observed by the physical separation methods was not identical to those observed by chemical separation techniques, although the analyses were somewhat hindered by substantial amounts of ferric oxide in the particulates obtained by chemical coagulation. These distributions are shown in Table 3.

D. Low Concentrations.

Although certain industrial waste waters may contain high concentrations of organics, the usual problem is that the organics are present in rather low concentrations. Domestic sewages for example, contain less than 0.1% organic material by weight. If physical separation procedures are employed, these concentrate the particulates during their removal, but the problem of a dilute soluble fraction remains.

If the required analytical methods are sufficiently sensitive, concentration procedures may not be required. In most cases, however, some type of concentration procedure which does not result in a loss of organic matter is necessary. Two techniques that have been used with advantage for this purpose are lyophilization and freeze concentration (6,12). In the former, the sample is frozen and

TABLE 3

Contribution of Physically and Chemically Separated Sewage Fractions to Sewage Organic Strength Parameters (as % of recovered strength parameter[a])

Strength Parameter	Separation Procedure			
	Physical		Chemical	
	Particulates %	Soluble %	Particulates %	Soluble %
Volatile Solids	57.6	42.4	63.0	37.0
Chemical Oxygen Demand	77.3	22.4	83.6	16.4
Organic Nitrogen	80.5	19.5	77.0	23.0

[a] Composited sewages from Highland Park, New Jersey, Fall-Winter, 1959-60.

the water is removed by sublimation, in the latter it is removed in the form of ice crystals. Which technique results in the least loss of organic matter is uncertain at present, but one advantage of lyophilization is that it results in a powder that can be kept dry and, thus, is not subjected to significant biological degradation.

E. Chemical Complexity.

Even after preliminary fractionation, waste waters represent complex mixtures of organic materials. To a certain extent, the problems encountered by the analyst are a function of the purpose of the analysis. If this is the determination of a specific compound, few conceptual problems exist. Analyses are selected on the basis of analytic sensitivity if no concentration is employed and insensitivity to suspected interferring substances. Even when total analyses are attempted, a knowledge of the materials present greatly simplifies the approach.

An example of this is found in the total analyses of domestic waste water. The contributions to waste water organic carbon of the various possible domestic sources were estimated as follows: feces = 17 g/capita/day, urine = 5 g/capita/day, dish washing and food preparation = 8 g/capita/day, and personal and clothes washing = 7 g/capita/day. The contributions to waste water organic nitrogen were estimated as follows: feces = 1.5 g/capita/day, urine = 1.7 g/capita/day, and dish washing and food preparation = 0.2 g/capita/day (4). Thus, domestic waste water analysis was based on what was known about the organic constituents expected from these sources, and what types of analytic methods and chemical separation procedures were available for their estimation (4, 6, 8, 11). Table 4 gives an example of the type of information used in deciding what analyses should be attempted (13).

The success of such estimations has been established reasonably well (4, 6), as most of the organic constituents of domestic waste water have been estimated or chemically described. For effluents from domestic sewage plants, the story is quite different. Using analytic approaches similar to those employed for domestic waste water analysis results in a recovery of organic matter of only a quarter to a third of the total (14, 15). Whether these organics represent the undetected constituents of the waste water before treatment, the materials formed during the treatment itself, or both, not enough information has been available to facilitate their analysis. If these organics have large molecular weights and complex structures, their identification and estimation will be a rather difficult task.

Industrial waste waters partake of both types of problems. Chemical wastes contain intermediates, final, and degradation products and, thus, a rationale at least exists for their analysis. Wastes from processes that do not markedly alter the structure of the organics (food processing, etc.) should allow analytic approaches based on a knowledge of the nature of the starting materials. Wastes produced by severe chemical alteration of the starting material (paper pulp production, etc.) do not allow such approaches and, therefore, represent a difficult and interesting analytic problem.

III. DOMESTIC WASTE WATER COMPOSITION

A large body of information exists on the distribution and nature of the organic constituents of domestic waste waters. To a certain extent this is due to their consideration as the "least common denominator" of waste waters and, therefore, most likely to produce information of general applicability.

TABLE 4

Organic Constituents of Human Excretion Products (13)

Compound	Origin	
	Urine mg/kg body wt.	Feces mg/kg body wt.
Creatinine	15(12-25)	
Hippuric acid	8(1-12)	
Urea	300(215-500)	
Uric acid	9(5-12)	
Alanine, total	0.55	
Arginine, free	0.31(0.15-0.5)	
Arginine, combined	0.1(0.0-0.2)	
Arginine, total	0.4(0.34-0.5)	3.8(2.9-5.0)
Aspartic acid, free	0.02(0.014-0.26)	
Aspartic acid, combined	2.3(1.2-3.7)	
Aspartic acid, total	2.32(0.37-3.7)	
Citrulline, free	0.58(0.26-0.7)	
Citrulline, total	(0.345-0.79)	
Cystine, free	1.3(0.65-2.0)	
Cystine, total	(1.5-2.4)	
Glutamic acid, free	0.52(0.0-1.07)	
Glutamic acid, combined	4.5(1.0-10.0)	
Glutamic acid, total	5.27(1.58-11.55)	
Glycine, free	10.1(9.0-12.0)	
Glycine, total	(2.3-18.0)	
Histidine, free	2.7(0.94-4.8)	
Histidine, combined	0.6(0.07-1.8)	
Histidine, total	3.0(0.98-6.59)	1.7(1.4-2.1)
Hydroxyproline, total	0.02	
Isoleucine, free	0.085(0.03-0.3)	
Isoleucine, combined	0.2(0.06-0.4)	
Isoleucine, total	0.3(0.11-0.6)	4.3(3.3-5.5)
Leucine, free	0.14(0.05-0.25)	
Leucine, combined	0.2(0.05-0.4)	
Leucine, total	0.32(0.20-0.52)	5.6(4.3-6.9)
Lysine, free	0.5(0.25-1.13)	
Lysine, combined	0.6(0.2-1.1)	
Lysine, total	1.04(0.48-2.0)	5.7(4.5-6.9)

TABLE 4 (Continued)

Compound	Origin	
	Urine mg/kg body wt.	Feces mg/kg body wt.
Methionine, free	0.11(0.05-0.18)	
Methionine, combined	0.03	
Methionine, total	0.14(0.12-0.17)	
Ornithine, free	0.15	
Phenylalanine, free	0.23(0.1-0.43)	
Phenylalanine, combined	0.1(0.04-0.2)	
Phenylalanine, total	0.33(0.21-0.6)	
Proline, free	0.12(0.05-0.21)	
Proline, combined	0.5(0.3-0.8)	
Proline, total	0.61(0.33-0.9)	
Serine, free	0.4(0.21-0.52)	
Serine, combined	0.25(0.0-0.5)	
Serine, total	0.65(0.35-1.4)	
Threonine, free	0.37(0.17-0.62)	
Threonine, combined	0.4(0.3-0.8)	
Threonine, total	0.77(0.36-1.2)	4.0(3.3-5.2)
Tryptophan, free	0.37(0.12-0.7)	
Tryptophan, combined	0.3(0.009-0.4)	
Tryptophan, total	0.7(0.23-1.3)	
Tyrosine, free	0.3(0.17-0.55)	
Tyrosine, combined	0.5(0.08-0.9)	
Tyrosine, total	0.79(0.35-1.45)	
Valine, free	0.065(0.04-0.125)	
Valine, combined	0.2(0.09-0.4)	
Valine, total	0.3(0.21-0.45)	4.6(3.6-6.2)
Citric acid	(3-17)	
Creatine	2.9(1.1-3.86)	
Guanidoacetic acid	(0.23-0.51)	
Formic acid	(0.42-2.0)	
Indoleacetic acid	(0.03-0.06)	
Lactic acid	40	
Oxalic acid	0.285(0.23-0.5)	
Phenols	4.0(0.19-6.6)	(0.0-3.0)
Cholesterol, total	(0.0-0.007)	8(10-20)

TABLE 4 (Continued)

Compound	Origin	
	Urine mg/kg body wt.	Feces mg/kg body wt.
Fat, total		56(30-100)
Fat, neutral		(10-45)
Fat, unsaponifiable		32(22-38)
Fatty acids, total		30(4-64)
Fatty acids, free		16(4-38)
Soaps		53(40-66)
Bilirubin, μg	70	
Coproporphyrin I and III, μg	(0.24-1.4)	
Porphyrins, μg	(0.0-0.4)	
Urobilin, μg	(7-20)	
Urobilinogen, μg	(0.6-3.0)	2.0(0.57-4.0)
Androgens, μg	(30-100)	
Epinephrine, μg	0.16(0.07-0.31)	
Estrogens, μg	(0.1-0.5)	
Formaldehydogenic steroids, μg	(3-140)	
17-Ketosteroids, μg	100 - 160	
Oxycorticosteroids, μg	(1.0-6.0)	
Noradrenaline, μg	0.41(0.18-0.9)	
Vitamins, A,D,K, μg	(0-trace)	
Ascorbic acid, μg	380(130-790)	70(60-70)
Biotin, μg	0.4(0.33-0.75)	1.9(0.63-6.64)
Carotenes, μg		(20-600)
Choline, μg	90(80-130)	
Cobalamin, μg	0.0004(0.00023-0.00079)	
Vitamin E, μg		308(226-391)
Folic acid group, μg	0.2(0.03-0.3)	4.3(1.8-7.7)
Inositol, μg	170(170-220)	
Nicotinic acid, μg	(11-105)	52(12-124)
N-methyl nicotinamide, μg	130(73-400)	
Pantothenic acid, μg	44(20-100)	31.4(3.85-63.4)
Para-aminobenzoic acid, μg	2.11(2.0-3.0)	3.5(1.01-8.2)
Pyridoxal, μg	3.0(0.7-5.4)	

TABLE 4 (Continued)

Compound	Origin	
	Urine mg/kg body wt.	Feces mg/kg body wt.
Pyridoxamine, μg	1.6(0.3-2.1)	
4-Pyridoxic acid, μg	51(9-160)	
Riboflavin, μg	14.3(0.2-22.5)	14.7(8.0-23.0)
Thiamine, μg	2.6(0.43-5.6)	7.8(0.67-18.0)
Trigonelline, μg	(30-300)	
Acetone bodies	0.285(0.03-0.7)	
Allantoin	0.27(0.18-0.36)	
Histamine	(0.2-1.0)	
Hydroxytyramine, μg	(1.4-2.8)	
Imidazole derivatives	(1.35-9.4)	(0.0-0.2)
Indican	0.14(0.06-0.45)	
Methionine sulfoxide	(0.0-0.31)	
Purine bases	0.41(0.18-0.92)	(2-3)
Reducing substances	(7-20)	
Sugars, as glucose	1.4	
Taurine	(0.105-0.2)	
Volatile acids, total, ml of 0.1 N NaOH		2.66(1.61-4.45)

A. Organic Matter Distribution.

Using the physical fractionation procedures previously mentioned (settling, centrifugation, and filtration) the distribution of domestic waste water organic matter among these fractions has been determined. A summary of these results is presented in Table 5. Thus, the ratio of particulate to soluble organic matter is 3:2, with the contributions of the particulates to the total organic matter decreasing with their particle size.

B. General Analyses.

Certain of the studies that contributed data on the distribution of

TABLE 5

Physical Distribution of Domestic Waste Water Organic Matter[a]

Physical Fraction	Domestic Waste Water Source				
	Bernardsville, N. J. 1966 (16)	Lambertville, N. J. 1966 (16)	Highland Park, N. J. 1954 (8)	Highland Park, N. J. 1959-60 (6)	English 1957-58 (4)
Settleable	23	25	31	30	30
Supracolloidal	18	19	24	19	19
Colloidal	9	9	14	10	13
Soluble	50	47	31	41	38

[a] Based on volatile solids determinations except for English results, which are based on volatile solids roughly calculated from the authors' reported organic carbon results.

organic materials in domestic waste waters also supplied considerable information on their general organic composition. A summary of these results for two composited American domestic sewages is presented in Table 6 (11). It is evident from these results that most of the waste water organics were detected and that these organics were mainly carbohydrates, amino acids, and fatty acid-esters. The same observations can be made from the results presented in Table 7 (4) for an English domestic waste water, even though the difference in concentration units makes direct comparison difficult.

C. Specific Analyses.

In addition to comprehensive analyses for major organic constituents of domestic waste waters, many analyses have been made to detect certain organic compounds or group compounds of interest to the investigators. This allows a more detailed look at the organic composition of domestic waste waters, but it should be noted that many of these were on grab rather than composited waste waters and, therefore, the reported concentrations are subject to wide variations.

1. Carbohydrates and sugars.

Particulate domestic waste water carbohydrates are in the form of polymers or polysaccharides, which differ in their behavior as well as in their composition. In general, carbohydrates in an American domestic waste water were found to be distributed as follows:

Pectin	12.7%
Hemicellulose	26.5%
Cellulose	60.8%

from a total of 35.4 mg/l carbohydrates. Hydrolysis of an English carbohydrate indicated the presence of glucose, galactose, arabinose, xylose, and rhamnose, with lesser amounts of mannose, maltose, ribose, and fructose (4).

Soluble carbohydrates in American domestic waste water analyzed about 10 mg/l, with a hexose/pentose ratio of 13:1. An English domestic waste water with a hexose/pentose ratio of 10-12:1 contained glucose, sucrose, and lactose with lesser concentrations of galactose, fructose, xylose, and arabinose (4,15).

TABLE 6

The Contributions of Physically Separated Sewage Fractions to Whole Sewage Organic Matter (11)

	Sampling Period									
	Winter-Spring, 1959					Fall-Winter, 1959-60				
Organic Constituent	Settle-able mg/l	Supra-Colloidal mg/l	Colloidal mg/l	Soluble mg/l	Constitu-ent Sum mg/l	Settle-able mg/l	Supra-Colloidal mg/l	Colloidal mg/l	Soluble mg/l	Constitu-ent Sum mg/l
Acids	0.89	1.64	1.48	22.56	26.57	1.14	1.66	1.82	33.95	38.57
Bases	-	-	-	3.24	3.24	-	-	-	3.55	3.55
Solvent Sol. Neutrals	9.45	6.64	3.58	13.59	33.26	8.99	5.64	3.54	17.31	35.48
ABS	0.08	0.14	0.10	3.94	4.26	0.11	0.13	0.09	4.02	4.35
Phenols	0.004	0.002	0.002	0.12	0.128	0.005	0.001	0.001	0.10	0.107
Cholesterol	0.04	0.02	0.03	0.03	0.12	0.02	0.03	0.04	0.05	0.14
Creatine-Creatinine	-	-	-	0.20	0.20	-	-	-	0.17	0.17
Uric Acid	-	-	-	0.33	0.33	-	-	-	0.34	0.34
Tannin-Lignin	2.02	1.05	1.15	-	4.22	2.12	0.64	1.34	-	4.10
Cellulose	11.50	3.15	2.43	-	17.08	21.58	2.05	2.34	-	25.97
Hemi-cellulose	3.53	4.32	1.33	-	9.18	3.21	5.36	0.94	-	9.51
Pectin	1.48	2.16	1.35	-	4.99	1.30	1.32	1.35	-	3.97

TABLE 6 (Continued)

	Sampling Period									
	Winter-Spring, 1959					Fall-Winter, 1959-60				
Organic Constituent	Settle-able mg/l	Supra-Colloidal mg/l	Colloidal mg/l	Soluble mg/l	Constitu-ent Sum mg/l	Settle-able mg/l	Supra-Colloidal mg/l	Colloidal mg/l	Soluble mg/l	Constitu-ent Sum mg/l
Hexoses	0.26	0.13	0.10	9.77	10.26	0.24	0.14	0.10	8.65	9.13
Pentoses	-	-	-	0.77	0.77	-	-	-	0.66	0.66
Amino Acids	8.59	12.84	5.37	9.05	35.85	12.41	12.26	4.76	9.01	38.44
Fraction Sum	37.84	32.09	16.92	63.60	150.46	51.12	29.23	16.32	77.81	174.49
Volatile Solids	52.25	41.29	20.64	72.19	186.37	63.89	36.86	18.79	87.86	207.40

TABLE 7

Organic Composition of English Domestic Waste Water[a]

Organic Constituent	Concentration[b]		
	Soluble mg/l	Particulate mg/l	Total mg/l
Total Carbohydrate	30.5	13.5	44.0
Free Amino Acids	3.5	0	3.5
Bound Amino Acids	7	21.25	28.25
Higher Fatty Acids	0	72.5	72.5
Soluble Acids	22.75	5	27.75
Esters	0	32.7	32.7
Anionic Surfactants	11.5	4	15.5
Amino Sugars	0	0.7	0.7
Amide	0	1.35	1.35
Creatinine	3.1	0	3.1
Fraction Sum	78.35	151	229.35
Present in Waste Water	94	211.5	305.5

[a] Average of two composite waste waters (4).
[b] In mg/l organic carbon.

2. Amino acids.

Whether in the soluble or particulate fractions, amino acids exist mainly as polymers of various molecular weights, compositions, and properties (peptides and proteins). To determine the amino acid contents of these materials requires hydrolysis. However, extended hydrolysis by boiling with 50% hydrochloric acid might be expected to degrade part of the organic material. Thus, the total amino acid nitrogen recovered after hydrolysis does not equal the Kjeldahl organic nitrogen concentration averaging only about 40 - 70% of this value (4, 3, 15, 17). In a nitrogen balance, it was observed that 10% of the organic nitrogen could not be explained by amino acids or their degradation products: humin and ammonia (8). It has also been reported that 8% of the organic nitrogen of waste water colloids was due to purenes and pyrmidines (18). Table 8 indicates what concentrations of free and combined amino acids may be found in domestic waste water. Table 9 gives a more detailed picture of the amino acid contents of domestic waste water particulates.

3. Non-amino acid nitrogenous compounds.

Those compounds having nitrogen but not classified as amino acids that have been detected in domestic waste water are presented in Table 10. In addition to these compounds, such enzymes as lipase, diastase, catalase, pepsin, and trypsin are probably present in a free form while others such as uricase are probably present in bacterial cell walls (32).

4. Organic acids and esters.

In domestic waste waters the higher fatty acid-fatty acid ester contents are in the range of 50-100 mg/l (4, 6). However, there is little agreement as to the distribution and nature of these acids and esters. In general, the following ranges have been determined (4, 6, 33, 34):

Free Acids	=	5 - 70% of total acids
Esterified Acids	=	30 - 95% of total acids
Saturated Acids	=	48 - 86% of total acids
Unsaturated Acids	=	14 - 52% of total acids

An example of the individual acid distributions for an Indian domestic waste water was found (33):

Saturated Acids	Concentration	Unsaturated Acids	Concentration
Lauric	0.12 mg/l	Oleic	17.0 mg/l
Myristic	0.24 mg/l	Linoleic	9.96 mg/l
Palmitic	11.7 mg/l	Linolenic	0 mg/l
Stearic	4.6 mg/l	Total Unsat.	17.96 mg/l
Total Saturated	16.66 mg/l		

The acids present in the soluble fraction of domestic waste water consists of the lower or volatile fatty acids and more polar soluble

TABLE 8

Amino Acid Contents of Domestic Waste Water

Constituent Amino Acid	Concentration Free mg/l	Concentration Total mg/l	References
Cystine	0 - trace	1.4-5.7	(19, 20, 21, 22)
Lysine and Histidine	Trace	5.1-9.7	(20, 21, 22)
Histidine	Present	Present	(19, 22)
Lysine	Absent Present	Absent Present	(19) (22)
Arginine	Trace	4.6-11.0	(19, 20, 21, 22, 23)
Serine, Glycine and Aspartic Acid	0.02-0.13	9.4-19.4	(20, 21, 22, 23)
Threonine and Glutamic Acid	0.01-0.18	4.5-24.8	(20, 21, 22, 23)
Alanine	0.02-0.09	5.1-11.9	(20, 21, 22)
Proline	0	0	(20, 21)
Tyrosine	0.06-0.09	1.7-6.4	(20, 21, 24)
Methionine and Valine	0.05-0.24	0.09-15.7	(20, 21, 22, 23)
Phenylalanine	0.02-0.33	4.7-16.8	(20, 21)
Leucine	0.06-0.28	4.2-13.1	(20, 21, 22, 23)
Tryptophane	Present	Present	(23, 25)

TABLE 9

Amino Acid Contents of Domestic Waste Water Particulates (8)
(Highland Park, N. J.)

Amino Acid	Settleable mg/l	Supra-colloidal mg/l	Colloidal mg/l	Total mg/l
Alanine	1.30	2.26	0.86	4.42
Cystine	0.63	1.27	--	1.90
Aspartic Acid	1.21	2.01	1.07	4.29
Glutamic Acid	1.88	2.50	0.79	5.18
Glycine	0.81	2.22	0.36	3.39
Histidine	1.31	--	0.72	2.03
Leucines and Phenylalanine	2.23	2.32	0.87	5.42
Lysine	--	1.44	0.04	1.48
Methionine and Valine	1.68	1.74	0.79	4.21
Serine	1.16	0.64	0.03	1.83
Threonine	0.78	0.78	0.19	1.85
Tyrosine	1.40	0.47	--	1.87
Unidentified[a]	1.03	1.87	0.72	3.64
Sum of Amino Acids	15.44	19.52	6.44	41.50
Nitrogenous Matter[b]	20.02	24.26	8.24	52.52

[a] As leucine.
[b] Total Kjeldahl nitrogen x 6.25.

TABLE 10

Non-amino Acid Nitrogenous Constituents of Domestic Waste Water

Compound	Concentration mg/l	References
Urea	2 - 16	(26)
Muramic Acid	0.5	(4)
Amino Sugars	1.2 - 2.2	(4)
Uric Acid	0.2 - 1.0	(6,15,19,27,28)
Hippuric Acid	Present	(4)
Xanthine	Trace	(19)
Indole	0.00025	(24)
Skatole	0.00025	(24)
Aliphatic Amines	0.1	(15)
Creatine-Creatinine	0.2 - 7.0	(4,6,19)
Organic Bases	3.4	(6)
Thiamine	0.029	(29)
Riboflavin	0.022 - 0.044	(29,30)
Niacin	0.135	(29)
Cobalamin	0.0008	(29)
Biotin	0.0003	(31)
Pantothenic Acid	Present	(29)
Folic Acid	Present	(29)

organic acids. The distribution of the volatile acids is presented in Table 11. The non-volatile soluble organic acids detected include lactic, citric, gallic, glutaric, glycollic, oxalic, pyruvic, succinic, benzoic, phenyllactic, phenylacetic, and hydroxybenzoic acids in concentrations ranging from 0.1 to 2 mg/l (15).

5. Anionic surfactants.

Due to the persistence of the original dodecylbenzene sulfonates (ABS) in the environment and the problems believed to be or actually associated with their presence, a large body of data exists on the concentration of these materials in domestic waste waters. For thirteen New Jersey waste waters, the ABS concentrations ranged from 1.2 to 20.1 mg/l, with an average of 6.9 mg/l (3,37). For eight American waste waters the concentration range was 2.8 to 39.0 mg/l with an average of 9.1 mg/l (7,37,38,39,40,41,42), and for ten English waste waters, the concentration range was 2.6 to 30.1 mg/l with an average of 12.0 mg/l (2,43,44,45,46,47,48).

6. Miscellaneous organic compounds.

A list of the organic compounds that have been detected in domestic waste water but which do not fall into the previous categories is presented in Table 12.

IV. WASTE WATER EFFLUENTS

Conventional domestic waste water treatment removes organic material by sedimentation and biological oxidation. It should be noted, however, that in addition to simple removal of organic matter, the organics remaining may be altered or even synthesized during the process.

A. Organic Matter Distribution.

A comparison of the organic matter distributions for waste water treatment plant influents and effluents is shown in Table 13. It is notable that the particulate waste water organics as measured by volatile solids undergo a much larger reduction than the soluble organics during secondary treatment.

TABLE 11

Volatile Acid Contents of Domestic Sewages

Volatile Acid	English Sewage (4) mg/l	American Sewage (35) mg/l	American Sewage (35) mg/l	American Sewage (25) mg/l	American Sewage (36) mg/l
Formic	0	0.1	1.3	0	0
Acetic	10.0	11.0	2.5	36	6
Propionic	2.6	2.8	2.2	8	1.2
Butyric	1.0	1.0	1.4	17	0.4
Valeric	0.4	0	0	0	0

TABLE 12

Miscellaneous Organic Compounds Present in Domestic Waste Water

Compound	Concentration mg/l	References
Lignin	1.5 - 5.0	(4, 6)
Aromatic hydrocarbons	1.3	(6)
Aliphatic hydrocarbons	4.0	(6)
Phenols	0.1 - 1.0	(6,15)
Nonionic surfactants	1 - 2	(49)
Cholesterol	0.3 - 0.26	(6, 50)
Coprostanol	0.1 - 0.75	(50)
Ascorbic acid	Present	(25)

B. General Analyses.

General analyses for effluents have been somewhat more limited than those for influents, i.e., waste waters. In the analyses of

TABLE 13

Distributions of Organic Matter for Activated Sludge Plants Influents and Effluents (9) (Winter-Spring 1965-66)

Waste Water Fraction	Bernardsville, N.J.				Lambertville, N.J.			
	Influent		Effluent		Influent		Effluent	
	mg/l	%	mg/l	%	mg/l	%	mg/l	%
Soluble	127	50	71	79	185	47	94	78
Colloidal	22	9	2	2	36	9	4	3
Supracolloidal	44	18	16	18	73	19	19	16
Settleable	57	23	1	1	97	25	4	3
Total	250	100	90	100	391	100	121	100

American waste water treatment plant effluents, the following results were obtained for the non-particulate organics (14):

Ether extractables	< 10% of total COD
Proteins	< 10% of total COD
Carbohydrates and Polysaccharides	< 5 % of total COD
Tannins and Lignins	< 5% of total COD
ABS (or MBAS)	~10% of total COD
Unidentified	~65% of total COD

This low recovery was also encountered by other investigators for an English waste water, where Table 14 indicates that an analytical method that had revealed 80% of the organic constituents of domestic waste water revealed now only 29% of the organic constituents of effluents (4). The reason for these low recoveries is not known at present. Only that it may be whatever organics are formed or altered during treatment or persist through treatment are sufficiently different from the organics in waste water influents so as to render similar analytic methods inappropriate.

C. Specific Analyses.

Even less is known about the detailed organic composition of

TABLE 14

Comparative Composition of a Domestic Sewage and Trickling Filter Effluent (for an English sewage in mg/l carbon)

Constituent	Influent		Effluent	
	Particulate	Soluble	Particulate	Soluble
Fat-acid	7.10	0	0.12	0
Fat-ester	28.2	0	0.12	0
Protein	23.0	8.0	2.74	0.25
Amino Acids	0	5.0	0	0.06
Carbohydrates	15.0	40.0	1.39	0.24
Soluble Acids	4.0	17.0	0.13	1.65
Amides	1.5	0	--	--
MBAS	3.0	11.0	0.05	1.40
Creatinine	0	3.5	--	--
Amino Sugars	1.8	0	0.38	0
Muramic Acid	0.2	0	0.05	0
Sum	147	85	5.0	3.6
Total Carbon	205	106	12.9	14.0

effluents than their general composition. Therefore, this section will simply discuss particulate and soluble organic constituents.

1. Effluent particulates.

Table 14 indicates what little is known about the organic composition of trickling filter effluent particulates. Little or nothing comparable to this is known about activated sludge effluent particulates. However, a certain amount of information is known about the organic composition of activated sludge itself. This information should be somewhat applicable to activated sludge treatment plant effluent organics. Table 15 describes the amino acid composition of activated sludge, and Table 16 its vitamin content.

The fatty acid and ester content of activated sludge is low, and the distribution comparable to the one previously discussed for influent fatty acids is as follows (33):

Saturated Acid	Concentration mg/l	Unsaturated Acid	Concentration mg/l
Lauric	0.11	Oleic	1.1
Myristic	0.13	Linoleic	2.1
Palmitic	1.3	Linolenic	0.06
Stearic	0.93	Total	2.26
Total	2.47		

In addition, there are such sugars as sucrose, lactose, arabinose, and glucose present (54). DNA is in the range of 3-7% (55).

2. Effluent solubles.

In the soluble fraction of the effluents, more direct information is available. Table 17 gives the average concentration of volatile (lower fatty) acids in secondary effluents (50). In addition to these, gallic, citric, and, perhaps, lactic acid have been detected also in effluents (35). Soluble amino acids are removed to a considerable extent during biological treatment and, thus, little has been done on their molecular composition (21, 22). It is possible that leucine, valine, serine, glycine, aspartic acid, glutamic acid, and threonine may be present (21, 22). Pyrene, a polynuclear hydrocarbon, has been detected also in municipal treatment plant effluents in concentrations of 0.4-1.0 μg/l (56).

Nonionic surfactants have been reported in concentrations of 0.5-1.0 mg/l (49). The sterols, cholesterol and coprostonol, have been detected in activated sludge effluents in concentrations of 15 and 8 μg/l, and in trickling filter effluents at 57 and 102 μg/l (16).

TABLE 15

Amino Acid Contents of Activated Sludge Hydrolysates

Amino Acid	Sludge Source		
	Switzerland (51) mg/g	Japan (52) mg/g	India (21) mg/g
Cystine	--	trace	14.8
Lysine	48.0	1.3	26.9
Histidine	5.0	--	
Arginine	20.0	8.7	18.1
Serine	16.0	2.0	36.4
Glycine	24.0	35.0	
Aspartic Acid	--	11.2	
Threonine	14.6	12.9	34.3
Glutamic Acid	--	30.4	
Alanine	--	20.5	22.1
Proline	--	trace	trace
Tyrosine	9.2	2.8	12.8
Methionine	--	--	31.7
Valine	20.0	23.5	
Phenylalanine	10.0	11.7	27.5
Leucine	22.0	72.6	29.9
Isoleucine	18.2		

TABLE 16

Vitamin Content of Activated Sludge (53)

Vitamin	Contents in mg/100 g	
	Centrifuged	Dried
Thiamine	2.05	0.63
Riboflavin	2.16	1.3
Pyridoxine	trace	0.7
Nicotinic Acid	13.6	10.0
Panthothenic Acid	4.6	4.4
Biotin	158[a]	83.4[a]
Folic Acid	182	140
B_{12}	159	100

[a] In micrograms/100 g.

TABLE 17

Volatile Fatty Acids in Secondary Effluents (50)
(average of 13 sewage effluents)

Acid	Concentration μg/l	Acid	Concentration μg/l
Formic	91.0	Butyric	30.7
Acetic	130.0	Isovaleric	73.4
Propionic	13.7	Valeric	8.1
Isobutyric	26.5	Caproic	47.9

Uric acid has been detected in concentrations ranging from 5-12 μg/l (28). At present, it would be expected that the ABS (anionic surfactant) concentration in effluents today would be lower than those recorded several years ago. In unpublished results by the author, the following was found for composited domestic waste water and an activated sludge treatment plant effluent from Bernardsville, N.J.:

	1963	1967
Influent ABS	6.4 mg/l	11.6 mg/l
Effluent ABS	5.6 mg/l	1.6 mg/l

These results should be considered only as indicative of the possibility that biological treatment plant effluents, at present, contain relatively minor amounts of anionic surfactants.

Finally, certain high molecular weight materials have been detected if not identified. A polysaccharide was precipitated from a glucose fed activated sludge (57), and a pigment called "Hestianic Acid" was recovered from a biological treatment effluent by acid precipitation (58). Although not pure, this material seems to resemble somewhat humic acids, differing mainly by having a much higher nitrogen content. As previous evidence indicated that as much as 40% of the soluble organics in effluents were non-dialyzable (14), the presence of such material is not surprising.

V. INDUSTRIAL WASTE WATERS

There are few generalizations that can be made about industrial waste water organic constituents. There are large variations of strength and composition with time, and the individual organic constituents vary from the relatively predictable (chemical industry) to the relatively unpredictable (pulp industry). Certain wastes, as those from the food processing industries, more or less resemble domestic sewage.

Due to the considerably more individualistic nature of industrial waste waters, data on organic matter distributions and general organic composition are lacking. Considerable results of analyses for individual organics have been published, but it is not truly possible to obtain a general picture of the organic composition of industrial waste waters from such data.

The description of individual organics in this section, therefore, is limited to Tables 18-24, noting those organics that have been detected and their source. This list is not intended to be exhaustive

TABLE 18

Organic Acids Found in Industrial Waste Waters

Compound	Source	Reference
Formic	Wood pulp production	(59, 60)
	Flax retting	(61)
Acetic	Wood pulp production	(59, 60)
	Flax retting	(61)
	Sugar refining	(62)
	Silage (grass)	(63)
Lactic	Silage (grass)	(63)
	Olive oil refining	(64)
Butyric	Wood pulp production	(59)
Oxalic	Wood pulp production	(59)
	Sugar refining	(62)
	Olive oil refining	(64)
Tartaric	Wine making	(65)
	Olive oil refining	(64)
Abietic	Wood pulp production	(66)
Dehydroabietic	Wood pulp production	(66)
p-Hydroxybenzoic	Wood pulp production	(59)
Pyruvic	Sugar refining	(62)
Adipic	Sugar refining	(62)
Malonic	Sugar refining	(62)
	Olive oil refining	(64)
Vanillic	Wood pulp production	(59)
Ferulic	Wood pulp production	(59)
Maleic	Olive oil refining	(64)
Fumaric	Olive oil refining	(64)

TABLE 19

Carbohydrates Found in Industrial Waste Waters

Compound	Source	References
Sucrose	Sugar refining	(62)
	Pineapple cannery	(67)
	Sugar beet processing	(68)
Raffinose	Sugar beet processing	(68)
Glucose	Silage (grass)	(63)
	Wood pulp production	(69)
Galactose	Silage (grass)	(63)
	Wood pulp production	(69)
Fructose	Wood pulp production	(69)
	Silage (grass)	(62)
Xylose	Silage (grass)	(62)
	Wood pulp production	(69)
Arabinose	Wood pulp production	(69)
	Silage (grass)	(62)
Rhamnose	Wood pulp production	(69)
Cellobiose	Wood pulp production	(69)
Glacturonic acid	Wood pulp production	(69)

TABLE 20

Organic Bases Found in Industrial Waste Waters

Compound	Source	References
Choline	Brewery	(70)
Melamine	Resin manufacture	(71)
Methylamine	Sugar refining	(62)
Ethanolamine	Sugar refining	(62)
Allylamine	Sugar refining	(62)
Amylamine	Sugar refining	(62)
Benzidine	Dye manufacture	(72)
Naphthylamine	Dye manufacture	(72)
β-Naphthylamine	Dye manufacture	(72)
Pyridine	Ammoniacal liquor (coal gas)	(73, 74)
	Coke plants	(75)
Methylpyridenes	Ammoniacal liquor (coal gas)	(74)
	Coke plants	(75)
Dimethylpyridenes	Ammoniacal liquor (coal gas)	(74)
Methylethylpyridenes	" " " "	(74)
Diethylpyridenes	" " " "	(74)
Trimethylpyridenes	" " " "	(74)
Aniline	" " " "	(74)
Methylaniline	" " " "	(74)
Dimethylaniline	" " " "	(74)
Quinoline	" " " "	(74)
Isoquinoline	" " " "	(74)

TABLE 21

Polynuclear Hydrocarbons Found in Industrial Waste Water

Compound	Source	References
3,4-Benzpyrene	Shale oil	(76)
	Coke production	(77,78)
	Oil refining	(79)
2,6-Dimethylnaphthalene	Oil refining	(80)
2,6-Dimethylanthracene	Oil refining	(80)
Benzpyrenes	Acetylene Production	(81)
Benzperylenes	" "	(81)
Pyrene	" "	(81)
Fluorene	" "	(81)
Anthracene	" "	(81)
Perylene	" "	(81)
Acenaphthylene	" "	(81)
Naphthalene	" "	(81)
Fluoranthene	" "	(81)
Coronene	" "	(81)
9,10-Dibenzpyrene	" "	(81)

TABLE 22

Phenols Found in Industrial Waste Water

Compound	Source	References
Dinitro-o-cresol	Pesticide manufacture	(82)
2,4-Dichlorophenol	Pesticide manufacture	(83)
p-Nitrophenol	Dye manufacture	(84)
p-Aminophenol	Dye manufacture	(84)
Phenol	Synthetic resin manufacture	(85)
	Coke manufacture	(86)
	Oil refining	(87)
o-Cresol	Coke manufacture	(86)
	Ammoniacal liquor (coal gas)	(73)
m-Cresol	Coke manufacture	(86)
	Ammoniacal liquor (coal gas)	(73)
p-Cresol	Ammoniacal liquor (coal gas)	(73)
	Coke manufacture	(86)
Xylenol	Ammoniacal liquor (coal gas)	(73)
Ethylphenol	" " " "	(73)
Guaiacol	Wood pulp production	(69)
Vanillin	" " "	(88, 69)
Vanillic Acid	" " "	(69)

TABLE 23

Miscellaneous Aromatics Found in Industrial Waste Water

Compound	Source	References
Divinylbenzene	Plastic manufacture	(89)
2, 4-Dinitrobenzene	Dye manufacture	(84)
Nitrotoluene	TNT manufacture	(90)
Dinitrotoluene	TNT manufacture	(90)
2, 4-Dinitrochlorobenzene	Dye manufacture	(91)
Nitrobenzene	Oil refining	(87)
Diphenyldisulfide	Oil refining	(80)

but merely indicative of the types of compounds detected in industrial waste waters. The division of organics listed here is more or less arbitrary, being roughly grouped into acids, bases, carbohydrates, polynuclear hydrocarbons, phenols, and miscellaneous aliphatics and aromatics. In addition, it should be noted that thiamine, biotin, pyridoxine, riboflavin, pantothenic acid, nicotinamide, p-amino-benzoic acid, inositol, and ergosterol have been detected in brewery waste waters (70), and fenchone, camphor, α terpineol, fenchyl alcohol, limonene, and terpinolene have been detected in Kraft (wood pulp) production wastes (100).

VI. CONCLUSIONS

The problems caused by organic materials entering the aquatic environment fall roughly into two categories. Soluble organics that undergo biological oxidation tend to reduce and, perhaps, deplete receiving water dissolved oxygen if present in sufficient concentrations.

TABLE 24

Miscellaneous Aliphatics Found in Industrial Waste Water

Compound	Source	References
Dimethyldisulfide	Wood pulp production	(92)
Methylmercaptan	" " "	(92)
Methanol	" " "	(92)
Methanol	Synthetic resin manufacture	(85)
Formaldehyde	Wood pulp production	(88)
Formaldehyde	Synthetic resin manufacture	(85)
Furfural	Wood pulp production	(69, 85, 93)
Betanine	Sugar beet processing	(94)
Methyl mercury chloride	Plastic manufacture	(95)
Isoprene	Synthetic rubber production	(96)
Dimerized isoprene	" " "	(96)
Dimethyldioxane	" " "	(96)
Dimethylformamide	Synthetic fiber production	(97)
Caprolactam	" " "	(98)
Aminocaprolactam	" " "	(99)

The biological population resulting from this degradation is frequently a nuisance and almost always aesthetically unattractive. Particulate organics settle to the bottom of water bodies forming deposits that exert a slow but extended oxygen demand on the receiving water, reduce sulfates to hydrogen sulfide, and are active in the release of organics and nutrients back into the overlying waters.

Organics that do not undergo biological degradation in the aquatic environment emphasize a different problem. This is the specific nuisances caused by organic materials,such as odor, color, foam, toxicity, and, possibly, carcinogenic properties. This is not intended to imply that only persistent organics cause such problems, only that they intend to center attention on them. This is exemplified by past concern about surfactants and present concern about dyes and pesticides. Both these categories stress only the more obvious reasons for an interest in waste water organics. There are also other possible roles for waste water organics, as in trace metal transport, etc., that have only recently come under detailed investigation.

In general, organic wastes originating from such activities as cooking, personal washing, urine, feces, industrial food processing, brewing, etc., are largely biodegradable and, therefore, are of immediate stream damage concern. Wastes originating from laundering, chemical industries, wood pulp production, plastic manufacture, etc., may contain organics that are resistant to biological degradation and are, therefore, of concern regarding their eventual effect on water quality and water reuse. Unfortunately, at present more is known about the nature and effects of the readily degraded organics than is known about the nature and eventual effects of the persistent organics.

ACKNOWLEDGMENT

Paper of the Journal Series, New Jersey Agricultural Experiment Station, Rutgers, The State University, Department of Environmental Sciences, New Brunswick, N. J.

REFERENCES

1. Report by the Subcommittee on Environmental Improvement, Cleaning Our Environment, The Chemical Basis for Action, American Chemical Society, Washington, D. C., 1969.

2. J. Murray, discussion of paper by D. Rayner, J. Proc. Inst. Sew. Purif., 27 (1960).

3. R. Manganelli, H. Orford, and C. Henderson, presented at the 32nd Annual Meeting, New York Sewage and Industrial Wastes Association, January 20-22, 1960.

4. H. H. Painter and M. Viney, J. Biochem. Microbiol. Technol. Eng., 1, 143 (1959).

5. P. McGauhey and S. Klein, Sewage Ind. Wastes, 31, 877 (1959).

6. J. V. Hunter and H. Heukelekian, J. Water Pollut. Contr. Fed., 37, 1142 (1965).

7. G. Malaney, W. Sheets, and J. Ayres, J. Water Pollut. Contr. Fed., 32, 1161 (1960).

8. H. Heukelekian and J. Balmat, Sewage Ind. Wastes, 31, 413 (1959).

9. D. Rickert and J. V. Hunter, J. Water Pollut. Contr. Fed., 39, 1475 (1967).

10. S. D. Faust and M. Manger, Int. J. Air-Water Pollut., 9, 565 (1965).

11. J. V. Hunter, Ph.D. Thesis, Rutgers, The State University, New Brunswick, N. J., 1962.

12. S. Kobayshi and G. F. Lee, Anal. Chem., 36, 2197 (1964).

13. W. Spector, Handbook of Biological Data, W. B. Saunders Co., Philadelphia, Pa., 1956.

14. R. J. Bunch, E. Barth, and M. Ettinger, J. Water Pollut. Contr. Fed., 33, 122 (1961).

15. H. H. Painter, M. Viney, and A. Bywaters, J. Inst. Sewage Purif., 302(1961).

16. J. Murtaugh and R. J. Bunch, J. Water Pollut. Contr. Fed., 39, 404 (1967).

17. B. Bai, C. Viswanathan, and S. Pillai, J. Sci. Ind. Res., 21C, 72 (1962).

18. O. Bolotina, Vodosnabzh. Sanit. Tekhnol., 4, (1962), C.A., 61, 4054J (1964).

19. A. Buswell and S. Neave, Bulletin No. 3, Dept. of Registration and Education, State of Illinois, 1930.

20. C. Sastry, P. Subrahanyam, and S. Pillai, Sewage Ind. Wastes, 30, 1241 (1958).

21. P. Subrahanyam, C. Sastry, A. Rao, and S. Pillai, J. Water Pollut. Contr. Fed., 32, 344 (1960).

22. L. Kahn and C. Wayman, J. Water Pollut. Contr. Fed., 36, 1368 (1964).

23. M. Aurich, W. Dummler, and D. Mucke, Wasserwirtech. Technol. (Germany), 8, 496 (1958).

24. W. Rudolfs and N. S. Chamberlin, Ind. Eng. Chem., 24, 111 (1932).

25. W. Rudolfs and B. Heinemann, Sewage Ind. Wastes, 11, 587 (1939).

26. A. Hanson and T. Flynn, Proc. 19th Ind. Waste Conf., Purdue Univ., 32 (1964).

27. G. Kupchik and G. Edwards, J. Water Pollut. Contr. Fed., 34, 410 (1965).

28. J. O'Shea and R. Bunch, J. Water Pollut. Contr. Fed., 37, 1444 (1965).

29. E. Srinath and S. Pillai, Curr. Sci., 35, 247 (1966).

30. L. Kraus, Sewage Ind. Wastes, 14, 811 (1942).

31. H. Neujahr and J. Hartwig, Acta. Chem. Scand., 13, 954 (1961).

32. M. Sridhar and S. Pillai, J. Sci. Ind. Res., 25, 167 (1966).

33. C. Viswanathan, B. Bai, and S. Pillai, J. Water Pollut. Contr. Fed., 34, 189 (1962).

34. H. Heukelekian and P. Mueller, Sewage Ind. Wastes, 30, 1108 (1958).

35. H. Mueller, T. Larson, and W. Lennarz, Anal. Chem., 30, 41 (1958).

36. E. Hindin, D. S. May, R. McDonald, and G. H. Dunstan, Water Sewage Works, 111, 92 (1964).

37. S. D. Faust, Water Sewage Works, 100, 242 (1953).

38. D. Croft and S. D. Faust, Water Sewage Works, 101, 286 (1954).

39. R. Culp and H. Staltenberg, J. Amer. Water Works Assoc., 45, 1187 (1953).

40. W. Russell, Sewage Ind. Wastes, 36, 1041 (1954).

41. C. Keefer, Water Sewage Works, 99, 84 (1952).

42. E. Hurwitz, J. Water Pollut. Contr. Fed., 32, 1111 (1960).

43. D. Rayner, J. Proc. Inst. Sewage Purif., 27 (1960).

44. W. Lockett, J. Proc. Inst. Sewage Purif., 225 (1956).

45. J. O'Neill, discussion of paper by D. Rayner, J. Proc., Inst. Sewage Purif., 27 (1960).

46. F. Roberts, Water Waste Treat. J., 302 (1957).

47. W. Smith, J. Proc. Inst. Sewage Purif., 153 (1957).

48. S. Jenkins. J. Inst. Public Health Eng., 59, 29 (1960).

49. Ministry of Technology (Brit.), Notes on Water Pollution, No. 34, 1966.

50. J. Murtaugh and R. Bunch, J. Water Pollut. Contr. Fed., 37, 410 (1965).

51. V. Corti, Schweiz. Z. Hydrol., 15, 152 (1953).

52. T. Akiyama and T. Sato, Trans. Faculty of Medicine, (Kagoshima University), 15, 366 (1963).

53. V. Kocher and V. Corti, Schweiz. Z. Hydrol., 14, 333 (1952).

54. T. Brown, J. Protozool., 14, 340 (1967).

55. E. Genetelli, J. Water Pollut. Contr. Fed., 39, R32 (1967).

56. P. Wedgewood, J. Proc. Inst. Sewage Purif., 20 (1952).

57. P. L. Busch and W. Stumm, Environ. Sci. Technol., 2, 49 (1968).

58. A. Fredericks and D. W. Hood, Atomic Energy Commission Publ. AD618932, Washington, D. C., 1965.

59. E. Eldridge, Trans. 2nd Seminar Biological Problems on Water Pollution; R. A. Taft Sanitary Engineering Center, Tech. Rept. W60-3, 255, 1959.

60. L. Ruus, Svensk Papperstid, 67, 221 (1964), C.A., 61, 6771d (1964).

61. K. Menzel and I. Thomas, Faserforsch. u. Textiltechnol., 8, 138 (1957), Chem. Zbl., 128, 13846 (1957).

62. E. Leclerc and F. Edeline, Centre Belge'etude Document. Eaux, 114-115, 201 (1960).

63. W. Moore, H. Walker, E. Gray, and E. Weir, Water Waste Treat. J., 8, 226 (1961).

64. J. Ursinos, Grasas y Aceites, 10, 30 (1959).

65. N. Rizaev and K. Merenkov, Teoriga i Prakt., donnoga Obmena, Akad. Nouk. Kaz. - USSR, Tr. Resp. Soveskch, 171 (1962).

66. R. Maenpaa, P. Hyminen, and J. Tikkai, Paperi Puu, 50, 143 (1965), C.A., 68, 117039q (1968).

67. N. Burbank and J. Kumagi, Proc. 20th Ind. Wastes Conf., Purdue Univ., 365, (1965).

68. J. Laughlin, Proc. 4th Ind. Wastes Conf., Texas Water Pollut. Contr. Assoc., 2 (1964).

69. T. Maloney, and E. Robinson, TAPPI, 44, 137 (1961).

70. F. Knorr, Brauwiss., 18, 191 (1965).

71. A. Koganovskii, T. Levchenko, and V. Kirichenko, Bum. Prom., 7 (1968), C.A., 68, 107741Y (1968).

72. N. Takemura, T. Okiama, and C. Nakajima, Int. J. Air-Water Pollut., 9, 665 (1965).

73. A. Elliot and A. Lafreniere, Water Sewage Works, 111, R325 (1964).

74. M. Hughes, J. Appl. Chem., 450 (1962).

75. M. Ettinger, R. Lishka, and R. Kroner, Ind. Eng. Chem., 46, 791 (1954).

76. I. Veldre, L. Lake, and I. Arro, Gigiena i. Sanit., 30, 104 (1965), C.A., 64, 4782h (1966).

77. Z. Fedorenko, Vapr. Gigiena Naselen. Mest. Kiev, Sb., 5, 101 (1964), C.A., 64, 15561f (1966).

78. Z. Fedorenko, Gigiena i. Sanit., 29, 17 (1964), C.A., 61, 4055 (1964).

79. K. Ershova, Gigiena i. Sanit., 33, 102 (1968), C.A., 68, 10774ly (1968).

80. S. Brady, Proc. Div. Refining, Amer. Petrol. Inst., 48, 556 (1968).

81. V. Livke and R. Vodyanik, Khim. Prom., 30 (1969), C.A., 70, 80677j (1969).

82. S. Jenkins and H. Hawkes, Int. J. Air-Water Pollut., 5, 407 (1961).

83. I. Oshina and N. Tyurina, Khimiz, Sel'sk. - Khaz. Bashkirii, Ufa, 5b., 19 (1964).

84. K. Papov, Khim. Ind., 37, 203 (1965), C.A., 64, 4783e (1966).

85. K. Singleton, Purdue Univ. Engineering Bulletin Extension Series, No. 121, 62 (1967), C.A., 68, 15886w (1968).

86. L. Semenchenko, V. Kaplin, and A. Potateuv, Koko Khim., 42 (1967), C.A., 67, 67425u (1967).

87. Y. Karelin, M. Ikramov, D. Zhukov, and D. Komarov, Khim i Teknol., Topl. i. Masel., 9, 29 (1964), C.A., 61, 11742c (1964)

88. K. Christofferson, Anal. Chim. Acta., 31, 233 (1969), C.A., 61, 14346d (1964).

89. N. Progressov, E. Prikliodko, and L. Ivanilova, Vodosnabzh. Sanit. Tekh., 8 (1968), C.A., 69, 80017y (1968).

90. H. Kurmeier, Wasser Luft., 8, 727 (1964), C.A., 65, 18317f (1966).

91. K. Papov, Khim. Ind., 37, 164 (1965), C.A., 64, 1803f (1966).

92. N. Kardos, Khim. Ind., 9, 4 (1960), C.A., 61, 2817e (1964).

93. Y. Tsirlin, V. Vasil'eva, and A. Yasinkaya, Gidroliz. Lesokhim. Prom., 19, 12 (1966), C.A., 66, 88495m (1967).

94. R. Pailthrop, J. Water Pollut. Contr. Fed., 32, 1201 (1960).

95. K. Irukayaina, Proc. 3rd Int. Conf. Water Pollut. Res., 3, 153 (1967).

96. V. Ivanov, Okhr. Priro. Tsentr. - Chernozemn. Polsy, 137 (1962), C.A., 61, 1599h (1964).

97. M. Thonke and D. Dittmann, Fortschr. Wasserchem. Ihrer Grenzgeb, 4, 272 (1966), C.A., 67, 36195g (1967).

98. V. Livke, V. Gabernatorova, and L. Golovanova, Vestn. Teklin. i Ekon. Inform. Nauchn. Issled. Inst. Teklin - Ekon. Issled Gas. Kom. Khim. i Neft. Prom. pri Gosplane, USSR, 25 (1963), C.A., 61, 15824g (1964).

99. J. Kaeding, Fortsch. Wasserchem. Ihrer Grenzgeb, 5, 258 (1967), C.A., 67, 63806p (1967).

100. L. Kieth, Presented before the Division of Water, Air, and Waste Chemistry, American Chemical Society, Minneapolis, Minn., April 15, 1969.

CHAPTER 5

TRACE ORGANICS IN SUBSURFACE WATERS

Sotirios G. Grigoropoulos
University of Missouri - Rolla

John W. Smith
ESSO Research and Engineering Company

I. INTRODUCTION

The growing population and expanding industrialization of the country continually require more water and of higher quality. To meet this demand, new sources of water need to be developed and the quality of all waters must be protected. Ground water constitutes one of the major sources of fresh water in the United States. Geraghty (1) recently estimated that 95 percent of all fresh water is underground and only 5 percent is accounted for by surface water in rivers, streams, lakes, and ponds. He further reported that of the total water used for all purposes in the country, only about 17 percent is ground water. Failure to utilize more of the underground reserves is due, in part, to the lack of knowledge of the quality of subsurface water; this is especially true with regard to trace organic pollutants.

Organic substances occur in the environment as the result of natural processes or are added by man in his efforts to improve specific parts of the ecosystem or to dispose his wastes. These substances, therefore, originate from several sources, including domestic, industrial and agricultural wastes, accidental spillage, runoff, and by-products of the natural biota. Regardless of source, many of these materials eventually end up in water supplies, surface and subsurface. Although organic contaminants are normally present in water in trace quantities, their presence is highly objectionable because of the aesthetic problems, primarily taste and odor, color, foaming, and the toxicologic problems which they impart to the water. Recognizing the importance of trace organics in drinking water, the U. S. Public Health Service (PHS), in the 1962 drinking water standards (2), set the maximum permissible concentration of chloroform soluble organics at 200 μg/l. Furthermore, the American Water Works Association (AWWA) in 1968 adopted (3) substantially more exacting water quality goals of 40 and 100 μg/l for chloroform and alcohol soluble organics. These levels have been established primarily because of aesthetic considerations; however, of greater importance is the health hazard represented by the long-term toxic effects of these substances, emphasized by the recovery of carcinogenic substances from surface and subsurface water (4, 5, 6, 7) and the recent massive fish kills in the United States and Europe. Borneff et al. (5) have proposed recently that toxic polycyclic aromatic hydrocarbons, whose presence in water they felt had been consistently demonstrated, are biosynthesized in plants, transferred to the soil by dead plants, and then carried to ground water by percolation. Walker (8) has reported the pollution with hydrocarbons of several Illinois ground water aquifers resulting from breaks in transmission lines and leaking fuel oil tanks; a strong taste and odor was imparted to the ground water which could persist for many years. Because subsurface waters are not usually treated, except for disinfection, a serious health threat exists to the consumer by the presence of trace organics.

In recent years, many investigations have been undertaken in the organic micropollutants area; however, most of this work has been primarily concerned with surface water organics and, consequently, little information is available in the literature on subsurface water materials. This paper presents information on the concentration of trace organic substances in a number of subsurface waters, including a large spring and several wells, and discusses the general characteristics and properties of these materials. Emphasis is placed on the toxic effects of these pollutants, both acute and long-term, and the mechanism of their toxic action. A method is also presented which enables the prediction of long-term toxicity on the basis of short-term experimental data.

II. RECOVERY AND CONCENTRATION OF TRACE ORGANICS

The minute concentrations at which trace organic substances are found in water necessitates that an appropriate method be employed to concentrate and recover workable quantities. Four general approaches have been followed for this purpose. These are briefly outlined in Table 1. The carbon adsorption method has been tentatively selected as a standard procedure for the determination of organic contaminants

TABLE 1

Methods for the Recovery of Trace Organics

Method	Characteristics	Application
Carbon Adsorption	Monitoring and concentrating device; large volumes of water can be sampled and sufficient material can be recovered for extensive characterization and identification studies; any solvent may be used; several days required for results; adsorption and/or elution may not be complete; composition of organics may be altered.	Chemical, biochemical, organoleptic characterization; chromatographic and spectrometic analysis; toxicity studies; PHS CCE limit and AWWA CCE & CAE goals established for drinking water.
Liquid-Liquid Extraction	Concentrating and monitoring device; limited volumes of water can be sampled with small amount of material recovered, except when continuous flow systems are employed; solvent must be immiscible with water; less possibility of chemical or biochemical alteration of organics.	Primarily chromatographic and spectrometric characterization; pesticide analysis; ASTM tentative method.
Freeze Drying	Concentrating device; limited volumes of water can be sampled with small amount of material recovered; mineral content of water can be a problem; solvent, when used, must be immiscible with water; little chance for chemical or biochemical alteration of organics.	Primarily chromatographic and spectrometric characterization; study of organic color in water.
Ion Exchange	Concentrating and monitoring device; large volumes of water can be sampled and sufficient material can be recovered for characterization and identification studies.	Study of organic color in water.

in water (9). It consists of passing a known volume of water through an activated carbon column over a period of several days, removing and drying the carbon, and eluting the organics from the carbon by sequential extraction with appropriate solvents. This method offers the advantage of enabling the convenient sampling of large volumes of water and the recovery of workable quantities of organics. Concern has been expressed, however, that the sorption and desorption of organics may not be complete and that their composition may be altered while they are adsorbed on the carbon. Liquid-liquid extraction can be performed in a batch or a continuous flow system and has been utilized to recover enough material for chromatographic and spectrometric characterization. It is less amenable to the sampling of large volumes of water, but appears to be more efficient and reduces the possibility of chemical or biochemical alteration of the organics. This procedure was recently approved by the American Society for Testing and Materials (ASTM) as a tentative method for the separation of solvent-extractable organic matter from water (10). Freeze drying has been employed also in the chromatographic and spectrometric characterization of organic micropollutants, as well as the study of organic color in water. The mineral content of the water can be a difficult problem with this method. Ion exchange has been primarily employed in the study of organic color.

The carbon adsorption method has been widely used in the United States. It was developed by Braus et al. (11) and later modified by several investigators to accomplish different experimental goals. Some of the major modifications are outlined in Table 2. These include the use of two large capacity filters and sequential elution with chloroform and ethanol employed by Middleton et al. (12); acidification before the second filter employed by investigators at Washington University (13, 14, 15); and adjustment of the pH of the elutant employed by Robinson et al. (16) in their study of organic materials in Illinois ground waters. Robinson et al. used two or three small capacity carbon filters in series, acidified the water before it entered the second filter, and extracted the carbon with chloroform, water, ethanol, ethanol plus ammonia, and ethanol plus hydrochloric acid.

Investigations of trace organic substances in Missouri subsurface waters have employed (17) three large capacity carbon filter units in series (Figure 1). Each unit contained a 0.75 cubic foot central layer of fine carbon (Nuchar C-190, +30 mesh) held between two 0.38 cubic foot end layers of coarse carbon (Cliffchar, 4 x 10 mesh). Water at its natural pH and without any pretreatment was passed through the three units in an upflow direction. After an appropriate volume of water had been filtered, the carbon was removed from the filter, dried

TABLE 2

Some Modifications of the Carbon Adsorption Method

Source of Water	Modification	Procedure Employed	Ref.
Ohio River	Standard filter	Sand filtration; PHS (0.073 cu ft) filter; elution with chloroform.	(11)
Ohio River	Pretreatment, size and number of filters; elutants used.	Sedimentation and sand filtration; two large capacity (1.2 cu ft) filters in series; elution with chloroform and ethanol.	(12)
Missouri River	Pretreatment; size and number of filters; pH of water; elutants used.	Sedimentation and diatomite filtration; two large capacity (1.3 cu ft) filters in series; pH adjustment to 2.0 before second filter; elution with chloroform, ethanol, acetone, and benzene.	(13) (14) (15)
Illinois wells	No pretreatment; number of filters; pH of water; elutants used; pH of elutants.	Three PHS filters in series; pH adjustment to 2.5-3.0 before second filter; elution with chloroform, water, ethanol, ethanol plus ammonia, ethanol plus hydrochloric acid.	(16)
Missouri spring and deep wells	No pretreatment; size and number of filters; elutants used.	Three large capacity (1.5 cu ft) filters in series; elution with chloroform, ethanol, acetone and benzene, or benzene and acetone.	(17)

FIG. 1

Activated carbon filter arrangement at Meramec Spring.

at 40°C. for 5 days, and sequentially was extracted with chloroform, ethanol, benzene and acetone, or acetone and benzene.

The concentrations of trace organics recovered from several subsurface water sources are summarized in Table 3 where the PHS limit and AWWA goals for drinking water are also given to provide reference values. Rather high concentrations of organic materials, in the mg/l range, were recovered from the Illinois well waters, especially those at Oakwood and Clinton where three carbon filters were employed in series. Robinson et al. (16) reported that the third filter recovered as much as 50 or 60 percent of the material obtained from the second filter. Significant concentrations of trace organics were found in Missouri's Meramec Spring and detectable concentrations were present in the deep well waters. Meramec Spring is one of the largest springs in Missouri with an average annual flow of 96 MGD and a maximum flow of 420 MGD; it was sampled at the point where the water emerged from the ground. The University of Missouri - Rolla (UMR) well was 1150 feet deep and had not been in operation at the time of sampling because it had shown evidence of bacteriological contamination; it has been

TABLE 3

Concentration of Trace Organics Recovered from Subsurface Waters by the Carbon Adsorption Method

Source of Water		Carbon Filters		Water Sampled			Organic Extract, μg/l			
Type	Location	Size cu ft	Number	Volume gal	Rate gpm	Solvents Used (in sequence)	Total[a]	CCE[b]	CAE[b]	Ref.
Well	Oakwood, Ill.	<0.1	3	1,073	1.0	chloroform, water, ethanol, ethanol plus ammonia	5,430	2,170	1,140	(16)
	Philo, Ill.		2	950			1,500	130	940	
	Clinton, Ill. Run 1		2	1,426			3,750	340	2,000	
	Clinton, Ill. Run 2		3	1,153			7,290	550	3,310	
	Atwood, Ill.		2	1,293			4,570	340	1,360	
Deep Well	Rolla, Mo.	1.5	3	262,000	6.2	chloroform, ethanol, acetone, benzene	3.2	0.1	2.1	(17,18)
	UMR			142,000	6.8		24.2	4.8	8.0	
Spring	Meramec Spring, Mo. Run 1			129,000	5.5		224.2	43.7	109.0	
	Meramec Spring, Mo. Run 2			133,000	4.9		324.7	92.1	195.5	
Deep Well	Rolla, Mo.			262,000	6.2	chloroform, ethanol, benzene, acetone	9.3	0.1	2.1	
	UMR			142,000	6.8		59.3	4.8	8.0	
Spring	Meramec Spring, Mo. Run 1			129,000	5.5		210.9	43.7	109.0	
	Meramec Spring, Mo. Run 2			133,000	4.9		332.5	92.1	195.5	
					PHS Drinking Water Standard		200	--		(2)
					AWWA Potable Water Goals		40	100		(3)

[a]The sum of CCE, CAE, and other extracts, as appropriate.

[b]CCE: Carbon chloroform extract; CAE: Carbon alcohol extract.

since sealed off. The Rolla well is 1745 feet deep and part of the municipal water supply. The concentrations of chloroform and alcohol soluble organics in Meramec Spring were less than the values reported by Sproul and Ryckman (19) for a Missouri unpolluted surface water (98 and 220 μg/l, respectively), but well exceeded the AWWA goals for potable water.

One carbon filter was sufficient to recover all the measurable CCE and CAE present in the two Missouri deep wells. However, significant quantities of CCE and CAE were recovered with each of the three filters used in the two runs which were made at Meramec Spring (Figure 2). Units 2 and 3 of Runs 1 and 2 recovered about 26 to 27 and 33 μg/l, respectively, less than their preceding unit. On this basis, it is estimated that the three units effectively recovered all these organics from the water in the first run, while approximately 10 percent of the CCE and CAE present in the water was not recovered in the second run. Both runs were essentially of the same duration, 17 and 18 days, and sampled about the same volume of water 129,000 and 133,000 gallons. Run 2 began on the same day that Run 1 was terminated, however, following a period of heavy rainfall, the spring water was highly colored during the final days of Run 2. A karstic terrain prevails in the infiltration area of Meramec Spring and it is suspected that the spring is supplied to a large extent by surface waters from the immediate area. It is possible then that rain water passing through the soil had picked considerable quantities of organic materials and carried them to the spring water.

III. CHARACTERIZATION OF TRACE ORGANICS

Chemical and biochemical determinations have been employed for the characterization of trace organics recovered from subsurface waters. Results of elemental chemical analyses performed on several extracts are summarized in Table 4, together with the corresponding empirical formulas for the various materials. In general, the alcohol soluble materials were found to be more oxygenated and contained less carbon and more nitrogen than the chloroform soluble organics. When determined, the sulfur and phosphorus contents were found to be low; exceptions were the sulfur content of the CCE from Unit 1 of the UMR well which was almost 28 percent, and the phosphorus content of the CAE from Unit 2 of Run 2 at Meramec Spring which was 2.8 percent.

Data on the organic carbon content and oxygen demand of subsurface water organics are presented in Table 5. The carbon content was determined by elemental chemical analysis and a total carbon analyzer and the values obtained by the two methods were quite similar.

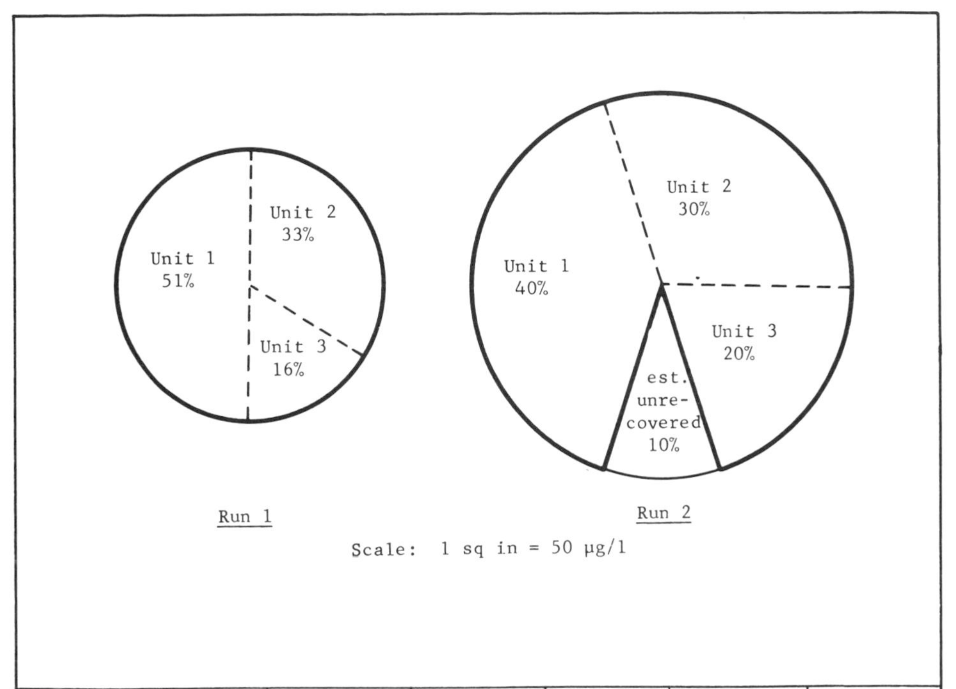

Source of Water	Run	Organic Extract	Unit 1		Unit 2		Unit 3		Unrecovered*		Total	
			μg/l	% of total#	μg/l	% of total	μg/l	% of total	μg/l	% of total	Recovered μg/l	Adjusted μg/l
Meramec Spring	1	CCE & CAE	78.0	51	50.4	33	24.3	16	0	--	152.7	152.7
	2		128.1	40	95.1	30	64.4	20	33	10	287.6	320.6

*Estimated on the basis of the relative recoveries of each of the three units.

#Based on adjusted total value.

FIG. 2

Relative efficiency of three activated carbon filters in series in recovering trace organics from subsurface water.

TABLE 4

Elemental Composition and Empirical Formulas of Subsurface Water Trace Organics

Source of Water			Organic	Elemental Composition, %						Empirical Formula	Ref.
Location	Run	Unit	Extract	C	H	O	N	S	P		
Meramec Spring	1	1	CCE	59.0	6.96	24.2	0.68	$<$0.1	0.36	$C_{4.5}H_{6.8}O_{1.5}N_{0.05}P_{0.01}$	(17,18)
			CAE	51.2	6.52	36.0	2.53	$<$0.1	0.35	$C_{4.2}H_{6.4}O_{2.4}N_{0.18}P_{0.01}$	
	2	2	CCE	62.5	7.40	24.0	0.54	0.20	0.22	$C_{5.2}H_{7.4}O_{1.5}N_{0.04}P_{0.01}S_{0.01}$	
			CAE	46.3	7.47	36.6	2.04	0.13	2.80	$C_{3.9}H_{7.5}O_{2.3}N_{0.14}P_{0.09}$	
UMR Well		1	CCE	52.5	6.74	14.7	0.32	27.8	0.22	$C_{4.2}H_{6.4}O_{0.92}N_{0.02}P_{0.01}S_{1.0}$	
			CAE	42.9	6.81	36.0	1.36	$<$0.1	0.24	$C_{3.6}H_{6.8}O_{2.3}N_{0.10}P_{0.01}$	
Clinton Well	2	1	CAE	44.1	5.41	32.1				$C_{3.7}H_{5.4}O_{2.0}$	(16)
		2	CCE	57.8	6.98	30.1				$C_{4.9}H_{7.0}O_{1.9}$	
			CAE	45.4	5.15	46.4				$C_{3.8}H_{5.2}O_{2.9}$	
Russian boring wells			Liquid-liquid extract w/isobutyl alcohol @pH 2-3 (composite)	53.0	6.90	36.7	3.20			$C_{4.4}H_{6.9}O_{2.3}N_{0.23}$	(20)

TABLE 5

Carbon Content and Oxygen Demand of Subsurface Water Trace Organics

Source of Water				Carbon Content %		TOD	COD		5 day BOD		
Location	Run	Unit	Organic Extract	ECA[a]	TCA[a]	mgO_2/mg extract	mgO_2/mg extract	% of TOD	mgO_2/mg extract	% of COD	Ref.
Meramec Spring	1	1	CCE	59.0	58.0	1.85	1.37	74.0	0.26	19.3	(17,18)
			CAE	51.2	49.6	1.52	1.49	98.0	0.23	15.3	
		2	CCE		55.0		1.48		0.13	8.7	
			CAE		51.6		1.30		0.10	8.0	
		3	CCE		57.0		1.35		0.15	11.1	
			CAE		53.0		1.40		0.14	10.0	
	2	1	CCE		60.5		1.32		0.18	12.9	
			CAE		50.5		1.31		0.22	16.5	
		2	CCE	62.5	62.0	2.01	2.00	99.5	0.16	8.2	
			CAE	46.3	46.0	1.48	1.40	94.5	0.19	13.6	
		3	CCE		56.2		1.72		0.14	8.4	
			CAE		61.2		1.94		0.13	6.7	
UMR Well	-	1	CCE	52.5	53.0	1.72	1.64	95.3	0.0		
			CAE	42.9	44.0	1.55	1.46	94.2	0.0		
Clinton Well	2	1	CAE	44.1		1.58	1.27	80.0			(16)
		2	CCE	57.8		1.99	1.67	83.9			
			CAE	45.4		1.20	1.23	100.2			

[a]ECA: Elemental chemical analysis; TCA: Total carbon analyzer.

The chemical oxygen demand (COD) of the extracts, determined experimentally through chemical oxidation (9), compared favorably with the theoretical amount of oxygen (TOD) required for the complete oxidation of the extracts to carbon dioxide and water, computed on the basis of the empirical formulas. The COD values ranged from 1.23 to 2.0 mg oxygen per mg organics and for many extracts represented over 94 percent of the TOD values. Biodegradability studies, using a Warburg respirometer at 20°C. with seed organisms from an activated sludge unit acclimated to domestic sewage, indicated that the Meramec Spring extracts exhibited sizeable oxygen uptakes, while the UMR well extracts were bioresistant. The 5 day biochemical oxygen demand (BOD) of the spring extracts ranged from 0.10 to 0.26 mg oxygen per mg extracts and represented from 7 to 19 percent of the COD values.

IV. TOXICITY OF TRACE ORGANICS

The acute and long term toxic effects of the organic micropollutants recovered from Missouri subsurface water were studied using batch-type bioassays with fish as the test animal. In addition to the gross lethal effect of the trace organics, their effect on specific fish organs was evaluated by means of enzyme studies using homogenized trout tissue.

Rainbow trout (*Salmo gairdneri*), blue green sunfish (*Lepomis cyanellus*), golden shiners (*Notemigonus crysoleucas*), and red shiners (*Notropis lutrensis*) were employed in acute toxicity studies which were performed as outlined in Standard Methods (9). The results are shown in Table 6.

The Missouri River CCE was a yearly composite of chloroform soluble materials in treated river water and is included to provide a basis for comparison between the toxic behavior of the subsurface and surface water organics. It was obtained using a standard carbon adsorption unit (9) and constituted a composite of 24 biweekly samples. The average concentration of this material during the year was 36 μg/l, while individual biweekly values ranged from 17 to 71 μg/l.

Individual subsurface water CCE and CAE were not toxic to the test fish, even at concentrations approaching their solubility limits (270 and 400 mg/l for a CCE and a CAE, respectively). However, the organics were found to exhibit strong synergistic effects and, when combined at their naturally-occurring ratio, the combined extract was frequently toxic to the test fish. Contrary to the subsurface water CCE or CAE materials, the surface water CCE was individually toxic and at a significantly lower concentration.

The acute toxicity studies demonstrated that some of the trace organics were toxic to fish at high concentrations over a relatively

TABLE 6

Acute and Predicted Long-Term Toxicity of Trace Organics

Test Fish		Source of Water			Organic	Acute TLm[a]; mg/l				Long-Term TLm	
Type	Length, cm/ Weight, g	Location	Run	Unit	Extract	24 hr	48 hr	96 hr	120 hr	mg/l	at days
Trout	10.3/16.1	Meramec Spring	2	1	CCE& CAE[b]	130	125	95	82	33.1	40
				2		88	75	61	56	3.9	70
				3		no effect up to 180					
	5.3/2.8	Missouri River	-	-	CCE	36	32	28	24	0.44	80
Sunfish	7.2/8.9	Meramec Spring	2	1	CCE& CAE[b]	166	141	115	103	25.4	50
				2		137	121	114	100	47.0	50
	7.7/11.7	Missouri River	-	-	CCE	56	49	45	39	30.1	30
Golden Shiners	6.5/4.1	Meramec Spring	2	2	CCE& CAE[b]	180	171	160	152	141	20
		Missouri River	-	-	CCE	59	52	39	33	7.5	40
Red Shiners	5.0/2.0	Meramec Spring	1	1	CCE& CAE[b]	no effect up to 240					
				2		no effect up to 200					
	2.8/1.2		2	1		no effect up to 305					
	5.7/1.3			2		195	170	148	120	13.3	50

[a]TLm: Median tolerance limit, or the concentration at which 50% of the test animals are able to survive for the specified period of exposure.

[b]Combined at naturally-occurring ratio.

short period of time. Realistically, some of the concentrations investigated may never be reached under normal conditions; yet, the long-term effects of these materials over an extended period of time could be significant.

Long-term toxicity studies were conducted by exposing the fish (trout or red shiners) to the test solution under static conditions for a period of 5 days, and then transferring them to a recovery solution for 5 days during which time the fish were fed daily; the procedure was repeated until at least 50 percent of the test animals had been killed. The recovery solution consisted of either fresh water or water containing one-tenth of the test concentration of the organics. The 5 day exposure and recovery times were selected on the basis of preliminary work and represented the maximum period of time the fish could exist without feeding and remain in an acceptable physical condition, and the minimum time required by the test fish to return to normal feeding conditions. The long-term toxicity data are shown in Table 7. When the exposure time was lengthened, the trace organics were toxic at concentrations well below the acute toxicity levels. Evaluation of the fresh water recovery and 10 percent concentration recovery data further indicated that the organics did have a cumulative effect and that there was a build-up of the toxicant in the fish.

The behavior before death of the test fish indicated that these substances were affecting the respiratory process and that the fish were dying from a shortage of oxygen. This apparent suffocation could have been caused by either a physical blockage of oxygen transfer at the gills or an internal disruption of respiratory enzyme activity. Macroscopic and microscopic examination of fish exposed to subsurface water organics revealed that the gills had lost their red color, a light colored material had accumulated on the gill lamella, and the gas bladder was empty, completely devoid of gas. On the contrary, examination of fish exposed to surface water organics showed extreme hemorrhaging around the heart and liver, while the gills and bladder appeared to be normal. In order to elucidate further the mode of action of the toxicants, respiratory enzyme studies and oxygen transfer studies were performed.

Enzyme studies were conducted using homogenized trout gill, heart, and liver tissue and measuring the uptake of the homogenates in a 0.1 M solution of succinic acid at 20°C. for a period of 2 hours in a Warburg respirometer. The test solution consisted of equal volumes of tissue homogenate suspended in a sterilized Robinson's EDTA isotonic solution (22), a 0.3 M solution of succinic acid, and a surfactant solution of trace organics of the desired concentration. The surfactant was a product (No. M-14019) of the Tretolite Co., St. Louis, and was used as a dispersant to aid in dissolving large quantities of organics. Typical results are shown in Figure 3, which has been developed for the gill

TABLE 7

Long-Term Toxicity of Trace Organics[a]

Test Fish: Type	Test Fish: Length, cm/ Weight, g	Source of Water: Location	Source of Water: Run	Source of Water: Unit	Organic Extract	Total Accumulated Test Time[b] for 50% Kill, days — Test Material Concentration, mg/l: 10		7.5		4.2		1.0		Control
						0[c]	1.0[c]	0	0.75	0	0.42	0	0.10	
Trout	8.2/11.0	Meramec Spring	2	2	CCE& CAE[d]	30	20	37	24	48	29	>54	>54	>54[e]
Trout	5.3/2.2	Missouri River	-	-	CCE	10	5	13	11	19	13	>19	>19	28[f]
						24		13.5		5.6				
						0	2.4	0	1.35	0	0.56			Control
Red Shiners	4.7/1.5	Meramec Spring	2	2	CCE& CAE[d]	30	20	39	30	>65	45			>65[f]

[a]Adopted from Smith and Grigoropoulos (21).
[b]Five day exposure and recovery periods were used.
[c]Test material concentration in recovery water, mg/l.
[d]Combined at naturally-occurring ratio.
[e]Fish died because of loss of temperature control.
[f]Fish showed signs of disease; test discontinued.

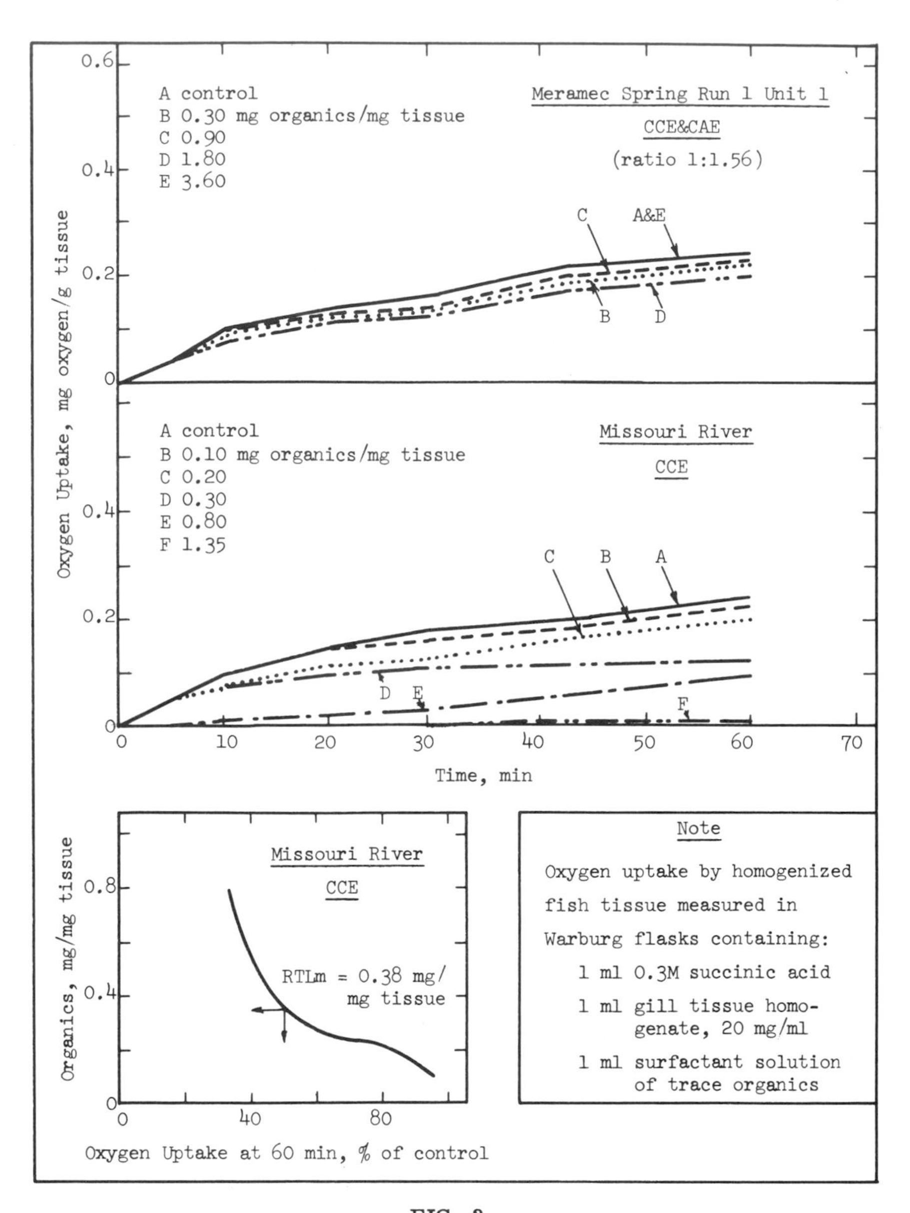

FIG. 3

Effect of trace organics on trout gill tissue enzyme activity.

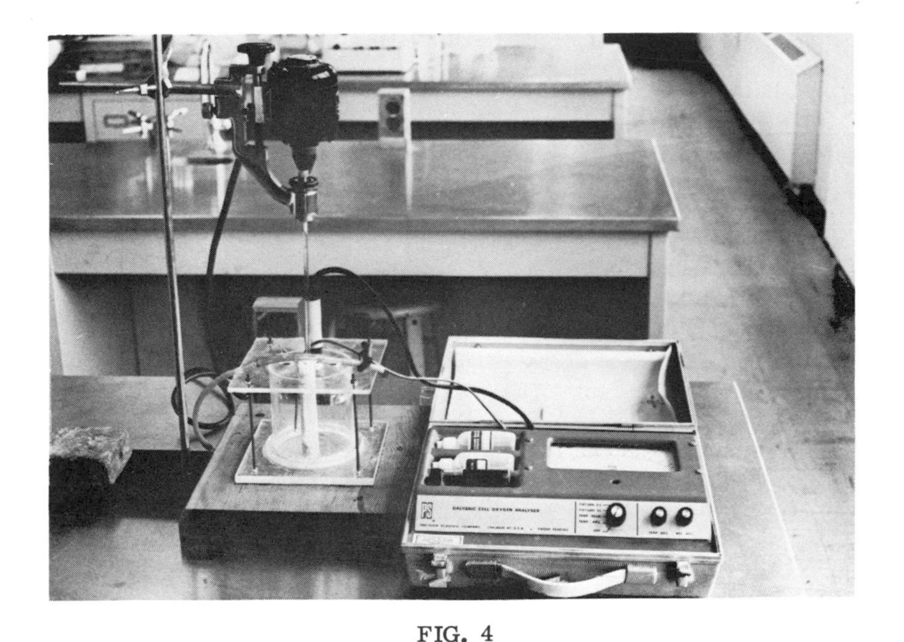

FIG. 4

Experimental arrangement for oxygen transfer studies.

tissue homogenate. The subsurface water CCE and CAE, individually or in combination, did not inhibit to a significant degree the respiratory activity of any of the three homogenates at the concentrations studied, which ranged from 0.1 to 3.6 mg organics per mg tissue. On the contrary, the surface water organics did inhibit the activity of all homogenates. The 60 minute median respiratory tolerance limits (RTLm) of the three trout homogenates, defined as the concentrations of the toxicant which reduced the activity of the test solutions to 50 percent of the control, were computed as shown in Figure 3 and found to be 0.38, 0.22, and 0.11 mg organics per mg tissue for gill, heart, and liver tissue, respectively.

Oxygen transfer studies were performed using an experimental model which was specifically developed for these studies and consisted of an air tight cylindrically shaped reaction vessel with facilities for mixing and for measuring and purging dissolved oxygen. The arrangement is shown in Figure 4. Oxygen transfer was evaluated using an oxygen-permeable silicone membrane (Cat. No. 40,824, Edmond Scientific, Inc., Barrington, N.J.). By bubbling nitrogen gas through the water inside the membrane, a differential oxygen concentration was established, oxygen was transferred from the test solution to the water inside the membrane, and a decrease in oxygen content of the test solution was measured. The effect of several subsurface and surface water organic micropollutants on the oxygen transfer across the membrane was determined. Typical results are shown in Figure 5. Neither the subsurface water CCE and CAE, when tested individually, nor the surface water CCE exerted a significant effect on the oxygen transfer efficiency. However, when the subsurface water organics were tested combined at their naturally-occurring ratio, they effectively blocked the transfer of oxygen across the membrane.

The mode of action of the subsurface and surface water organics was, therefore, different. The combined spring water chloroform and alcohol soluble materials blocked the physical transfer of oxygen across the gills but did not affect the enzymatic activity; on the contrary, the treated river water chloroform soluble organics disrupted the enzymatic utilization of oxygen.

V. PREDICTION OF LONG-TERM TOXICITY

Because of the length of time involved in performing a long-term bioassay, the amount of material required, and the difficulties encountered in maintaining the test fish in the laboratory for extended periods, the development of a method which could be employed to estimate the long-term toxic levels from a short-term or acute toxicity study was highly desirable. The mathematical model represented by the toxicity equation described in Figure 6 was developed for this purpose and was found to provide very satisfactory results. It utilized a toxicity factor which was employed to normalize the effects of the toxicant's mode of action and the fish's physical characteristics on the TLm concentration. In defining the toxicity factor it was considered highly desirable to keep the data required for this relationship as simple as possible without, however, impairing the accuracy of the model. Since the mode of action of the subsurface water organics was a physical blockage of oxygen transfer across the gills, the toxicity factor for these materials was defined as the relationship between the TLm value and the fish characteristics which were expressed as the gill surface area divided by the condition factor of the fish. This factor is defined (23) as the weight of the fish divided by its length cubed. On the other hand, since the gills were not physically affected by the surface water extracts, the gill surface area was not included in the toxicity factor relationship for these materials. The constants for the toxicity equation can be determined for a given fish and toxicant from an acute bioassay and a general knowledge of the mode of action.

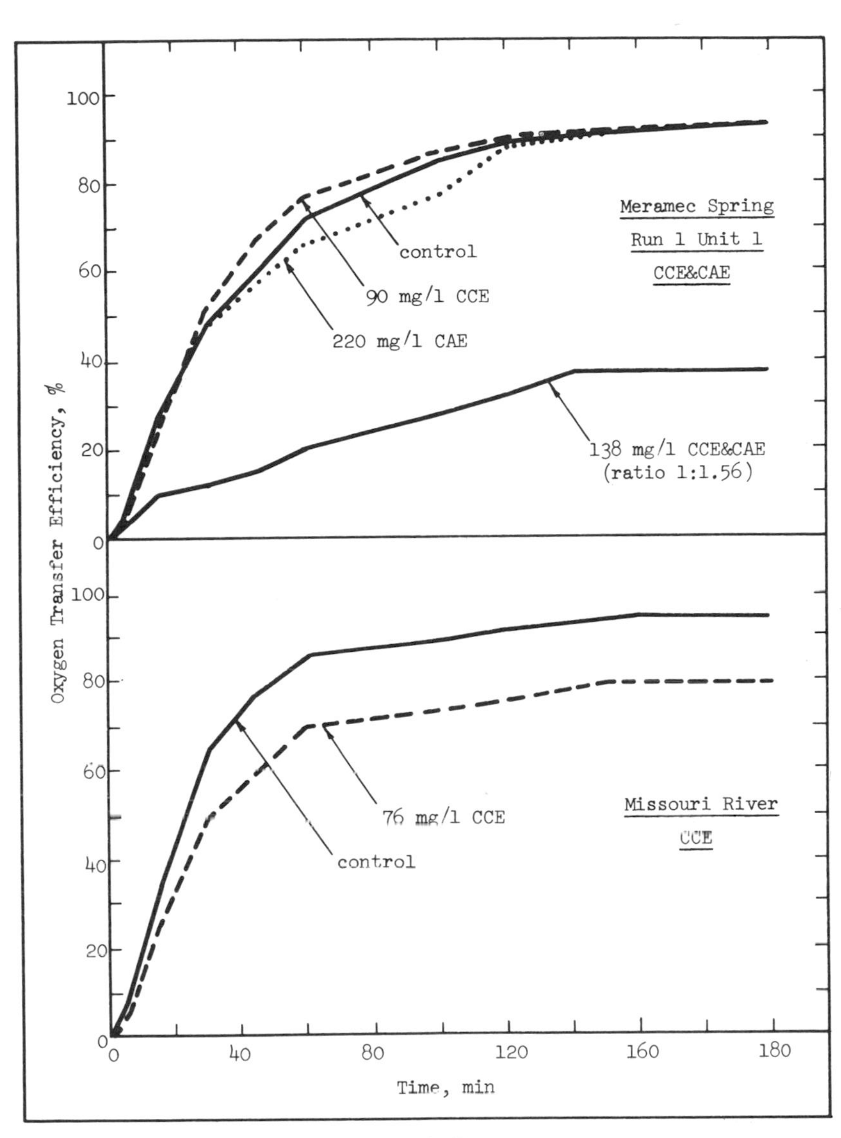

FIG. 5

Effect of trace organics on oxygen transfer efficiency.

TOXICITY EQUATION

enables prediction of long-term toxic levels from short-term data

$$y_t = y_c + (y_o - y_c)\, e^{-bt}$$

where: y_t = toxicity factor at time t
y_c = toxicity factor corresponding to the long-term TLm value
y_o = toxicity factor corresponding to the immediate TLm value
b = a constant depending on the test material and test fish

TOXICITY FACTOR

relates the toxicant concentration with the physical characteristics of the test fish and the mode of action of the toxicant

for <u>Subsurface Water Organics</u>

$$y_t = C_t \left[\frac{GSA}{K}\right]^{1.25}$$

for <u>Surface Water Organics</u>

$$y_t = C_t/K$$

where: C_t = TLm value at time t, mg/l
GSA = gill surface area, mm^2
K = condition factor, weight of test fish divided by its length cubed, g/cm^3

<u>Note</u> 1. y_o can be determined from short-term toxicity data by plotting y_t against t on semilogarithmic paper, fitting a straight line to the various points, extending the line to the y_t axis, and determining the intercept

y_o can be taken approximately equal to the 2 hour TLm

2. b can be determined by simultaneous trial and error solution of the toxicity equation at time t_1 and t_2

FIG. 6

Prediction of long-term toxic levels of trace organics.

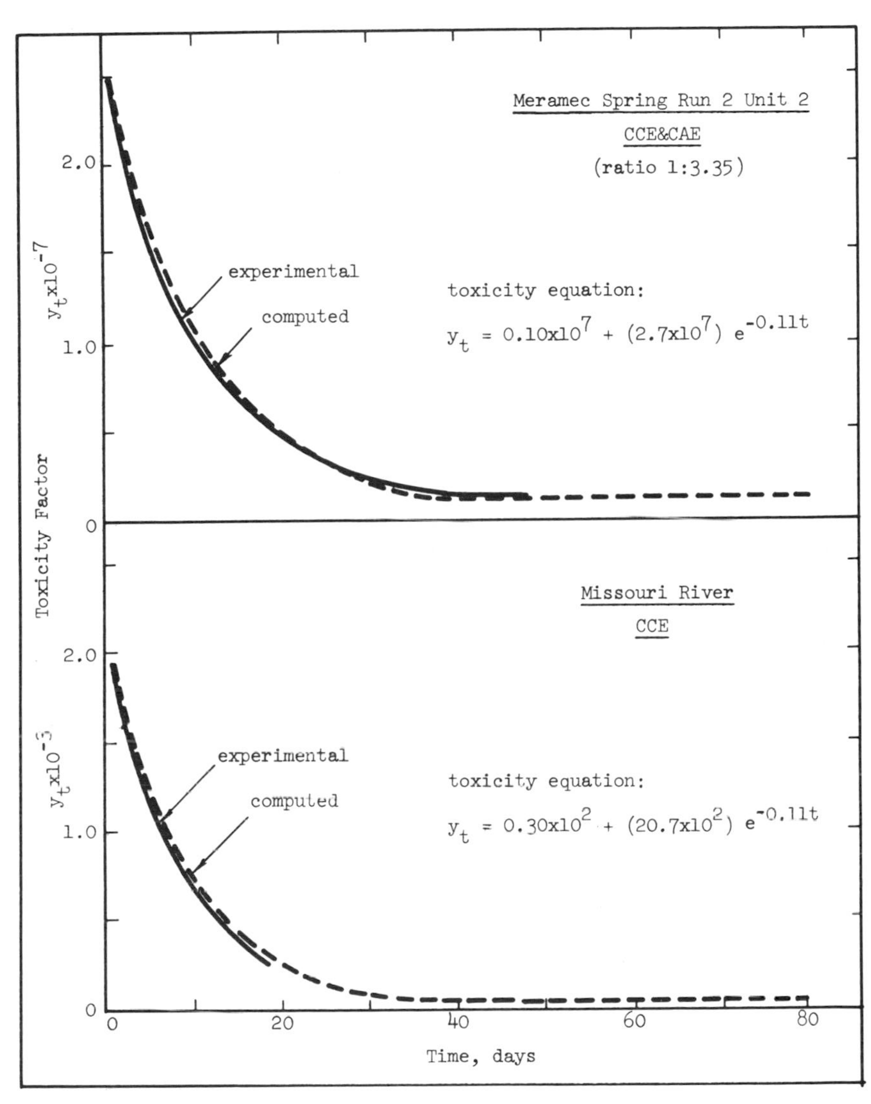

FIG. 7

Toxicity curves for trout exposed to trace organics.

The theoretical model was evaluated with long-term toxicity values determined by the authors or reported in the literature and was found to provide satisfactory results. Typical findings are presented in Figure 7. Similar data were obtained for other materials, including the pesticide malathion and acrylonitrile, and other types of test fish. The computed and experimental values were close, with the computed value deviating from the experimental in most cases by less than 10 percent within a range from a low of 2 percent to a high of 40 percent.

It should be emphasized that the toxicity factor is not the TLm concentration, but rather it is the relationship between this concentration and appropriate physical characteristics of the test fish. However, the TLm value at any time can be computed from the corresponding toxicity factor. Long-term TLm values determined for different types of fish are summarized in Table 6 together with the acute TLm values to facilitate comparison. Although the acute TLm values may not be reached under natural conditions, some of the long-term levels could be approached; these concentrations would have to persist for a long period of time before the health of aquatic life is endangered if the organics were acting alone. Synergistic action between the various trace organics and other materials in the water could significantly magnify the effect of the organic micropollutants.

VI. SUMMARY

This paper has presented information on the concentration of trace organics in a number of subsurface waters and has discussed the characteristics of these materials, emphasizing their acute and long-term toxicity. A method has also been outlined which enables the reliable estimation of the long-term or eventual median tolerance limit on the basis of acute toxicity data. Considerable work remains to be done in the trace organics area, especially in establishing the identity of the organic micropollutants and in developing effective and practical methods for their destruction or removal from water. Because of the potential threat to the consumer represented by the presence of these substances in water, there is an urgent need to intensify the study of these contaminants and to establish trace organics as an integral component of human ecologic investigations.

ACKNOWLEDGMENTS

This investigation was supported in part by PHS Research Grant No. ES-00082#6 and FWPCA Training Grant No. 5T1-WP-86. The authors wish to express their appreciation to the James Foundation,

St. James, Mo., the City of Rolla, and the University of Missouri - Rolla for providing sampling locations; the St. Louis County Water Company for providing samples of Missouri River trace organics; and the Division of Fisheries, Missouri Department of Conservation, and Huzzah Fishery, Salem, Mo., for providing test fish.

REFERENCES

1. J. J. Geraghty, J. Amer. Water Works Assoc., 59, 820 (1967).

2. Public Health Service, Drinking Water Standards - 1962, Public Health Service Publication No. 956, U. S. Government Printing Office, Washington, 1962, p. 31.

3. Anon., J. Amer. Water Works Assoc., 60, 1317 (1968).

4. J. Borneff, Das Gas-und Wasserfach, 110 (2), 20 (1969) (in German).

5. J. Borneff, F. Selenka, H. Kuntc, and A. Maximos, Environ. Res., 2, 22 (1968).

6. W. C. Hueper and W. W. Payne, Amer. J. Clin. Pathol., 39, 475 (1963).

7. N. Takemura, T. Akiama, and C. Nakajima, Int. J. Air Water Pollut., 9, 665 (1965).

8. W. H. Walker, J. Amer. Water Works Assoc., 61, 31 (1969).

9. Standard Methods for the Examination of Water and Wastewater, 12th ed., Amer. Public Health Assoc., New York, 1965.

10. Report of Committee D-19 on Water, 72nd Annual Meeting, Amer. Soc. Testing Materials, Atlantic City, N. J., June, 1969.

11. H. Braus, F. M. Middleton, and G. Walton, Anal. Chem., 23, 1160 (1951).

12. F. M. Middleton, H. H. Pettit, and A. A. Rosen, Proc. 17th Purdue Univ. Ind. Wastes Conf., 112, 454 (1962).

13. H. N. Myrick and D. W. Ryckman, J. Amer. Water Works Assoc., 55, 783 (1963).

14. J. N. Dornbush and D. W. Ryckman, J. Water Pollut. Contr. Fed., 35, 1325 (1963).

15. R. G. Spicher and R. T. Skrinde, J. Amer. Water Works Assoc., 55, 1174 (1963).

16. L. R. Robinson, J. T. O'Conner, and R. S. Engelbrecht, J. Amer. Water Works Assoc., 59, 227 (1967).

17. S. G. Grigoropoulos and J. W. Smith, J. Amer. Water Works Assoc., 60, 586 (1968).

18. J. W. Smith, Ph.D. Dissertation, Univ. of Missouri - Rolla, Rolla, 1968.

19. O. J. Sproul and D. W. Ryckman, J. Water Pollut. Contr. Fed., 33, 1188 (1961).

20. I. A. Goncharora and V. G. Datsko, Gidrokhim. Materialy, 33, 166 (1961) (in Russian); through CA, 57, 9592c (1962).

21. J. W. Smith and S. G. Grigoropoulos, J. Amer. Water Works Assoc., 60, 969 (1968).

22. W. W. Umbreit, R. H. Burris, and J. F. Stauffer, Manometric Techniques, 4th ed., Burgess, New York, 1964, p. 133.

23. K. F. Lagler, J. E. Bordach, and R. R. Miller, Ichthyology, John Wiley, New York, 1962, p. 173.

CHAPTER 6

ORGANICS IN LAKES

R. F. Christman and R. A. Minear

University of Washington

I. INTRODUCTION

The whole history of our organic evolution has been intimately connected to aqueous systems, dating back three to four billion years to water's first appearance on earth. Subsequent photochemically induced formation and association of organic fragments under shallow layers of water ultimately led to formation of life (1). Considering this plus the fact that current life forms in and surrounding water environments are diverse and complex, it is not surprising that the composition and structure of natural organics in water systems should reflect this complexity. Also, since the natural aquatic environment is a highly dynamic system, it is reasonable to expect a rapid turnover of the simpler organic structures and that the steady-state buildup of appreciable quantities of dissolved organic material is the result of long term processes leading to the formation of complex, chemically and biologically stable compounds. This buildup has been estimated to be the result of several thousand years accumulation for deep sea water and one to two orders of magnitude less for surface and shallow marine waters (2).

Certainly, short term variations exist due to changes in environmental and biological conditions, plus the influences of man. Thus it would be in order at this point to distinguish between naturally-occurring and man sponsored compounds.

A. Natural Product Organics.

Many of these compounds result from the dissolution of soil and less ossified living tissue like wood, either directly from living tissue or via the refractory humus pool. Some of these compounds are in true solution while others exist as microsuspensions of large molecules whose true solubility is questionable. In addition, the class of organic molecules must be considered which derive from cell lysis or excretion by living biological material within the system.

B. Man Sponsored Organics.

The variable existence of many organic structures is recognized as a consequence of man's activities, from pesticides and detergents to a host of exotic organics resulting from industrial effluents. If concentrations are considered, we find that these can also be variable. In Table 1 which is to be treated as exemplary rather than as rigorous fact, we intend to paint a picture showing that only in rare cases, and man-made at that, are organics present in appreciable concentrations. Also, natural product organics can be much more concentrated than the usual things which come to mind, e.g., pesticides or detergents. However, we do not mean to diminish the importance of these man-made products, particularly in view of their ecological impact, but simply to emphasize that they usually represent a small fraction of the soluble-colloidal organic matter present in natural waters.

Since fresh water is being considered, we should like to point out that although admittedly not as chemically complex overall as sea water, fresh water may have organic concentrations an order of magnitude greater than sea water and may be subject to greater composition variation (3).

A more detailed discussion will be developed of only two groups of organics: natural product organics responsible for natural water color and the soluble organic phosphorus compounds of fresh water. While by no means clearly resolved, much information has been gained about the former group of compounds through efforts in the authors' laboratory and by others in the past eight years or so. On the other hand, very little specific information exists regarding the nature of the latter group of compounds, particularly in fresh water systems, and, in contrast to color, these compounds seem to come from living systems.

II. COLOR PRODUCING ORGANIC SUBSTANCES

Many natural waters exhibit distinct yellow to brown coloring due solely to their natural product organic contents. In fact these substances appear to be present in most waters, though not always at

TABLE 1

Some Representative Concentrations of Organics in Water Systems

Source	Concentration	Moles/Liter
Petrochemical wastes		
(Direct)	400-500 mg/l as phenol.	$\sim 5 \times 10^{-3}$
(After treatment)	1-2 mg/l as phenol.	$\sim 1 \times 10^{-5}$
Pesticides	10^{-12} to 10^{-8} mg/l	$\sim 10^{-14}$ to 10^{-10}
Natural Color (of ~100 Pt. units)	10-30 mg/l as C	$2\text{-}6 \times 10^{-6}$ [a]

[a]Assuming ~50% carbon and an average molecular weight of 10,000.

levels sufficient to impart coloration to the water. The specific structure of these compounds remains unresolved but their macromolecular dimensions and phenolic nature are well established. The known characteristics of these color substances are summarized below and the methods used in obtaining some of this information are mentioned.

A. Molecular Size Distribution.

Recently there have been several studies on the molecular size characteristics of natural color using Sephadex gel filtration techniques (4-7). Each of these has used vacuum evaporation for sample concentration, which is of some importance since Ghassemi (8) found the ion exchange technique to be selective for higher molecular weight material.

Since the principle of gel filtration is essentially a sieving and fractionation according to molecular size (9) when conducted with ionic eluants, selection of the proper gels allows the determination of the range and distribution of molecular size for constituents of a mixture.

The exclusion limits and fractionation ranges for the Sephadex gels are given in Table 2. The values refer to the molecular weight of dextrans but the relationship between molecular weight and molecular size (the separation criterion) will depend upon the type of compound. In addition, this relationship may also be a function of pH (illustrated below) or ionic strength for a given compound.

The elution profiles obtained in the authors' laboratory for six different colored waters on Sephadex G-25 gel are given in Figure 1. (Note: molecular size decreases approximately logarithmically from the exclusion limit as the elution volume increases linearly). The exclusion and inclusion limits were determined with Blue Dextran 2000 (molecular weight $\sim 2 \times 10^6$) and glucose, respectively. These are indicated by the horizontal bars between the two profiles. The tailing beyond the glucose elution position indicates reversible adsorption has occurred, characteristic of lower molecular weight aromatic compounds (10), particularly on the G-25 gel. However, these materials were completely excluded from the G-10 gel as illustrated for two colored waters in Figure 2. This indicates that the lower molecular size limit is greater than or equal to that of dextrans with a molecular weight of 700. Elution profiles for these waters on G-50 and G-75 gels (8) indicated that most of the color producing substances had equivalent dextran molecular weights between 700 and 10,000, with small fractions exceeding 50,000. These ranges are similar to those found by Shapiro (5).

The variation in molecular size ranges and distributions for different colored waters is indicated in Figure 1 and has been reported also by other authors (4, 5, 6). The much higher upper limits reported in two of these studies (4, 6) may partly reflect experimental differences since much coarser filtration was used. Gjessing and Lee (6) indicated that waters with higher color values seemed to contain more high molecular weight components but this pattern does not appear to fit the waters in Figure 1.

It was mentioned earlier that pH can influence the relationship between molecular size and molecular weight. It is clearly demonstrated in Figure 3 that pH does significantly influence the molecular size (but not necessarily the molecular weight) of the color substances. This behavior is consistent with the macromolecular nature of these substances in view of the functional groups now known to be contained in the structure. As the pH value is increased, the acidic phenolic and carboxylate groups become more ionized causing polymer extension from the greater charge density, thus a larger molecular size and a larger apparent molecular weight based on elution position are possible.

The well known indicator effect, i.e., an increase in color with

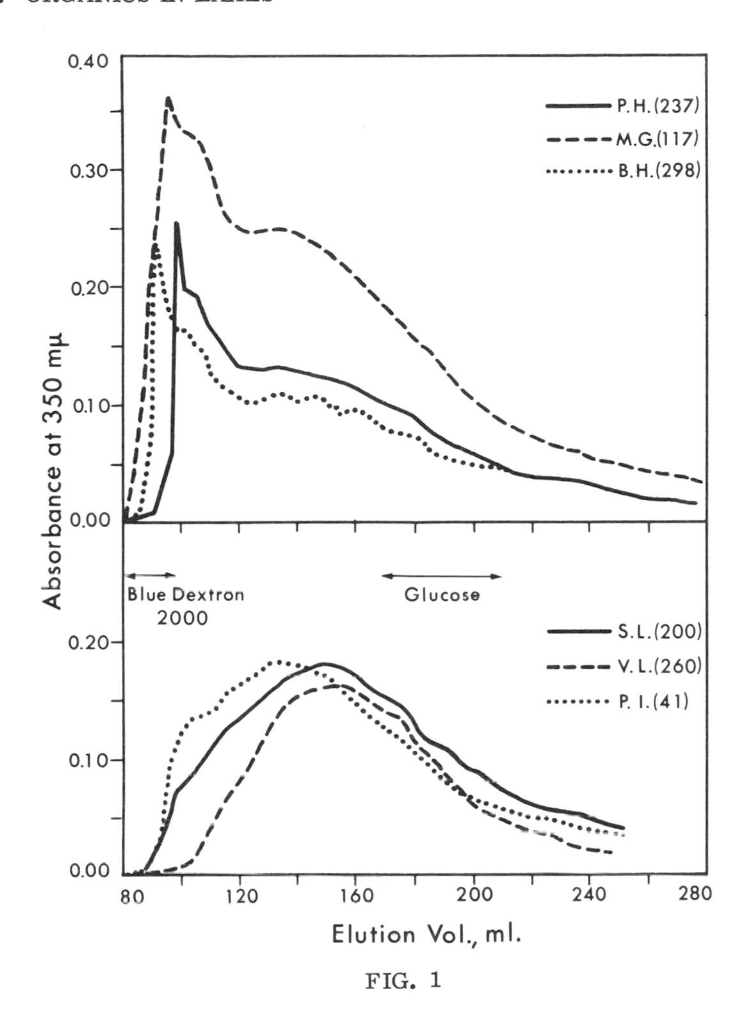

FIG. 1

Elution profiles for six different colored waters on Sephadex G-25 (2.3 x 46 cm.). Eluent: pH6 Clark and Lub buffer.

increased pH, has long been noted (11,12). This was attributed earlier to a decrease in molecular weight as pH value was increased due to the breaking-up of agglomerates of smaller molecules by the charge repulsion resulting from ionization at higher pH. In view of the behavior in

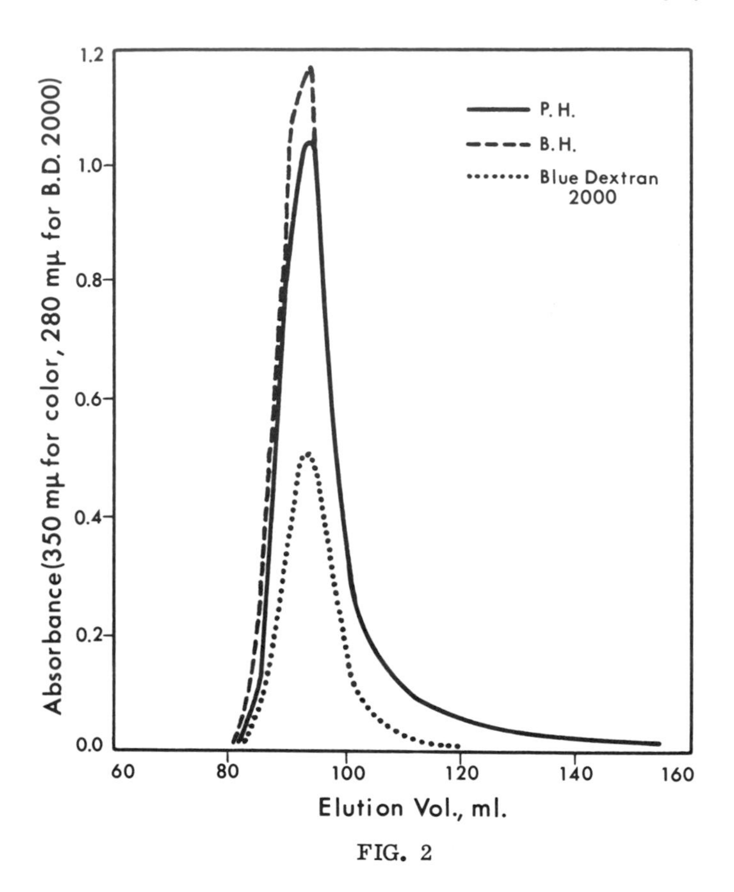

FIG. 2

Exclusion of color by Sephadex G-10 (2.3 x 48 cm.). Eluent: pH6 Clark and Lub buffer.

Figure 3 it would seem that the color intensity change must be accounted for by the well known phenol-phenoxide absorption behavior since the molecular size is increasing instead of decreasing.

Another characteristic, brought to light during investigations, was the ability of borates to form complexes with color substances (7, 8). This is illustrated in Figure 4 where a dramatic shift to higher molecular weights is exhibited in the presence of a pH 7 borate buffer compared to

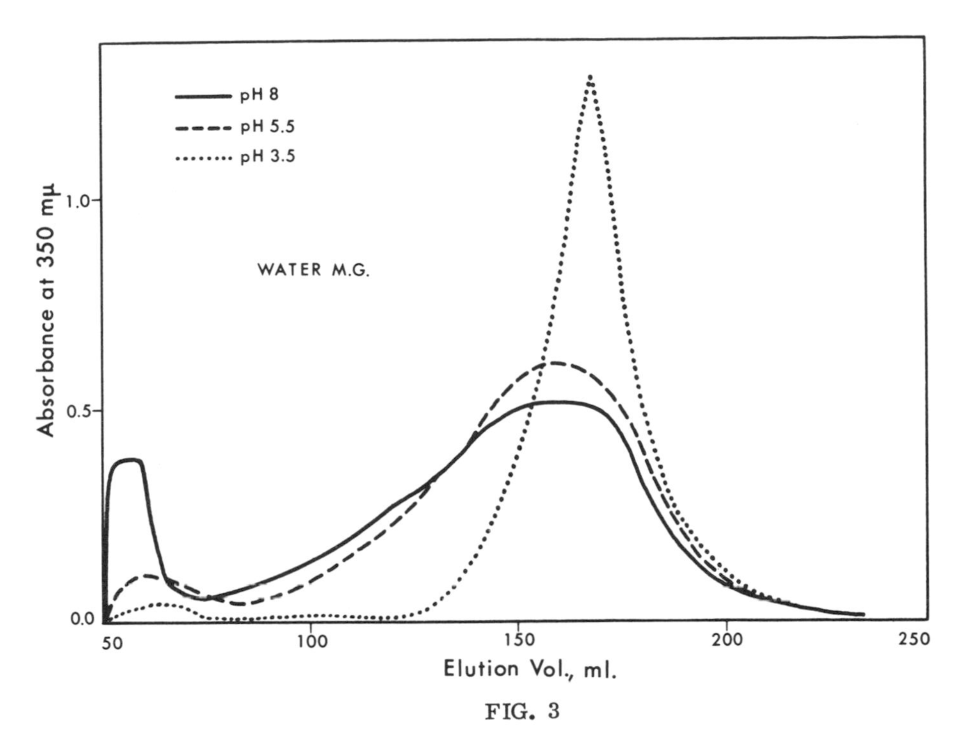

FIG. 3

Effect of pH on molecular size on Sephadex G-75 (2.3 x 46 cm.). Eluent: 0.01 N NaCl, pH adjustment with HC1 or NaOH.

a similar sample eluted with a pH 7, non-borate buffer. Other data (7) showed that even though the pH value was increased incrementally from 8 to 9 to 10 with buffers of decreasing borate content, the molecular size distributions on Sephadex G-50 decreased with decreased borate concentration (increasing pH). Control studies with Blue Dextran 2000 and Dextran 10 eliminated variations in the Sephadex gel's properties as a causative factor for this behavior.

B. Degradation Studies.

These studies are primarily responsible for the firm establishment of the aromatic-polyphenolic nature of color substances. Two techniques have been used: (a) mild oxidative degradation with alkaline

TABLE 2

Sephadex Gel Properties

Gel Type	Exclusion Limit[a] (MW)	Fractionation Range[a] (MW)
G-10	700	0 to 700
G-15	1,500	0 to 1,500
G-25	5,000	100 - 5000
G-50	10,000	500 - 10,000
G-75	50,000	1,000 - 50,000
G-100	100,000	5,000 - 100,000

[a]Based on behavior of polydextrans of these molecular weights.

copper oxide in a sealed bomb, under nitrogen at 180°C. (the details of this procedure are given by Christman and Ghassemi (13)) , and (b) reductive degradation with sodium amalgam under total reflux at atmospheric pressure (14).

Since the degradation studies required gram quantities of color substances, collection of these materials was achieved in the earlier oxidative studies (13) by collection on macroreticular anion exchange resins. This method, patterned after that of Packham (15), exploits another well known property of color, its negative charge in solution (12,16), yet its removal is permitted from the resin. The elution and extraction scheme leading to the isolation of the color solids is shown in Figure 5. More recent studies on both oxidative and reductive degradation (14) used a combination of vacuum concentration, dialysis, and freeze drying which is outlined in Figure 6.

The compounds identified as degradation products are shown in Table 3. These compounds are not likely to be produced during the degradative processes and are not present in the untreated water. Identification of products from the earlier oxidative studies (13) was achieved by thin-layer chromatography of the products on silica gel G using various solvent systems, different sprays, and comparison with known compounds. Confirmation was obtained by ultraviolet absorption spectra of the eluted spots. The specific details of this work are given elsewhere (13). It should be pointed out that the identification of products from the recent oxidative and reductive degradations (14) is only

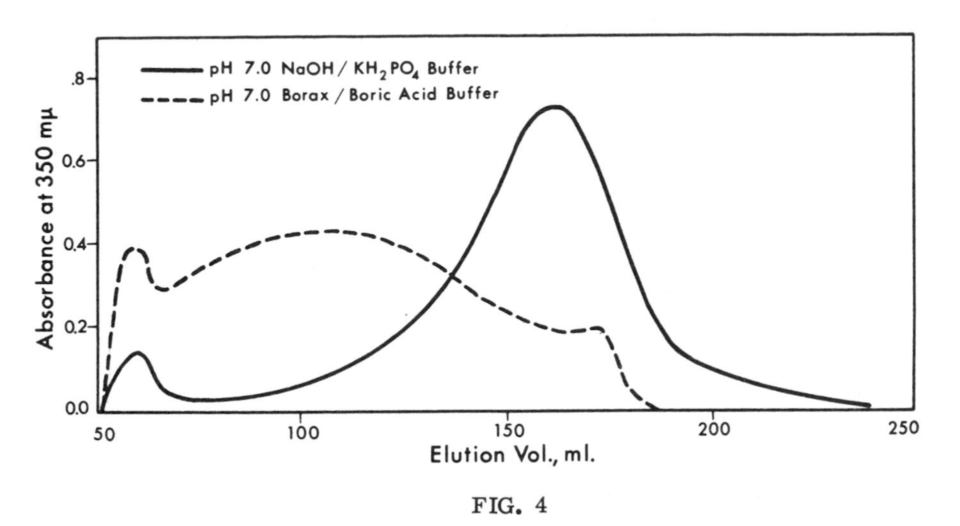

FIG. 4

Effect of borate on the molecular size distribution of color on Sephadex G-75 (2.3 x 46 cm.). Eluents: as indicated above.

tentative since they are based solely on comparative retention data for gas chromatography of their trimethylsilyl derivatives on only one column. (These derivatives were formed from BSA, (N, O-Bis (Trimethylsilyl Acetamide)) as described in (14)).

Degradation studies with model compounds (7,8,13,14) provided the required knowledge for interpretation of the color degradation data. From the alkaline copper oxide oxidation data, the following conclusions can be made:

(a) Aromatic structures remain intact. Thus the isolated aromatic compounds are evidence of related structures in the color molecules.

(b) Alkyl side chains are oxidized to aromatic acid functional groups only if the ring system is adequately activated by substituents. Since none of the isolated aromatic products possesses substituents of greater than one carbon atom, unsubstituted rings are not abundant, if present at all, in the parent molecules. This is not to say that alkyl side chains are not present in the parent molecules but only that when present on activated rings, they will be oxidatively cleaved to aromatic acid functional groups.

(c) The aromatic acid group may also be lost under certain conditions of ring activation. Therefore, the phenols, catechol and

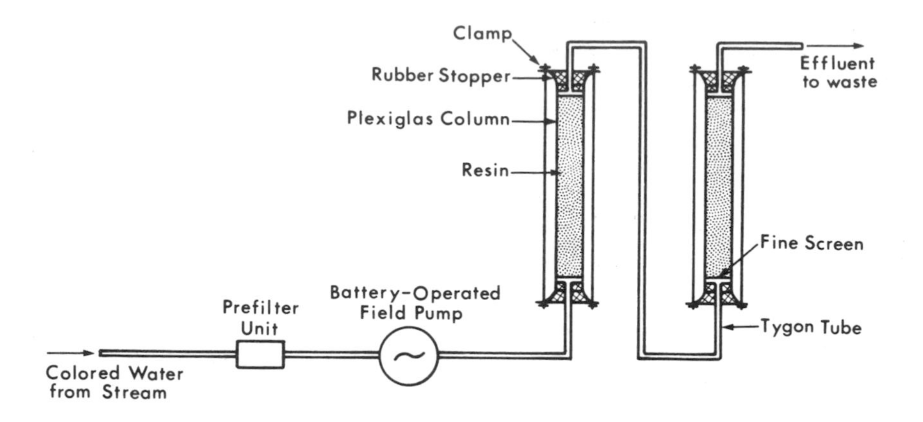

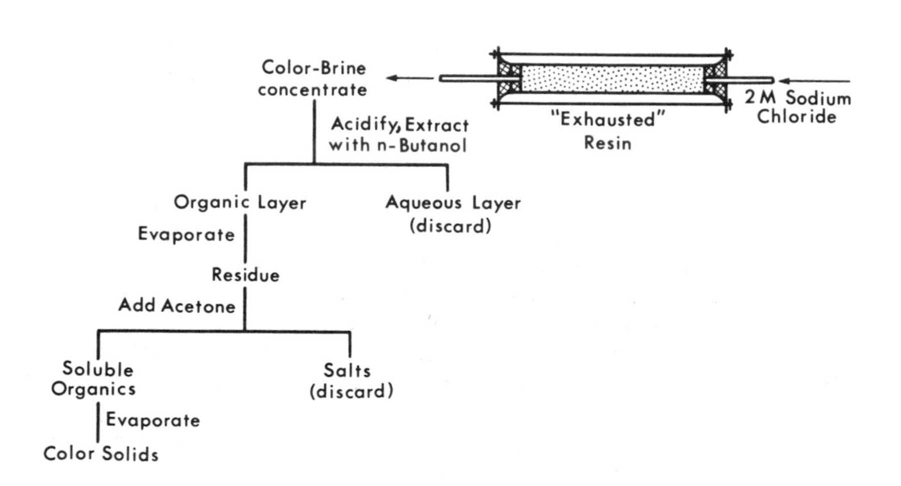

FIG. 5

Ion exchange method used for isolation of color solids.

TABLE 3

Components of Oxidative and Reductive Degradation Mixtures of Color Solids (Tentative Identification of Trimethylsilyl Derivatives by G. C.)

Compound	Alkaline CuO Oxidation	Na-Amalgam Reduction
Catechol[a]	+	+
Resorcinol[a]	+	
p-Methylphenol	+	+ +[b]
o-Methoxylphenol	+	+ +
Benzoic acid	+	+ +
3,4-Dihydroxybenzoic acid[a]	+	
3,5-Dihydroxybenzoic acid[a]	+	
Vanillin[a]	+	+
Vanillic acid[a]	+	
Syringic acid[a]	+	
p-Hydroxybenzoic acid	+	+

[a]Compounds identified in earlier oxidative studies by TLC and UV spectroscopy.
[b]+ + Indicates greater concentration than +.

resorcinol, most likely result from protocatechuic acid and 2,4- or 2,6-dihydroxybenzoic acid intermediates, respectively. Their presence in simple ether or ester linkages is precluded by their absence after simple acid hydrolysis of the unoxidized solids.

(d) The absence of other than single aromatic acid structures in the products indicates that any ring structure in the parent molecule containing a free acid functional group is likely to be bound through an ether linkage and also that each ring system contains only one alkyl-aryl, carbon-carbon bond.

For the sodium amalgam reduction, model compound studies showed no reduction of carboxylic acids and no alkyl carbon-carbon bond cleavage. The evolved hydrogen serves to maintain anoxic conditions during the alkaline hydrolysis.

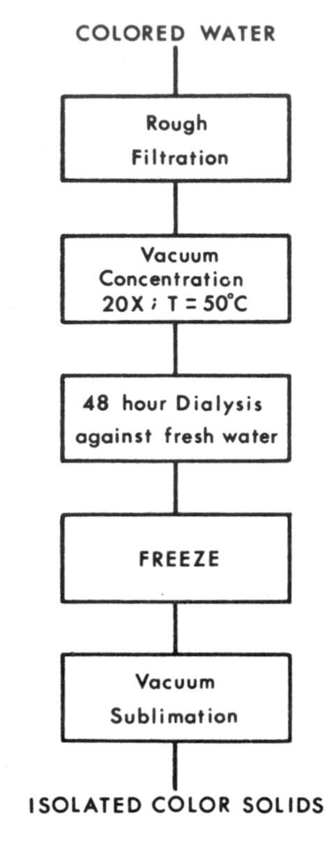

FIG. 6

Concentration and isolation of color solids used in more recent degradation studies.

On the basis of this degradative work, Christman (14) has proposed the symbolic representation of the unoxidized color macromolecule shown in Figure 7.

C. Origin and Genesis of Color.

The inverse relationship between the color of a water and its productivity reduces consideration of aquatic metabolic by-products as sources or precursors of color. It has been suggested (13, 18) that color in water is derived from: (a) the aqueous extractable material of

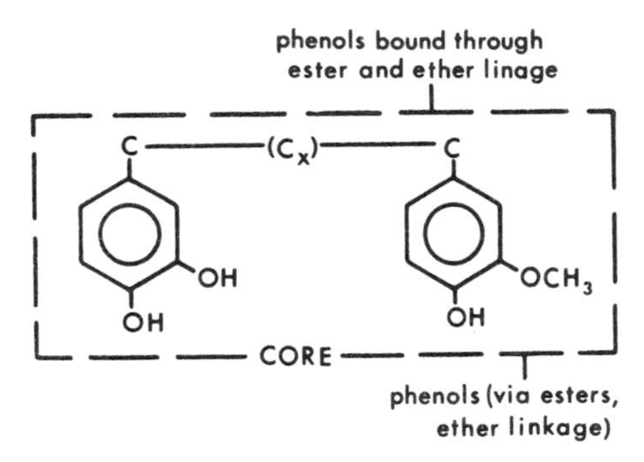

FIG. 7

A symbolic description of the unoxidized color macromolecule.

living woody tissue, (b) the dissolution of the decomposition products of decaying wood, (c) the dissolution of soil organic material, or (d) some combination of these.

Many similarities between the characteristics of wood and soil water extractables and color substances, i.e., such as pH-color behavior, fluorescence characteristics, and the similarity of degradation products (7, 8, 14, 17) support these materials as probable sources of color. More so for soil extractives since the similarities are greater between the soil matter and color substances. That lignin is not the sole precursor of colored substances is well established by the presence of meta hydroxylation patterns in the color degradation products (13, 14). These patterns do not occur in similar degradation studies on lignin (18).

It is possible that natural extractive products of wood, such as the Flavan-3,4-diols and Flavan-3-ols which are known to undergo condensation polymerization reactions, perhaps aided by biological activity during degradation, may contribute in part to the color macromolecule.

Significant variations are observed in the chemical properties of color substances from different waters. For waters of equal color at pH 7.0, different pH-color variations have been observed. Waters with similar mineral composition and molecular size distribution may display entirely different coagulation behavior and have quite different color to carbon ratios (8). Shapiro (5) was unable to relate his molecular size distributions to location or overall chemical characteristic for the lakes that were studied. Yet, different colored waters appear to exhibit significantly different size distributions (4-7) and it is likely that the compounds contained in a particular water reflect the surrounding environment and the age of the compounds themselves, but the nature of this relationship, if any, is not established.

D. Metal Ion-Color Association.

Another well recognized property of organic color is its association with iron. This has been demonstrated by: (a) the ability of iron to remain in solution in the presence of color while being rapidly removed from solution in its absence (19), (b) the elution behavior of iron on Sephadex gels in the presence of color (7,8), and (c) the fluorescence behavior of color in the presence and absence of iron (8).

The nature of this association, whether primarily peptization of ferric hydroxide colloidal particles or actual chelation by color molecule functional groups is in dispute. Christman (20) recently submitted a model showing that data provided as proof of a non-chelation could, with certain speculative assumptions, also support a chelation mechanism.

Studies on the association of color and iron in terms of their behavior on Sephadex gels (7) showed the degree of association to be highest at either lower pH or very high pH and lowest in the intermediate, pH 7 to 8, region contradicting other work (19). Use of strong chelating agents as eluant (e.g., 0.5% Versenex 80) showed that much of the iron could be removed from the color but not all of it, particularly that associated with the higher molecular weight material. This and further work by Ghassemi and Christman (unpublished) indicate that the color-metal ion interaction proceeds on a purely chemical basis.

Since the association of metal ions and color increases with pH at higher pH values, another or concurrent explanation is possible for the molecular size increase of color with increasing pH. Under highly alkaline conditions the color molecules rather than or in addition to extension may increase their size by association with iron or other metal-hydroxo complexes.

E. Soluble Organic Phosphorus Compounds.

In terms of the total organic compounds present in many lake waters, those compounds containing phosphorus certainly represent a very small proportion. The percent phosphorus ranges from ~2% for vitamin B_{12} to ~28% for phytic acid. Therefore, if 20 μg/l of organic phosphorus are taken as representative of a lake with a high soluble organic phosphorus content and 15% is used as an average value for percent phosphorus content, this would only represent 0.13 mg/l of organic matter, a small portion of the total soluble organic content of many lakes. But, the importance of phosphorus to living systems is certainly not in dispute as life requires this element.

The dynamic nature of phosphorus in the aquatic environment has been well established through radioactive phosphate tracer studies (21-24) and the turnover rate of soluble phosphorus forms can be very high in the summer. Rigler (25) has shown that most of the soluble phosphorus in lakes during the summer months is present as organic phosphorus. Recent data in our laboratory support this observation, with values of 70 to 80% organic obtained in early June for a highly productive lake in central Washington. The corresponding low orthophosphate level during the summer period is a well-known phenomenon (26, 27, 28). Thus, these soluble organic phosphorus compounds may play a significant role in the overall phosphorus chemistry of a water system, particularly during the summer. Yet the chemical nature of these compounds in fresh water is totally unknown. Knowledge of the chemical nature of these compounds would surely advance the understanding of the phosphorus cycle in natural waters and perhaps clarify a recent observation made by Rigler (29) that the soluble reactive phosphate measured by the heteropolyblue phosphate method may be 10 to 100 times greater than the actual soluble orthophosphate concentration.

Somewhat more work has been done for sea water (30-34), but, to date, no major component compounds have been identified, although several organic phosphorus esters have been resolved by two dimensional paper chromatography after isolation from marine algal cultures initially containing radiophosphate (34).

Considerable work has been done regarding the release of phosphate and organic phosphorus by various organisms and by different mechanisms (31, 32, 35-48). Which of these is responsible for the steady-state soluble organic phosphorus pool is still conjecture. Figure 8 depicts a summation of the possible sources suggested by several authors. General conclusions to be drawn from these studies are:

(a) Soluble organic phosphorus compounds apparently result from biological activity.

(b) Several organisms are capable of releasing soluble organic phosphorus.

(c) Whether soluble organic phosphorus is principally the result of active metabolism or death and decay is not clearly established.

(d) Which source represents the major contribution to the natural soluble organic phosphorus pool is not known.

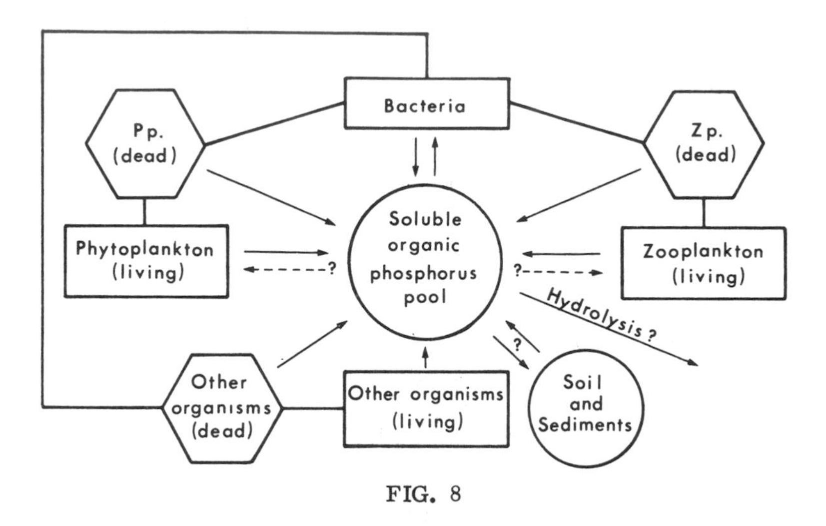

FIG. 8

A summary of possible contributors to and losses from the soluble organic phosphorus pool in natural waters.

(e) The role of bacteria is not clearly defined.

(f) In the case of death, certain cellular compounds are thought to be released by autolysis, some by enzymatic release and others by bacterial action only.

1. Preliminary work.

Some data are being acquired on this subject and, in some preliminary work with an axenic, unialgal, batch cultures of Chlamydomonas reinhardtii grown in a completely synthetic medium, soluble organic phosphorus levels of 0.5 mg/l as P have been obtained. This represented 25% of the total soluble phosphorus 12 days after inoculation. Our general solubility criterion has been filtration with 0.45 μ Millipore filters but occasional additional filtration through 0.10 μ Gelman filters yielded identical values.

2. Isolation of soluble phosphorus.

The culture system mentioned above was centrifuged at less than 1000 rpm for 5 to 10 minutes and the supernatant filtered in succession through a glass fiber prefilter at less than 10 inches of vacuum and then through a 0.45 μ membrane filter. The resultant solution was quickly frozen and then freeze dried. Roughly 85% of the organic phosphorus compounds in the freeze dried solids was found to be directly water extractable. This ready extractability was certainly enhanced by the absence of calcium in the culture medium but it is anticipated that cation exchange prior to freeze drying of lake water samples should aid in recoveries by direct water extraction. Greater total phosphorus extraction was accomplished by 0.5N HCl but there was some evidence of hydrolysis.

Extraction of the culture solids with a lipid solvent system, $CHCl_3$-CH_3OH-HCl, 200:100:0.2 (v/v), was capable of picking up part of the phosphorus containing compounds from the solids. However, a saline wash of the organic phase removed nearly all the phosphorus into the aqueous phase, indicating the absence of appreciable quantities of phospholipids.

3. Sephadex gel studies.

Sephadex G-10, G-15, and G-25 gels were evaluated with regard to their abilities to separate some known phosphorus compounds. The G-10 gel was found to be inadequate since the orthophosphate was nearly excluded from the column. The utility of the other two gels is illustrated in Figures 9 and 10. The results of more than one experiment are superimposed for convenience. The G-15 gel was capable of resolving sodium glycerophosphate from orthophosphate in the same sample. In a separate experiment phytin was dissolved in 0.01N HCl and was separated into two major peaks, hexaphosphate and possibly the pentaphosphate structure in the excluded peak and orthophosphate released by hydrolysis as the other. The smaller peaks most probably are due to the intermediate inositol phosphates with the one roughly corresponding to the position of glycerophosphate likely being the monophosphate. Similarly the G-25 gel separates orthophosphate from sodium DNA (which is excluded as expected) and phytate which is now in the fractionation volume of the gel (less hydrolysis had occurred with this sample as evidenced by the greater ratio of the main organic P peak to the ortho peak).

Extraction of $\sim$0.7g of the culture freeze dried solids with 11 ml of distilled water yielded a solution containing 89 mg P/l, $\sim$39% of which is organic. This solution was applied to each of the columns previously described. The results are shown in Figures 11 and 12.

In each case, a significant portion of the organic P is excluded from the column, indicating that some of the organic phosphorus material is possibly of large molecular size. The material excluded

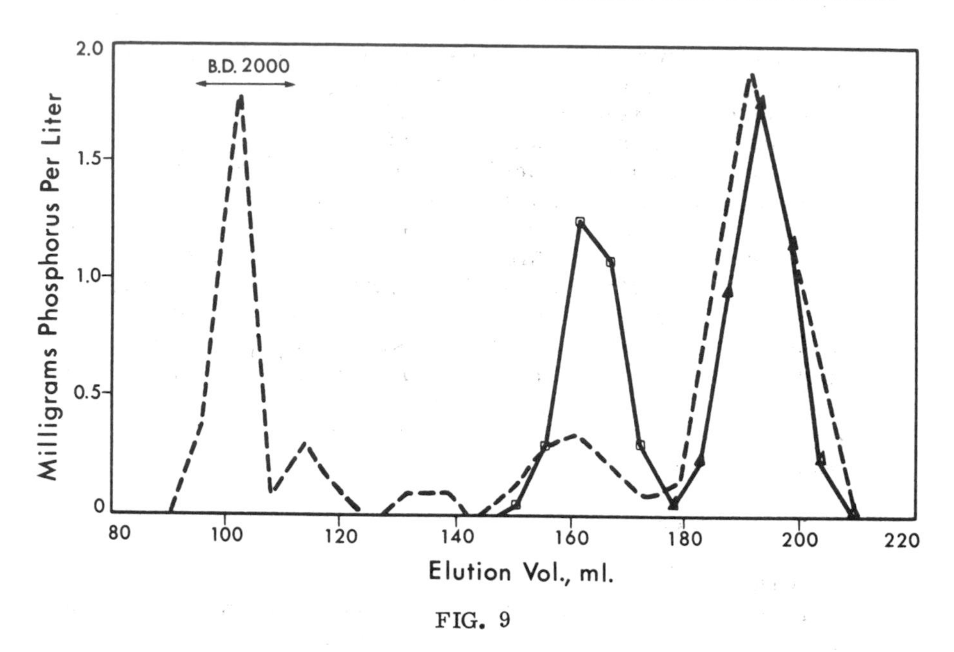

FIG. 9

Elution profile of some known phosphorus containing compounds on Sephadex G-15 (2.5 x 55 cm.). Eluent: 0.01 N HCl, 0.05 N NaCl.
□ - Na glycerophosphate, ▲ - KH_2PO_4, ----- CaMg phytate (Total P determined only).

from the G-25 gel was collected in a composite sample and its ultraviolet spectrum was determined. Although exhibiting only end absorption, significant absorption was shown in the region (240-290 mμ) of the DNA absorption peak. But, since many other compounds may well be in this solution and in the absence of a distinct peak corresponding to that of the nucleic acids, little significance can be attached to the spectrum. It only suggests that nucleic acid material could be present in this fraction and this is currently under investigation.

A close examination of both the G-25 and G-15 profiles for the culture extract material shows that the orthophosphate position has shifted considerably. It appears that this is due to a change in eluant from the HCl - NaCl system to NH_4HCO_3 (NH_4HCO_3 was used to allow desalting of excluded material by freeze drying and removal of the volatile salt) rather than a real change in molecular size of the culture extract orthophosphate (responding) material. Later studies on the

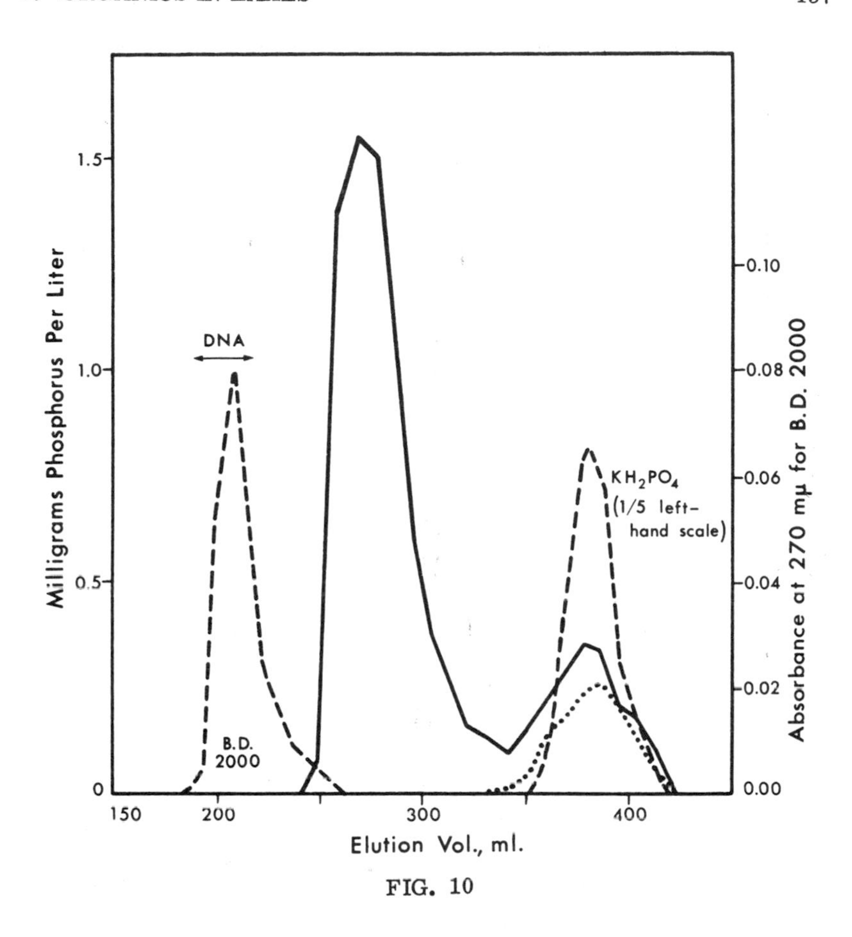

FIG. 10

Elution profile of some known phosphorus containing compounds on Sephadex G-25 (2.5 x 90 cm.). Eluent: 0.01 N HCl, 0.05 N NaCl except for DNA which used 0.05 N NH_4HCO_3. ——— CaMg phytate sample, Total P, •••• CaMg phytate sample, ortho P, others as labeled.

G-15 gel showed this to be probably a pH effect rather than difference in ionic content as 0.05 N NaCl produced the same shift for KH_2PO_4 as the 0.05 N NH_4HCO_3 eluant.

In each profile there is evidence of lower molecular weight organic phosphorus material. Significant amounts coelute with the

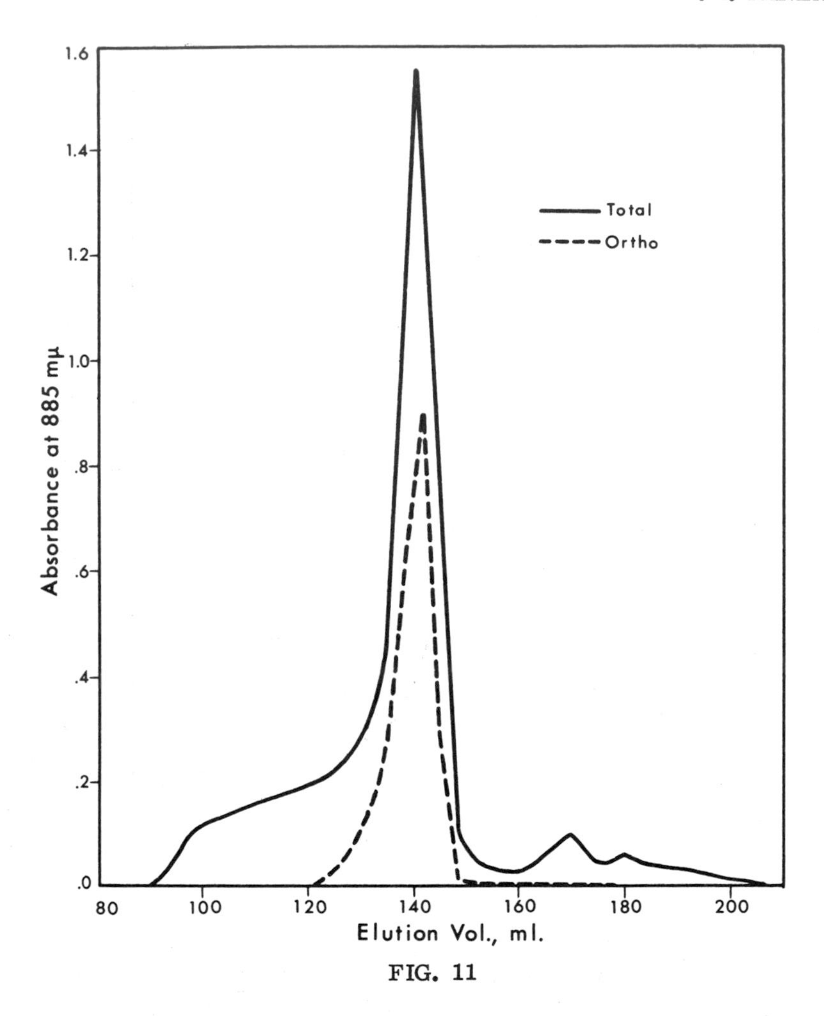

FIG. 11

Elution profile of water extractable phosphorus compounds from culture freeze dried solids on Sephadex G-15 (2.5 x 55 cm.). Eluent: 0.05 N NH_4HCO_3.

orthophosphate material and additional compounds elute after the orthophosphate peak. The identify of these compounds has not been established yet.

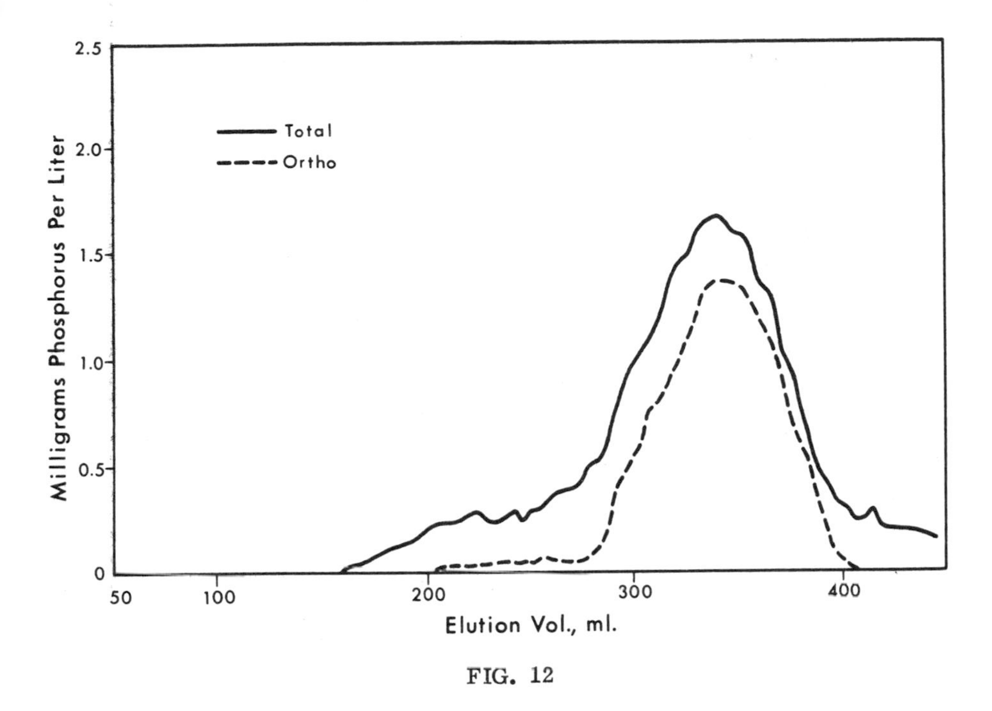

FIG. 12

Elution profile of water extractable phosphorus compounds from culture freeze dried solids on Sephadex G-25 (2.5 x 90 cm.). Eluent: 0.05 N NH_4HCO_3.

III. SUMMARY

Two separate natural organic types present in lake waters have been considered: natural product organics responsible for color and soluble organic phosphorus compounds which are more likely to have their origin from living, dying, or dead aquatic organisms. Some of the properties elucidated for color that have been considered are:

1. The macromolecular nature of color substances as defined by elution behavior on Sephadex gel columns.

2. The polyphenolic nature of color resulting from identification of compounds resulting from mild oxidative and reductive degradation studies on isolated color solids.

3. The ability of color to complex with borates and other metal ions as illustrated by its interaction with iron.

4. The variable nature of color from different sources as indicated by both chemical and physical differences.

5. The exclusion of lignin as the sole precursor of color.

A few comments regarding Sephadex data are in order at this point. Such things as the interaction of color acids with the gels and the effect of sample concentration have not been evaluated. Therefore, until the validity of Sephadex data is confirmed by other methods of molecular size analysis, these data can not be taken as conclusive. It is quite likely that the actual size distribution could be entirely different from that indicated by Sephadex.

The absence of specific knowledge on the chemical nature of soluble organic phosphorus compounds has been pointed out. It is quite likely that these compounds result from a quasi steady-state build-up during periods of high biological activity but the primary sources and mechanisms responsible in the natural environment are not well defined.

Some preliminary work on these compounds in fresh water systems has shown that appreciable build-ups of the compounds are possible in pure algal cultures and these compounds appear in part to be of higher molecular weight. The absorption of the higher molecular weight fraction in the ultraviolet region makes nucleic material a possible source.

ACKNOWLEDGMENT

Special thanks are extended to Dr. Masood Ghassemi of Atomics International, Canoga Park, California, who read and provided helpful comments on the first draft of this manuscript.

REFERENCES

1. J. Keosian, The Origin of Life, 2nd ed., Reinhold, New York, 1968, 120 pp.

2. T. R. Parsons and H. Seki, Symposium on Organic Matter in Natural Waters, University of Alaska, College, Alaska, September, 1968.

3. I. A. Breger, Symposium on Organic Matter in Natural Waters, University of Alaska, College, Alaska, September, 1968.

4. E. T. Gjessing, Nature, 208, 1091 (1965).

5. J. Shapiro, The Symposium of the Hungarian Hydrological Society, Budapest-Tihang, Hungary, September, 1966.

6. E. T. Gjessing and G. F. Lee, Environ. Sci. Technol., 1, 631 (1967).

7. M. Ghassemi and R. F. Christman, Limnol. Oceanog., 13, 583 (1968).

8. M. Ghassemi, Ph. D. Thesis, University of Washington, Seattle, 1967.

9. H. Determann, Gel Chromatography, Gel Filtration, Gel Permeation, Molecular Sieves. A Laboratory Handbook, Springer-Verlag, New York, 1968.

10. B. J. Gelotte, J. Chromatogr., 3, 330 (1960).

11. A. S. Behrman, R. H. Kean, and H. Gustafson, Paper Trade J., 92, 1 (1931).

12. A. P. Black and R. F. Christman, J. Amer. Water Works Assoc., 55, 753 (1963).

13. R. F. Christman and M. Ghassemi, J. Amer. Water Works Assoc., 58, 723 (1966).

14. R. F. Christman, Symposium on Organic Matter in Natural Waters, University of Alaska, College, Alaska, September, 1968.

15. R. F. Packham, Proc. Soc. Water Treat. Exam., 13, 316 (1964).

16. A. Saville, J. New Eng. Water Works Assoc., 31, 78 (1917).

17. R. F. Christman, Trend, 16, 10 (1964).

18. K. V. Sarkanen, in The Chemistry of Wood (B. L. Browning, ed.), J. W. Wiley, New York, 1963, Chap. 10.

19. J. Shapiro, J. Amer. Water Works Assoc., 56, 1062 (1964).

20. R. F. Christman, Environ. Sci. Technol., 1, 302 (1967).

21. C. C. Coffin, F. R. Hayes, L. H. Jodrey, and S. G. Whiteway, Can. J. Res., Zool Sci., 27D, 207 (1949).

22. F. H. Rigler, Ecol., 37, 550 (1956).

23. E. Harris, Can. J. Zool., 35, 769 (1957).

24. L. R. Pomeroy, Sci., 131, 1731 (1960).

25. F. H. Rigler, Limnol. Oceanog., 9, 511 (1964).

26. V. B. Scheffer and R. J. Robinson, Ecol. Monogr., 9, 95 (1939).

27. G. E. Hutchinson and V. T. Bowen, Proc. Nat. Acad. Sci., 33, 145 (1947).

28. F. Ruttner, Fundamentals of Limnology, 3rd ed., University of Toronto, 1963, 295 pp.

29. F. H. Rigler, Limnol. Oceanog., 13, 7 (1968).

30. L. M. Jeffry, Symposium on Organic Matter in Natural Waters, University of Alaska, College, Alaska, September 1968.

31. W. D. Watt and F. R. Hayes, Limnol. Oceanog., 8, 276 (1963).

32. J. E. Phillips, in Principles and Applications in Aquatic Microbiology (H. Heukelekian and N. C. Dondero, eds.), J. W. Wiley, New York, 1964, p. 61.

33. E. J. Kuenzler and J. P. Perras, Woods Hole Oceanographic Institute Reference No. 65-59, 1965, (Unpublished Manuscript).

34. E. J. Kuenzler, ESE Notes, 4, 2 (1967).

35. A. C. Redfield, H. P. Smith, and B. H. Ketchum, Biol. Bull., 73, 421 (1937).

36. D. M. Pratt, J. Mar. Res., 9, 29 (1950).

37. F. R. Hayes and J. E. Phillips, Limnol. Oceanog., 3, 459 (1958).

38. J. D. H. Strickland and K. H. Austin, J. Fish. Res. Board, Can., 17, 337 (1960).

39. H. L. Golterman, Acta. Bot. Neer., 9, 1 (1960).

40. F. H. Rigler, Limnol. Oceanog., 6, 165 (1961).

41. E. J. Kuenzler, Limnol. Oceanog., 6, 400 (1961).

42. N. S. Anita, C. D. McAllister, T. R. Parsons, K. Stephens, and J. P. H. Strickland, Limnol. Oceanog., 8, 184 (1963).

43. A. D. Ansell, J. E. G. Raymont, K. F. Lander, E. Crowley, and P. Shackley, Limnol. Oceanog., 8, 184 (1963).

44. L. R. Pomeroy, H. M. Mathews, and H. S. Min, Limnol. Oceanog., 8, 50 (1963).

45. R. E. Johannes, Limnol. Oceanog., 9, 225 (1964).

46. R. E. Johannes, Limnol. Oceanog., 9, 235 (1964).

47. R. E. Johannes, Limnol. Oceanog., 10, 434 (1965).

48. B. T. Hargrave and G. H. Geen, Limnol. Oceanog., 13, 332 (1968).

CHAPTER 7

THE DISTRIBUTION AND CYCLING OF ORGANIC MATTER IN THE OCEAN

P. M. Williams

Institute of Marine Resources
University of California, San Diego

"The amount of organic matter dissolved in lake and sea water depends upon the nature of the water, the plankton organisms, and the higher forms of plant and animal life inhabiting lake or sea."

S. A. Waksman, 1936 (1)

I. INTRODUCTION

The basic question in an understanding of the organic carbon cycle in the oceans is: what are the magnitudes of the living and dead organic carbon reservoirs and what is the flux of organic carbon and its direction through these reservoirs? The inter-relationships between phytoplankton, zooplankton, bacteria, dissolved and particulate organic matter, and organisms higher in the food chain are complicated and only partially resolved. In this report, emphasis will be directed towards the distribution, chemical nature, and stability of the dissolved and particulate organic matter in the sea.

II. DEFINITIONS AND METHODS

The dissolved organic matter is generally considered to be that which passes a 0.45 μ millipore HA® filter, a Whatman GF/C® glass fiber filter (nominal pore size: 1-2 μ), or Selas Flowtronics® silver filters of varying porosities. Thus the dissolved organic matter includes colloidal as well and truly dissolved material. The particulate organic matter is that retained on the above filters.

The analytical method for previously unpublished dissolved organic carbon results reported in this paper is that of Menzel and Vaccaro (2), and for the particulate organic carbon, a modification of Menzel and Vaccaro (3). Vitamin B_{12} determinations were done by A. F. Carlucci using the method of Carlucci and Silbernagel (4), and dissolved oxygen was determined by the Winkler titration (3).

III. THE DISTRIBUTION OF DISSOLVED AND PARTICULATE ORGANIC CARBON

The relative amounts of dissolved and particulate organic carbon (hereafter designated as DOC and POC, respectively) are depicted schematically in Figure 1, where the mean depth of the ocean is taken as 3800 m. The average concentration of DOC from 0-300 m is 1.0 mg C/l (range: 0.3-2.0 mg C/l) and from 300-3800 m, 0.5 mg C/l (range: 0.2-0.8 mg C/l). For POC from 0-300 m, the average concentration is 0.1 mg C/l (range: 0.03-0.3 mg C/l), and from 300-3800 m, 0.01 mg C/l (range: 0.005-0.03 mg C/l). Recent determinations of DOC and POC and references to earlier work can be found in (5, 6, 7, 8, 9, 10, 11, 12, 13). The results of Skopintsev and co-workers for DOC are several times higher than those reported here but the general distribution is similar. Thus, in the surface layer the DOC is ten times that of POC, and in the deep water 50 times higher. In comparison, the total inorganic carbon from 0-3800 m is of the order of forty times the total organic carbon. This transition depth of 300 m may vary from 100-400 m, depending upon surface production of organic matter and physical mixing processes, but the most outstanding observation is that the concentrations of DOC and POC may vary widely above this transition depth while remaining relatively constant below it down to the sea floor. This vertical homogeneity in the deep water occurs even though other non-conservative properties, such as dissolved oxygen and inorganic phosphate, nitrate, and silicate have well-defined maxima and minima and are themselves interrelated. Vertical profiles of DOC and POC at stations in the Atlantic and Pacific Oceans (Figs. 2, 3, and 4) illustrate this homogeneity although the absolute concentrations of DOC and POC may vary spatially.

Duursma (5), Menzel (6), and Barber (10) have suggested that the distribution of DOC may be used to trace subsurface water movements. Menzel (6) observed that variations of DOC from 0.2 to 2.0 mg C/l occurred at specific density surfaces in the Indian Ocean, and that the concentration of DOC was predictable from salinity. Why

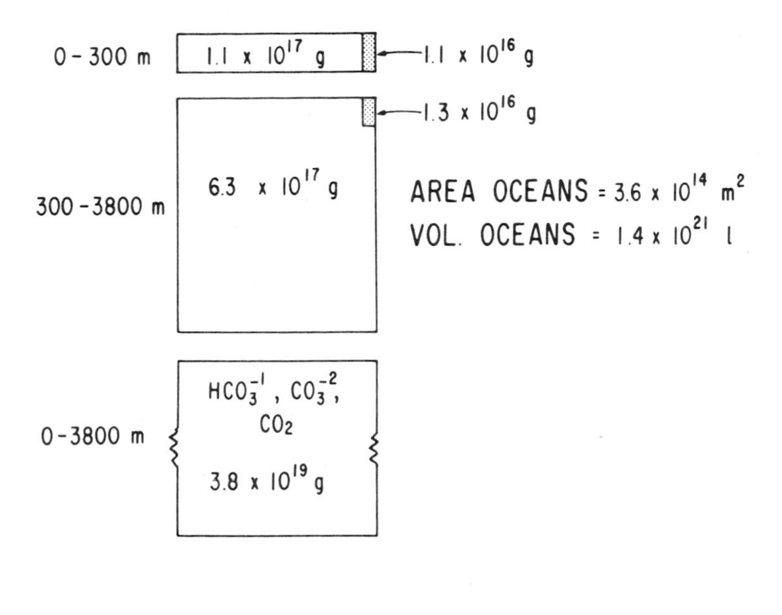

FIG. 1

The amounts of dissolved and particulate organic carbon and inorganic carbon in the various oceanic reservoirs.

these differences in DOC can be maintained in the Indian Ocean and not in the Atlantic or Pacific is as yet obscure.

There have been no extensive studies on the secular variation of DOC in subsurface waters at a single oceanic location. However, DOC (including POC, which is insignificant in amount relative to DOC in deep water) was determined for samples taken on cruises which traversed an identical transect along the 155° meridian from the Hawaiian to the Aleutian Islands in Aug., 1965 and Jan., 1966 (Fig. 5). Integration of the total DOC per m^2 from 0-300 m and 300-1500

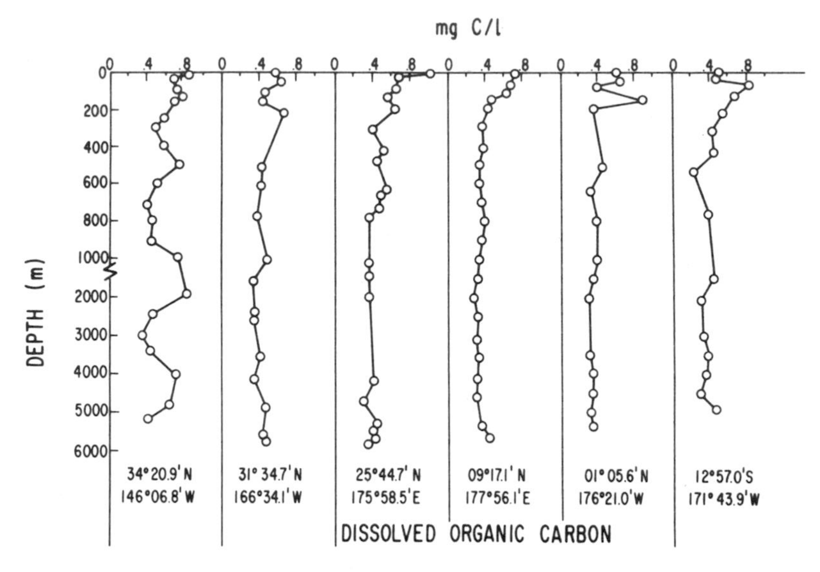

FIG. 2

Vertical profiles of dissolved organic carbon at stations from California to Samoa. Samples collected on expedition "Nova," May 1967, by M. Koide.

m shows that the organic carbon content of the deep water was significantly higher in Jan., 1966 than in Aug., 1965 between the latitudes 40-45°N, whereas in the surface waters there were no real differences. Integrated dissolved oxygen values are included but it is not certain that the differences in deep water are real or due to methodology which was changed between cruises. It appears that at 40-45°N, the transition zone between the northern subarctic water and the subtropical central Pacific water, there occurred an influx of water containing a high concentration of DOC between Aug., 1965 and Jan., 1966, or that high surface productivity resulted in the sinking and solubilization of POC at depth. A random, seasonal variation of POC in subsurface waters in the Sargasso Sea has been reported (14) although at any one time the vertical distribution was constant.

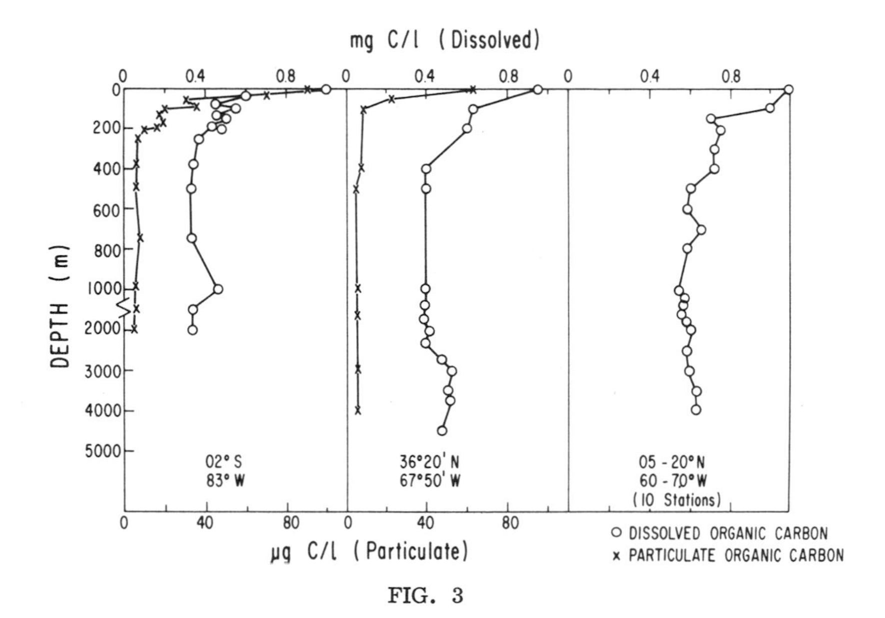

FIG. 3

Vertical profiles of dissolved and particulate organic carbon at stations in the Atlantic Ocean. Redrawn from Ref. (7) by courtesy of Pergamon Press Ltd.

IV. THE RELATION OF DISSOLVED AND PARTICULATE ORGANIC CARBON TO PRODUCTIVITY

In the open ocean it has been demonstrated that DOC is not related to the primary productivity at the sea surface, and in fact, may be inversely related to primary organic production in certain areas (6, 11). A coastal plankton survey, Apr. - Sept., 1967, off southern California (15) indicated no correlation of DOC with chlorophyll a in surface waters 3 to 6 nautical miles offshore, and only a slight correlation ($P < 0.05$) at the inshore (1 nautical mile) station. Earlier data (Fig. 6) for the concentration of DOC off the Scripps Institution pier does demonstrate that when massive "red tide blooms" of dinoflagellates are present, the DOC increases markedly due to lysis of dead cells and then is rapidly reduced by bacterial mineralization

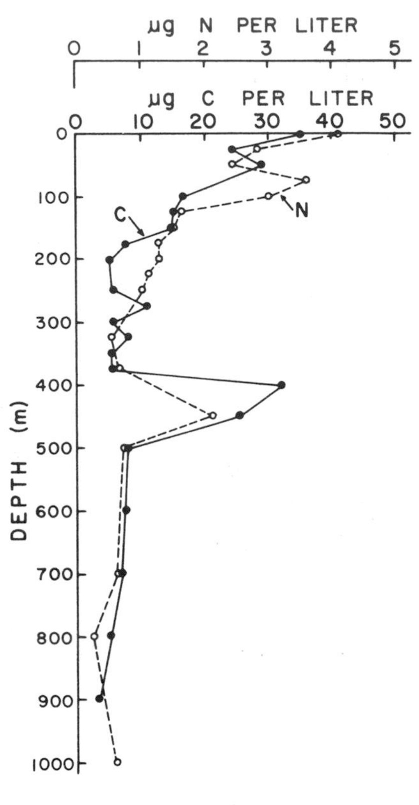

FIG. 4

Vertical profile of particulate organic carbon and nitrogen in the northeast Pacific Ocean. Reprinted from Ref. (45) by courtesy of Allen Press, Inc.

such that the resultant concentration of DOC is about 1 mg C/l, the normal, baseline value (16).

In subsurface waters, DOC is not related to its concentration at the surface or to organic production in the euphotic zone (17). There has been considerable speculation, and some indirect experimental evidence, that dissolved organic carbon may be transformed to particulate organic matter at depth (14, 18, 19), or that organic aggregates, which form at the sea surface due to wave-induced bubble

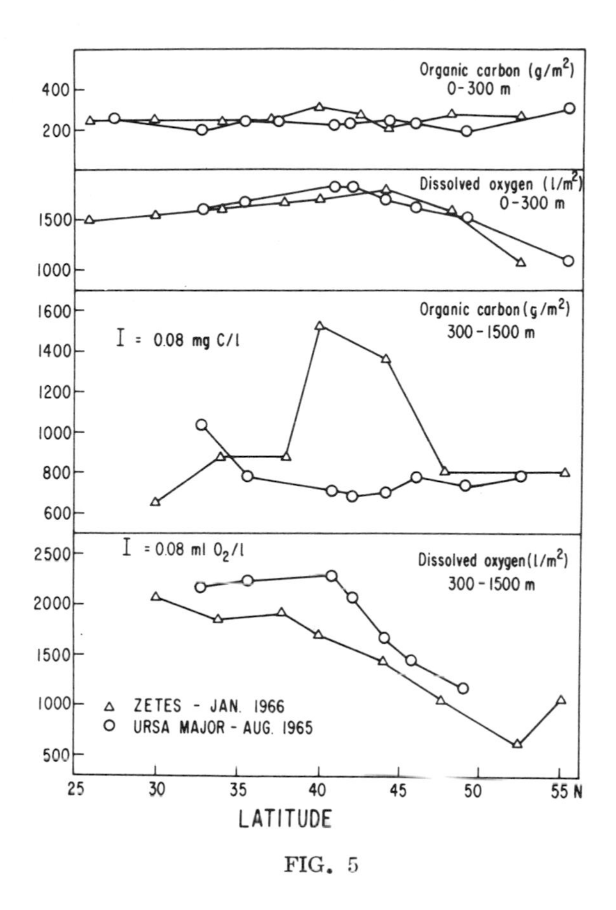

FIG. 5

Secular variation of dissolved organic carbon and dissolved oxygen, Hawaii to the Aleutian Islands. The ordinate represents the integration of the amounts of oxygen and DOC per m^2 per depth interval given. The DOC samples were collected by C. Miller on expeditions "Ursa Major" and "Zetes."

action, sink and then may become solubilized. This remains an open question, however.

The concentration of POC is highly variable in surface waters and is directly related to phytoplankton productivity (15, 17, 20) as determined by chlorophyll a measurements or the uptake of radiocarbon by

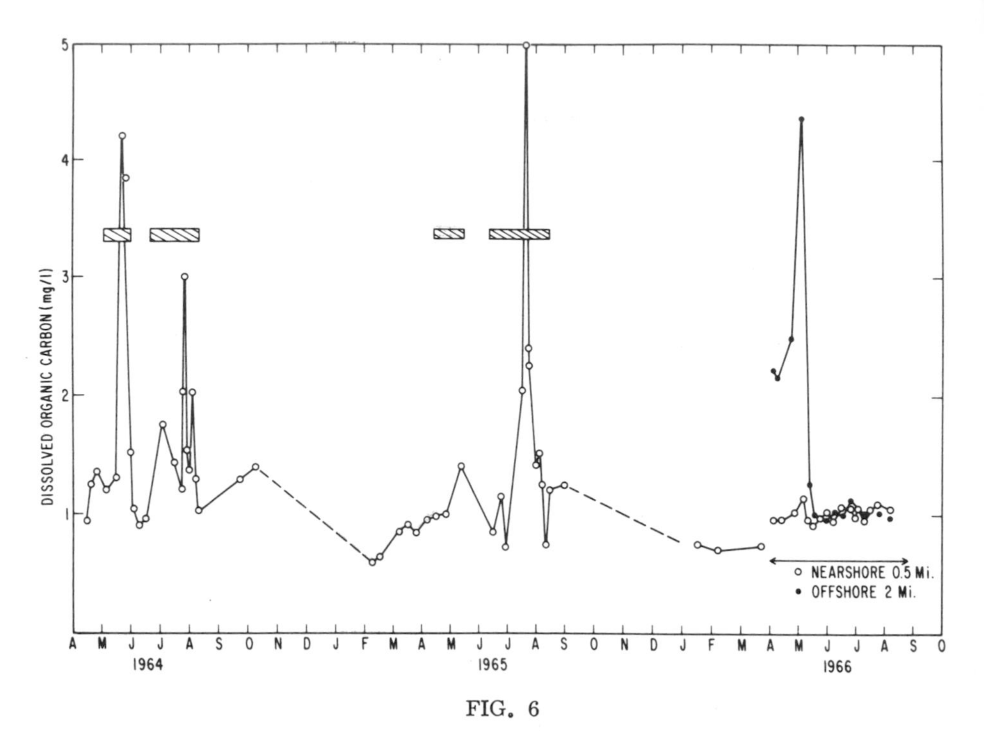

FIG. 6

Secular variation of DOC for surface samples taken off Scripps Institution Pier and at offshore stations. The hatched bars represent "red tide blooms" of dinoflagellates. Reprinted from Ref. (16) by courtesy of Allen Press, Inc.

phytoplankton. As with the DOC, there is no evidence that the concentration of subsurface POC is in any manner a reflection of the organic production at the sea surface (7). Menzel and Ryther (17) have suggested that, in the deep water, the amount of POC may represent the lower limit that can be grazed by zooplankton or that the POC is chemically inert and not utilizable by grazing organisms or bacteria. The steady state transformation of DOC into POC by adsorption of DOC onto inorganic or biologically inert surfaces (14, 21) with subsequent utilization of these conglomerates may also maintain the low and constant concentration of POC at depth.

In effect, there does not appear to be a simple mechanism which can explain the distribution and concentration of the dissolved and particulate organic matter in the deep sea with relation to its production, utilization, and decomposition in the euphotic zone.

V. DISSOLVED ORGANIC CARBON AND THE OXYGEN MINIMUM

In most areas of the oceans, dissolved oxygen exhibits a minimum concentration at a depth anywhere from 100 to 1500 meters, and in some cases double minima are present (22). The oxygen concentration at the minimum in the northeast Pacific Ocean is of the order of 0.2-0.5 ml O_2/l, and in the eastern equatorial Pacific concentrations are less than 0.1 ml/l at the minimum. An oxygen concentration of 0.1 ml/l is equivalent to the oxidation of approximately 0.04 mg C/l assuming an oxidative ratio of 2.6:1 (O:C) by atoms (23). Since the concentration of DOC at the depth of the oxygen minimum is 0.2-0.6 mg C/l at these above locations, it is obvious that large sections of the deep oceans would become anoxic should heterotrophic bacteria readily utilize the DOC over short time periods or should chemical oxidation of DOC occur. This is not the case. The relationship between DOC and the oxygen minimum in the Atlantic Ocean has been discussed by Menzel and Ryther (11) and by Bubnov and co-workers (24, 25). These authors conclude that oxygen utilization and the decomposition of organic matter occur in the upper 300-400 m of the ocean, and that this low oxygen water then sinks. Subsequent mixing of opposing water masses then determines the concentration of dissolved oxygen and DOC. In this model, the DOC does not undergo biochemical utilization once the water has left the surface.

VI. THE CHEMICAL NATURE OF THE DISSOLVED AND PARTICULATE ORGANIC MATTER

The problem of the molecular nature of the dissolved organic matter in the ocean, and especially in the deep sea, is similar to, and more difficult than, the identification of "humus" in soils. It is necessary to extract without alteration and identify 0.5 to 1.0 mg C/l or less of organic matter from an aqueous solution containing 3.5 x 10^4 times as much inorganic salts.

The more easily identifiable fatty acids, amino acids, and sugars have been reported for surface waters and for several deep sea profiles (for a summary of work prior to 1965 see Duursma, (26)). In addition to these more ubiquitous compounds, trace amounts of C_1-C_4 hydrocarbons (27), C_{20}-C_{33} n-alkanes (28), and pristane (29) have been identified in sea water. In all the above cases, however, there were no total DOC determinations made on the same water samples from which the organic compounds were isolated. Without such total DOC values, it is difficult to assess contamination problems and total recoveries. Table 1 lists the total amounts and molecular nature of the DOC in the northeast Pacific Ocean. The data for the amino acids, fatty acids, sugars, etc., were compiled from different sources (30, 31, 32, 33) for samples taken at different times in the same general area. The main point is to illustrate that only about 10% of the total DOC has been identified in surface and subsurface waters, leaving some 90% which has been termed variously as "humic acids" or "lignin-type" material. This is the organic matter which may be refractory to microbial degradation or chemical oxidation. In fact, Barber (34) has demonstrated that the *in situ* bacterial population will not utilize the DOC in the deep water, even though it is concentrated five-fold. Co-precipitation of the DOC in deep water in the Pacific Ocean (31) yielded a mixture of hydroxylated organic acids having an average molecular weight of 395, an equivalent weight of 208, and an empirical formula $C_{18}H_{24}O_{12}$. Khailov and Finenko (35), however, report the presence of high molecular weight compounds in surface sea water. In sea surface films containing high concentrations of organic carbon (36), long chain alcohols have been found in addition to fatty acids (37), and there is indirect evidence that copper, zinc, and iron may be complexed with organic ligands in the sea (38, 39, 40).

It is evident that little is really known concerning the molecular nature of the DOC in the sea, and such identification may be as intractable as the identity of "humus" in soils.

The molecular nature of the particulate organic matter in the sea is again limited to fatty acids (41), amino acids (30), and sugars (30, 42, 43). Most analyses of POC have been by non-specific colorimetric techniques, where acid hydrolysis precedes the analysis. There is indirect evidence for the presence of uronic acids in the POC (44), but no identification of the individual acids has been achieved. Adenosine triphosphate (ATP) and deoxyribonucleic acid (DNA) have been determined in the POC taken in vertical profiles off southern California (45). Both ATP and DNA show no gradient in concentration below 200-300 m. The small quantities of POC in the deep water (0.005-0.03 mg C/l) necessitates filtration of large amounts of sea water in order to obtain sufficient material for identification, and this is only just becoming a practical routine.

TABLE 1

The Molecular Nature of the Dissolved Organic Matter in the Northeast Pacific Ocean

	μg C/l (mean)	
	0-300 m	300-3000 m
Total organic carbon	1000	500
Amino acids (free and combined)	25	25
Sugars (free)	10	10
Fatty acids (free and combined)	40	10
Urea (free)	20	<2
Aromatics (substituted phenols)	1	-
Vitamins (B_{12}, B_1, biotin)	10^{-2}	10^{-2}
$\Sigma \approx$	100	50
% Identified of Total $\approx$	10	10

VII. CARBON ISOTOPES AND THE STABILITY OF THE DISSOLVED ORGANIC MATTER

A. Carbon-13: Carbon-12 Ratios.

The stable carbon isotopic composition of the dissolved and particulate organic carbon in the Pacific Ocean has a very narrow range of values compared to the composition of the plankton from which it was ultimately derived (46, 47). These $^{13}C/^{12}C$ ratios are between -22.0‰ and -24.4‰ relative to the PDB_1 standard. In Figure 7, the $\delta^{13}C$ values are given for various marine and terrestrial carbon reservoirs. The more minus the value, the more it is depleted in C^{13}:

$$\delta^{13}C = \frac{(^{13}C/^{12}C)\text{ sample} - (^{13}C/^{12}C)\text{ standard}}{(^{13}C/^{12}C)\text{ standard}} \times 1000 \qquad (7\text{-}1)$$

Included in Figure 7 are $^{13}C/^{12}C$ ratios for the DOC and POC in the Amazon River. Since the discharge of the Amazon is approximately

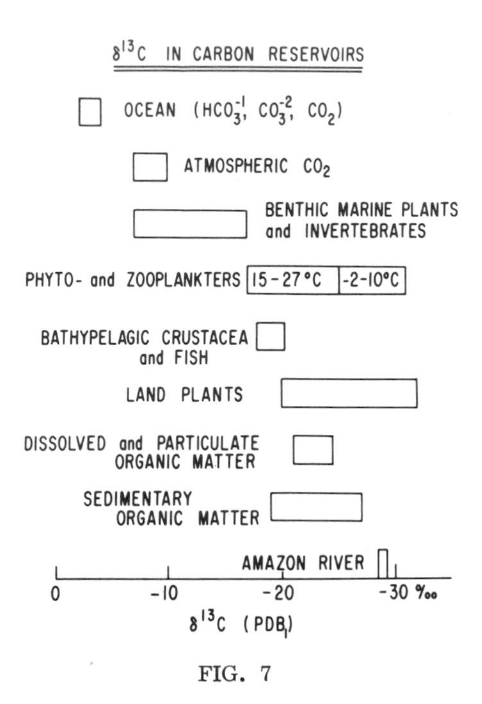

FIG. 7

The range of $\delta^{13}C$ for marine and terrestrial sources of organic and inorganic carbon. See Ref. (47) for source material. The temperature ranges given for the phyto- and zooplankters are the surface water temperatures at the collection locations.

20% of the entire world-wide river runoff (48), then if the DOC in the oceans was derived largely from terrestrial runoff, its $\delta^{13}C$ value should be nearer to -28‰ than to -22‰.

The vertical profile of the $^{13}C/^{12}C$ ratios in the DOC (Fig. 8) off southern California shows that the $\delta^{13}C$ values are essentially constant with depth, that the POC has $^{13}C/^{12}C$ ratios similar to the DOC, and that there is no correlation of $\delta^{13}C$ with the concentration of dissolved oxygen (or with total DOC or total POC). Deuser and Hunt have recently shown (49) that the stable carbon isotope ratios of the inorganic carbon in the Atlantic Ocean correlate with the dissolved oxygen distribution; the inorganic carbon being lighter in carbon-13 at the oxygen minimum. This depletion in carbon-13 was

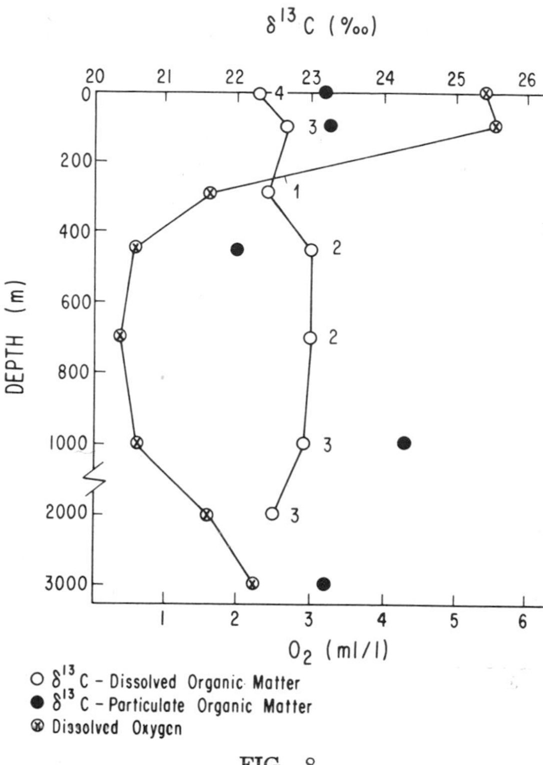

FIG. 8

The vertical distribution of $\delta^{13}C$ in the DOC and POC at stations off southern California, and the dissolved oxygen profile at these locations. The numbers to the right of the open circles designate the number of samples analyzed.

attributed to oxidation of marine organic matter, itself light in carbon-13, to carbon dioxide in the euphotic zone, with resultant sinking of this low oxygen, carbon-13 depleted water.

The range of $\delta^{13}C$ for the DOC is close to that for the so-called cellulose and "lignin" fractions of marine phyto- and zooplankters (50). This suggests that the DOC has a singular and common source and that this material is stable to biochemical fractionation.

B. The Natural Radiocarbon Activity of the Dissolved Organic Carbon in the Deep Sea.

The apparent "age" of the dissolved organic matter at 2000 m in the northeast Pacific Ocean is approximately 3400 years (b. p.) (51). Calculation of the residence time of the DOC in the deep-sea from input and output fluxes can be made assuming a steady-state situation (Table 2). The net primary production figures are taken from Ryther (52). For rain (organic content = 1 mg C/l) it was assumed that 2.23 x 10^{17} l/yr falls on the ocean (53). For rivers (organic content = 2 mg C/l) a total discharge of 1.58 x 10^{16} l/yr (48) was assumed. Sedimentation rates were taken as 1 mm/10^3 yrs for 2.68 x 10^8 km^2 of pelagic sediments and 40 mm/3 yrs for 9.2 x 10^7 km^2 of nearshore sediments (54, 55). The organic content of the pelagic and nearshore sediments is 1% and 2.5% on a dry weight basis, where the density of the sediments is 2 g/cm^3 and the water content is 50%. The output (9.5 x 10^{13} g/yr) does not take into account microbial oxidation of the organic matter. Considering the uncertainty in estimating all of these fluxes, the apparent "age" of the DOC in the deep sea is of the order of what could be predicated on the basis of steady-state conditions assuming some 0.4% of the carbon fixed by photosynthesis enters the deep sea as DOC. The calculation of the residence time of DOC from output rates is about twice the 3400 years found experimentally, but it is still of the same order of magnitude. Bruyevich (56) and Romankevich (57) have calculated that some 0.01% of the total particulate carbon produced in the euphotic zone by photosynthesis passes into the sediments, and that the turnover time of POC is 3000 years.

VIII. SUMMARY

The following facts, enumerated above, support the hypothesis originally put forth by Krogh (58), that the dissolved organic matter, and possibly the particulate organic matter is chronologically old, chemically and biochemically inert, and is not significant in the marine food chain in the deep sea: (a) the vertical and spatial distribution of DOC and POC is relatively constant below 200-400 m; (b) indigenous microorganisms cannot utilize the bulk of the DOC in the deep-sea; (c) the chemical composition and carbon-13: carbon-12 ratios suggest that the DOC is a "humic" type material; and (d) the radiocarbon "age" of the DOC indicates that its turnover time is some 3400 years.

TABLE 2

Input and Output Rates and Balances for a Steady-state Distribution of Dissolved Organic Matter

		g/yr
Input:		
Net primary productivity (100 g C fixed/m^2/yr)		3.6×10^{16}
Rain (1 mg C/l)		2.2×10^{14}
Rivers (2 mg C/l)		3.1×10^{13}
	Σ =	3.6×10^{16}
Output:		
Sedimentation		
Nearshore		2.7×10^{12}
Pelagic		9.2×10^{13}
	Σ =	9.5×10^{13}

Residence time (T):

$$T = \frac{\text{amount in reservoir (g)}}{\text{input or output (g/yr)}}$$

Input:

$$T = \frac{6.3 \times 10^{17}}{3.6 \times 10^{16}} \approx 2 \times 10^{1} \text{ yrs}$$

If 0.36% of primary production enters deep ocean,

$T = 3.4 \times 10^{3}$ yrs (experimentally determined value)

Output:

$$T = \frac{6.3 \times 10^{17}}{9.5 \times 10^{13}} \approx 7 \times 10^{3} \text{ yrs}$$

ACKNOWLEDGMENT

This work was supported in full by the United States Atomic Energy Commission, Contract No. AT(11-1)GEN 10, P. A. 20.

REFERENCES

1. S. A. Waksman, Humus, Williams and Wilkins, Baltimore, 1936, p. 289.

2. D. W. Menzel and R. F. Vaccaro, Limnol. Oceanogr., 9, 138 (1964).

3. J. D. H. Strickland and T. R. Parsons, A Practical Handbook of Seawater Analysis, Fish. Res. Bd. Can. Bull. 167, Ottawa, 1968.

4. A. F. Carlucci and S. B. Silbernagel, Can. J. Microbiol., 12, 175 (1966).

5. E. K. Duursma, Neth. J. Sea Res., 1, 1 (1961).

6. D. W. Menzel, Deep-Sea Res., 11, 757 (1964).

7. D. W. Menzel, Deep-Sea Res., 14, 229 (1967).

8. O. Holm-Hansen, J. D. H. Strickland, and P. M. Williams, Limnol. Oceanogr., 11, 548 (1966).

9. P. M. Williams, unpublished reports from the Institute of Marine Resources, University of California, San Diego, report No.'s 65-6, 65-7, 65-30, 67-9, and 68-8.

10. R. T. Barber, Ph. D. Thesis, Stanford Univ., Palo Alto, 1967.

11. D. W. Menzel and J. H. Ryther, Deep-Sea Res., 15, 327 (1968).

12. B. A. Skopintsev and S. N. Timofeyeva, Trudy morsk. gidrofiz Inst., 25, 110 (1962) (in Russian).

13. B. A. Skopintsev, S. N. Timofeyeva, and O. A. Vershinina, Okeanologiia Akad. Nauk SSSR, 6, 251 (1966) (in Russian).

14. G. A. Riley, D. Van Hemert, and P. J. Wangersky, Limnol. Oceanogr., 10, 354 (1965).

15. J. D. H. Strickland, L. Solorzano, and R. W. Eppley, Scripps Institution Oceanogr. Bull. 1970 (1969).

16. R. W. Holmes, P. M. Williams, and R. W. Eppley, Limnol. Oceanogr., 12, 503 (1967).

17. D. W. Menzel and J. H. Ryther, Symposium on Organic Matter in Natural Waters, Univ. of Alaska, College, Alaska, Sept. 1968 (to be published).

18. G. A. Riley, Limnol. Oceanogr., 8, 372 (1963).

19. P. J. Wangersky, Amer. Sci., 53, 358 (1965).

20. D. W. Menzel and J. J. Goering, Limnol. Oceanogr., 11, 333 (1966).

21. K. E. Chave, Science, 148, 1723 (1965).

22. F. A. Richards, in Treatise on Marine Ecology and Paleoecology, (J. Hedgepeth, ed.), Mem. Geol. Soc. Amer., 67, 185 (1957).

23. A. C. Redfield, B. H. Ketchum, and F. A. Richards, in The Sea (M. N. Hill, ed.) II, Interscience, New York, 1963, p. 26.

24. V. A. Bubnov, Okeanologiia Akad. Nauk SSSR, 6, 240 (1966) (in Russian).

25. V. A. Bubnov, O. A. Gushchin, and L. M. Krivelevich, Okeanologiia Akad. Nauk SSSR, 8, 605 (1968) (in Russian).

26. E. K. Duursma, in Chemical Oceanography (J. F. Riley and G. Skirrow, eds.) I., Academic Press, London, 1965, p. 433.

27. J. W. Swinnerton and V. J. Linnenbomm, Science, 156, 1119 (1967).

28. E. Peake and G. W. Hodgson, J. Amer. Oil Chem. Soc., 43, 215 (1966).

29. M. Blumer, Symposium on Organic Matter in Natural Waters, Univ. of Alaska, College, Alaska, Sept., 1968 (to be published).

30. E. T. Degens, J. H. Reuter, and K. N. E. Shaw, Geochim. Cosmochim. Acta, 28, 45 (1964).

31. P. M. Williams, Nature, 189, 219 (1961).

32. A. F. Carlucci, private communication, 1969.

33. J. J. McCarthy, private communication, 1969.

34. R. T. Barber, Nature, 220, 274 (1968).

35. K. M. Khailov and Z. Z. Finenko, Okeanologiia Akad. Nauk SSSR, 8, 980 (1968) (in Russian).

36. P. M. Williams, Deep-Sea Res., 14, 791 (1967).

37. W. D. Garrett, Deep-Sea Res., 14, 221 (1967).

38. P. M. Williams, J. Fish Res. Bd. Can., 23, 575 (1966).

39. J. F. Slowey and D. W. Hood, Ann. Rept. AEC Contract No. AT-(40-1)-2789, Texas A & M Rept. 66-2F (1966).

40. P. M. Williams, Limnol. Oceanogr., 14, 156 (1969).

41. P. M. Williams, J. Fish, Res. Bd. Can., 22, 1107 (1965).

42. N. Handa, Rec. Oceanogr. Works Jap., 9, 65 (1967).

43. T. R. Parsons and J. D. H. Strickland, Science, 136, 313 (1962).

44. P. M. Williams and J. S. Craigie, Symposium on Organic Matter in Natural Waters, Univ. of Alaska, College, Alaska, Sept. 1968 (to be published).

45. O. Holm-Hansen, Limnol. Oceanogr., 14, 740 (1969).

46. P. M. Williams, Nature, 219, 152 (1968).

47. P. M. Williams and L. I. Gordon, Deep-Sea Res. (in press, 1970).

48. R. J. Gibbs, Bull. Geol. Soc. Amer., 78, 1203 (1967).

49. W. G. Deuser and J. M. Hunt, Deep-Sea Res., 16, 221 (1969).

50. E. T. Degens, M. Behrendt, B. Gotthardt, and E. Reppmann, Deep-Sea Res., 14, 11 (1968).

51. P. M. Williams, H. Oeschger, and P. Kinney, Nature, 224, 256 (1969).

52. J. H. Ryther, in The Sea (M. N. Hill, ed.), Interscience, New York, 2, 347 (1963).

53. G. Neumann and W. J. Pierson, Jr., Principles of Physical Oceanography, Prentice-Hall, Engelwood-Cliffs, N.J., 1966.

54. K. O. Emergy, W. L. Orr, and S. C. Rittenberg, in Essays in the Natural Sciences in honor of Captain Allan Hancock, Univ. of Southern California, 1955, p. 299.

55. P. H. Kuenen, Marine Geology, J. Wiley and Sons, New York, 268 pp. (1950).

56. S. W. Bruyevich, Geokhimiya, 3, 329 (English translation) (1963).

57. E. A. Romankevich, Okeanologiia Akad. Nauk SSSR, 8, 825 (1968) (in Russian).

58. A. Krogh, Ecol. Monogr., 4, 421 (1934).

CHAPTER 8

SOLUTION ADSORPTION THERMODYNAMICS FOR ORGANICS ON SURFACES

A. C. Zettlemoyer and F. J. Micale

Lehigh University

I. INTRODUCTION

The term adsorption is generally used to describe the concentration of a particular component at an interface between two phases. A number of possibilities arises as a result of this definition. For the solid/gas interface, theoretical developments depend upon the adsorption of one component, i.e., the gas molecules, onto the solid surface. The major problem in the development of a thermodynamic theory is concerned with the definition of the usually uncertain, heterogeneous solid surface in terms of a mathematical model. The liquid/liquid and liquid/gas interfaces, however, are more susceptable to theoretical development and interpretation because of the homogeneous nature of the adsorbing surface. The solid/liquid interface, which is the subject matter of this paper, is complicated by the heterogeneous nature of the solid surface as well as the fact that at least two components are involved in competitive adsorption at the solid surface. Although theoretical progress has been relatively slow for the case of the solid/liquid interface, the application of adsorbent carbons to remove coloring matter in solution is one of the oldest technical processes. The technique was noted in the fifteenth century and used by Lowitz, about 1791, to purify raw sugar solutions.

A vast technical literature has grown out of the many studies of adsorption from solution by charcoals and carbon blacks. For the most part the carbon blacks possess rather complicated surfaces because of a multiplicity of residual oxygenated functional groups. These groups produce quite heterogeneous surfaces so that adsorption data are often difficult, if not impossible, to interpret. Beginning in the early 1950's, surface chemists and physicists were provided with carbons possessing very uniform surfaces, but with sufficient surface areas to yield adequate adsorption data, i.e., the so-called graphitized carbon blacks. For a number of investigators, the path, thus, was opened toward meaningful theoretical interpretations.

Adsorption from solution can be divided into two categories. One concerns adsorption, e.g., surfactants or polymers, from dilute solution where there is high specificity for the solute. Binary mixtures of liquids over the entire concentration range form another important type. There are three main types of solution adsorption isotherms as shown in Figure 1. Type I, often called the Langmuir isotherm, falls into the category of adsorption from dilute solutions and is exemplified by limiting or monolayer adsorption. Type II and III isotherms are typical of composite isotherms of two components over their entire concentration range and are distinguishable, as will be seen shortly, by the relative degree of interaction of one component over the other.

Heats of solution adsorption, measured calorimetrically, provide an added approach to the measurement of the isotherms which could lead to much information about the adsorption process. Little has been done, however, with direct calorimetric measurements of heats of interaction of solutions or with the measurements of multi-temperature isotherms.

II. ADSORPTION FROM BINARY MIXTURES

Examples of two types of composite isotherms from binary mixtures over the entire concentration range are given in Figure 1, b and c. Solution adsorption is ordinarily followed by measuring the change in mole fraction of component 1, Δx, from its initial value, x_o, to its equilibrium value, x_1. Therefore:

$$\Delta x = x_o - x_1 \tag{8-1}$$

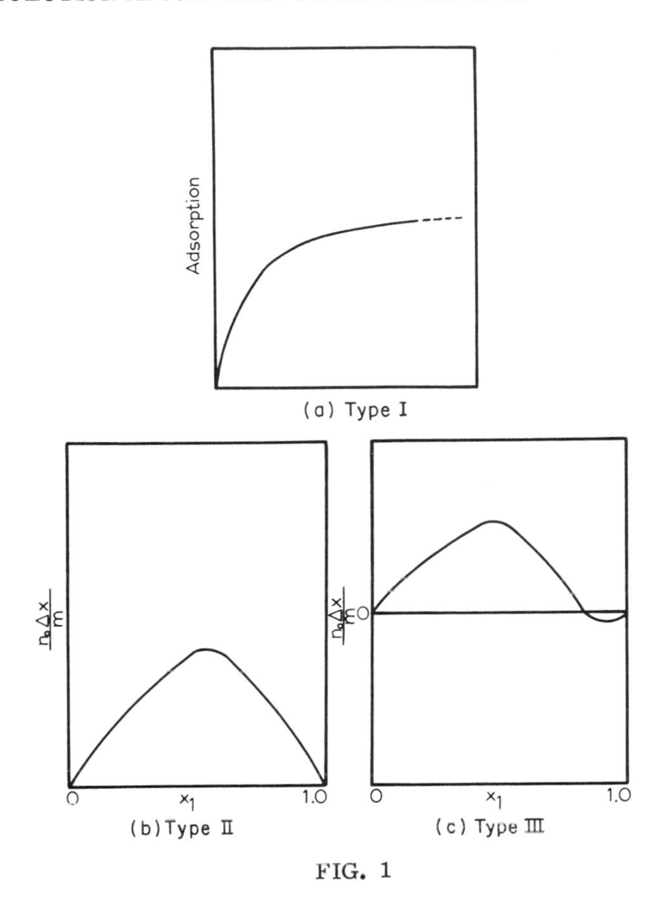

FIG. 1

Solution adsorption isotherms.

The total number of moles in the system may be defined:

$$n_o = n_1 + n_2 + n_1^s m + n_2^s m \tag{8-2}$$

where n_1 and n_2 are the number of moles of components 1 and 2 in the liquid phase and n_1^s and n_2^s are the number of moles of components 1 and 2 adsorbed per gram of solid. The value of m represents the weight of the added solid adsorbent in grams. Furthermore, by definition:

$$x_o = \frac{n_1 + n_1^s m}{n_o} \qquad (8\text{-}3)$$

$$x_1 = \frac{n_1}{n_1 + n_2} \qquad (8\text{-}4)$$

and, combining Equations (8-1) through (8-4):

$$\Delta x = \frac{n_1 + n_1^s m}{n_1 + n_2 + n_1^s m + n_2^s m} - \frac{n_1}{n_1 + n_2} \qquad (8\text{-}5)$$

or:

$$\Delta x = \frac{m\,(n_2 n_1^s - n_1 n_2^s)}{n_o\,(n_1 + n_2)} \qquad (8\text{-}6)$$

Since the mole fraction of component 2, x_2, may be defined as:

$$x_2 = (1 - x_1) = \frac{n_2}{n_1 - n_2} \qquad (8\text{-}7)$$

$$\frac{n_o \quad \Delta x}{m} = n_1^s \;(1 - x_1) - n_2^s \; x_1 \qquad (8\text{-}8)$$

Equation (8-8) is the basic equation of solution adsorption theory. When the left hand side of Equation (8-8), $n_o \;\Delta x/m$, is plotted as a function of x_1, this plot is referred to as the composite isotherm. The Types II and III isotherms given in Figure 1 now having meaning in terms of Equation (8-8). When $n_o \;\Delta x/m$ is positive over the entire mole fraction range, then n_1^s must be large with respect to n_2^s, or component 1 adsorbs much more strongly than component 2. When negative values of $n_o \;\Delta x/m$ appear, e.g., a Type III composite isotherm, the adsorption of components 1 and 2 are comparable, although one may be larger than the other. Furthermore, the value of

$n_o \Delta x/m$ must be zero at $x_1 = 0$ (where $n_1^s = 0$) and at $x_1 = 1$ (where $n_2^s = 0$); the composite isotherm over its entire range must begin and end at zero apparent adsorption.

Although a composite isotherm gives limited information about the relative adsorption of each component, the individual isotherms, i.e., n_1^s and n_2^s as a function of concentration, cannot be obtained readily from Equation (8-8) because it has two unknowns. Several approaches, based on different models, have been advanced in order to evaluate n_1^s and n_2^s. One approach, which might be called the monolayer hypothesis, was suggested by Williams (1) and developed by Elton (2) and Kipling and Tester (3). This theory is based on the assumption that the adsorbed layer is monomolecular and that the surface of the adsorbent is completely covered at all concentrations. This means, in effect, that both components completely wet the surface and that the presence of the surface is not effective beyond the first layer. These assumptions lead to the relationship:

$$\frac{n_1^s}{(n_1^s)_m} + \frac{n_2^s}{(n_2^s)_m} = 1 \qquad (8\text{-}9)$$

where $(n_1^s)_m$ and $(n_2^s)_m$ are the number of moles of components 1 and 2, respectively, required to form a complete monolayer per unit weight of solid. The values of $(n_1^s)_m$ and $(n_2^s)_m$ can be estimated from individual vapor adsorption isotherms. If vapor adsorption measurements are not possible, the monolayer moles can be calculated from the specific surface of the adsorbent and from molecular dimensions assuming a suitable orientation of the adsorbed species. By combining Equations (8-8) and (8-9), the individual adsorption isotherms can be calculated from the composite isotherms. Figures 2 and 3 give examples from the literature (4, 5) of Type II and Type III isotherms along with the individual isotherms calculated from the monolayer hypothesis.

Schay and Nagy (6, 7), on the other hand, noted that, on many porous adsorbents such as carbon blacks, the composite isotherms often exhibit a long linear decreasing portion extending over a wide range of concentrations. They have suggested that the composition of the adsorbed phase is the same over this entire linear portion. This condition is readily seen by rewriting Equation (8-8) as:

$$\frac{n_o \Delta x}{m} = n_1^s - (n_1^s + n_2^s)\, x \qquad (8\text{-}10)$$

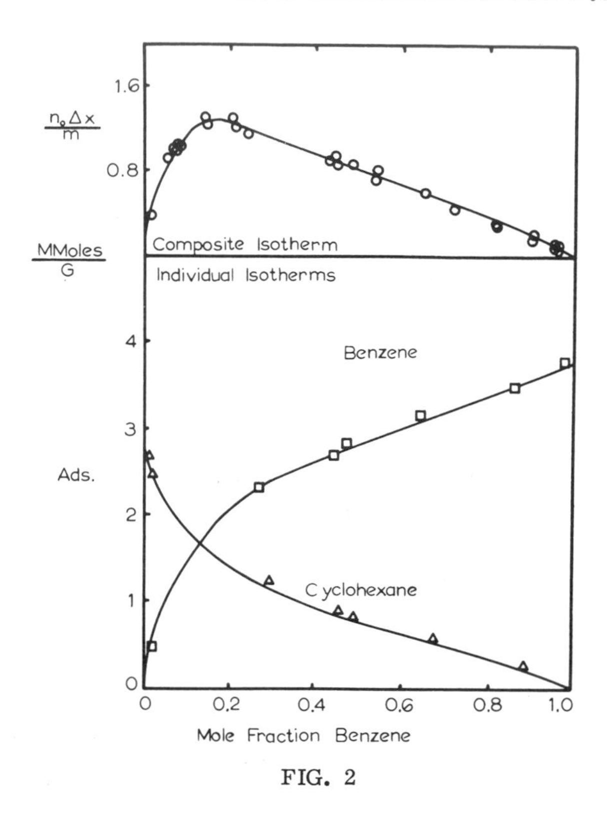

FIG. 2

Composite and individual isotherms of benzene-cyclohexane on charcoal. Reprinted from Ref. (4), p. 3819, by courtesy of J. Chem. Soc.

where n_1^s and n_2^s could both be considered to be constant to produce a straight line. Extrapolation of the linear portion of the isotherm to $x = 0$ and $x = 1$ gives values of $(n_1^s)_c$ and $(n_2^s)_c$ which define the composition of the adsorbed phase in the linear portion. From the monolayer values, $(n_1^s)_m$ and $(n_2^s)_m$, for the two components, Schay and Nagy evaluated the mean thickness (number of monolayers) of the adsorbed layer t:

$$t = \frac{(n_1^s)_c}{(n_1^s)_m} + \frac{(n_2^s)_c}{(n_2^s)_m} \qquad (8\text{-}11)$$

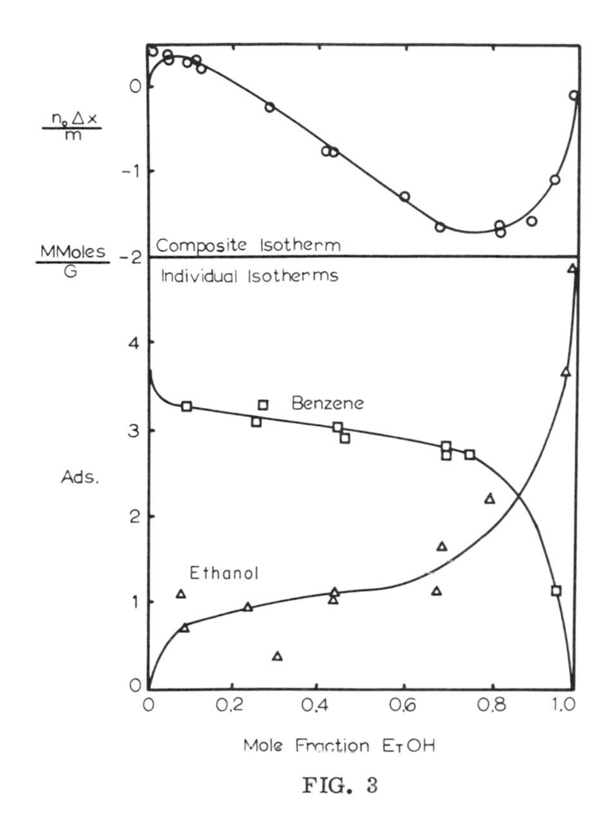

FIG. 3

Composite and individual isotherms of benzene-ethanol on charcoal. Reprinted from Ref. (5), p. 4123, by courtesy of J. Chem. Soc.

The value of t, which is given as the number of molecular layers, was found to be approximately unity for a number of porous adsorbents. An example of application of this analysis to a Type II and Type III composite isotherm is given in Figure 4. Curve a represents the Type II isotherm where component 2 undergoes very little adsorption, i.e., $(n_2^s)_c \cong 0$, and component 1 adsorbs strongly. Curve b represents a Type III isotherm where both components undergo comparable adsorption over the entire composition range.

Cornford, Kipling, and Wright (8) have applied the concepts of Schay and Nagy to the adsorption of several binary mixtures on

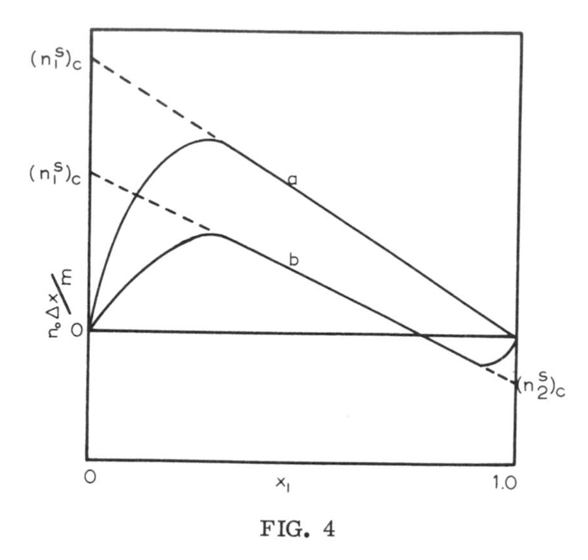

FIG. 4

Composite isotherms exhibiting linearity over wide concentration range.

Spheron 6 and Graphon. They found that the mean thickness of the adsorbed layer was unity even for these nonporous adsorbents. This finding lends support to the assumption that the adsorption from binary mixtures is often confined to a monolayer.

Cornford and co-workers (8) pointed out that the existence of an adsorbed layer of constant composition is a possible but not the only condition for the linearity of the isotherms. By using a virial equation for the individual isotherms, they showed that three specific cases satisfy the condition for linearity of the isotherm. The individual isotherms are given by:

$$n_1^s = a_1 + b_1 x + c_1 x^2 + d_1 x^3 + \ldots\ldots$$

$$n_2^s = a_2 + b_2 (1 - x) + c_2 (1 - x)^2 + d_2 (1 - x)^3 + \ldots\ldots \tag{8-12}$$

The terms beyond the first two can be neglected sometimes.

Case 1. If both b_1 and b_2 are zero, $n_1^s = a_1$ and $n_2^s = a_2$, which is the condition of Schay and Nagy. This condition has to satisfy an important thermodynamic criterion. The chemical potentials of component 1 in the solution phase and the adsorbed phase are given by:

$$\mu_1^s = (\mu_1^s)^o + RT \ln (f_1^s x_1^s), \text{ and}$$

$$\mu_1^1 = (\mu_1^1)^o + RT \ln (f_1^1 x_1), \quad (8\text{-}13)$$

where μ_1 and $(\mu_1)^o$ are the chemical potentials of component 1 in the mixture and standard state, respectively; f is the activity coefficient, and the superscripts s and 1 represent the adsorbed and liquid phase, respectively. At equilibrium $\mu_1^s = \mu_1^1$. The composition of the adsorbed phase will be constant only if changes in $f_1^1 x_1^1$ are compensated by changes in f_1^s. This condition is unlikely to occur in practice.

Case 2. If $b_1 = b_2 = b$, Equation (8-12) reduces to the form:

$$n_1^s = a_1 + bx \quad (8\text{-}14a)$$

$$n_2^s = a_2 + b - bx \quad (8\text{-}14b)$$

Equations (8-14a, b) represent a pair of parallel straight lines and the composite isotherm, which is a resultant of the two, will also be a straight line.

Case 3. If b_1 and b_2 are both small, the composite isotherm will approximate a straight line. Cornford et al. (8) have suggested that this situation is likely to occur in the case of systems which are far from being ideal and that possibly the majority of the systems investigated by Schay and Nagy are of this type.

Although extensive experimental work on solution adsorption has been in progress for over 50 years, no satisfactory physical model which enables a vigorous thermodynamic treatment of the process is available. The main reason for the lack of theoretical development in the field is the complexity of the process due to the heterogeneity of most solid surfaces and the lack of knowledge of the various interactions both between the two components in solution and on the surface, and between the individual components and the adsorbent. Recently, however, attempts have been made to define an ideal picture for solution adsorption and to evaluate the thermodynamic properties of real systems in terms of deviations from this ideal picture.

Such an idealized model for the adsorption from binary mixtures has recently been examined in detail by Everett (9), and the analysis has been extended to adsorption from imperfect solutions (10). In this idealized model, the solution is described in terms of a quasi-crystalline array of plane lattices. The solid is composed of a uniform array of adsorption sites. The adsorption is restricted to a monolayer and the components of the binary mixture are assumed to be of the same size. Everett's theory appears to imply, furthermore, that there should be preferential adsorption for one of the components throughout the entire concentration range for a perfect system. The adsorption is envisaged as an exchange process in which a small amount of component 1 in the adsorbed phase is replaced by an equal amount of the component 2 with change in concentration:

$$(1)^s + (2)^l \rightleftharpoons (1)^l + (2)^s \qquad (8\text{-}15)$$

With these assumptions, Everett (9) has described the adsorption from binary mixtures by means of an equation analogous to the Langmuir adsorption isotherm:

$$\frac{x_1^s \; x_2}{x_2^s \; x_1} = K \qquad (8\text{-}16a)$$

or:

$$n_1^s = \frac{K \, n^s \, x_1}{1 + (K-1)\, x_2} \qquad (8\text{-}16b)$$

where n^s is the total number of moles adsorbed, i.e., $n^s = n_1^s + n_2^s$, x_1^s is the mole fraction of component 1 on the surface or $x_1^s = \frac{n_1^s}{n^s}$, x_1 and x_2 are the mole fractions of components 1 and 2 in the liquid phase, and K is a constant given by:

$$K = \exp - \left(\frac{\Delta Ha}{RT} - \frac{\Delta Sa}{R} \right) \qquad (8\text{-}17)$$

where ΔHa is the differential heat of adsorption and ΔSa is the standard differential entropy of adsorption. The value of K determines the strength of adsorption of component 1 by the surface. The differential heat and entropy of adsorption are actually heat and entropy of the phase exchange process represented by Equation (8-15).

The composite isotherm equation may now be rewritten by substituting Equation (8-16b) into Equation (8-10) and rearranging to put it in its linear form:

$$\frac{x_1 \; x_2}{n^o \; \Delta x/m} = \frac{1}{n^s} \left[x_1 + \frac{1}{(K-1)} \right] \tag{8-18}$$

A plot of the left hand side of Equation (8-18) as a function of x_1 results in a straight line and n^s and K can be calculated from the slope and intercept. The value of obtaining n^s lies in the fact that solution adsorption data can be used to determine the surface area of the adsorbent by assuming a cross-sectional area for the adsorbate.

Everett (9) has shown that the heat of immersion ΔH of a solid by the solution is given most simply by:

$$\Delta H = x_1^s \; \Delta H_1^o + x_2^s \; \Delta H_2^o \tag{8-19}$$

where ΔH_1^o and ΔH_2^o are the heat of immersion values of the absorbent in the pure liquids 1 and 2, respectively. Equation (8-19) may now be combined with Equation (8-10) and (8-16b) to yield:

$$\Delta H - (x_1 \; \Delta H_1^o + x_2 \; \Delta H_2^o) = \frac{x_1 \; x_2}{\left[x_1 + \frac{1}{(K-1)} \right]} \; (\Delta H_1^o - \Delta H_2^o) \tag{8-20}$$

To test Equation (8-20), it is necessary to know K, already determined from Equation (8-18), and to measure ΔH over the entire concentration range. From Equation (8-20), therefore, a plot of the left hand side against $x_1 \, x_2 / [x_1 + (K-1)^{-1}]$ should be linear with a slope $(\Delta H_1^o - \Delta H_2^o)$.

Although the number of systems likely to approximate the conditions imposed by this theory is small, a few authors (9-13) have attempted correlations with varying degrees of success. The most significant contribution has come from Wright (12,13) who studied solution adsorption of benzene in a number of different solvents and on a variety of carbon surfaces. The results are presented in Figure 5 in the form of composite isotherms where the amount absorbed,

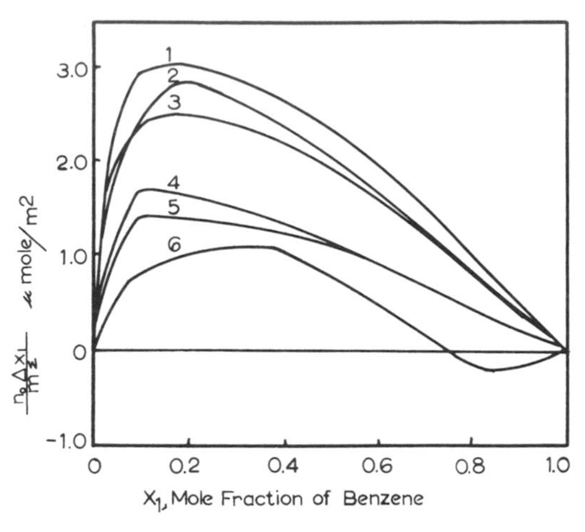

FIG. 5

Composite isotherms for benzene in different solvent-adsorbent systems (12, 13).

Curve No.	Solvent	Adsorbent
1	ethylene dichloride	graphon
2	cyclohexane	coconut shell charcoal
3	cyclohexane	decolorizing charcoal
4	carbon tetrachloride	decolorizing charcoal
5	carbon tetrachloride	coconut shell charcoal
6	ethylene dichloride	spheron 6

$n_o \Delta x/m$, has been reduced to per unit area, i.e., μmoles/ (meter)^2, based on BET nitrogen area measurements, so that all six isotherms could be presented on the same graph. The experimental points have been eliminated for the sake of clarity. The results are plotted according to Equation (8-18) and are given in Figure 6. The solid lines represent the best fit of the experimental points and the dotted lines the deviation of the data from linearity. Table 1 identifies the curves in Figures 5 and 6 and gives the results of calculations according to Equations (8-18) and (8-20).

The results of the benzene-ethylene dichloride adsorption on Graphon and Spheron 6, curves 1 and 6, respectively, indicate that Graphon obeys the requirements for a perfect system while Spheron

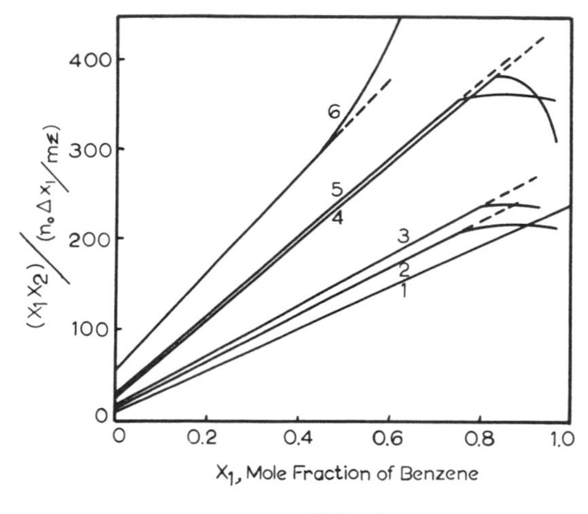

FIG. 6

Composite isotherms according to Equation (8-18) (12, 13). (Identification of curves is given in Fig. 5 and Table 1.)

6 does not. This conclusion is arrived at by observing the deviation from linearity according to Equation (8-18), Figure 6, at a mole fraction of 0.5 for the Spheron 6 while the Graphon system is linear for the entire concentration range. Furthermore, the surface area determinations from the n^s values show reasonably good agreement with the BET nitrogen area for Graphon and poor agreement for Spheron 6. The (ΔH_1^o - ΔH_2^o) values in the last column of Table 1 are too small, practically within experimental error, either to prove or to disprove Equation (8-20). As explained heretofore, we should expect good agreement with the theory for the homogeneous Graphon surface; the heterogeneous surface of the carbon black provided a poor test adsorbent.

Curves 2-5 represent the results for benzene in cyclohexane and carbon tetrachloride on decolorizing charcoal and coconut shell charcoal. Water adsorption results on these adsorbents (14) indicate that the coconut shell charcoal surface was partly covered with a moderate amount of hydrophilic groups and that the decolorizing charcoal had a much higher percentage of surface hydrophilic groups. Figure 6 shows that all four of these systems obey Equation (8-18)

TABLE 1

Results of Calculations for Benzene in Different Systems

Curve No.	Solvent	Adsorbent	K	Area of Adsorbent, m^2/g Eq.(8-18)	BET N_2	$(\Delta H_1^o - \Delta H_2^o)$ cal/g Eq.(8-20)	Measured	Ref.
1	Ethylene DiChloride	Graphon	18.5	119	89	0.54	$0.97 \pm .4$	(12)
2	Cyclohexane	Coconut Shell Charcoal	19.9	677	593	12.0	10.2 ± 1.5	(13)
3	Cyclohexane	Decolorizing Charcoal	16.4	596	582	7.1	5.9 ± 1.5	(13)
4	Carbon Tetrachloride	Decolorizing Charcoal	16.1	374	582	4.0	3.4 ± 1.5	(13)
5	Carbon Tetrachloride	Coconut Shell Charcoal	18.2	350	593	3.8	3.0 ± 1.5	(13)
6	Ethylene Dichloride	Spheron 6	5	57	119			(12)

up to a benzene mole fraction of approximately 0.8. These deviations, however, are expected in view of the chemically heterogeneous nature of the adsorbent surfaces and the slight differences in the sizes of the adsorbing molecules. The surface areas calculated according to Equation (8-18) show very good agreement with the BET nitrogen areas for the cyclohexane systems and less agreement for the carbon tetrachloride systems. The (ΔH_1^o - ΔH_2^o) values show generally good agreement with the directly measured differences.

The number of ideal systems (and not all of the above approached this state, particularly due to surface heterogeneities) will always remain small. What is more, the technologically important solution is not likely to be nearly ideal. Instead, strong preferential adsorption of one of the components is likely to occur. Such cases will be discussed next.

III. ADSORPTION FROM DILUTE SOLUTIONS

In the case of dilute solutions, adsorption is treated in a manner similar to gas adsorption with the restriction that the adsorption is confined to a monolayer, i.e., Type I isotherm. Multilayer adsorption has been observed in a few cases (15), but generally the solute-adsorbent interactions are not strong enough to overcome the forces extended by the solvent or solution except in the first layer. Equations representing the Langmuir adsorption isotherm have been widely applied to adsorption from solution, but they must be regarded as empirical equations without thermodynamic validity for such multicomponent solutions.

For dilute systems, with high specificity for adsorption of the solute, it is generally assumed for simplicity that the solvent has no influence on the adsorption of the solute. However, even in the case of dilute systems the role of the solvent can be a complicating factor. It is not possible sometimes to exclude the adsorption of the solvent, and it is rather difficult to assess the extent of interaction of the adsorbed solute and solvent. These difficulties are minimized by choosing systems in which the adsorption of the solute from the solution is quite dominant. Accordingly, a polar solvent such as water can be used for adsorption on a low- or medium-energy surface such as Graphon and a nonpolar hydrocarbon solvent for studies on high-energy surfaces such as a metal oxide.

Another problem that confronts the investigator in solution adsorption concerns the particular orientation of the adsorbed solute molecules. In general, it can be said that on high-energy surfaces a molecule is adsorbed through its polar head group, and on a low-energy surface the nonpolar end of the molecule lies on the surface. This

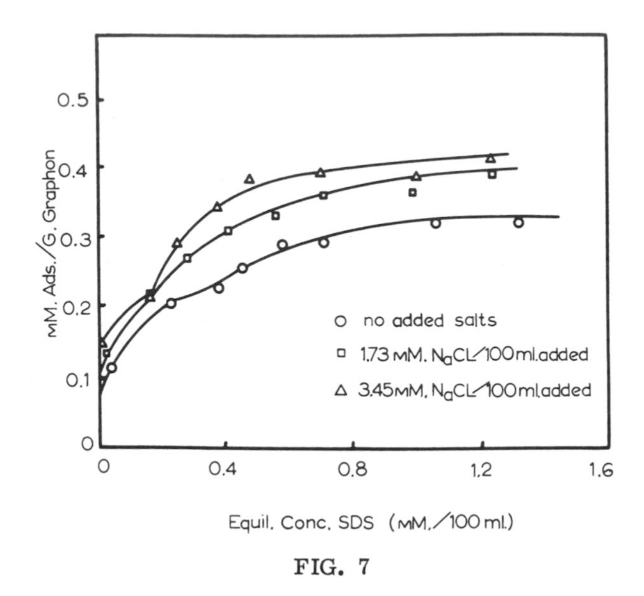

FIG. 7

Adsorption of sodium dodecylsulfate onto graphon. Reprinted from Ref. (16), p. 280, by courtesy of J. Amer. Oil Chemist Soc.

picture, however, is oversimplified; the orientation of the adsorbed molecule depends to a large extent on the nature of the polar head group and also on the concentration of the adsorbed species on the surface. For example, it has been shown that on rutile, n-propyl alcohol, which can interact with the surface by hydrogen bonding, is adsorbed through its hydroxyl, the hydrocarbon tail extending into the bulk phase. However, n-butyl chloride on the same surface lies flat.

The change in orientation of the same adsorbate, depending on the surface concentration of the adsorbed species, has been demonstrated in the adsorption of sodium dodecyl sulfate (NaDS) on Graphon (16) and is shown in Figure 7. At low surface concentrations, the isotherm reaches an apparent plateau followed by an increase to a higher plateau with increasing concentration. The interpretation given is that, at low concentrations, the NaDS molecule adsorbs at some equilbrium orientation with a longer length of the hydrocarbon chain in contact with the surface. At some higher concentration, the orientation changes to some new equilbrium position which leads to increased adsorption to a second plateau.

Figure 7 also shows the effect of Na^+ ions on the adsorption of NaDS on Graphon. Sodium ions increase the adsorption of the surfactant. This increase is caused by a decrease in mutual repulsion between head groups. Similar effects have been noted for sodium dodecyl benzene sulfonate on Graphon (16). Based on these results, a model for the Graphon/solution interface has been developed in which the adsorbed surfactant ions have their hydrocarbon chains oriented toward the Graphon surface and the polar head groups oriented toward the bulk aqueous phase. This situation is very similar to spreading at liquid interfaces. The packing densities in such an arrangement can be estimated by applying Davies' equation (17,18) to the data. This equation takes into account the interionic repulsion and the double-layer formation.

An interesting aspect of solution adsorption is the existence of a relationship between the solubility of a series of compounds and the adsorption at an interface (19). Hansen and Craig (20) have demonstrated that this relationship is indeed quantitative for the adsorption of a series of fatty acids and alcohols on Graphon. When the amount adsorbed per gram of Graphon was plotted against the reduced concentration C/C_0, where C_0 represents the solubility, the isotherms for a homologous series of compounds were superimposed. This ratio is analogous to relative pressures used in gas adsorption. Two examples are given in Figure 8 where curve a is indicative of multilayer adsorption and curve b is the Langmuir type isotherm usually found for porous solid. The Langmuir and BET equations, both of which were developed for gas adsorption, have sometimes been used to analyze isotherms of this type. The equations, however, can only be used as empirical equations since the constants do not have any physical significance in the case of solution adsorption. Hansen et al. (15) have demonstrated multilayer formation for a number of long chain alcohols on both carbon and graphite surfaces, although the number of adsorbed layers at saturation concentration was found to be small, i.e., three to four layers; in gas adsorption, five to ten layers are common at saturation pressures.

IV. HEATS OF IMMERSION INTO SOLUTIONS

When a solid surface is immersed in a solution which exerts no chemical or solvent action, some of the solute is adsorbed thereby diluting the solution. Generally, the solute is not adsorbed to the complete exclusion of the solvent. The surface enthalpy of the solid, thus, is replaced by the interfacial enthalpies arising from adsorption of solvent and solute. In addition, interfaces are established between

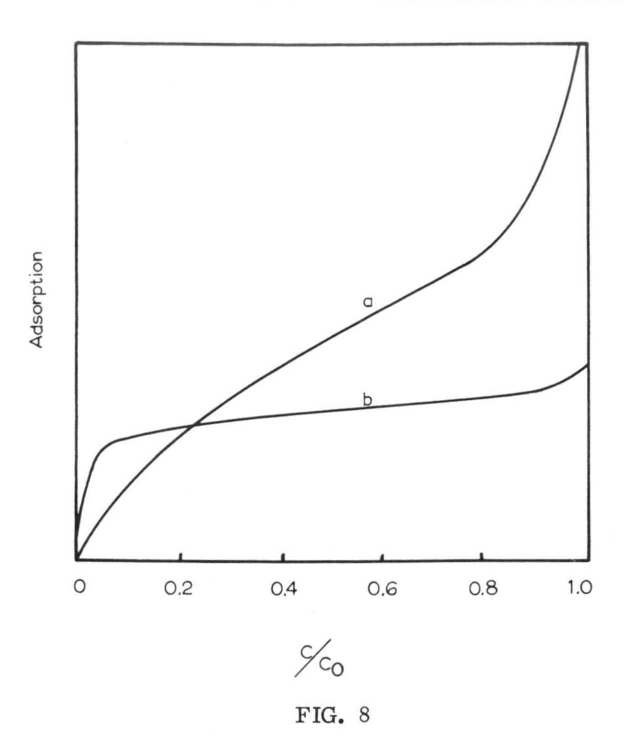

FIG. 8

Adsorption isotherms at reduced concentrations.

these new adsorbed phases and the remaining solution. If it is assumed that the interactions between molecules in the adsorbed phase are of the same order as for those in solution, the heat of wetting for the process can be formulated in terms of surface and interfacial enthalpy changes. The total enthalpy change for the wetting process is given (21) by the equation:

$$h_w = \theta(h_s - h_{s1}) + (1 - \theta)(h_s - h_{s2}) + \theta(h_{s1} - h_{1\sigma}) + (1 - \theta)(h_{s2} - h_{2\sigma}) \qquad (8\text{-}21)$$

where $h_w = \Delta Hw/\Sigma$ is the heat of wetting per unit area, θ is the surface coverage, $\theta(h_s - h_{s1})$ is the enthalpy contribution for the formation of the first adsorbed layer, $\theta(h_s - h_{1\sigma})$ is the enthalpy contribution for the formation of the interface between adsorbed solute and solution, and $(1 - \theta)(h_s - h_{s2})$ and $(1 - \theta) \cdot (h_{s1} - h_{2\sigma})$ are the enthalpy changes for the adsorption of first layer of water molecules and the formation of the interface between these molecules and the solution.

On simplification, Equation (8-21) reduces to:

$$h_w = [\theta h_{s1} + (1 - \theta) h_{s2} + \theta h_{1\sigma} + (1 - \theta) h_{2\sigma}] \tag{8-22}$$

The total heat of immersion obtained by calorimetry, on the other hand, is given by:

$$\Delta H_1 = \Delta H_W + \Delta H_D + \Delta H_M + \Delta H_B \tag{8-23}$$

where the subscripts W, D, M, and B refer to the separate contributions of wetting, dilution of the solution with respect to one component, demicellization when it occurs, and bulb-breaking plus accessory events. The contribution due to demicellization arises only in the case of surfactants above their critical micelle concentration. ΔH_D, ΔH_M, and ΔH_B can all be determined by separate experiments, allowing the evaluation of h_w.

The above concepts have been verified for the system Graphon-n-butanol from aqueous solutions (21). The value of h_{s1} was determined as the difference between the heat of immersion of a clean Graphon surface and one covered with a monolayer of butanol. The value of $h_s + h_{2\sigma}$ was assumed to be the same as for the heat of immersion in water. $h_{1\sigma}$ was obtained from the heat of immersion of a Graphon surface covered with a monolayer of adsorbed butanol molecules into a solution in equilibrium with this surface. ΔH_D and ΔH_M were determined experimentally and were quite small. The calculated and experimental heats of immersion are plotted in Figure 9. The excellent agreement between the calculated and experimental values provides some verification of the validity of the simplified assumptions employed in deriving Equation (8-22). Specifically, there appears to be no special interactions between the adsorbed butanol and water molecules, and the suggestion is made that the individual adsorbates occur in patches.

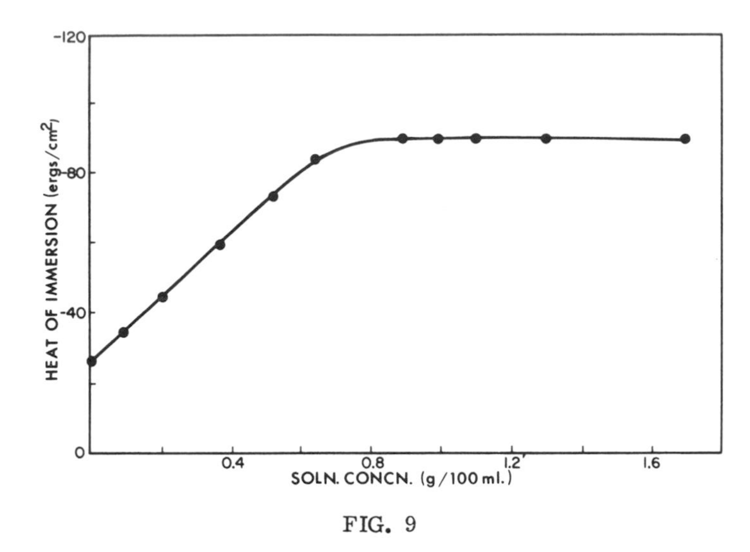

FIG. 9

Heats of immersion of graphon in aqueous n-butyl alcohol solutions. Reprinted from Ref. (21), p. 396, by courtesy of J. Phys. Chem.

The heat of immersion technique deserves further exploitation in the study of solution adsorption.

REFERENCES

1. A. M. Williams, Medd K. Vetenskapsakad, 2: No. 27 (1913).

2. G. A. H. Elton, J. Chem. Soc., 2958 (1951).

3. J. J. Kipling and D. A. Tester, J. Chem. Soc., 4123 (1952).

4. A. Blackburn and J. J. Kipling, J. Chem. Soc., 3819 (1954).

5. See reference 3.

6. G. Schay and L. Nagy, J. Chem. Phys., 58, 149 (1961).

7. G. Schay, L. Nagy, and T. Szekrenyesy, Periodica Polytech., 4, 95 (1960).

8. P. V. Cornford, J. J. Kipling, and E. H. M. Wright, Trans. Faraday Soc., 58, 1 (1962).

9. D. H. Everett, Trans. Faraday Soc., 60, 1803 (1964).

10. D. H. Everett, Trans. Faraday Soc., 61, 2478 (1965).

11. B. C. Y. Lu and R. F. Lama, Trans. Faraday Soc., 63, 727 (1967).

12. E. H. M. Wright, Trans. Faraday Soc., 62, 1275 (1966).

13. E. H. M. Wright, Trans. Faraday Soc., 63, 3026 (1967).

14. C. Pierce, R. N. Smith, J. W. Wiley, and H. Cordes, J. Amer. Chem. Soc., 73, 4551 (1951).

15. R. S. Hansen, Y. Fu, and F. E. Bartell, J. Phys. Colloid Chem., 53, 769 (1949).

16. A. C. Zettlemoyer, J. D. Skewis, and J. J. Chessick, J. Amer. Oil Chemists Soc., 39, 280 (1962).

17. J. T. Davies, Trans. Faraday Soc., 48, 1052 (1952).

18. J. T. Davies, Proc. Roy. Soc. (London), A245, 417 (1958).

19. F. E. Bartell and Y. Fu, J. Phys. Chem., 33, 676 (1929).

20. R. S. Hansen and R. P. Craig, J. Phys. Chem., 58, 211 (1954).

21. G. J. Young, J. J. Chessick, and F. H. Healey, J. Phys. Chem., 60, 394 (1956).

CHAPTER 9

SELECTIVE ADSORPTION OF ORGANICS ON INORGANIC SURFACES

Thomas W. Healy

University of Melbourne

INTRODUCTION

In natural aquatic environments, dissolved or suspended organic materials range from such simple soluble organic molecules as amino acids or phenols, to long-chain organic surfactants such as the alkylbenzene sulphonate (ABS) detergents, to complex natural "colour" in the form of tannates or humates, to synthetic macromolecules, i.e., polymer flocculants, to natural macromolecules, i.e., proteins, and finally colloidal organic aggregates, i.e., organic sols. This list is not meant to be comprehensive but rather it serves to highlight the diversity of organic species found to a greater or lesser extent in aquatic environments.

These materials may come in contact with inorganic substances during their lifetime in the environment and, to varying degrees, will be taken up or absorbed by these inorganic materials. Once again there is a wide range of inorganic material types from the authigenic clays, hydrous sediments, precipitated metal hydroxides, and metal carbonates to the inorganic additives such as filter aids, inorganic pollutants and sediments; the actual modification of each of these materials will clearly be a function of the nature of any particular environment.

The aim of this paper is to consider the principles that control the selective adsorption of organics onto inorganic surfaces. The complex and variable nature of both inorganic and organic substances, that may appear in the natural environment, force us to consider the governing principles of the adsorption process. The task of applying these principles to individual situations, therefore, may be made a little simpler. Recent advances in environmental science, in water renovation science, in oceanography, and in geochemistry, that relate to organic adsorption onto inorganics, have forced the colloid and surface chemist to take a fresh look at his understanding of the solid/aqueous solution interface and at the process of adsorption at that interface. New advances are being made, but much remains unfinished. During this paper it will be necessary to consider new areas of research that have only just begun.

Since we will be concerned with principles of adsorption, the accompanying bibliography will not be comprehensive but selected to include data that substantiates generally accepted principles that are currently proving to be of analytical value.

In the following sections, the nature of inorganic surfaces will be considered together with a generalized treatment of the thermodynamics of adsorption. With this background, the adsorption of simple organics, surfactants, polymers, and, finally, organic sols will be considered in detail.

II. INORGANIC SURFACES

A. The Nature of the Inorganic Solid-Aqueous Solution Interface.

While it is true that there is a great variety of inorganic materials present in aquatic environments, it is possible to make a simplifying assumption that appears to be useful in many situations, viz., we shall consider the surface or interfacial region of the inorganics in water to have the properties of a simple inorganic oxide. If this is the case we can write the surface region as:

$$-\overset{|}{\underset{|}{M}}-O^- \underset{}{\overset{OH^-}{\rightleftharpoons}} -\overset{|}{\underset{|}{M}}-OH \overset{H^+}{\rightleftharpoons} -\overset{|}{\underset{|}{M}}-OH_2^+ \qquad (9\text{-}1)$$

where MO^- and MOH_2^+ represent negative and positive surface sites, respectively. Thus, pH is that fundamental variable which controls the nett or effective surface charge on the inorganic and which, therefore, controls the nature of the electrical charge distribution or electrical double layer at the solid/aqueous solution interface (1, 2).

Support for this assumption comes from a variety of sources: for example, kaolinite and other layer alumino silicate clay minerals tend to dissolve or leach in water to yield a hydrated silica and/or hydrated alumina surface (3,4), the electrical surface properties of which are controlled by pH. Again, several authors have shown that pH is the fundamental control variable for the surface properties of hydrated iron oxide (5,6), apatite (7), hydrated manganese minerals (8,9), beryl (10), soils (11), and natural sediments (12).

In terms of Equation (9-1) it is, therefore, possible to define a surface equilibrium constant (K_S) for the process:

$$MO^- + 2H^+ = MOH_2^+ \quad (9\text{-}2)$$

$$K_S = \frac{a_{MOH_2^+}}{a_{MO^-}} \cdot a_{H^+}^{-2} \quad (9\text{-}3)$$

i.e., if $a_{H^+} = a^o_{H^+}$ at the condition or activity of protons in solution for which there are equal numbers of positive and negative surface sites, then:

$$K_S = (a^o_{H^+})^{-2} \quad (9\text{-}4)$$

This condition is referred to as the zero-point-of charge (Z.P.C.) of the solid/liquid interface.

The Z.P.C. values for oxide-aqueous solution interfaces range from pH 1-2 for silica gels, pH 2.5-3.5 for α quartz, pH 6-8 for hydrated iron oxides, pH 8-9 for hydrated aluminum oxides, and pH 12 for magnesium oxide. The variation of Z.P.C. values determined for oxides and silicates has been reviewed extensively by Parks (1,13). Examination of the data tabulated by Parks reveals that the Z.P.C. value of an oxide (or silicate) varies with the purity of the sample, its thermal and chemical pretreatment history, the degree of surface hydration, and the crystal modification of the material. The quantitative analysis of the variation of Z.P.C. values of oxides with their crystal structure has been considered in detail by Healy and Fuerstenau (2,9). In outline, the electrostatic field of the crystal lattice is considered to polarize and induce dissociation of water molecules at the oxide surface. The greater the field, the greater is the dissociation, and, therefore, the greater is the pH of the Z.P.C. Thus, Z.P.C. values range from around pH 1.0 for essentially non-polar solids (14), to pH 2-3 for such covalent solids as SiO_2 and WO_3, to pH values around 7 for oxides whose metal-oxygen bond character is equally ionic and covalent to such ionic crystals as MgO where the Z.P.C. is around pH 12. If we now consider the Z.P.C. data in the literature in terms of this field strength model an important result emerges, viz., if the surface of a crystalline material is allowed to hydrate to the extent that the hydrolysed surface so formed is essentially amorphous, then the effective crystal field is lowered and the Z.P.C. is lowered concomitantly. The properties of quartz, for example, that has been in contact with water, in general, reflect a silica gel surface rather than a quartz surface (2).

It has not proved possible, to date, to describe the electrical properties of the solid-liquid interface in terms of this negative-site, positive-site model. Rather, it becomes necessary to regard the nett zero charge of the surface at the Z.P.C. as a condition of, in fact, zero charge; above the Z.P.C. the surface is negative and below the Z.P.C. the surface is positive. We can define, therefore, a surface potential (ψ_o) such that:

$$\psi_o = \frac{RT}{F} \ln \frac{a_{H^+}}{a^o_{H^+}} \tag{9-5}$$

where R is the gas constant, F is the Faraday, and T is the absolute temperature. The interface is considered to be a region where the potential ψ_o decays to zero potential in the bulk liquid. The oxide surface has a negative surface potential above the pH of the Z.P.C. and positive below the pH of the Z.P.C.

The most satisfactory technique, from a theoretical viewpoint, for Z.P.C. determination is that of direct titration where one essentially performs a potentiometric titration of the surface acid/base sites. The course of the titration is expressed in terms of ($\Gamma_{H^+} - \Gamma_{OH^-}$) moles cm^{-2}, i.e., the nett surface charge density with respect to potential-determining ions. At the Z.P.C.,($\Gamma_{H^+} - \Gamma_{OH^-}$) is zero. From an experimental point of view, the method is less satisfactory and great care must be taken in the interpretation of results (15). Some representative data on Z.P.C. determination is shown in Figure 1 compiled from the data on Onoda and de Bruyn (15) and Yopps and Fuerstenau (16). The role of the supporting, surface-inactive electrolyte is shown in Figure 1, where increased salt concentration diminishes the potential determining ion adsorption density by simple compression of the electrical double layer; however, the pH of the Z.P.C. is unaffected by the surface inactive salt. In the next section it will be necessary to distinguish surface active electrolytes as ones which cause an apparent shift in the Z.P.C. with increasing concentration.

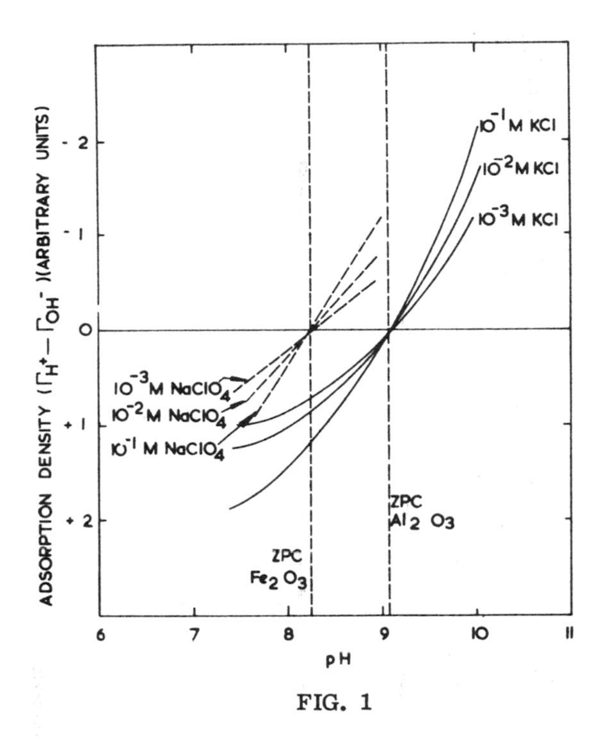

FIG. 1

Titration curves of α- Fe_2O_3 (15) and α-Al_2O_3 (16) showing the Z.P.C. values for each oxide.

An alternative technique for Z.P.C. determination is that of electrokinetics in which the double layer is sheared by an applied potential (electrophoresis) or by an applied hydrodynamic force (streaming potential). The movement of charge affected by this applied force represents a current (or potential) in streaming potential determination or a moving colloidal particle in electrophoresis. The reader is referred to earlier review articles for details of electrokinetic techniques (17).

The electrokinetic technique of electrophoresis provides data in the form of mobilities, e.g., microns/sec per volt/cm. In many instances, particularly if one is interested only in a Z.P.C. value, i.e., zero mobility of a material, no further reduction of the measured electrokinetic quantity is required. However, it is often necessary

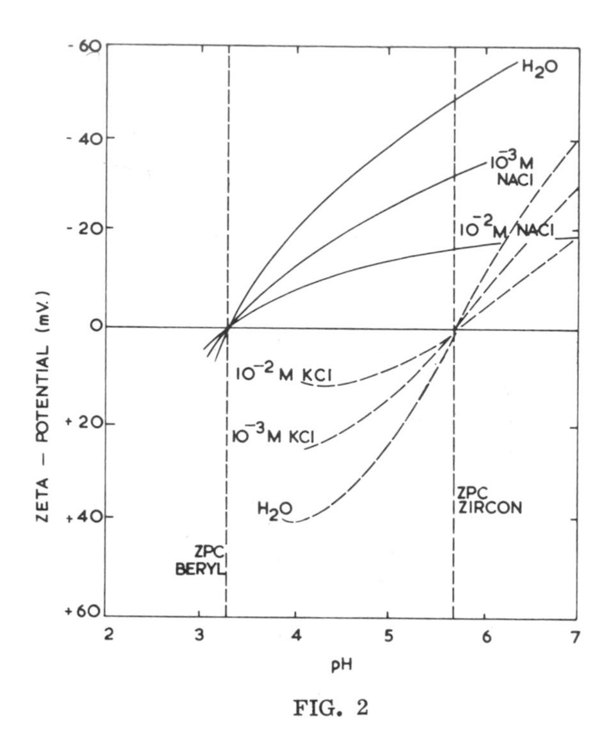

FIG. 2

Zeta-potential determinations of the minerals beryl (10) and zircon (58) showing Z. P. C. values.

to convert the mobility or streaming potential to an equivalent electrical double layer potential, i.e., the potential in the double layer at the plane of shear. This so-called zeta-potential and its use were challenged extensively in the discussion contributions at the Fourth Rudolfs Research Conference (17). It is sufficient here to point out that the use of zeta-potentials in double layer calculations can be shown to be of considerable value. This will be illustrated in subsequent sections.

Electrokinetic data are used extensively for Z. P. C. determination and a selection of such data is shown in Figure 2. Again the surface inactive role of the supporting electrolyte is shown by these sets of data.

A final technique for Z. P. C. determination depends on changes in stability of colloidal dispersions or suspensions of the material induced by a change in the potential-determining ion concentration. At the Z.P.C., the colloid has no double layer and coagulation is rapid. Above and below the Z.P.C., the double layer is formed, double layers repel each other upon collision, and coagulation is prevented.

The technique typically (16) consists of a glass vessel containing the suspension through which is passed an intense beam of light. By monitoring changes in transmitted light intensity with a photomultiplier or simple photocell, the pH of maximum coagulation can be determined.

An example of such coagulation data is shown in Figure 3 for a manganese oxide colloid (18). The electrophoretic mobility and sodium ion adsorption data for the same oxide are included also. The Z.P.C. value is discerned readily.

B. General Aspects of Adsorption at the Inorganic Solid-Aqueous Solution Interface.

In an earlier review (19) and again in the present volume (20) the general thermodynamics of adsorption has been discussed. It is sufficient for present purposes to abstract the special descriptions for organic adsorption.

The total free energy of adsorption is expressed conveniently as:

$$\Delta G_{ads.} = \sum_{all\ i} \Delta G_i \qquad (9\text{-}6)$$

where ΔG_i refers to each contribution to the total process of adsorption. This concept has been used by Grahame (21) and by Fuerstenau et al. (23) and more recently by James and Healy (22). The general form appropriate for the present discussion is:

$$\Delta G_{ads.} = \Delta G_{el.} + \Delta G_{solv.} + \Delta G_{hb.} + \Delta G_{VdW.} + \Delta G_{chem.} \qquad (9\text{-}7)$$

where: (a) $\Delta G_{el.}$ is the coulombic interaction and is given by:

$$\Delta G_{el.} = z e \psi_{\delta} \qquad (9\text{-}8)$$

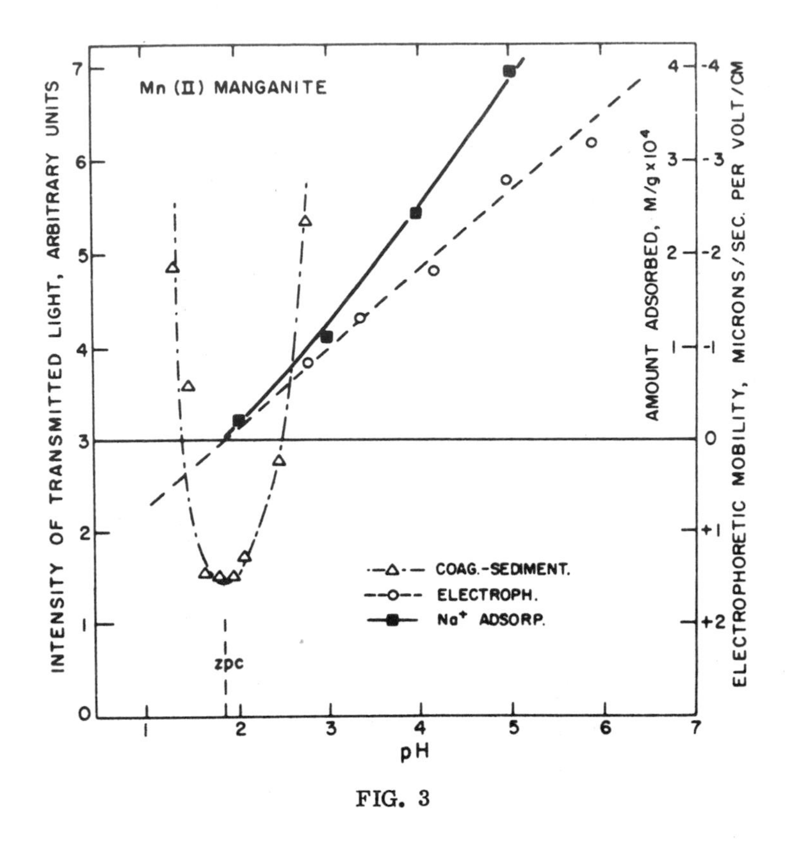

FIG. 3

The Z. P. C. of manganese (II) manganite as determined by the adsorption of sodium ions, electrophoresis, and coagulation. Reprinted from Ref. (18), p. 74, by courtesy of the American Chemical Society.

where z is the valence of the ion including sign, e is the electronic charge, and ψ_δ is the potential in the double layer where the ion resides,

(b) $\Delta G_{solv.}$ is the work required to replace all or part of the secondary solvation sphere with water molecules associated with the interface; to a first approximation:

$$\Delta G_{solv.} = \frac{z^2 e^2}{8} \left(\frac{1}{\epsilon_i} - \frac{1}{\epsilon_o}\right)\left(\frac{1}{x}\right) \quad (9\text{-}9)$$

where ϵ_i and ϵ_o are the dielectric constants of water in the interface and bulk, respectively, and x is the distance between the center of the ion and the first layer of interfacial water,

(c) $\Delta G_{hb.}$ is the free energy of formation of hydrogen bonds between surface species and the adsorbing entity,

(d) $\Delta G_{VdW.}$ is the free energy change associated with removal of alkyl chains from an aqueous environment into association with themselves or into a non-polar liquid of the same chain length,

(e) $\Delta G_{chem.}$ is the free energy change associated with surface chemical compound formation as evidenced by fatty acid adsorption-reaction with calcium minerals.

Parts of this general breakdown of the free energy of adsorption have been considered previously, principally by Fuerstenau et al. (23, 24). Anderson and Bockris (25), and James and Healy (22) have considered in detail the solvation energy term summarized to a first approximation by Equation (9-9).

For ions oppositely charged to the surface, $\Delta G_{el.}$ is negative and spontaneous adsorption is favored. For cations adsorbing on negative surfaces, if the ion is small and highly charged, $\Delta G_{solv.}$ will be large and positive and ($\Delta G_{el.} + \Delta G_{solv.}$) may be positive $\Delta G_{solv.}$ will decrease with decreasing charge and increasing size. $\Delta G_{hb.}$ will generally be small but negative and of significance for adsorption of molecular or neutral species only. $\Delta G_{VdW.}$, which shall be discussed in detail in the section on surfactant adsorption, will be negative for the adsorption process and will increase in negative value as the chain length is increased. Finally, $\Delta G_{chem.}$ generally will be restricted to small anion (i.e., low solvation) adsorption onto ionic crystals containing a large divalent or trivalent ion as a lattice ion such that chemical compounds will form at the surface.

It is the intricate balance of at least these five energy terms that will control selective adsorption. The specific application of these contributions to small organic molecules, surfactant ions, and macromolecule adsorption will be considered below.

III. ADSORPTION OF ORGANICS

A. Adsorption of Simple Organics.

The uptake of such simple organics as phenol, sucrose, or resorcinol onto graphitic surfaces has been the subject of many studies. Snoeyink and Weber (26) have reviewed this field in considerable detail. There are a number of scattered references to the adsorption of such molecules on inorganic surfaces, but to date no systematic study of the effect of molecular structure on adsorption has been made. In the extensive classification of adsorption isotherms given by Giles (27) such simple organics adsorbing on Al_2O_3 or SiO_2 out of water follow simple Langmuir or simple S-shaped isotherms; in either case, evidence of monolayer saturation is observed. From such studies, Giles and Nakhama (28) have recommended that p-nitrophenol adsorption out of water provides a simple and reliable method of surface area determination.

The principal contribution to the free energy of adsorption for such solutes as phenols and related molecules, appears to be due to hydrogen bond formation. For example, silica tends to adsorb p-nitrophenol extensively but nitrobenzene or anthraquinone show only minor adsorption out of water (29). Again, Lotse (30) in a study of the adsorption on lake sediments of the insecticide, lindane, concludes that hydrogen bond or van der Waals interactions control the uptake.

It is important to point out that the above discussion and in fact all of the discussion in this paper refers to adsorption from dilute or moderately dilute aqueous solutions of organic solutes. The problem of adsorption from a binary solution onto an inorganic solid where the mole fraction of solute is varied from 0 to 1 will not be considered. The reader is referred to the work of Everett whose school has made many contributions in this area (31).

It is surprising that so little work has appeared on such relatively simple systems as phenols, etc., adsorbed on inorganic oxides. The many advances in physical organic chemistry on the effect of substituent groups on reaction mechanisms should provide an excellent foundation for the interpretation of adsorption of simple organics in which substitution was considered systematically.

Somewhat more work exists on the adsorption of organic dyes at oxide-aqueous solution interfaces. Fairly direct evidence is available from the optical properties of such adsorbed molecules that charge-transfer reactions occur (32). Again the enhanced reaction of simple organics with aqueous suspensions of such photoconducting solids as ZnO in the presence of ultraviolet light (33) would appear to be fairly direct evidence of charge transfer reaction of adsorbed organics. However, in such cases, it is important to consider the $ZnO-H_2O$ interface as reactive rather than as a true photo-catalyst (33). Once again such studies usually have been directed at the specific optical or photochemical properties of the adsorbed molecule and little attention has been given to the nature of the adsorption step itself. In fact, in these kind of systems, the effect of changes in the organic molecule or substrate structure on the adsorption step are swamped by the overriding energies involved in electron transfer from substrate to adsorbed molecule or between adsorbed molecules (33).

B. Adsorption of Surfactants.

There is a formidable body of literature on the adsorption of organic surfactants on a variety of inorganic substrates. Much of this literature is associated with mineral flotation where the adsorbed surfactant renders first one mineral then another hydrophobic and selective separation is possible. Again, in detergency, the lifting of inorganic stains from garments and other materials with the aid of surfactants represents an important application of organic surfactant adsorption on inorganic surfaces. Finally in more recent years there has been a growing body of literature concerning the removal of detergents from polluted water systems by soil or sand filters.

To illustrate the principles of surfactant adsorption that control selective uptake, only simple alkyl and simple alkylaryl surfactants shall be considered. As previously, we shall consider that H^+ and OH^- ions are potential-determining and, therefore, pH is the single variable with which the surface charge can be controlled most appropriately.

For surfactants with polar groups: $-NH_3^+$, $-N(Me)_3^+$, $-SO_3^-$, and $-SO_4^-$, it has been shown that specific chemical interaction (i.e., so-called specific or super-equivalent adsorption where $\Delta G_{chem.}$ is large and negative) is absent. That is, the anionics do not adsorb extensively above the Z.P.C. and cationics do not adsorb below the Z.P.C. This is shown dramatically in Figure 4 that was taken from the data of Iwasaki (34). In the upper half of Figure 4, it can be seen that the Z.P.C. of the goethite substrate is clearly defined at pH 6.7, and is independent of the surface inactive supporting electrolyte concentration. The data in the lower half of Figure 4 are the flotation recovery results - the percentage of the mineral that floats under standard conditions - and it can be assumed reasonably that recovery is related simply to extent of adsorption.

The flotation data illustrate the control on surfactant adsorption imposed by the Z.P.C. condition. In all oxide or silicate-aqueous solution interfacial studies it is essential to refer adsorption to this Z.P.C. condition.

The energy term, $\Delta G_{solv.}$, has not been considered, to date, for surfactant adsorption. It is probable that it is small if one considers monovalent ions for which $\Delta G_{solv.}$ will be small and long chain

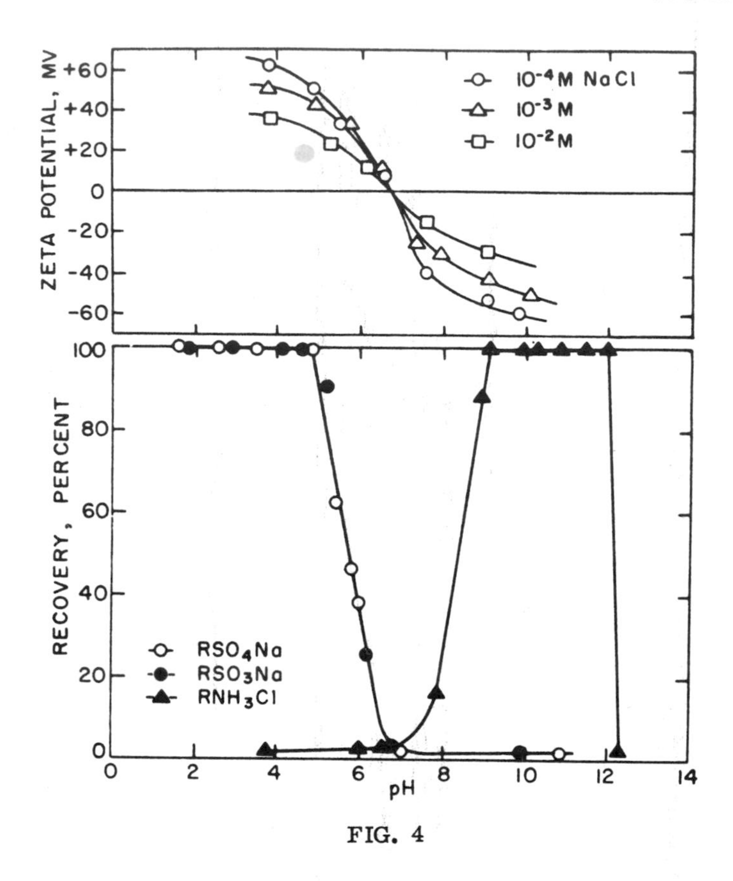

FIG. 4

The zeta-potential in salt solutions and flotation recovery behaviour in surfactant solutions of the mineral goethite. Reprinted from Ref. (34), p. 192, by courtesy of Mining Engineering.

surfactants for which $\Delta G_{VdW.}$ is large. Thus, for nonspecifically adsorbed surfactants of chain lengths above about C_5, a good approximation may be made:

$$\Delta G_{ads.} = \Delta G_{el.} + \Delta G_{VdW.} \qquad (9\text{-}10)$$

$$\Delta G_{ads.} = ze\psi_{\delta} + n\,\phi_{CH_2} \qquad (9\text{-}11)$$

where ϕ_{CH_2} is the van der Waals contribution per CH_2 group and n is the effective number of carbons removed into the interface and into association with carbons from adjacent absorbed surfactants. The same description applies to micelle formation where, in aqueous systems, almost all of the alkyl chain is removed out of the water structure into the non-polar hydrocarbon liquid - like core of the micelle. The obvious similarity between micelle formation in bulk and the formation of two dimensional aggregates of surfactants at the solid-liquid interface has led Gaudin and Fuerstenau (35) to describe interfacial association as hemi-micelle formation. The hemi-micelle concentration (H.M.C.) is defined, therefore, as that solution concentration at which $\Delta G_{VdW.}$ operates and drastically modifies the adsorption process. It is significant that the H.M.C. is approximately 100 times lower (i.e., more dilute) than the critical micelle concentration (C.M.C.).

Hemi-micelle formation is illustrated in Figure 5, for alkyl sulphonate adsorption onto well-characterized α-Al_2O_3 (36). The low concentration region (Region 1.) has been shown for several substrate-surfactant systems to involve either:

(a) simple diffuse layer compression for variable ionic strength systems (37) and, therefore, controlled by a Guoy-Chapman expression for cation adsorption:

$$\Gamma_d^+ = 6.1 \times 10^{-11} \sqrt{C}\left[(\exp - 19.46\psi_\delta) - 1\right] \tag{9-12}$$

at 25° C., where Γ_d^+ is the adsorption density of the cation in the diffuse layer, C is the equilibrium (bulk) concentration, and ψ_δ is the potential at plane δ in the double layer. In several sets of separate calculations, substitution of the zeta-potential for ψ_δ has proved to be successful. The log Γ - log C isotherm, therefore, has a slope of +0.5, or

(b) simple ion exchange in the diffuse layer for constant ionic strength systems. It appears that this is essentially a 1 to 1 exchange, e.g., alkyl sulphonate ions for chloride ions (24, 36), with some preferential tendency for concentration of the surfactant ion in the diffuse layer.

The H.M.C. is usually well-characterized by extrapolating the Region 1 and 2 lines. The magnitude of the van der Waals contribution can be obtained by considering the variation of the H.M.C. with chain

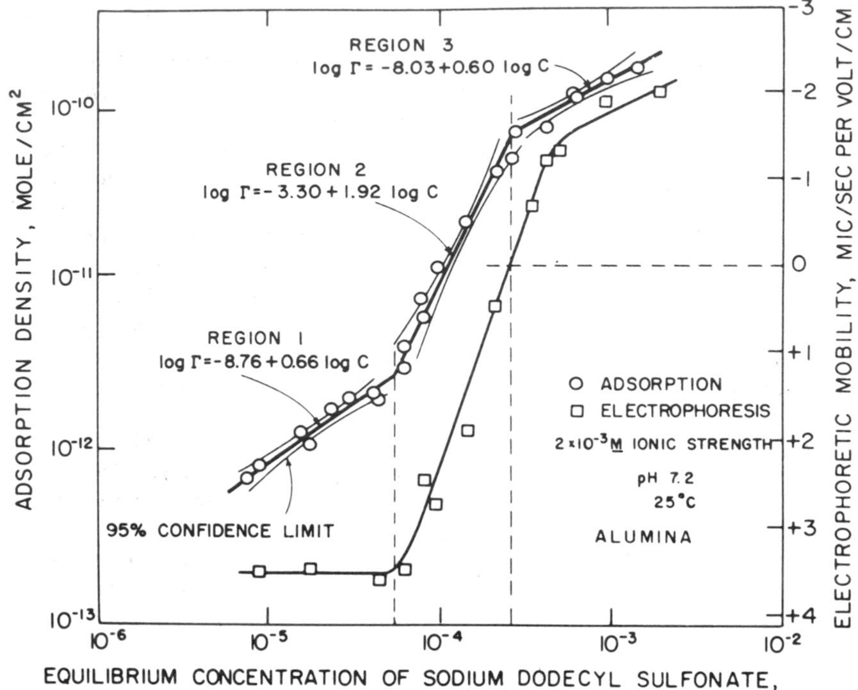

FIG. 5

The electrophoretic mobility and adsorption density of α-Al_2O_3 as a function of the concentration of sodium dodecyl sulphonate. Reprinted from Ref. (36), p. 161, by courtesy of the American Chemical Society.

length of the surfactant for an homologous series. To do this, it is necessary to postulate that upon the operation of $\Delta G_{VdW.}$, the adsorption is specific and into the Stern layer of the double layer system. Thus, following Grahame (21), for such specific adsorption, again for cationic surfactants:

$$\Gamma_\delta^+ = 2rC \exp(-\Delta G_{ads.}/RT) \tag{9-13}$$

where $\Delta G_{ads.}$ is given by Equation (9-11) and r is the size of the polar head of the adsorbed surfactant. In logarithmic form:

$$\ln \Gamma_\delta^+ = \ln C + \ln 2r - (z e \psi_\delta + n \phi_{CH_2})/RT \qquad (9\text{-}14)$$

If we further assume that 2r and ϕ_{CH_2} are independent of chain length and concentration, and that, by considering adsorption in Region 2, n can be regarded as the stoichiometric chain length, then:

$$\partial \ln \Gamma_\delta^+ / \partial \ln C = 1 - ze(\partial \psi_\delta / \ln C)/RT \qquad (9\text{-}15)$$

$$\partial \ln (\Gamma_\delta^{+\prime}) / \partial n = \partial \ln C' / \partial n - ze(\partial \psi_\delta' / \partial n)/RT - \phi_{CH_2}/RT \qquad (9\text{-}16)$$

where the prime indicates adsorption density, concentration, and ψ_δ-potential at the H.M.C.

It has been shown that if one makes the highly probable assumptions that:

$$\partial \psi_\delta / \partial \log C \equiv \partial \zeta / \partial \log C \qquad (9\text{-}17)$$

and/or that at $\zeta = 0$, $\psi_\delta = 0$, where ζ is the zeta-potential, then both ψ_δ terms in Equations (9-14) and (9-15) can be evaluated. Since the ψ_δ term in Equation (9-15) is negative, the steep slope in Region 2 can be evaluated and compared to experiment (24).

To obtain ϕ_{CH_2}, there are two possible routes. First, if one compares adsorption isotherms for various chain lengths, the adsorptive density at the H.M.C. is altered only slightly with change in chain length (23). Thus, if the concentration in solution at which the zeta-potential is reduced to zero (i.e., $\psi_\delta = 0$) is plotted on a log scale as a function of n, a straight line of slope ϕ_{CH_2}/RT is obtained. A set of such data is shown in Figure 6B taken from the zeta-potential data of Figure 6A for alkyl ammonium salt adsorption onto negative silica

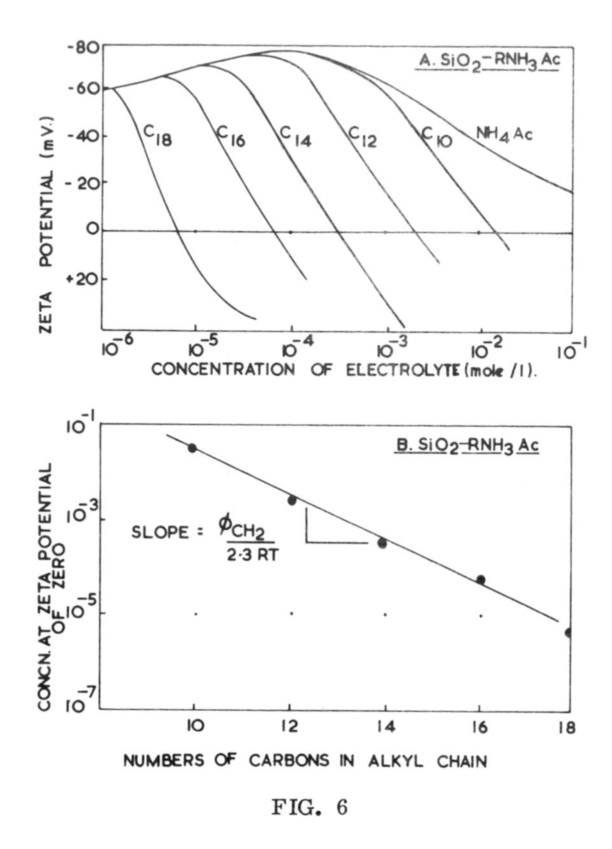

FIG. 6

The effect of hydrocarbon chain length on the zeta-potential of quartz for a series of alkylammonium acetate surfactants illustrating hemi-micelle formation (23).

(23). A similar set of data exists for alkyl sulphonate (24) and alkylaryl sulfonate (38) adsorption onto positive alumina.

The value of ϕ_{CH_2} obtained from micelle formation studies is approximately 1.2 kT or 625 cal./mole; for adsorption at oxide-water interfaces using the hemi-micelle model, ϕ_{CH_2} is 0.9-1.0 kT or approximately 580 cal./mole (23, 36) which for a 12 carbon surfactant is a respectable 6 kcal./mole.

It is important to point out that this $\Delta G_{VdW.}$ term, when applicable, i.e., at and above the H.M.C., will be negative and will assist the $\Delta G_{el.}$ term for surfactant ion adsorption onto an oppositely charged surface. However, when sufficient surfactant ions are adsorbed to bring the surface charge to zero, $\Delta G_{VdW.}$ still operates and will continue to promote adsorption even though $\Delta G_{el.}$ is now positive and, therefore, unfavorable. The characteristic reversal of charge shown by surfactants, typified by the results of Figure 6A, is due to the operation of $\Delta G_{VdW.}$.

The adsorption of alkyl benzene sulphonate surfactants at inorganic surfaces has been considered in several recent studies (40). Similar hemi-micelle behaviour is exhibited and, in two very recent papers (38, 39), the effects of alkyl chain length and alkyl chain branching on the H.M.C. have been considered in detail. The effect of branching can be considered for the C 12 benzene sulphonate isomers where the ring is attached to the alkyl chain at carbons 1, 2, 3, 4, 5, and 6 of the alkyl chain. For adsorption at the alumina-water interface at pH 6.9 (i.e., positive surface), the H.M.C. increases as the ring is moved from position 1 to position 6; in other words, branching hinders hemi-micelle formation in that effective removal of the C 1 isomer from water can be achieved at a lower concentration than for the branched C 6 isomer.

It has been shown from micelle formation studies that the benzene ring is "equivalent" to approximately 3.2 CH_2 groups. This same equivalence is observed when one compares alkyl and alkylaryl surfactants (38). It seems reasonable, therefore, to ascribe a simple hydrophobic bonding role to the aromatic nucleus in the same way as one ascribes a hydrophobic bonding role to the alkyl chains. To date, no specific chemical interaction between two adsorbed benzene rings or between a benzene ring and a surface has been identified.

While subsequent papers will consider organic adsorption onto organic surfaces, it is of interest to relate this A.B.S. adsorption work to Swisher's study (41) of the effect of chain length and chain branching on biodegradation of A.B.S. materials. For example, it was found that, for a C 12 A.B.S. surfactant mixture of the six isomers, there was preferential degradation of the straight chain isomer, the C 2 material, compared with the C 6 or branched isomer. In view of the fact that the sulphonate group appears to be anchored adjacent to the active enzyme site, an adsorption step may control the resultant kinetics of degradation.

C. Adsorption of Macromolecules.

Most standard colloid chemistry texts describe the classical experiment for the determination of the isoelectric point of a protein by measuring the electrophoretic mobility of, say, quartz particles that possess an adsorbed coating of the protein. Again it has been known for many years that moderately large concentrations of polyelectrolytes will stabilize a colloidal dispersion of an inorganic material. A reawakening of interest in polymer adsorption on inorganics came with the finding that, at low concentrations, a polyelectrolyte can cause extensive aggregation rather than stabilization of the colloidal dispersion. This aggregation by polymers was considered by LaMer, Healy, and others (42, 43) to result from the bridging action of the extended polymer molecules in binding the particles into a loose aggregate or floc. This aggregation has been termed <u>flocculation</u> in contrast to <u>coagulation</u> through simple double layer adsorption of electrolytes (44, 45).

Since the first use of macromolecules as flocculants in the mineral industry and later in water treatment practice, a series of detailed reviews have appeared on polymer adsorption and polymer flocculation (43,46, 47, 48, 49). It is, therefore, unnecessary at this time to add to this topic by way of review.

There are two aspects of polymer flocculation that need more detailed study. The first concerns the long-accepted bridging model of polymer flocculation, while the second concerns the recent, exciting findings on so-called bioflocculation where an organism is able to provide its own polymer flocculant.

The generally accepted model of polymer flocculation is shown schematically in Figure 7. The extent of flocculation is followed by light scattering, sedimentation rate, filtration rate, and related parameters. There is a pressing need to relate the structure of the polymer molecule to its ability to flocculate a given dispersion. By careful control of polymer chain length, stereoregularity, flexibility (radius of gyration), partial cross-linking, etc., a critical test of the rather qualitative bridging theory would be possible. That such rigorous testing has not been forth-coming, is not surprising when it is remembered that polymer adsorption-flocculation is a process controlled by chemical as well as physical variables. By careful optimization of physical variables (degree and intensity of agitation, etc.) La Mer, Smellie, and Healy (50, 44, 45) were able to establish the validity of a simple, yet, general theory of polymer flocculation and filtration. A recent attempt by Slater and Kitchener (51) to highlight the deficiencies of the theory of LaMer, Smellie, and Healy must be regarded with suspicion; no attempt was made by these workers to survey and then control the physical conditions of polymer adsorption and flocculation. Their result that floc size and strength did not correlate with the refiltration rate simply highlights the fact that their conditions of agitation and mixing were far from an optimum value, so that polymer usage

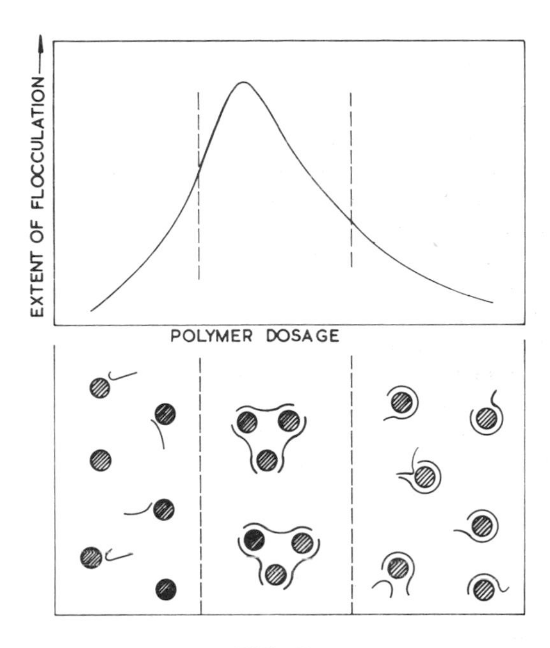

FIG. 7

Schematic representation of the effect of polymer concentration on the extent of flocculation of a dispersion and of the "bridging" model below, at and above the optimum polymer concentration.

was inefficient. When such incomplete flocculation is observed, the refiltration technique does detect it. However, had these workers used conditions of optimum agitation, their maximum refiltration rate would have been increased markedly and floc size and strength would have been optimized similarly. This optimum condition corresponds to surface coverage of the particles by adsorbed polymer segments of 0.5 (45); flocculation may be observed at higher coverages but such flocculation will be incomplete.

Finally it is to be hoped that the recent work of Stumm et al. (49, 52) on bioflocculation will be continued. It is clear that a wide range of colloidal dispersions in natural aquatic environments and in vivo are concerned with what we might naively call "polymer flocculation".

D. Adsorption of Organic Sols.

In this section it is proposed to consider briefly the possible interactions between inorganic surfaces and organic material in the form of a sol or as discrete particles of colloidal size. It is perhaps more appropriate to refer to this interaction as coagulation (strictly hetero- or mutual coagulation) rather than adsorption.

The author suggests that this coagulation of organic sols onto inorganic surfaces or of organic sols with other organic surfaces warrants closer examination. It is possible that coagulation of this kind may be a rate-determining process for many reactions of biological colloids with sediments, synthetic polymer surfaces, and in vivo organic surfaces. Living organisms, blood platelets, protein aggregates, etc., may represent the organic sol component; it should be pointed out that subsequent reaction may involve deformation or breakdown of the organic. The present discussion concerns the first step in the overall process.

The presence of charged groups on the inorganic surface and on the organic sol gives rise to electrical double layers on each component. Thus, as the sol approaches the inorganic surface, repulsion will result if the double layers are of the same charge and attraction if they are of opposite charge. In addition, the van der Waals attraction interaction assists the process of aggregation. The sum of the double layer-repulsion and van der Waals attraction yields the curve of the total potential energy of interaction as a function of distance apart of the interacting bodies.

The form of this energy curve is given elsewhere for particles of the same charge (53) and for particles of opposite charge (54). The curve has characteristically a so-called secondary minimum at large distances, a barrier to coagulation, and a deep minimum at close approach. Usually the secondary minimum is shallow and unable to provide an effective coagulation well. The barrier to coagulation due to double layer repulsion is reduced by adsorbing electrolytes into the double layer. The critical coagulation concentration (C. C. C.) represents a reduction of the coagulation barrier such that van der Waals attraction results.

If one considers coagulation of a spherical colloidal sol particle with a flat surface (sphere-plate coagulation), the secondary minimum at large distances becomes much deeper than for coagulation of, say, equal sized spheres.

The conditions which lead to coagulation are summarized (55) by the domain diagram of Figure 8. This graph summarizes a large number of calculations for materials of equal surface potential (ψ_o)

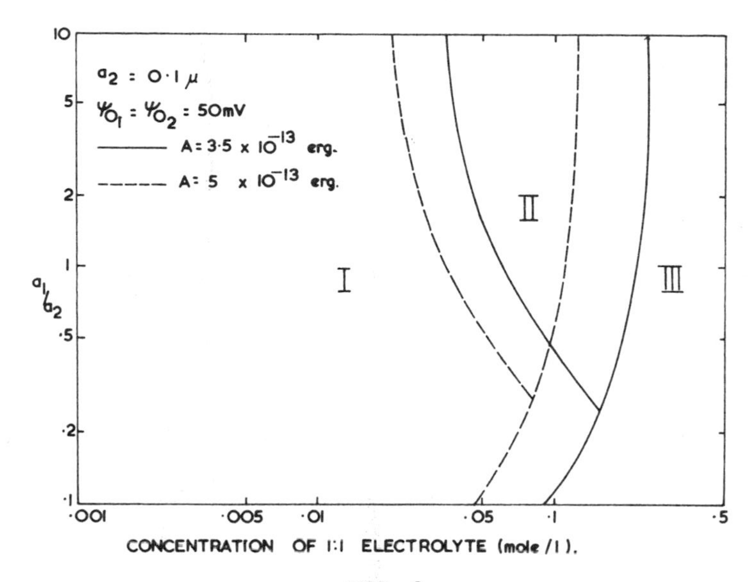

FIG. 8

Coagulation domains for Hamaker constants appropriate for organic sol-inorganic surface interactions. In Region I, $V^*_{max.} > 5kT > V_{sec.}$; in Region II, $V^*_{max.} > 5kT < V_{sec.}$ and in Region III, $V^*_{max.} < 5kT$.

but whose particle radii have been allowed to vary. Type 2 particles of radius, a_2, have been fixed at one value of radius, while particles of Type 1, radius, a_1, have been allowed to vary from $0.1a_2$ to $10a_2$. This latter case closely approximates the sphere-plate situation with Type 1 being the plate and Type 2 being the sphere. In Domain I, the electrolyte concentration is not sufficient to cause coagulation and the sol is dispersed. In Domain II, the secondary minimum ($V_{sec.}$) is deepened to greater than the average kinetic energy of the particles, but the barrier to coagulation ($V^*_{max.}$) is high and prevents particles from going over into the primary minimum. In Domain III, the coagulating electrolyte concentration is such that there is no significant barrier to prevent coagulation into the primary coagulation well.

The data of Figure 8 are for a system of an inorganic surface and an organic sol. The Hamaker constant (A) values represent values that are appropriate for inorganic-organic interactions (56). The Hamaker constant is the parameter which controls the magnitude of the van der Waals attraction between particles.

It is significant that, for organic sol coagulation onto flat inorganic surfaces ($a_1/a_2 > 10$), there is a wide coagulation region, a region much wider than predicted by conventional coagulation theory. It is suggested that this wide zone may be responsible for and would control, the initial uptake of live or dead organisms on natural sediments, soils, and sand filters. Recent work by Ives and Gregory (57) on the uptake of dispersed inorganic colloids on sand filters uses some of the above concepts but does not consider the important coagulation into the secondary minimum.

It is hoped to extend, theoretically and experimentally, this new approach to coagulation of organics with one another or with inorganics as all situations are of considerable interest in a variety of natural aquatic environments.

IV. SUMMARY

The principles that are considered to control the selective adsorption of organic materials present in the natural aquatic environment onto inorganic surfaces have been reviewed. The various contributions to the total free energy of adsorption, coulombic, solvation, hydrogen bond, van der Waals and chemical, have been considered for small organic molecules, surfactants, polymers, and organic sols. The way in which these contributions vary with molecular structure of the organic and surface structure of the inorganic material is shown to control selective uptake. Profitable new areas of research and application have been explored.

ACKNOWLEDGMENTS

The author wishes to thank the Fifth Rudolfs Research Conference. Several research students: Messrs. James, Wiese, Dick, and Dixon, of the colloid and surface chemistry group of the Department of Physical Chemistry provided some of the more recent data used in several sections of the paper. Their contributions are gratefully acknowledged.

REFERENCES

1. G. A. Parks, Chem. Revs., 65, 177 (1965).

2. T. W. Healy and D. W. Fuerstenau, J. Colloid Sci., 20, 376 (1965).

3. A. S. Buchanan and R. C. Oppenheim, Aust. J. Chem., 21, 2367 (1968).

4. J. S. Smolik, A. Harman, and D. W. Fuerstenau, Trans. A.I.M.E., 235, 367 (1966).

5. G. A. Parks and P. L. de Bruyn, J. Phys. Chem., 66, 967 (1962).

6. R. J. Atkinson, A. M. Posner, and J. P. Quirk, J. Phys. Chem., 71, 550 (1967).

7. P. Somasundaran, J. Colloid Interf. Sci., 27, 659 (1968).

8. J. J. Morgan and W. Stumm, J. Colloid Sci., 19, 347 (1964).

9. T. W. Healy, A. P. Herring, and D. W. Fuerstenau, J. Colloid Interf. Sci., 21, 435 (1966).

10. M. C. Fuerstenau, Bull. Inst. Min. Met. (Lond.), 74, 381 (1965).

11. D. Mulfadi, A. M. Posner, and J. P. Quirk, J. Soil Sci., 17, 212 (1966).

12. E. A. Jenne, in Adv. in Chem. Series No. 73, Amer. Chem. Soc., Washington, D. C., 1968, p. 337.

13. G. A. Parks, in Adv. in Chem. Series No. 67, Amer. Chem. Soc., Washington, D. C., 1967, p. 121.

14. T. W. Healy and D. W. Fuerstenau, in preparation, (1969).

15. G. Onoda and P. L. de Bruyn, Surf. Sci., 4, 48 (1966).

16. J. A. Yopps and D. W. Fuerstenau, J. Colloid Sci., 19, 61 (1964).

17. M. Bier and F. C. Cooper, in "Principles and Applications of Water Chemistry", (S. D. Faust and J. V. Hunter, eds.), John Wiley, New York, 1967, p. 217.

18. D. J. Murray, T. W. Healy, and D. W. Fuerstenau, Adv. in Chem. Series No. 79, Amer. Chem. Soc., Washington, D. C., 1968, p. 74.

19. See Ref. 17, 1967, p. 241.

20. A. C. Zettlemoyer and F. J. Micale, Chapter 8 this publication.

21. D. C. Grahame, Chem. Revs., 41, 441 (1947).

22. R. O. James and T. W. Healy, in preparation, (1969).

23. D. W. Fuerstenau, T. W. Healy, and P. Somasundaran, J. Phys. Chem., 68, 3562 (1964).

24. P. Somasundaran and D. W. Fuerstenau, J. Phys. Chem., 68, 3562 (1964).

25. T. N. Anderson and J. O'M. Bockris, Electrochim. Acta, 9, 347 (1964).

26. V. L. Snoeyink and W. J. Weber Jr., Adv. in Chem. Series No. 79, Amer. Chem. Soc., Washington, D. C., 1968, p. 112.

27. C. H. Giles, J. Chem. Soc., 3973 (1960).

28. C. H. Giles and S. N. Nakhama, J. Appl. Chem., 12, 266 (1962).

29. C. H. Giles in "Hydrogen Bonding", Pergamon Press, New York, 1959, p. 449.

30. E. G. Lotse, Environ. Sci. Tech., 2, 353 (1968).

31. D. H. Everett, A.C.S. National Colloid Symposium, Cleveland, June 1969, and J. Colloid Interf. Sci., in press.

32. A. H. Hertz, R. P. Danner, and G. A. Janusonis, Adv. in Chem. Series No. 79, Amer. Chem. Soc., Washington, D. C., 1968, p. 173.

33. D. R. Dixon and T. W. Healy, submitted for publication, 1969.

34. I. Iwasaki, U. S. Bureau of Mines Report No. 5593, (1960).

35. A. M. Gaudin and D. W. Fuerstenau, Trans. A.I.M.E., 202, 958 (1955).

36. T. Wakamatsu and D. W. Fuerstenau, Adv. in Chem. Series No. 79, Amer. Chem. Soc., Washington, D. C., 1968, p. 161.

37. P. L. de Bruyn, Trans. A.I.M.E., 200, 291 (1955).

38. D. W. Fuerstenau, and T. W. Healy, in preparation, 1969.

39. S. D. Dick and T. W. Healy, in preparation, 1969.

40. C. Wayman, see Ref. 17, 1967, p. 127.

41. R. D. Swisher, J. Water Pollut. Contr. Fed., 35, 877 (1963).

42. R. A. Ruehrwein and D. W. Ward, Soil Sci., 73, 485 (1952).

43. V. K. LaMer and T. W. Healy, Revs. Pure Appl. Chem. (Australia), 13, 112 (1963).

44. V. K. LaMer and T. W. Healy, J. Phys. Chem., 67, 2417 (1963).

45. T. W. Healy and V. K. LaMer, J. Colloid Sci., 19, 323 (1964).

46. F. Rowlands, R. Bulas, E. Rothstein, and F. R. Eirich, Ind. Eng. Chem., 57, 46 (1965).

47. V. K. LaMer, Disc. Faraday Soc. No. 42, p. 248 (1966).

48. W. Heller, Pure Appl. Chem. (I.U.P.A.C.), 12, 249 (1966).

49. M. W. Tenney and W. Stumm, J. Water Pollut. Contr. Fed., 37, 1370 (1965).

50. R. H. Smellie Jr. and V. K. LaMer, J. Colloid Sci., 13, 589 (1958).

51. R. M. Slater and J. A. Kitchener, Disc. Faraday Soc. No. 42, p. 267 (1966).

52. P. L. Busch and W. Stumm, Environ. Sci. Tech., 2, 49 (1968).

53. E. J. Verwey and J. Th. Overbeek, "Theory of the Stability of Lyophobic Colloids", Elsevier Publishing Co., Inc., Amsterdam, Holland, 1948.

54. R. Hogg, T. W. Healy, and D. W. Fuerstenau, Trans. Faraday Soc., 62, 1638 (1966).

55. G. R. Wiese and T. W. Healy, Trans. Faraday Soc., 66, 490 (1970).

56. F. M. Fowkes, Ind. Eng. Chem., 56, 40 (1964).

57. K. J. Ives and J. Gregory, Proc. Soc. Water Treat., 15, 93 (1966).

58. J. M. Cases, J. Chim. Phys., 64, 1101 (1967).

CHAPTER 10

INTERACTIONS RESPONSIBLE FOR THE SELECTIVE ADSORPTION OF ORGANICS ON ORGANIC SURFACES

Richard L. Gustafson
and
John Paleos

Rohm and Haas Company

I. INTRODUCTION

The recent syntheses of high surface area organic polymers, commonly termed "macroreticular" resins, have led to the development of a new class of adsorbents in which organic species may be adsorbed, via van der Waals' interactions, from either aqueous or non-aqueous environments. The present discussion will be limited to studies of the distribution of water-soluble organics between an aqueous phase and the polymer surface. The role of variables such as: (a) chemical structure of the adsorbate and adsorbent, (b) surface area of the adsorbent, (c) concentration of adsorbate, (d) ionic strength, and (e) temperature upon the extent of adsorption will be discussed.

Kauzmann (1) has shown that the positive values of the unitary free energies found in such processes as the transfer of methane or ethane from organic phases to water are produced by large negative entropy changes. These decreases in free energy overcome small enthalpy changes which actually favor dissolution of the hydrocarbon in the aqueous phase. Similar unfavorable entropy changes have been observed in cases of dissolution in water of polar organic molecules such as diethylketone, diethylether, ethyl acetate, n-butanol, ethyl bromide, and heptaldehyde.

Frank and Evans (2) have shown that this unfavorable entropy of solution of non-polar molecules in water is produced by the orientation of water molecules around the organic species. This tendency toward greater crystallinity of the aqueous phase has been termed "iceberg" formation. Frank has shown that, as the size of the organic molecule increases, the size of the "iceberg" increases and the entropy loss required in dissolving the organic molecule increases.

In the adsorption of organic species on the surfaces of macro-reticular polymers or on a high surface area carbon, the "icebergs" in the aqueous phase are broken up with an accompanying large entropy gain. This entropy effect is the main driving force in physical adsorption from aqueous solution. Schneider et al. (3) have shown that the entropies of adsorption of low molecular weight fatty acids by a high surface area styrene-divinylbenzene copolymer become more positive as the molecular weights of the acids increase. Sixty percent of the free energy of adsorption of butyric acid is contributed by a favorable entropy change.

II. EXPERIMENTAL

Most of the adsorbents were commercially available polystyrene-divinylbenzene (DVB) copolymers (1) or polymethacrylate materials (2) cross-linked with a suitable non-aromatic material.

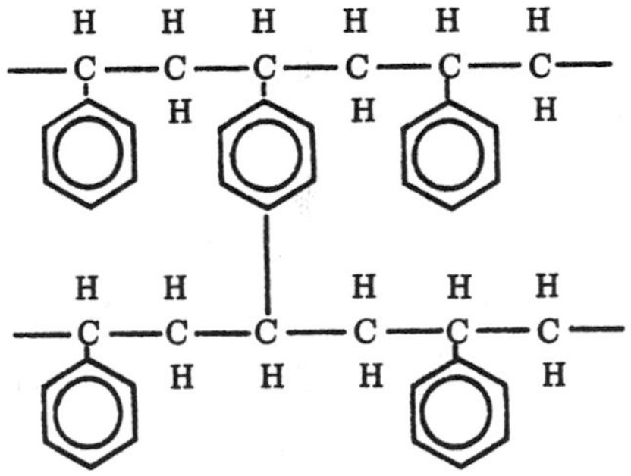

```
   H    H    H    H    H    H
 --C----C----C----C----C----C--
   |    H    |    H    |    H
  (O)       (O)       (O)
             |
   H    H    |    H    H    H
 --C----C----C----C----C----C--
   |    H    H    H    |    H
  (O)                 (O)
```

(1)

STYRENE-DVB COPOLYMER

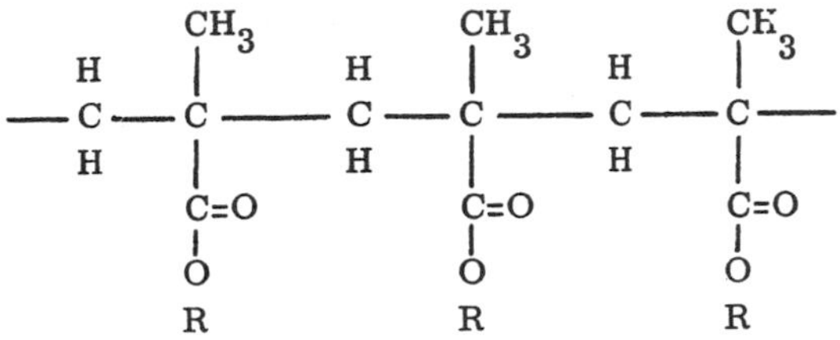

```
         CH3            CH3            CH3
    H    |         H    |         H    |
  — C —— C ——————— C —— C ——————— C —— C ——
    H    |         H    |         H    |
         C=O            C=O            C=O
         |              |              |
         O              O              O
         R              R              R
```

(2)

METHACRYLATE-BASED COPOLYMER

(R = polyfunctional aliphatic residue or methyl group)

Typical properties of the dried resins and Pittsburgh activated carbon (CAL 12 x 40) are given in Table 1. Surface areas were determined by the conventional BET method. Prior to use in the equilibration experiments, weighed quantities of the dry adsorbents were solvated with acetone and rinsed with deionized water until acetone-free. The resins were transferred to volumetric flasks and equilibrated with appropriate volumes of solutions of known solute concentration. After equilibrium was attained, the amount of solute adsorbed by each resin sample was determined by difference following analysis of the solution phase.

III. RESULTS

A. Adsorption Equilibria-General Trends.

Freundlich isotherms representing the adsorption of a variety of organic species by a styrene-divinylbenzene copolymer (Amberlite XAD-2) which had a surface area of 313 m^2/g, a porosity of 0.430 ml/ml, and an average pore diameter of 91 Å, are shown in Figure 1. As the aromaticity of analogous sodium sulfonates increases in the order: benzenesulfonate < naphthalenesulfonate < anthraquinonesulfonate, the affinity of the resin for these materials increases in the same order. Although tannic acid contains ten aromatic rings, it is not adsorbed much more extensively than sodium anthraquinonesulfonate because of the presence of a large number of hydrophilic hydroxy groups on the molecule. As the size of the hydrophobic group increases in the order: propionic acid < butyric acid (data not shown in Fig. 1)

TABLE 1

Typical Properties of Various Adsorbents

Resin	Porosity ml/ml	Surface Area m^2/g	Ave. Pore Diam. Å
		STYRENE-DVB ADSORBENTS	
XAD-1	0.352	100	200
XAD-2	0.430	313	91
XAD-5	0.434	415	68
EXP-500	0.387	525	45
XAD-4	0.552	860	51
		METHACRYLATE-BASED ADSORBENTS	
XAD-7	0.532	445	82
XAD-8	0.513	212	160
		ACTIVATED CARBON	
CAL(12 x 40)	0.685	1045	38

< valeric acid, the extent of adsorption also increases. The binding of resorcinol is much less than that of phenol because of the presence of an additional hydroxy group in the former compound, thus favoring its retention by the aqueous phase. The high affinity of Amberlite XAD-2 for sodium dodecylbenzenesulfonate (ABS) is produced by the strong interaction of the aromatic resin matrix with the twelve membered, branched, aliphatic chain of the adsorbate coupled with the incompatibility of these chains with the aqueous phase.

In general, for a homologous series of adsorbates, the degree of adsorption from aqueous solutions by aromatic adsorbents increases as the molecular weight of the adsorbate increases or as the aqueous solubility of the organic species decreases (4).

In many cases, Langmuir isotherms are non-linear over a wide concentration range. An example is shown for the cases of low

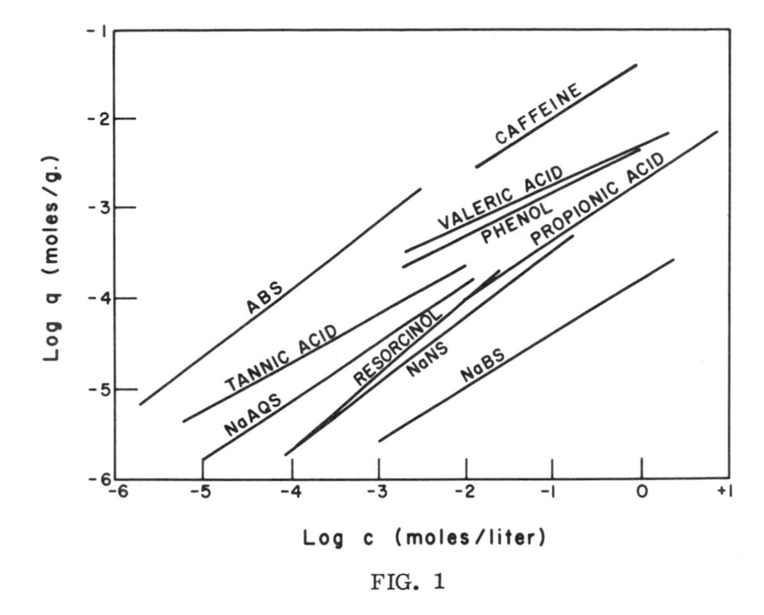

FIG. 1

Freundlich isotherms for adsorption of organic compounds by Amberlite XAD-2 at 25°C.

molecular weight fatty acids in Figure 2. Here values of c/q (in which c is the molar concentration of fatty acid in solution and q is the number of moles of acid adsorbed per gram of Amberlite XAD-2) have been plotted against c according to the relation:

$$\frac{c}{q} = \frac{1}{Kb} + \frac{c}{b} \qquad (10\text{-}1)$$

where b is the number of moles of adsorbate adsorbed per gram of resin in forming a monolayer on the adsorbent surface and K is the equilibrium constant for the resin-sorbate interaction.

However, at low concentrations, linear Langmuir isotherms may be obtained. For instance, a linear plot was obtained for the

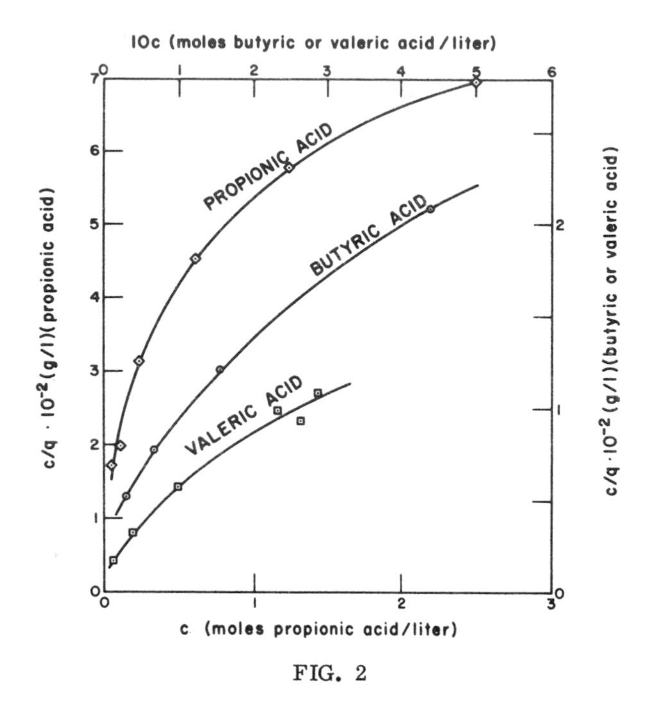

FIG. 2

Langmuir plots for adsorption of low molecular weight fatty acids by Amberlite XAD-2 at 25°C.

butyric acid-Amberlite XAD-2 system provided that the amount of the acid adsorbed did not exceed 9×10^{-4} mole/g, or 60% of monolayer coverage.

In Figure 3, the column capacity of Amberlite XAD-5 for the aromatic species, crystal violet, is compared with that of the polymethacrylate resin, Amberlite XAD-7. Although both resins have comparable surface areas and average pore diameters, the capacity of the aromatic resin, Amberlite XAD-5, is far superior, presumably because of the interaction of the π electrons of the solute with those of the adsorbent. Another illustration of $\pi - \pi$ interaction may be seen in a comparison of the results of binding of phenylalanine and glycine by Amberlite XAD-2. Seven times as much of the aromatic solute was adsorbed by the resin in a 0.1 M amino acid solution.

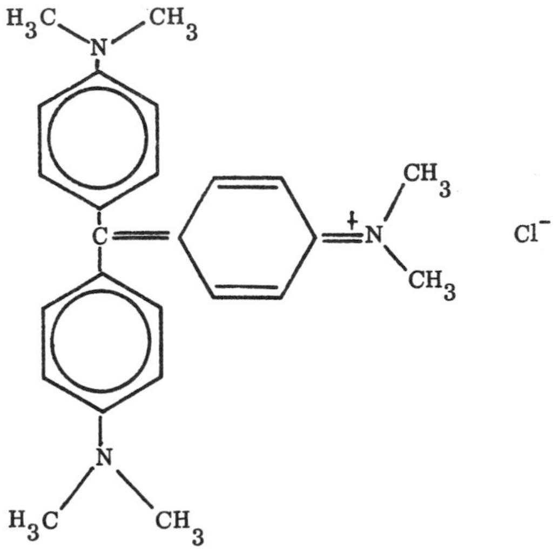

CRYSTAL VIOLET

The amount of phenylpropanolamine bound by Amberlite XAD-2 decreased by an order of magnitude when the amine was converted to its hydrochloride salt. Advantage may be taken of this behavior by removing an amino species via adsorption from aqueous solution and subsequently eluting it from the column by regeneration with a mineral acid. Similarly, phenolic compounds and carboxylic acids may be eluted with ammonia or alkali metal hydroxides. Schneider et al. (3) have shown previously that high surface area styrene-DVB copolymers have virtually no affinity for the sodium salts of low molecular weight fatty acids.

B. Relation Between Surface Area and Adsorption Efficiency.

The effect of increasing the surface area of styrene-DVB adsorbents may be seen in Figure 4 in which Freundlich isotherms describe the adsorption of sodium benzenesulfonate on Amberlite XAD-1 (100 m^2/g), Amberlite XAD-2 (354 m^2/g), EXP-500 (526 m^2/g), and granular carbon (CAL 12 x 40, 1045 m^2/g). The degree of adsorption increases with increasing surface area. Similar results have been found for a wide variety of solutes including sodium anthraquinonesulfonate, alkylbenzenesulfonate, and caffeine. Analysis of adsorption

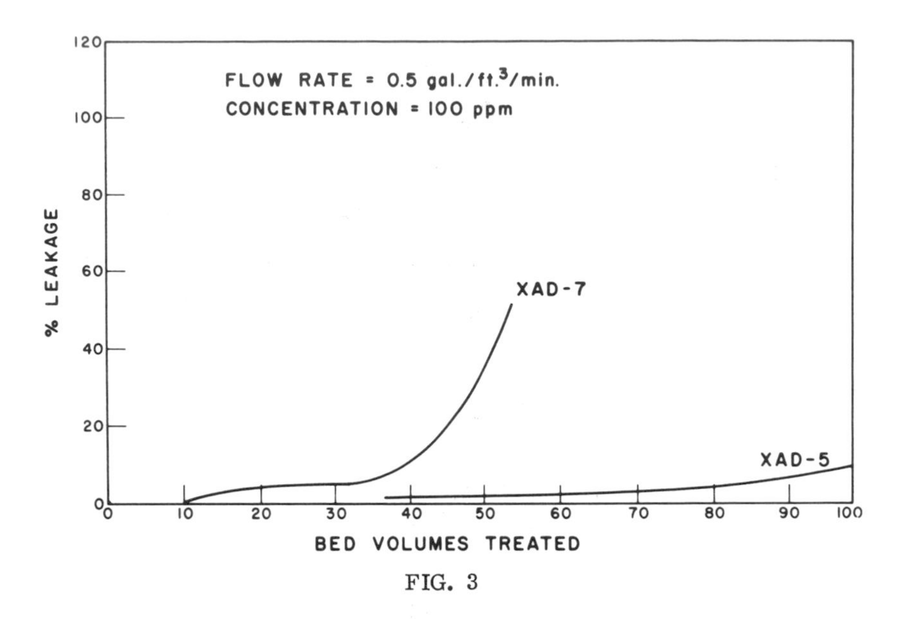

FIG. 3

Comparison of the efficiencies of adsorption of crystal violet from a 100 mg/l solution by Amberlite XAD-5 and Amberlite XAD-7 at a flow rate of 4 bed volumes per hour in column operations.

data for these three systems shows that the amount of solute which is adsorbed at a given solution concentration is not proportional to the surface area of the styrene-DVB adsorbent. The numbers of moles of AQS, ABS, and caffeine adsorbed per m^2 of surface of Amberlite XAD-1, Amberlite XAD-2, and EXP-500 at a solute concentration of 4×10^{-4} M are shown in Table 2. The efficiency of utilization of the available surface area in the adsorption of AQS decreases in the order: EXP-500 ~ XAD-1 > XAD-2. The adsorptive efficiency of ABS decreases in the order: XAD-1 > EXP-500 > XAD-2, while that of binding of caffeine decreases in the order: EXP-500 > XAD-2 ~ XAD-1.

C. Effect of Resin Composition Upon Adsorption Efficiency.

Data pertaining to the equilibrium adsorption of phenol and p-cresol by an aromatic resin, Amberlite XAD-2, and an aliphatic resin, Amberlite XAD-7, are shown in Table 3. The extent of binding

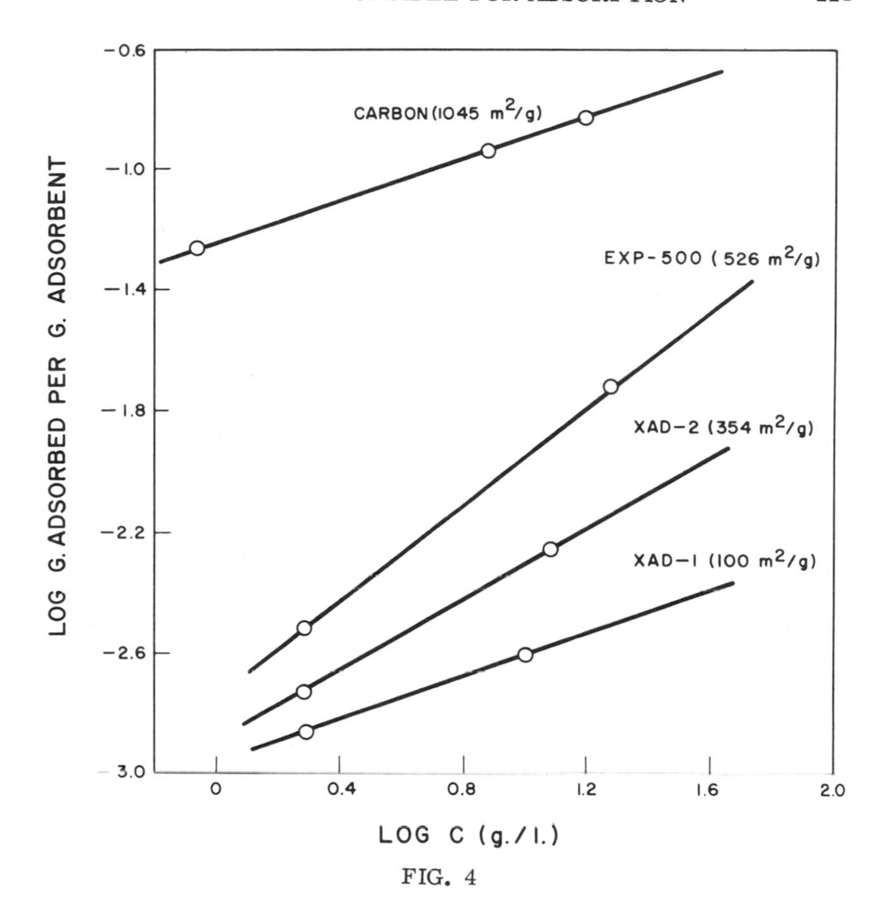

FIG. 4

Freundlich isotherms for adsorption of sodium benzenesulfonate by high surface area styrene-DVB resins and activated carbon at 25°C.

of p-cresol is greater than that of phenol in both cases. In addition, the adsorption of both solutes by Amberlite XAD-7 is greater than that by the aromatic resin. However, other data show that the extent of binding of phenol by the high surface area Amberlite XAD-4 (860 m^2/g) is approximately 20% greater than that by Amberlite XAD-7 (surface area = 445 m^2/g). The efficiency of utilization of available surface area in the adsorption of phenol is somewhat greater in the case of the poly(methacrylate) resin.

TABLE 2

Efficiency of Utilization of Surface Areas of Amberlite XAD-1, Amberlite XAD-2, and EXP-500 in the Adsorption of AQS[a], ABS[b], and Caffeine

Solute	Resin	Surface Area m^2/g	Moles Solute Adsorbed per Square Meter
AQS	XAD-1	100	7.30×10^{-8}
AQS	XAD-2	354	5.08×10^{-8}
AQS	EXP-500	526	7.81×10^{-8}
ABS	XAD-1	100	1.59×10^{-6}
ABS	XAD-2	354	0.86×10^{-6}
ABS	EXP-500	526	1.00×10^{-6}
Caffeine	XAD-1	100	2.13×10^{-7}
Caffeine	XAD-2	354	2.24×10^{-7}
Caffeine	EXP-500	526	4.01×10^{-7}

[a]AQS = sodium anthraquinonesulfonate
[b]ABS = sodium dodecylbenzenesulfonate

TABLE 3

Adsorption of Phenol and p-Cresol by Amberlite XAD-2 and Amberlite XAD-7 at 25°C.

Resin	Solute	Equilibrium Solute Concentration moles/l	Grams Solute Adsorbed per Gram Resin
XAD-7	Phenol	0.0152	0.0764
XAD-2	Phenol	0.0306	0.0511
XAD-7	p-Cresol	0.0146	0.1431
XAD-2	p-Cresol	0.0265	0.0924

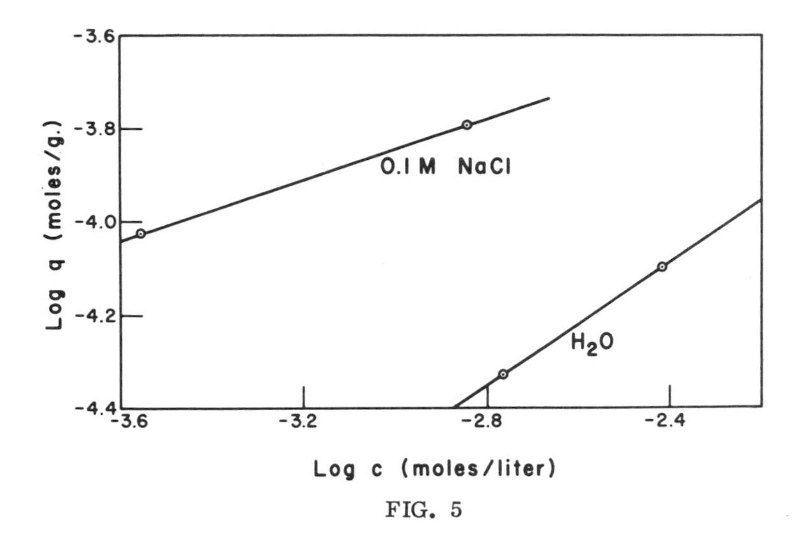

FIG. 5

Effect of NaCl addition upon adsorption of sodium anthraquinone-sulfonate by Amberlite XAD-2 at 25°C.

D. Influence of Ionic Strength.

The effect of the addition of sodium chloride upon the binding of sodium anthraquinonesulfonate by Amberlite XAD-2 is shown in Figure 5. There is approximately a four-fold increase in the binding capacity of Amberlite XAD-2 when sufficient NaCl is added to produce a 0.1 M solution. The sodium ions of the added salt distribute themselves in such a way as to minimize the electrostatic repulsions of neighboring negative sulfonate groups, with the result that the resin has an increased capacity for the organic species.

E. Influence of Solvent Upon Adsorption.

The most efficient regenerants for styrene-DVB adsorbents are water-soluble organic solvents such as low molecular weight alcohols and ketones. The solvent that will function best is the one which has a solubility parameter closest to that of the adsorbent. Polystyrene and polymethylmethacrylate have solubility parameters of approximately 9.1 and 9.3, respectively (5). It is expected that the cross-linked polymers will have solvent compatibilities which are similar to those of the linear polymers. It has been found that the efficiency of regeneration of styrene-DVB copolymers increases in the order: methanol < ethanol < propanol < acetone. This is in agreement with the anticipated order of regeneration efficiency predicted by consideration of the following solubility parameters:

Solvent	Solubility Parameter $cal^{0.5}/cm^{1.5}$
2-Butanone	9.3
2-Propanone	10.0
1-Butanol	11.4
1-Propanol	11.9
Ethanol	12.7
Methanol	14.5
Water	23.2

F. Adsorption of Organic Anions by Anion Exchange Resins.

The trends observed in the adsorption of organic species by high surface area organic polymers are similar to those found in the binding of organic ions to anion exchange resins (6). In the latter case, extremely high selectivities for organic species are found because of the combined effect of electrostatic interactions and hydrophobic bonding.

Selectivity coefficients, $K^{O^-}_{Cl^-}$, for the reaction:

$$R^+ Cl^- + O^- = R^+ O^- + Cl^- \qquad (10\text{-}2)$$

$$K^{O^-}_{Cl^-} = \frac{\text{Mequiv. } O^- \text{ in resin}}{\text{Mequiv. } Cl^- \text{ in resin}} \cdot \frac{[Cl^-]_s}{[O^-]_s} \qquad (10\text{-}3)$$

which describes the exchange of chloride ions and organic ions (O^-) by either a quaternary ammonium ion exchange resin or the protonated form of a weakly basic resin, have been determined for such organic species as sodium ethanesulfonate (ES), sodium benzenesulfonate (BS), sodium naphthalenesulfonate (NS), and sodium anthraquinonesulfonate

TABLE 4

Selectivity Coefficients for Adsorption of Organic Ions by Exchange Resins

Organic Ion	Log $K^{O^-}_{Cl^-}$ Resin (1)[a]	Log $K^{O^-}_{Cl^-}$ Resin (2)[a]
ES	-0.21	-0.52
BS	0.86	0.16
NS	2.18	1.11
AQS	3.14	2.31

[a]The Log $K^{O^-}_{Cl^-}$ values were measured under conditions in which the resin contained equivalent amounts of chloride and organic ions.

(AQS). Values of the selectivity coefficients as determined in 0.1 N solutions at 25°C. in the presence of poly(N, N, N-trimethylvinylbenzyl-ammonium)-DVB (1) and protonated poly (3-N, N-dimethylaminopropyl-acrylamide) (2) resins are shown in Table 4.

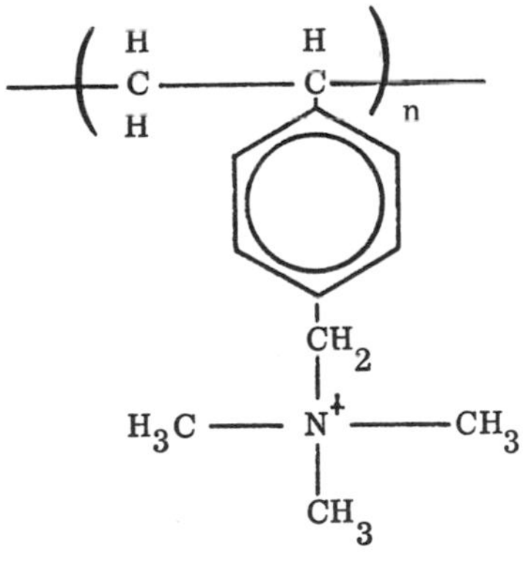

(1)

$$\left(\begin{matrix} H \\ -C- \\ H \end{matrix} - \begin{matrix} H \\ C- \\ | \end{matrix}\right)_n$$

$$C = O \qquad\qquad CH_3$$

$$N - CH_2 - CH_2 - CH_2 - N{:}$$

$$H \qquad\qquad\qquad\qquad CH_3$$

(2)

As in the case of styrene-DVB copolymers, the selectivities of both resins (1) and (2) for organic species increase markedly as the number of aromatic rings in the organic ion increases. In addition, the selectivity of the aliphatic resin for each of the organic ions is less than that of the styrene-based resin. Additional experiments have shown that the selectivity coefficients, $K^{O^-}_{Cl^-}$, for gallate, tert-butylcatecholsulfonate, and dodecylbenzenesulfonate ions are: 150, 400, and 32,000, respectively, in the presence of a type (1) resin.

The following values of $K^{O^-}_{Cl^-}$ were determined by equilibration of sodium naphthalenesulfonate with a poly-(3-N, N, N-trimethylammoniumpropylacrylamide) resin (3) in 50% H_2O-50% organic solvent systems at an ionic strength of 0.10 at 25°C.

$$\left(\begin{matrix} H \\ -C- \\ H \end{matrix} - \begin{matrix} H \\ C- \\ | \end{matrix}\right)_n$$

$$C = O \qquad\qquad CH_3$$

$$N - CH_2 - CH_2 - CH_2 - N^+ - CH_3$$

$$H \qquad\qquad\qquad\qquad CH_3$$

(3)

Solvent	$K^{O^-}_{Cl^-}$ [a]
H_2O	11.3
50% CH_3OH	1.86
50% C_2H_5OH	0.82
50% n-C_3H_7OH	~0.4
50% $(CH_3)_2CO$	~0.4

[a]Measured at a point at which resin contains equivalent amounts of chloride and organic ions.

Values of $K^{O^-}_{Cl^-}$ for binding of sodium napthalenesulfonate by a type (1) resin in methanol-water mixture are as follows:

Volume Percent CH_3OH	$K^{O^-}_{Cl^-}$
0	133
25	39
50	8.6
75	3.0

These results are similar to those obtained in the binding, via hydrophobic bonding only, of organic species by non-ionic copolymers. As water is replaced gradually by methanol, which has a solubility parameter much closer to that of the adsorbent, the affinity of the resin for the solute rapidly decreases. The resin-sorbate interaction decreases even more when methanol is replaced by a solvent having a lower solubility parameter.

G. Thermodynamics of Adsorption of Organics by Styrene-DVB Copolymers.

Enthalpies of adsorption were obtained by measuring the variation of the equilibrium binding constant with temperature. The appropriate relations are:

$$K = \frac{\theta}{(1-\theta)\ c} \tag{10-4}$$

$$K^{*} = 55.5\ K \tag{10-5}$$

$$\Delta F_u^o = -RT\ln K^{*} \tag{10-6}$$

$$\Delta S_u^o = \frac{\Delta H^o - \Delta F_u^o}{T} \tag{10-7}$$

where θ represents the fraction of sites occupied by sorbate molecules and ΔF_u^o and ΔS_u^o are the unitary free energy and entropy, respectively, of adsorption.

A plot of log K vs. 1/T (Fig. 6) for data obtained upon adsorption of sodium naphthalenesulfonate by Amberlite XAD-2 at 25.0, 37.8, and 50.1°C. in the 4×10^{-4} to 2×10^{-3} M concentration range is linear. A value of $\Delta H^o = -4.4$ kcal/mole was calculated. This value, while small, is larger than many values of ΔH^o which have been obtained for the adsorption of organic species on activated carbon. For instance, a value of -1.4 kcal/mole was calculated for the adsorption of sodium benzenesulfonate on granular Columbia LC activated carbon (7). At 25°C., the unitary free energy of adsorption of sodium napthalenesulfonate by Amberlite XAD-2 is -5.7 kcal/mole while ΔS_u^o is +4.4 entropy units. Thus, less than one fourth of the total free energy change is contributed by a favorable entropy change in the case of adsorption of this relatively soluble (in water) compound.

Reasonably linear plots of log K vs. 1/T were obtained upon adsorption of sodium anthraquinonesulfonate by Amberlite XAD-2 in the

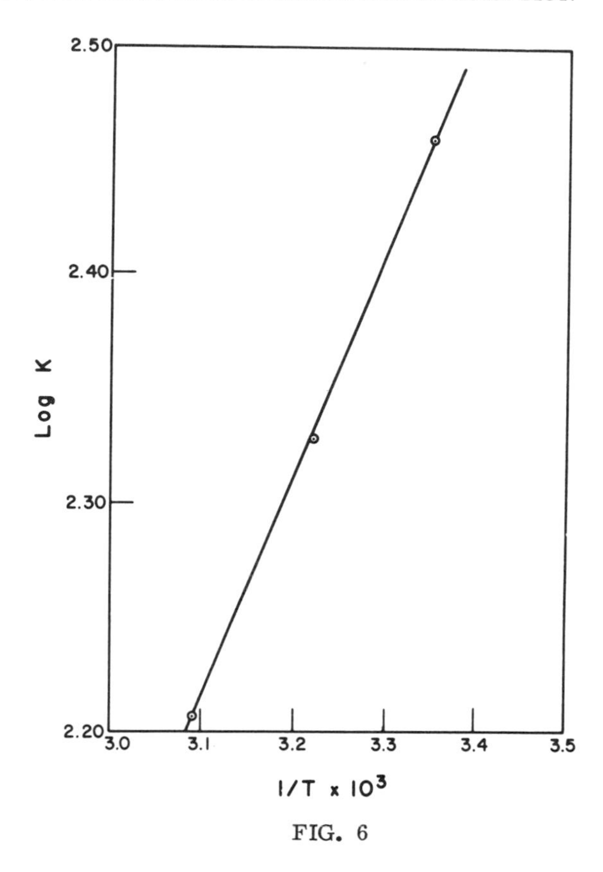

FIG. 6

Plot of log K vs. 1/T for binding of sodium naphthalenesulfonate by Amberlite XAD-2.

5×10^{-4} to 1.5×10^{-3} M concentration range at 4.7, 25.0, 37.8, and 50.1°C. At 25°C. the following quantities were determined:

$$\Delta F_u^o = -5.4 \text{ kcal/mole}$$

$$\Delta H^o = -1.8 \text{ kcal/mole}$$

$$\Delta S_u^o = +12 \text{ entropy units}$$

TABLE 5

Properties of 90% DVB - 10% Styrene Adsorbents

Adsorbent	Surface Area m^2/g	Porosity ml/ml	Ave. Pore Diameter Å
A	478	0.604	117
B	536	0.573	90
C	518	0.468	63
Carbon CAL(12 x 40)	1045	0.685	38

In this case of more favorable adsorption, two thirds of the total free energy of adsorption is provided by a favorable entropy change.

Measurements of binding of caffeine by activated carbon and three 90% divinylbenzene-10% styrene copolymers were made at temperatures of 25.0, 37.8, and 50.1°C. The characteristics of the adsorbents are listed in Table 5.

Values of the enthalpies of adsorption, as calculated at various loading levels, are listed in Table 6. Examination of these data and the corresponding adsorption isotherms shows that there is no correlation among the values for the extent of adsorption, the enthalpy of adsorption, and the average pore diameter.

For a series of eight 90% DVB - 10% styrene copolymers, the amount of caffeine adsorbed per unit surface area of resin increases, in general, as the average pore diameter decreases, as may be seen in Figure 7. This comparison was made by the use of values of the amount of caffeine adsorbed by the resin in equilibrium with a 0.10 M sorbate solution. It is possible that, as the pores decrease in size, the extent of surface to surface contact between the sorbate molecule and the adsorbent surface increases, thus producing greater binding energies.

H. Adsorption in Column Operations.

The application of a variety of styrene-DVB copolymers to the removal of phenol from a 474 mg/l aqueous solution is illustrated in

TABLE 6

Enthalpies of Adsorption of Caffeine by Activated Carbon and Macroreticular Divinylbenzene (90%)-Styrene (10%) Copolymers

Resin	Ave. Pore Diam. Å	-log q^a	$-\Delta H^o$ kcal/mole
A	117	3.4	6.8 ± 0.4
A	117	3.6	6.8 ± 0.3
A	117	3.8	6.8 ± 0.2
A	117	4.0	6.7 ± 0.1
B	90	3.4	7.0[b]
B	90	3.6	7.0[b]
B	90	3.8	6.8[b]
B	90	4.0	6.4[b]
C	63	3.4	4.9 ± 0.3
C	63	3.6	4.8 ± 0.4
C	63	3.8	4.7 ± 0.6
C	63	4.0	4.7 ± 0.8
Carbon			
CAL(12 x 40)	38	2.9	8

[a]q = moles caffeine adsorbed per gram of dry adsorbent.
[b]Calculated from data obtained at only two temperatures.

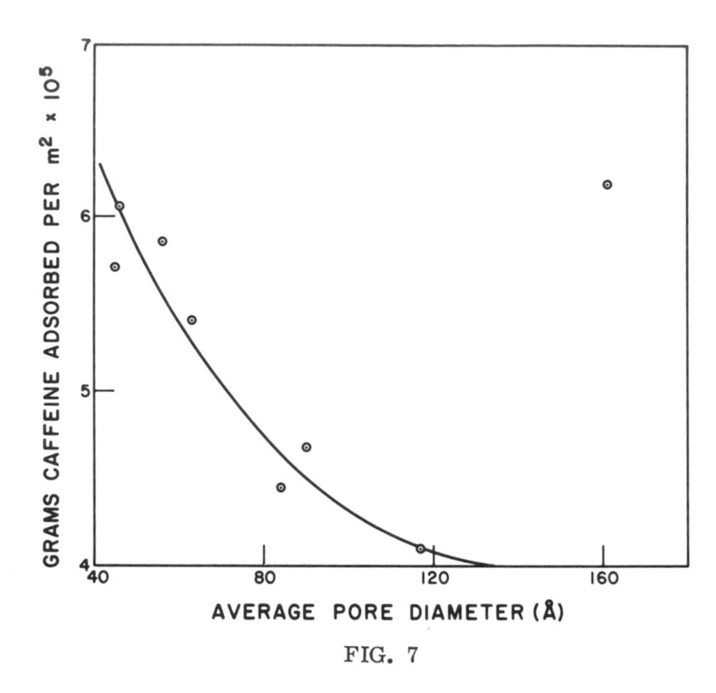

FIG. 7

Influence of average pore diameter upon efficiency of utilization of available surface area in adsorption of caffeine by 90% DVB-10% styrene resins.

Figure 8. The efficiency of removal increases in the order of increasing surface area: i.e., XAD-1 < XAD-2 < XAD-5 < EXP-500 < XAD-4. A separate experiment in which 230 mg/l phenol was passed through CAL (12 x 40) at a 16 bed volume per hour flow rate at 25°C. showed that approximately 190 bed volumes of solution were treated prior to the attainment of a 10% leakage level. No phenol was detected in the effluent until 160 bed volumes of solution had been treated. In a similar experiment phenol leakage was obtained after the passage of 22 bed volumes of solution through a column of EXP-500 and a 10 % leakage level was realized at the 36 bed volume mark.

The efficiency of removal of phenol and its chlorinated derivatives by Amberlite XAD-4 is illustrated in Table 7. The observed

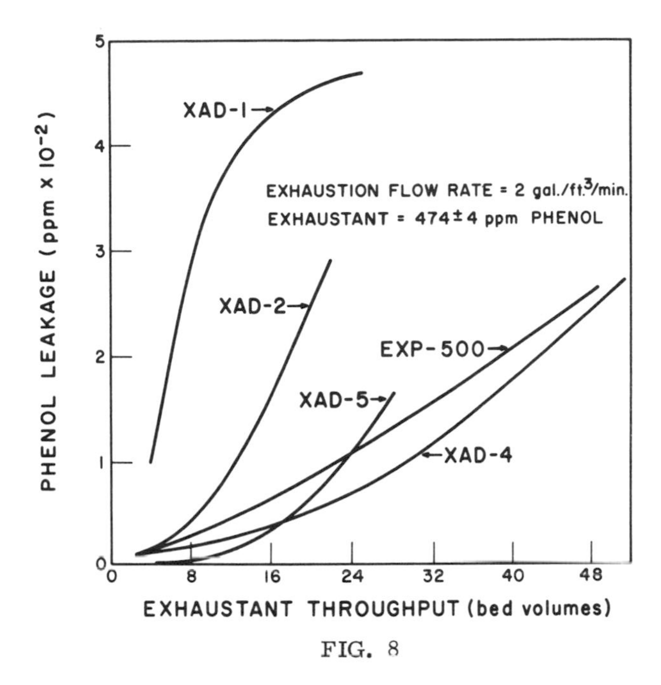

FIG. 8

Adsorption of phenol from a 474 mg/l aqueous solution by styrene-DVB copolymers. Flow Rate = 2.0 gal./ft.^{3}/min. Temperature = 25°C.

trend is a reflection of the decreased water solubilities of the solutes as the number of chlorine atoms per solute molecule is increased.

Curves showing the removal of ABS by various adsorbents are shown in Figure 9. The column capacity of Amberlite XAD-2 is greater than that of Amberlite XAD-1, because of the greater equilibrium capacity of the former resin. However, the performance of XAD-2 is also superior to that of CAL (12 x 40), despite the fact that the activated carbon has the greatest affinity for ABS of all of the adsorbents studied. The poor kinetics of ABS adsorption is produced undoubtedly by the small pore diameters of this adsorbent. A column of EXP-500 showed virtually no capacity despite the fact that the equilibrium capacity of this resin for ABS is greater than that of XAD-2.

TABLE 7

Adsorption of Phenol and Substituted Phenols by Amberlite XAD-4

Solute	Solubility of Solute in Water g/100 ml	Solute Concentration moles/l	Bed Volumes Treated Prior to 10 mg/l Leakage[a]
Phenol	6.7 at 16°C.	2.65×10^{-3}	54
m-Chlorophenol	2.60 at 20°C.	2.71×10^{-3}	116
2,4-Dichlorophenol	0.46 at 20°C.	2.67×10^{-3}	203
2,4,6-Trichlorophenol	0.08 at 25°C.	2.60×10^{-3}	431

[a]Flow rate = 4 bed volumes/hour. Temperature = 25°C.

The efficiency of removal of fulvic acid by Amberlite XAD-2 from an aqueous solution which was 0.002 M in this material and 0.02 M in HCl is illustrated in Table 8. Seventy bed volumes of solution were passed through the resin column at a flow rate of 8 bed volumes per hour and at a temperature of 23 ± 2°C. The resin bed was regenerated with 5 bed volumes of 1% NaOH at the same flow rate. The results show complete regeneration of the resin by the alkaline solution.

Delaware River water which contained 19 mg/l COD (chemical oxygen demand) was acidified by passage through a hydrogen form cation exchange resin, after which it was passed through a column of Amberlite XAD-2 at a flow rate of 8 bed volumes per hour at a temperature of 15-20°C. The effluent COD was 6 and 10 mg/l after the passage of 45 and 135 bed volumes of solution, respectively.

Other organic materials, such as color bodies in raw cane and wood sugar, have been removed effectively by Amberlite XAD-2. Additional experiments have shown the utility of this resin in the

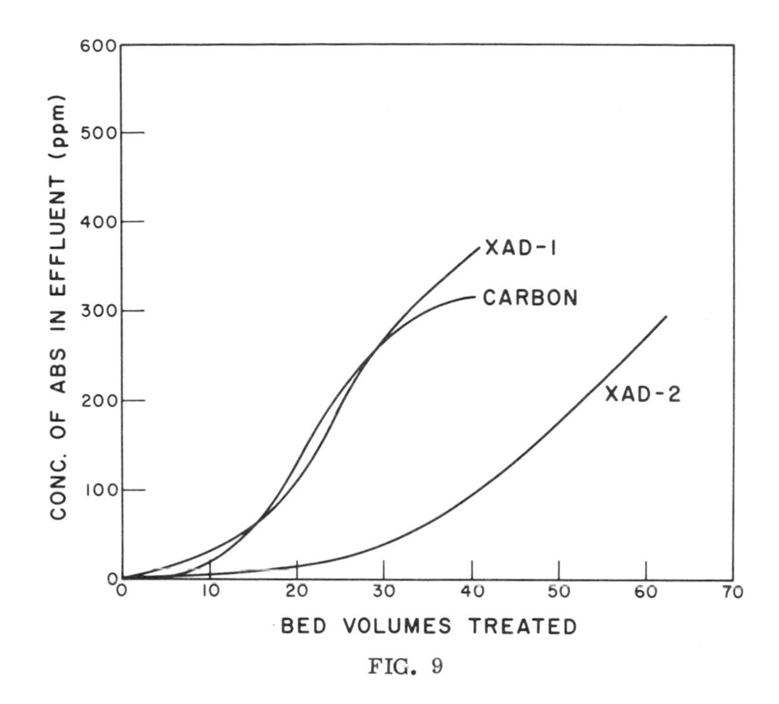

FIG. 9

Adsorption of ABS by styrene-DVB copolymers and activated carbon at room temperature. Influent concentration = 680 mg/l. Flow rate = 2.0 gal. /ft.^{3}/min.

separation and purification of steroids, vitamin B_{12}, tetracycline, caffeine, theophylline, and a number of other compounds of pharmaceutical interest.

IV. SUMMARY

The degrees of physical adsorption of a number of water-soluble organic species by several high surface area styrene-divinylbenzene and polymethacrylate resins have been measured at various concentrations and temperatures. For resins of similar chemical composition, the extent of adsorption increases with increasing surface area. In general, adsorption increases as the water solubility of the organic

TABLE 8

Removal of Fulvic Acid by Amberlite XAD-2

Cycle No.	Percent of Fulvic Acid Adsorbed	Percent Adsorbed Fulvic Acid Regenerated (with 1% NaOH)
1	76	102
2	77	97
3	78	100
4	78	94
5	72	108
6	76	101
7	75	98
Average	76	100

compound decreases. In a series of related compounds, adsorption via hydrophobic bonding increases as the length of a hydrocarbon chain increases, as the number of hydrocarbon substituents increases, or as the number of aromatic rings increases.

Similar effects are noted in the binding of aromatic sulfonates by anion exchange resins, i.e., as the number of aromatic rings increase in the order benzenesulfonate < naphthalenesulfonate < anthraquinonesulfonate, the selectivity of a poly(N, N, N-trimethyl-vinylbenzylammoniumdivinylbenzene) resin for the organic species increases markedly. The efficiency of displacement of organics from both high surface area adsorbents and anion exchange resins by water-miscible organic solvents increases with the decreasing solubility parameter of the solvent.

REFERENCES

1. W. Kauzmann, Adv. Protein Chem., 14, 1 (1959).

2. H. S. Frank and M. W. Evans, J. Chem. Phys., 13, 507 (1945).

3. H. Schneider, G. C. Kresheck, and H. A. Scheraga, J. Phys. Chem., 69, 1310 (1965).

4. R. L. Gustafson, R. L. Albright, J. Heisler, J. A. Lirio, and O. T. Reid, Jr., Ind. Eng. Chem. Prod. Res. Develop., 7, 107 (1968).

5. J. L. Gardon, in "Encyclopedia of Polymer Science and Technology", (H. F. Mark, N. G. Gaylord, and N. M. Bikales, eds.), Interscience Publ., New York, 1966.

6. R. L. Gustafson and J. A. Lirio, Ind. Eng. Chem. Prod. Res. Develop., 7, 116 (1968).

7. W. J. Weber, Jr. and J. C. Morris, J. Sanit. Eng. Div., Proc. Amer. Soc. Civ. Eng., 90, 79 (1964).

CHAPTER 11

PRINCIPLES OF COMPLEX FORMATION

Arthur E. Martell

Texas A&M University

I. INTRODUCTION

The purpose of this paper is to outline the principal factors that determine the affinities and interactions between metal ions and organic substances in solution, particularly in dilute solution. Knowledge of the degree of combination of metal ions with organic compounds is useful in understanding and interpreting the effects of metal ions on the reactions that organic compounds undergo. Therefore, examples of typical metal-catalyzed reactions of organic substances will be presented after the nature of the acid-base interactions and degree of formation of metal-organic complexes have been reviewed.

A. Types of Donor Groups.

The types of organic functional groups that may serve as effective electron donors in the binding of metal ions to organic compounds are:

aliphatic amino	$R-\ddot{N}H_2$, $RR'\ddot{N}H$, $RR'R''\ddot{N}$
aromatic amino	$\gtrdot N:$
carboxylate	$-COO^-$
enolate	$>C=\overset{\vert}{C}-O^-$
alkoxide	$R-O^-$

phenoxide	$\gt C-O^-$
mercaptide	$R-S^-$
phosphate	$-O-P(=O)(O^-)-O^-$
phosphonate	$-CH_2-P(=O)(O^-)-O^-$

Also, there are a number of donor groups that have relatively weak affinities for metal ions, and do not, by themselves, lead to the formation of stable complexes in solution. They may be important, however, as auxiliary donor groups in complexes that are held together by one or more of the more effective donor groups listed above. Examples of these weakly-coordinating groups are:

carbonyl	$\gt C=O$
ether	$\gt O$
ester	$R-C(=O)-OR'$
amide	$R-C(=O)-NHR'$
thioether	$R-S-R'$
hydroxyalkyl	$R-OH$

The presence of one or more of these functional groups in organic compounds renders them a more or less effective complexing agent for metal ions, with the affinities generally increasing with the number of donor groups present, as well as with the affinities of these groups for metal ions.

B. Metal Chelate Compounds.

When a metal ion is coordinated simultaneously to more than one donor group of an organic complexing agent, called an organic "ligand", the resulting compound contains a heterocyclic ring which includes the metal ion, and is called a metal chelate compound. A typical aqueous metal complex and an analogous metal chelate are illustrated schematically as follows:

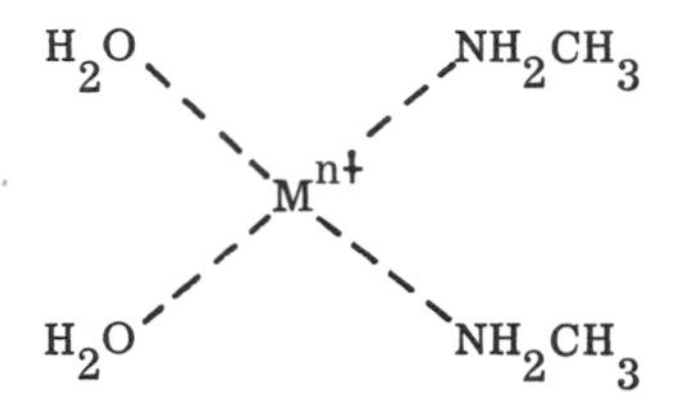

I. Diaquobismethylamine metal complex

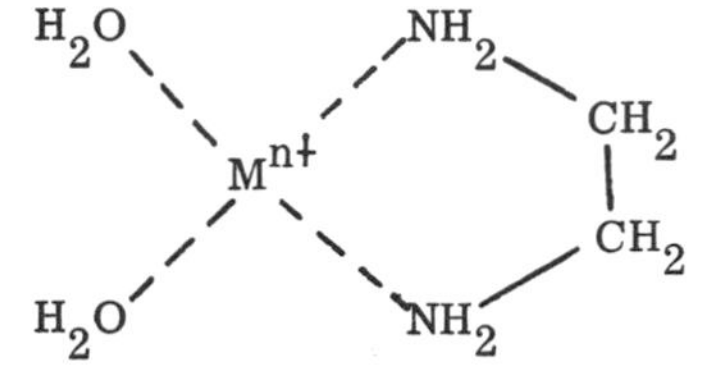

II. Diaquoethylenediamine metal chelate

In these compounds, if the amino donor groups have the same affinity for the metal ion in both the complex and the chelate compound, it would seem that the distinction between these two compounds is merely a formal one, and that the metal chelate would not be expected to have any special properties not found in the complex. While this general conclusion is certainly valid, it will be shown that when the metal chelate ring meets certain requirements, the chelate compound will differ markedly from the complex in stability, and in other ways.

C. Metal Ions.

The metal ions of significance for this paper will consist mainly of those frequently found in water and in biological systems, plus those that may be introduced in various ways as contaminants. The former group includes the alkaline earths (mainly Ca^{2+} and Mg^{2+}), Zn^{2+}, and the common transition metal ions: Cu^{2+}, Fe^{2+}, Fe^{3+}, Mn^{2+}, and VO^{2+}. To these may be added other common metal ions such as: Sr^{2+}, Ba^{2+}, Cd^{2+}, Co^{2+}, and Ni^{2+}, that are related closely to the biologically essential metals and are introduced frequently along with them into water as contaminants.

In dealing with metal ions in solutions it is understood that they are always fully coordinated (i.e., solvated) by water molecules.

Therefore, the combination of a metal ion with the donor group of an organic ligand, or any ligand, will involve the displacement of one or more water molecules, depending on the number of donor groups of the ligand that effectively simultaneously coordinate the metal ion. In most examples, to be illustrated below, the coordination and displacement of water molecules about metal ions will be understood and in most cases not specifically indicated.

The number of water molecules and the number of donor groups of a ligand, that may become coordinated to a metal ion, are dependent on the ionic dimension and the charge of the metal ion. This number is called the coordination number. The coordination number of the metal ions listed above is six; coordination numbers of larger and more highly-charged ions, such as the lanthanide and actinide ions, are considerably larger (8, 10, and 12 are common). The coordination numbers of metal ions in solution are somewhat variable and depend to some extent on the nature and size of the coordinated donor groups of the solvent and attached ligands.

II. METAL-ORGANIC COMPLEX EQUILIBRIA

The stabilities of metal complexes in solution are measured usually by the equilibrium constants for their formation. If L represents a ligand, with an indefinite number of donor groups and ionic charge, a series of complex formation reactions occurs which may be described in terms of the following chemical equilibria and equilibrium constants:

$$M^{n+} + L \rightleftharpoons ML^{n+} \qquad K_1 = \frac{[ML^{n+}]}{[M^{n+}][L]} \qquad (11\text{-}1)$$

$$ML^{n+} + L \rightleftharpoons ML_2^{n+} \qquad K_2 = \frac{[ML_2^{n+}]}{[ML^{n+}][L]} \qquad (11\text{-}2)$$

$$ML_2^{n+} + L \rightleftharpoons ML_3^{n+} \qquad K_3 = \frac{[ML_3^{n+}]}{[ML_2^{n+}][L]} \qquad (11\text{-}3)$$

$$ML_{i-1}^{n+} + L \rightleftharpoons ML_i^{n+} \qquad K_i = \frac{[ML_i^{n+}]}{[ML_{i-1}^{n+}][L]} \qquad (11\text{-}4)$$

$$M^{n+} + iL \rightleftharpoons ML_i^{n+} \qquad \beta_i = \frac{[ML_i^{n+}]}{[M^{n+}][L]^i} \qquad (11\text{-}5)$$

Considerable research has occurred over the past twenty years on the determination of the stability constants, K_i and β_i, of metal ion-organic complexes in aqueous solution. At the present time, quantitative data exist for the combination of about 1500 organic compounds with a wide variety of metal ions.

A. The Chelate Effect.

The high stability of chelate compounds is illustrated by reactions such as the following

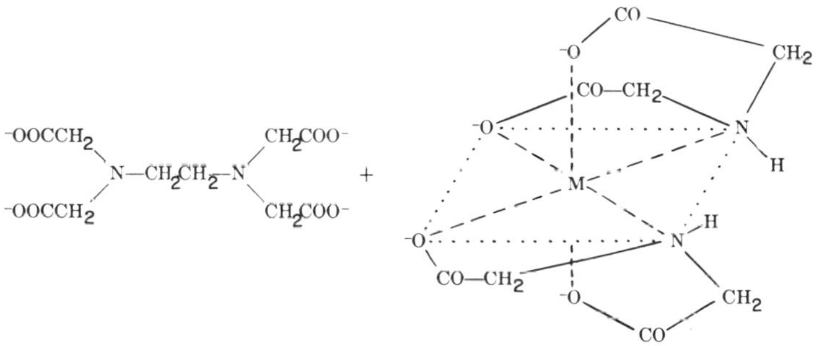

$$({}^-OOCCH_2)_2N{-}CH_2CH_2{-}N(CH_2COO^-)_2 \; + $$

Metal chelate of iminodiacetete ligand

$$\rightleftharpoons \qquad + \; 2\,HN(CH_2COO^-)_2 \qquad (11\text{-}6)$$

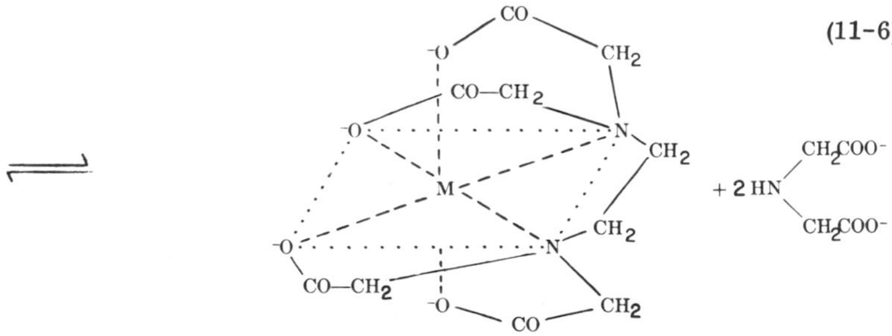

Metal chelate of ethylenediaminetetraacetate ligand

in which stabilities are compared for two similar chelates or complexes containing the same coordinated groups, but differing in the number of metal chelate rings present. The stability constant of the Ca(II)-EDTA chelate is about 10^5 greater than that of the corresponding 2:1 chelate of iminodiacetic acid, so that the equilibrium constant of the preceding replacement reaction, when M = Ca^{2+}, is 10^5.

Similar effects have been observed (1) for a large number of reactions that involve an increase in the number of metal chelate rings present. The stabilizing effect of metal chelate ring formation in solution has long been accepted as a general principle in coordination chemistry.

A partial explanation of this effect, advanced by Schwarzenbach (2), is related to the entropies of the complexes in solution, and may be visualized as related to the number of molecules or ions formed in solution. Since the solutions are dilute and solute molecules have freedom of motion, an increase in the number of particles in solution results in the entropy increase which is responsible for driving the reaction forward. Thus, for a reaction involving the formation of an additional metal chelate ring, of the type:

$$MA_2^{n+} + A\text{-}A \longrightarrow M\langle{}^{A}_{A}\rangle + 2A \quad (11\text{-}7)$$

The entropy change may be expressed as the difference in the solution entropies of three moles of product and two moles of reactant. The entropy increase is:

$$\Delta S^o = \bar{S}^o_{M(A\text{-}A)} + 2\bar{S}^o_A - \bar{S}^o_{MA_2} - \bar{S}^o_{A\text{-}A} \quad (11\text{-}8)$$

$$= R\ln 55.5$$

For the formation of n chelate rings, the entropy increase is

$$\Delta S^o = nR\ln 55.5$$

$$= 7.9\ n \text{ entropy units (at } 25^oC.) \quad (11\text{-}9)$$

This increase is roughly equivalent to a factor of 100 in the stability constant, for each metal chelate ring formed.

This effect is also inherent in the equilibrium constant expressions for the formation of metal complexes and chelates, which predict increasing disparity between the stabilities of complexes and chelates as the solutions become more and more dilute. This principle is illustrated in Table 1 for three hypothetical chelates which differ in the number of chelate rings present, but not in the number of groups coordinated. These examples assume that ΔH^o of complex formation is the same for all these complexes. For purposes of comparison, an arbitrary value of 10^{24} is selected for the stability constant of the metal chelate formed from the quadridentate ligand. The translational entropy effect given above predicts approximately 10^{22} for the complex with one less chelate ring and about 10^{18} for the complex with no chelate rings.

The data in Table 1 clearly show that the chelate stability is maintained in a very dilute solution relative to the corresponding analogous complexes, and that the stabilizing effect increases rapidly with the number of metal chelate rings present. From these examples, it is seen that even complexes with quite high stability constants will dissociate extensively in dilute solution, while the corresponding chelate compounds are maintained at high dilution without appreciable dissociation. Nature takes advantage of this principle through the use of chelating donor groups in metalloenzymes. Without metal chelate ring formation, such enzymes could not remain in the active associated form at the very dilute concentrations at which they must function.

The stability constant expression itself, however, relates the formation of the complex or chelate to the free metal ion, as indicated by the following reaction:

$$M(H_2O)_x^{n+} + L^{-m} \rightleftharpoons ML(H_2O)_{x-y}^{(n-m)+} \qquad (11\text{-}10)$$

where the ligand L has y available donor groups. Here the principle of metal chelate and complex formation is similar to that observed above for the entropy changes in chelate displacement reactions, since the donor groups of the ligand replace water molecules that are released to the solvent, resulting in an increase in entropy. To this is added still another stabilizing factor resulting from the combination of positive and negative ions. The charges of both metal ion and negative donor groups of the ligand place constraints on the surrounding solvent molecules that are largely overcome when the chelate is formed, thus giving rise to an entropy increase and stabilization of the metal chelate or complex thus formed.

TABLE 1

Comparison of Degrees of Dissociation of Complexes and Chelates in Dilute Solution

Ligand	*Complex*	*No. of Chelate Rings*	*Quotient*	*Value*	*1.0 M Complexes*		*1.0 × 10⁻³ M Complexes*	
					Free [M]	*% Dissociation*	*Free [M]*	*% Dissociation*
L	L L M L L	0	$\frac{[ML_4]}{[M][L]^4}$	10^{18}	1×10^{-5}	1×10^{-3}	1×10^{-4}	10
L—L	L L M L L	2	$\frac{[M(L{-}L)_2]}{[M][L{-}L]^2}$	10^{22}	3×10^{-8}	3×10^{-8}	5×10^{-9}	5×10^{-4}
L—L—L—L	L L M L L	4	$\frac{[M(L{-}L{-}L{-}L)]}{[M][L{-}L{-}L{-}L]}$	10^{24}	1×10^{-12}	10^{-10}	3×10^{-14}	3×10^{-9}

TABLE 2

Structural Analogs of NTA and EDTA

(t = 25°C.; μ = 0.10)

η^{a}		Ligand	Log Formation Constants of 1:1 Chelates			
			Ca^{+2}	Cu^{+2}	La^{+3}	Th^{+4}
0	NTA	O N—O O	6.57	12.96	10.47	12.4
1	EDTA	O O N—N O O	10.4	18.7	15.2	23.2
2	DTPA	O O O N—N—N O O	10.7	21.1	19.5	$\tilde{>}$27
3	TTHA	O O O O N—N—N—N O O	9.9	20.3	23.1	≫27
4	TPHA	O O O O O N—N—N—N—N O O	~9.0	~20	>27	≫27

$$^{-}OOCCH_2\left(\begin{array}{c}CH_2COO^{-}\\ |\\ N-CH_2CH_2\end{array}\right)_{\eta}N\begin{array}{l}CH_2COO^{-}\\ \\ CH_2COO^{-}\end{array}$$

XVIII. *Analogs of NTA and EDTA*

An interesting series of synthetic chelating agents that illustrate these principles are the analogs of NTA and EDTA, listed in Table 2. Inspection of these stability data (1) shows that as the number of donor groups and negative charges of the ligands increase, the stabilities increase up to a point and then level off or decrease slightly. This behavior is characteristic of the increasing number of coordinate bonds formed, which reach a limit when the number of donor groups matches the characteristic coordination number of the metal ion. An increase in donor groups beyond this point does not increase the number of groups coordinated or the number of metal chelate rings formed. The first two metal ions listed, Ca^{2+} and Cu^{2+}, have the lower coordination numbers (i.e., 6), while the coordination numbers of the La^{3+} and Th^{4+} ions are believed to be much higher, as is indicated by the fact that increasing the number of donor groups of the ligand beyond 6 greatly increases the stabilities of the chelates formed. Studies of this type provide considerable insight into the structures of the corresponding metal chelates, and the probable number of coordinate bonds formed in solution.

B. Size of Chelate Ring.

From the above discussion, it is seen that the number of donor groups of the ligand controls the number of water molecules displaced from the metal ion, and the amount of charge neutralization that occurs. Since these effects are reflected in the entropy of reaction, the determination of entropy changes on chelate formation should give us an indication of the number of groups coordinated and the number of metal chelate rings formed. Some new data (3), illustrating these principles, are presented in Table 3, which shows heats and entropies of reaction for the formation of some representative metal chelates of homologs of EDTA in which the number of carbon atoms between the nitrogen atoms of the ligand increases from two to five. Thus, it is seen that the main differences in the metal chelates formed are due to an increase in the size of the central metal chelate ring from five to eight atoms.

While space does not permit a complete discussion of the implications of these data, two relatively significant facts can be indicated. Firstly, the constancy of the entropy increase for a given metal ion in the formation of most of the chelates indicates that the structures of the chelates in solution are very similar, involving the same number of donor groups and the same number of metal chelate rings as the size of the central chelate ring increases (only for the eight-membered rings is there a decrease in the entropy of reaction showing a lower degree of coordination and consequent inability of the larger ring to form). Second, considerable change is observed in the values of ΔH, which become less favorable (more endothermic) as the size of the ring increases. Here is an illustration of a general principle of metal chelate ring formation that the smaller chelate ring is the more stable. This stability variation with ring size is manifested to a large extent in the heat of formation of metal chelates in solution, and must be inherent in the heats of formation of the ligands, in which strongly polar donor groups are spaced close together in the organic structure.

TABLE 3

Thermodynamics of Formation of Metal Chelates of EDTA Homologs

Ligand	Mg^{2+}	Ca^{2+}	La^{3+}	Ni^{2+}	Cu^{2+}	Zn^{2+}	Thermodynamic Quantity
$(^{-}OOCCH_2)_2N(CH_2)_2N(CH_2COO^{-})_2$	8.8	10.8	15.6	18.7	18.6	16.3	Log K
	3.5	-6.6	-2.8	-7.6	-8.2	-4.8	ΔH^o
	51	27	61	59	58	59	ΔS^o
$(^{-}OOCCH_2)_2N(CH_2)_3N(CH_2COO^{-})_2$	6.3	7.4	11.3	18.3	19.1	15.3	Log K
	9.1	-1.7	3.8	-6.8	-7.7	-2.3	ΔH^o
	59	27	64	60	60	62	ΔS^o
$(^{-}OOCCH_2)_2N(CH_2)_4N(CH_2COO^{-})_2$	6.3	5.8	9.2	17.5	17.5	15.2	Log K
	8.5	0.9	0.1	-7.0	-6.5	-3.5	ΔH^o
	54	30	42	56	57	57	ΔS^o
$(^{-}OOCCH_2)_2N(CH_2)_5N(CH_2COO^{-})_2$	4.8	4.6	--	13.6	--	12.8	Log K
				-8.5		-4.0	ΔH^o
				34		44	ΔS^o

TABLE 4

Factors Influencing Solution Stabilities of Complexes

Enthalpy Effects	Entropy Effects
Variation of bond strength with electronegativities of metal ions and ligand donor atoms.	Number of chelate rings.
Ligand field effects.	Size of the chelate ring.
Steric and electrostatic repulsions between ligand donor groups in the complex.	Changes of solvation on complex formation.
Enthalpy effects related to the conformation of the uncoordinated ligand.	Arrangement of chelate rings.
Other coulombic forces involved in chelate ring formation.	Entropy variations in uncoordinated ligands.
	Effects resulting from differences in configurational entropies of the ligand in complex compounds.

C. Summary of Factors Influencing Stabilities of Complexes in Solution.

Space does not permit a more detailed discussion of all the factors involved in the formation of metal complexes and chelates in aqueous solution. In Table 4, there is a list of the more important properties of the metal ions, organic ligands, and the resulting complexes, that determine their stabilities and degrees of formation in solution. For a more complete discussion of the principles of complex and chelate formation in solution, the reader is referred to a review of this subject by the author (4).

D. Coordination Interactions of Metal Complexes in Solution.

Once a stable metal chelate or complex is formed, it undergoes very little or no further interaction when the number of donor groups provided matches the coordination number of the metal ion. This is not the case, however, when some of the coordination positions of the metal ion are not combined with the ligand and remain in a relatively active "aquo" form. Under these conditions, several types of reactions have been observed. The main possibilities are:

(a) Combination of one of the ligand donor groups with more than one metal ion. Such groups act as bridges binding two metal ions together, thus leading to the formation of "polynuclear complexes".

(b) Hydrolysis of one or more of the water molecules still coordinated to a metal ion will produce the hydroxide ion as an additional ligand. Frequently the hydroxide ion acts as a bridging group for the formation of polynuclear complexes. This process is known as "olation".

(c) Hydroxo metal complexes frequently disproportionate with the precipitation of the metal hydroxide and the formation of more stable complexes having a greater number of ligands coordinated to the metal ion.

(d) A secondary ligand may be added to satisfy the coordination requirement of the metal ion, leading to the formation of mixed-ligand complexes containing two or more different ligand molecules or ions.

When the coordination number of a metal ion is less than the number of donor groups on a ligand, or when some of the donor groups of a ligand are sterically oriented so as to be unavailable to the coordinated metal ion, the ligand may combine simultaneously with more than one metal ion, leading to a different type of polynuclear complex, in which the whole ligand molecule acts as a bridging group. When the two metal ions are different, interesting "mixed-metal" polynuclear complexes may be formed.

III. CATALYTIC EFFECTS OF METAL IONS

Now that the general principles governing organic complex formation in solution have been reviewed, it is important to consider the reactions of the organic ligand that may occur as the result of the influence of the coordinated metal ion. When the interaction of a metal ion and an organic ligand is formulated in the following way:

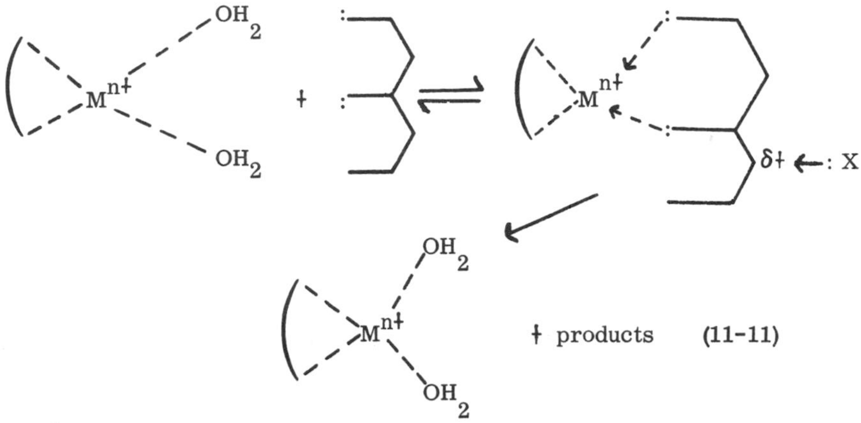

M^{n+} + products (11-11)

it can be seen that the primary effect of the metal ion must always involve the shifting of electrons toward the metal ion. The donor atoms themselves and the adjacent portions of the organic molecule are the most strongly polarized. The effect of this interaction with the metal ion is to make the organic molecule more reactive toward electron-rich (nucleophilic) reagents. Such interactions may lead to solvolysis, rearrangement of the organic molecule, and even decomposition reactions resulting from the change of electron distribution.

When the metal ion is capable of changing its valence state, it may actually remove one or more electrons from the organic molecule, thus effecting the first step in the oxidation of the substrate. Usually this first step is the most difficult, so that the metal acts as a catalyst for speeding up the rate-determining step. Subsequent steps in the oxidation of the compound usually occur relatively rapidly so that changes occur in which more than one electron is removed. When an oxidant is present that can reoxidize the metal to the initial state, a catalytic system is set up whereby a trace of metal ion may catalyze the oxidation of a large amount of an organic substrate.

In the following discussion of the catalytic effects of metal ions, both the acid base (i.e., coordination-initiated) reactions and the redox (electron-transfer initiated) reactions will be considered.

A. Metal Ions as Acid Catalysts.

As an electron acceptor (i.e., a Lewis acid), the metal ion reacts with electron donors (complexing agents) to form coordination compounds in which the basic nature of the ligand is reduced greatly. When the electron shift that occurs is sufficiently strong, the metal may lend enough acid character to the ligand to render it reactive towards other bases that may be present.

1. Solvolysis reactions.

An excellent example of this type of reaction is catalysis of the hydrolysis of amino acid esters, amides, and peptides by metal ions. A typical reaction would be:

$$H_2N{-}CH_2{-}C(=O){-}OC_2H_5 + M(H_2O)_x^{n+} \rightleftharpoons [H_2N{-}CH_2{-}C(OC_2H_5){=}O \cdots M^{n+}(H_2O)_{x-2}] \xrightarrow{^{-}OH} [H_2N{-}CH_2{-}C(OC_2H_5)(OH){-}O^{-} \cdots M^{n+}(H_2O)_{x-2}] \rightleftharpoons [H_2N{-}CH_2{-}C(=O){-}O^{-} \cdots M^{n+}(H_2O)_{x-2}] + C_2H_5OH \qquad (11\text{-}12)$$

Metal ions such as transition metals bind strongly to the amino group and rather weakly to the carbonyl oxygen, which is nevertheless polarized by an electron shift toward the metal ion, greatly increasing the acceptor activity of the carbonyl carbon atom. Accelerated attack of the carbonyl carbon by hydroxide ion permits the shift of a negative charge to the coordinated oxygen, greatly increasing the stability of the complex. A further electron shift in the coordinated ligand then releases the proton and the ethylate group (as ethanol), completing the hydrolysis. Thus, it is clearly seen that the donor-acceptor reaction between the metal ion and the ligand catalyze the hydrolysis reaction. The driving force for this reaction is the formation of a more stable metal chelate compound.

Examples of this type of metal-catalyzed hydrolysis have been described (5, 6). Similar reaction mechanisms have been advanced for the hydrolysis of peptides (7).

2. Decarboxylation.

Another interesting example of the catalytic effect of metal ions on organic compounds is the decarboxylation of alpha keto acids in which there is a second carboxyl group beta to the carbonyl group, such as the oxaloacetate ion (8, 9, 10, 11, 12). The reaction mechanism for decarboxylation may be described by the following reaction sequence:

Oxaloacetate

$\longrightarrow$ CH_3COCOO^- $+ M^{n+}$

$Cu^{II} > Zn^{II} > Ni^{II} > Co^{II} > Mn^{II} > Ca^{II}$

$-H^+$

H^+

Inactive forms

(11-13)

It is interesting to note that the oxaloacetate ion may form a very stable metal chelate involving a carboxylate and an enolized β-keto group, which is inactive with respect to decarboxylation. When the adjacent carboxylate and alpha keto group are simultaneously coordinated to the metal ion, leaving the beta-carboxylate free, an electron shift toward the metal ion breaks the carbon-carbon bond of the beta-carboxylate, and carbon dioxide is released, whilst a more stable chelate is formed.

These reactions have nearly exact counterparts in enzymic systems, and are good models for metallo-enzyme catalyzed decarboxylations.

3. Transamination.

Interesting examples of the metal ion catalysis of rearrangement reactions of organic substances are the transamination reactions that occur in the Schiff bases of alpha amino acids and pyridoxal (vitamin B_6). The following reaction scheme compares the metal ion and hydrogen ion catalysis of rearrangement of a Schiff base of an alpha amino acid and pyridoxal to the Schiff base of the corresponding alpha keto acid and pyridoxamine (13). The influence of the metal ion increases the rate of the proton-catalyzed reaction, so that the metal chelate branch of the mechanism is much more important when such metal ions as Cu^{2+} and Zn^{2+} are present. It is seen that the catalytic effect of the metal ion may be explained by an electron shift toward the metal ion, which releases a proton and forms an intermediate carbanion. Return of a proton to a different part of the molecule accomplishes the rearrangement reaction.

Transition state

Transition state

(11-14)

As the above reaction scheme indicates, a system containing a relatively small amount of pyridoxal and a catalytically active metal ion may equilibrate relatively large amounts of amino acids and keto acids, bringing them to an equilibrium state corresponding to their relative activities in the medium under consideration. It should be noted that metal ions are not required for the activation of transaminase itself, but are sometimes essential for the activation of other enzymes for which pyridoxal is a cofactor.

B. Redox Mechanisms.

1. Oxidation of oxalate ion.

A classic example of the oxidation of an organic substrate by a metal ion is the reaction of the Mn(III) ion with oxalate (14,15,16):

OXALATE

$$\text{(oxalate)}Mn^{3+} \longrightarrow \text{(oxalate}^{\bullet}\text{)}Mn^{2+} \longrightarrow Mn^{2+} + {}^{\bullet}O{-}CO{-}CO{-}O^{-} \xrightarrow{Mn^{3+}} \text{(oxalate}^{\bullet}\text{)}Mn^{3+} \longrightarrow Mn^{2+} + 2CO_2$$

(11-15)

The suggested mechanism involves a single electron transfer to the metal ion forming a free radical ligand intermediate which is a poorer donor than the original ligand, and, therefore, dissociates. Reaction of the intermediate with a second Mn(III) ion completes the oxidation of the organic molecule to carbon dioxide.

2. Ascorbic acid oxidation.

Another example of this type of reaction is given in a recent report by Taqui Khan and Martell (17) on the oxidation of ascorbic acid to dehydroascorbic acid by the Fe(III) ion. The following mechanism, involving two successive one-electron transfers, has been proposed:

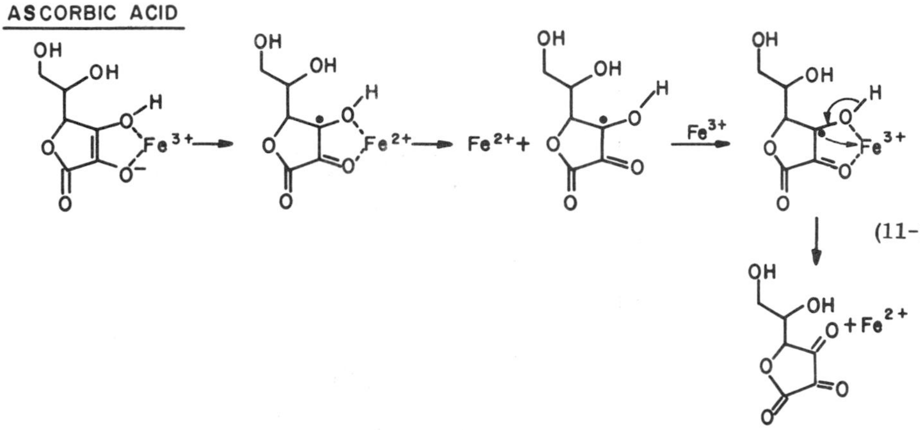

ASCORBIC ACID

(11-16)

The final product has two less electrons than the original ligand and is a very poor electron donor, so that the metal chelate dissociates in solution. When molecular oxygen is present in such a system, the metal ion (e.g., Fe^{2+}, Cu^{+}, V^{3+}) is reoxidized to the higher valence state, so that a small amount of metal ion may catalyze the oxidation of a relatively large amount of organic substrate. The following is the mechanism proposed (18) for the metal ion-catalyzed oxidation of ascorbic acid by dissolved molecular oxygen. In this case a substrate-metal-oxygen mixed ligand complex is proposed because the reaction rate was found to increase linearly with oxygen concentration.

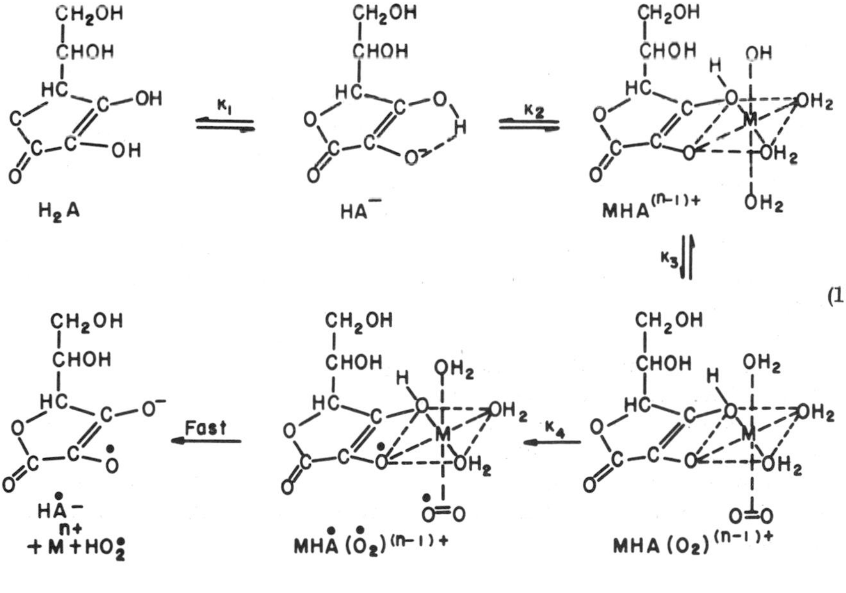

(11-17)

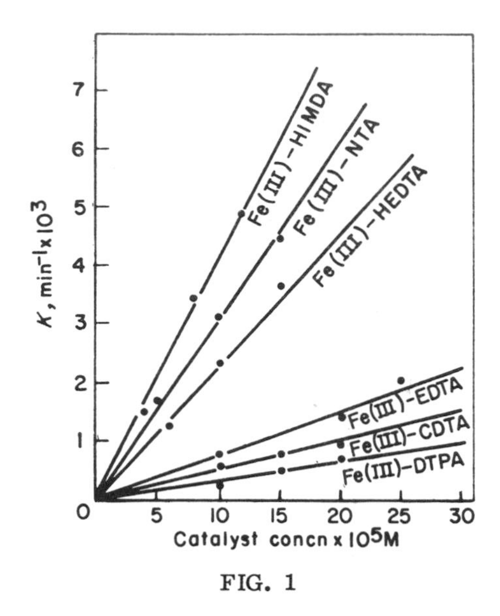

FIG. 1

Dependence of specific rate constant k on metal chelate concentration for the Fe(III) chelate-catalyzed oxidation of ascorbic acid at 25°C., ionic strength = 0.10M with KNO_3.

In the above mechanism, the metal ion may be considered as providing a pathway for the flow of electrons from the substrate to the oxidant.

Although such chelating agents as EDTA have been suggested and used for the protection of ascorbic acid from metal ion-catalyzed oxidation, they also catalyze the oxidation of this substrate, although at a much lower rate than does the metal ion. The linear increase in the first order rate of oxidation of ascorbic acid by molecular oxygen as the metal chelate concentration is increased, illustrated in Figure 1 (19), shows that the activated complex involved in the rate-determining step contains both the ascorbate and the chelating ligand coordinated to the metal ion. In this case, the reaction was found to be independent of the concentration of molecular oxygen, so that the activated complex does not contain a coordinated oxygen molecule. This is reasonable in view of the fact that the coordination sphere of the metal ion is occupied by other ligands which exclude the weakly-coordinated oxygen molecule. In these systems, the catalytic action must be achieved by dissociation of the lower-valence metal chelate followed by reoxidation in the body of the solution.

The relative catalytic effects of the metal chelates illustrated in Figure 1 decrease as the stabilities of the metal chelates increase. This is apparently a reflection of the tendency of the ligands to stabilize the higher valence state of the metal ion and thus inhibit electron transfer from the substrate. This tendency increases with the stability of the chelate formed.

3. Oxygen insertion reactions.

An important and sometimes difficult reaction to achieve in the removal of organic industrial wastes from water is the oxidation of phenol and phenol derivatives. A way to do this chemically by metal ion catalysis is illustrated by a system first described by Udenfriend et al. (20, 21, 22). Metal ions such as Fe(II), Mn(II), Cu(II), and Co(II) are combined with a chelating agent as a carrier, and undergo initial reaction with oxygen to form a more reactive free radical oxidant, which then substitutes on the aromatic ring. A typical reaction mechanism for the system containing Fe^{2+}, EDTA, salicylic acid as substrate, and ascorbic acid as reductant, is illustrated in Equation (11-18)(23).

The reaction sequence is pictured as a free-radical substitution of the aromatic ring by $HO_2^{\bullet}$, generated by a single electron transfer from the Fe(II) chelate to oxygen. The Fe(II)-chelate is regenerated by the two-electron reducing agent. Thus, we have the first step in the conversion of a phenol to the corresponding catechol derivative. This is the most difficult step in the oxidative chain that results in the destruction of the aromatic ring.

$$Fe^{II}\text{-EDTA} + O_2 + H_2O \longrightarrow Fe^{III}\text{-EDTA} + OH^- + HO_2^{\bullet}$$

COO⁻⁻ / OH (salicylate) $\xrightarrow{HO_2^{\bullet}}$ COO⁻ / O / • H $\xrightarrow{O_2}$ COO⁻ / O / • O–OH $\xrightarrow{Fe^{II}\text{-EDTA}}$ COO⁻ / OH / OH

$(+ H_2O_2)$ $(+OH^-)$

Regeneration of Fe (II)

$$Fe^{III}\text{-EDTA} + H_2A + OH^- \longrightarrow Fe^{II}\text{-EDTA} + HA^{\bullet} + H_2O$$
$$Fe^{III}\text{-EDTA} + HA^{\bullet} + OH^- \longrightarrow Fe^{II}\text{-EDTA} + A + H_2O$$
$$Fe^{III}\text{-EDTA} + H_2O_2 \longrightarrow Fe^{II}\text{-EDTA} + OH^{\bullet} + OH^-$$

(11-18)

(where H_2A is a two-electron reductant such as ascorbic acid).

C. Amine Oxidase Models.

An interesting example of the metal-catalyzed oxidation of an amine, which provides a model of amine oxidase enzymic systems, is illustrated in the following reaction sequence:

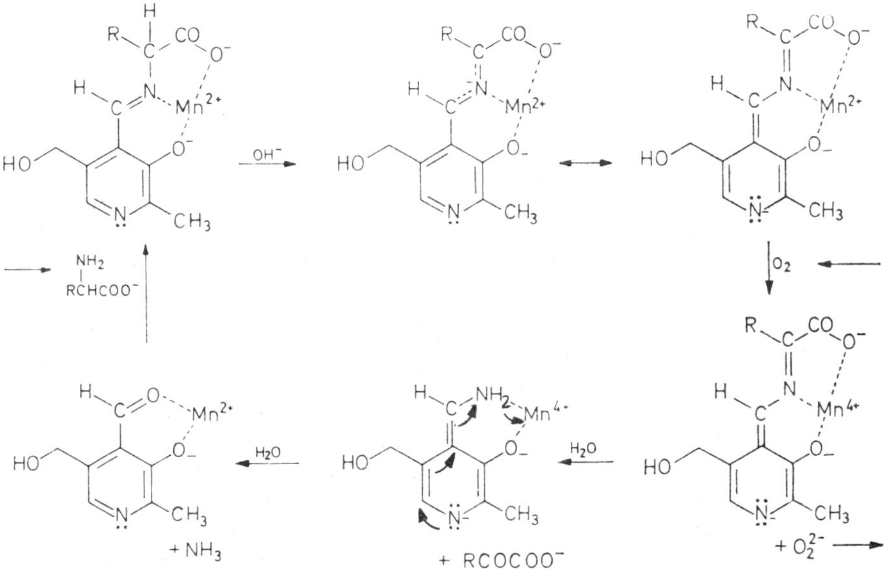

Net reaction

$$RCH(NH_2)COO^- + O_2 + H_2O \longrightarrow RCOCOO^- + NH_3 + H_2O_2$$

$$Mn^{II} > Co^{II} > Cu^{II} >> Ni^{II} \sim 0 \qquad (11\text{-}19)$$

This mechanism is a modification of the suggestion of Hamilton and Revesz (24) for the Mn(II)-catalyzed oxidative deamination of amino acids in the presence of pyridoxal. Initial combination of Mn(II) with the Schiff base of pyridoxal and an amino acid results in the formation of a metal ion-stabilized carbanion intermediate similar to the one already illustrated above for the transamination reaction. An internal electron shift results in conversion of the amino acid moiety to a keto acid residue of a pyridoxamine Schiff base, which can dissociate then to the keto acid and the metal chelate of pyridoxamine. At some point in the reaction sequence, the ligand is oxidized by molecular oxygen, reacting through the metal ion, regenerating the pyridoxal which may then again combine with a new amino acid to start the reaction sequence anew.

Recent work on this system (25) indicates the relative catalytic activities of metal ions to be: $Mn^{2+} > Co^{2+} > Cu^{2+} \gg Ni^{2+} \sim 0$. Thus, it is seen that the function of the metal ion cannot involve merely the transamination step (for which Cu^{2+} is the best catalyst), but also the ability of the metal ion to undergo oxidative-reductive transformations.

The above are only a few representative examples of the many metal-catalyzed reactions of coordinated organic ligands described in the literature. For a more complete treatment the reader is referred to a review of this type of catalysis by the author (26).

IV. SUMMARY AND CONCLUSIONS

1. The principles involved in the combination of metal ions with organic ligands have been discussed, and the factors influencing the stabilities of metal-organic complexes in solution have been described.
2. In the case of metal chelate compounds, the high stabilities of metal chelate compounds in solution have been explained on the basis of the entropies and enthalpies of chelate formation.
3. The acid-base reactivities of metal complexes and metal chelates in solution have been explained on the basis of the degree to which the donor groups of the ligand match the coordination requirements of the metal ions.
4. The Lewis acid character of metal ions provides an understanding of their catalytic effects on the reactions of organic ligands with nucleophilic reagents, and of rearrangements of organic ligands resulting from the distortion of electron distribution by coordinated metal ions.
5. The ways in which metal ions oxidize organic compounds by removal of electrons have been described. This concept is used as the basis for the reaction mechanisms of catalytic systems in which the metal is reoxidized continuously by a secondary oxidant.
6. Metal ions that undergo valence change in solution are catalytic agents for the insertion of oxygen into refractory aromatic compounds such as phenol and phenol derivatives, thus providing the first step of oxidative reaction sequences resulting in the destruction of aromatic rings.

REFERENCES

1. L. G. Sillen and A. E. Martell, Stability Constants, The Chemical Society, London, 1964.

2. G. Schwarzenbach, Helv. Chim. Acta, 35, 2344 (1952).

3. G. Anderegg, Proceedings of the 8th International Conference on Coordination Chemistry, Vienna, September 7-11, 1964, p. 34.

4. A. E. Martell, The Chelate Effect, in Advances in Chemistry Series, No. 62, Amer. Chem. Soc., 1966, p. 272.

5. H. Kroll, J. Amer. Chem. Soc., 74, 2036 (1952).

6. M. L. Bender and B. W. Turnquest, J. Amer. Chem. Soc., 77, 4271 (1955).

7. L. Meriwether and F. H. Westheimer, J. Amer. Chem. Soc., 78, 5119 (1956).

8. R. W. Hay, J. Chem. Educ., 42, 413 (1965).

9. A. Kornberg, S. Ochoa, and A. H. Mehler, J. Biol. Chem., 174, 159 (1947).

10. J. F. Speck, J. Biol. Chem., 178, 315 (1948).

11. R. W. Hay and E. Gelles, J. Chem. Soc., 3673 (1958).

12. E. Gelles and A. Salama, J. Chem. Soc., 3683, 3689 (1958).

13. A. E. Martell, Proceedings of Symposium on Chemical and Biological Aspects of Pyridoxal Catalysis, Rome, 1962, Pergamon Press, New York, 1963, p. 13.

14. H. Taube, J. Amer. Chem. Soc., 69, 1418 (1947).

15. H. Taube, J. Amer. Chem. Soc., 70, 1216 (1948).

16. F. R. Duke, J. Amer. Chem. Soc., 69, 2885 (1947).

17. M. M. Taqui Khan and A. E. Martell, J. Amer. Chem. Soc., 89, 5585 (1967).

18. M. M. Taqui Khan and A. E. Martell, J. Amer. Chem. Soc., 89, 4176 (1967).

19. M. M. Taqui Khan and A. E. Martell, J. Amer. Chem. Soc., 89, 7104 (1967).

20. S. Udenfriend, C. T. Clark, J. Axelrod, and B. B. Brodie, Fed. Proc., 11, 301 (1952).

21. S. Udenfriend, C. T. Clark, J. Axelrod, and B. B. Brodie, J. Biochem., 208, 731, 741 (1954).

22. J. Axelrod, S. Udenfriend, and B. B. Brodie, J. Pharmocol. Expt. Therap., III, 176 (1954).

23. R. R. Grinstead, J. Amer. Chem. Soc., 82, 3472 (1960).

24. G. A. Hamilton and A. Revesz, J. Amer. Chem. Soc., 88, 2069 (1966).

25. M. Haruta and A. E. Martell, to be published.

26. A. E. Martell, Pure Appl. Chem., 17, 129 (1968).

CHAPTER 12

METAL-ORGANIC INTERACTIONS IN THE MARINE ENVIRONMENT

Alvin Siegel

Southampton College of Long Island University

I. INTRODUCTION

The general nature of organic matter in the marine and fresh water environments has been previously reviewed (1-4). Specific areas of organic type and function which have been discussed in the literature include: nitrogen compounds (5), carbohydrates (6), lipids (7), particulate (8), volatile (9), surface active (10), and phosphorus-carbon organics (11). A good starting point for trace metals is Hogdahl's review (12) and for coordination chemistry in the oceans, Martin's work (13).

The macroscopic picture of the organic chemistry of the oceans is rather simple, although on a microscopic level it must be very complicated. There is an upper euphotic zone of photosynthesis of some 500m in depth where production, release, and recycling of organic matter occurs. In this zone the organic content is both geo graphically and seasonally variable, ranging from 0.00 to 0.13 mg/l and 0.4 to 4.0 mg/l, for the particulate and dissolved fractions, respectively (2). Below this zone the vertical and geographical distribution of these organic fractions is remarkably homogeneous. The particulate organic carbon varies from 3 to 10 μg C/l and the dissolved organic carbon ranges from 0.35 to 0.70 mg C/l (8). Further simplifying the picture is the apparent unreactive nature of the dissolved organic carbon at depth (14).

To the list of those organic species known to be present in the oceans, every organic metabolite and its degradation products should be added. The chief question, of course, is the concentration and functionality of the organic compound. What effect can its presence produce in the biosphere and in the inorganic realm? Needless to say some compounds may be extremely important even at low concentrations -- vitamins, antibiotics, and enzymes, for example.

There has been considerable speculation on the importance in the ocean of the group of organic materials which possess the ability of associating with metal ions in solution, whether by complexing, chelating (ring formation), or peptizing. This interest has arisen primarily because of certain questions, which, as of this date, have not yet been answered:

(a). The state of trace metals in the oceans. The concentration of iron in sea water is greater by some 8 or 9 orders of magnitude than would be predicted on the basis of the solubility product of ferric hydroxide (10^{-36}) and the pH value of sea water (pH = 8.0).

(b). The interaction of chelation phenomena with phytoplankton productivity. In the laboratory it is often necessary to add a known chelator to a culture medium which, in all (?) measurable respects, is essentially sea water if growth of phytoplankton is to occur.

An excellent introduction to this area is provided by the papers of Saunders (15) and Johnston (16,17). Johnston concluded that chelation was an important mechanism in determining trace metal availability to phytoplankton, particularly in the case of iron. If the organic solubilizers possessed chelating powers of the order of ethylenediaminetetra-acetic acid (EDTA) they need constitute only 0.01 to 0.1% of the naturally-occurring organic pool. This figure was based on the need to provide iron at the 10^{-14}M level to meet the average instantaneous requirement for iron. The competitive ionic situation that exists in sea water was indicated. The great quantities of magnesium and calcium present in sea water may associate with any chelator, making it unavailable to the trace metals. The ratio of any trace metal complex to magnesium complex could be calculated from Equation 12-1:

$$(ML)/(MgL) = (M)/(Mg) \times K_{ML}/K_{MgL} \qquad (12\text{-}1)$$

where (ML) is the trace metal complex concentration, (MgL) is the magnesium complex concentration, (M) is the concentration of free trace metal, (Mg) is the concentration of free magnesium, K_{ML} is the equilibrium constant for the trace metal complex formation, and K_{MgL} is the equilibrium constant for the magnesium complex formation. For most trace metals, considering their low concentrations relative to magnesium and calcium, the stability constant ratio must favor the trace metal by at least 6 orders of magnitude in order to have any appreciable bonding of the trace metal.

Duursma (18) introduced a second major factor by considering the importance of competition among the trace metals themselves for the available organic ligands. Using EDTA as an example of a typical chelator, it was calculated that, given the concentrations of the metal ions in sea water, virtually all of the trace metals would be chelated at an EDTA concentration of 1×10^{-6}M (0.12 mg C as EDTA/l) whereas at a concentration lower than 1.7×10^{-7}M (0.02 mg C as EDTA/l) all the trace metals would be completely free, except for Fe^{3+} and Cr^{3+}. That "iron is completely regulating the chelation of all other metals at low concentrations due to its relatively high stability constant" was stated by Duursma.

While this conclusion is valid as far as the defined system is concerned, i.e., EDTA as a chelator and the trace metals at their known total concentrations, this picture of the actual competition between the trace metals for available ligands may be somewhat distorted for several reasons:

(a) The use of the figure of 1.79×10^{-7}M as the concentration for Fe^{3+}ion may be far too large since the bulk of the iron may be unavailable to the EDTA. Duursma and Sevenhuysen (19) have shown that freshly added Fe^{3+}ion in sea water is reactive to EDTA but the actual state and reactivity of "aged" iron is unknown (17). Barber (20) found that freshly added ferric chloride greatly increased the rate of growth of phytoplankton in contrast to the "aged" iron originally present in the sea water. Thus the "aged" iron may be present in some unavailable form, possibly colloidal, in which case it may react with organic chelators only to a slight degree. The amount of iron which reacts may be sufficient to raise the level of the iron in true solution to 10^{-14}M, i.e., sufficient to meet Johnston's criterion for biological activity. This situation would then allow for the bulk of the EDTA or natural chelator to be available to the other trace metals.

(b) The choice of EDTA as an example of a chelator may not reflect the natural chelator's selectivity for various metal ions. Table 1 compares the log of the equilibrium constant for the formation of a series of metal-organic complexes (21). The variation in the ratio of stability constants for ferric ion versus copper ion and chromium ion

TABLE 1

Log of the Equilibrium Constants of Metal-Chelator Complex[a]

Metal	EDTA	Glycine	Aspartic Acid	QCA[b]	NTA[c]
Mg^{2+}	8.69	3	2.43	1.37	7.0
Ca^{2+}	10.59	1.38	1.60	1.42	6.41
Sr^{2+}	8.63	0.9	1.48	1.2	4.90
Fe^{3+}	25.0	10.0	11.4	$\beta_2 = 7.7$ $K_1 = ?$	8.2
Zn^{2+}	16.58	5	5.84	5	10.45
Cu^{2+}	18.38	8.22	8.4	5.91	12.96
Ni^{2+}	18.45	5.97	7.12	4.19	11.54
Cr^{3+}	24.0	--	3.62	--	>10
Pb^{2+}	18.2	5.47	--	3.95	11.8
K_{Fe}/K_{Cu}	$4x10^6$	60	10^3	(1)	$1.6x10^{-5}$
K_{Cr}/K_{Cu}	$3x10^6$	--	$6x10^{-3}$	--	$(1x10^{-1})$

[a] Selected metal-organic complexes from Ref. (21).
[b] QCA = Quinoline-2-carboxylic acid.
[c] NTA = Nitrilotriacetic acid.

versus copper ion is obvious. EDTA shows a considerable selectivity for Fe^{3+} and Cr^{3+} ions. The concept of ion competition is one to keep in mind but until there is a clearer idea of the actual nature of the chelators present in sea water and their specific metal affinities, quantitative relationships are interesting but somewhat speculative. As Duursma says ". . . the problem of chelation in the sea and also in fresh waters, between trace metals and dissolved organic matter cannot be solved in a simple way and that many efforts are necessary to study, on the one hand, the complete chemical composition of the dissolved organic matter, and on the other hand, their chelating stability constants in relation to all cations present" (18).

At this point, an important distinction between thermodynamic stability and lability should be emphasized. The trace metal-organic compounds present in sea water may be of two types:

(a) Those metal-organic species existing in equilibrium with their free components in the sea; the entire system can be described by reversible reactions, each with an appropriate equilibrium constant.

(b) Those compounds whose relatively constant concentration in sea water is not the result of equilibrium processes but more accurately of steady state conditions and which are thermodynamically unstable relative to the concentrations of their components in sea water. Kinetic barriers may exist which prevent the system from reaching equilibrium (22). Metal-organic species of this second type may be produced as a result of metabolic activity or decay of marine organisms.

Compounds of the second type would be non-labile, i.e., no exchange of added labeled metal ion with the incorporated metal ion would occur over a long period of time. What is not equally well recognized is the possibility that longer time periods than are usually used in laboratory experiments may be required before equilibrium is reached even for complexes of type (a). For example, Tyree (23) cites the slow reaction of chelating agents with isopolyhydroxometal ions.

II. NATURE OF ORGANIC CHELATORS IN SEA WATER

The chief source of natural supply of organic chelators would appear to be the biosphere in the ocean. Many papers have reviewed the uptake and release of organics from marine organisms: phytoplankton (24, 25) and invertebrates (26). Nicol (27), in a review of the free amino acids (FAA) released by marine invertebrates, reported that the FAA constitute the largest fraction of the dissolved organic nitrogen released by a number of invertebrate groups. Among these, the most common are glycine, serine, ornithine, and alanine. These are also the FAA found in greatest abundance in sea water (28, 29, 30). The rate of excretion of amino acids by zooplankton is sufficient to replace the natural levels of FAA within one month (31).

The fraction of the organic matter present in the sea which can interact with trace metals has been the most difficult fraction to investigate, due to its polar nature and tendency to remain with the salt fraction on isolation. Amino acids are a group whose concentration and species nature is slowly being brought to light. Concentrations in near-shore waters have been determined by Siegel and Degens (28), using ligand exchange chromatography (32). The method is based on the isolation of the FAA from sea water by chelation to copper ion, itself bound by chelate formation to Chelex-100 ion-exchange resin. Near shore waters had levels of FAA averaging 60 μg/l with glycine and serine being the predominant types. The bulk of the dissolved amino acids was present in polymeric form (240 μg/l) which, on acid hydrolysis could be analyzed for amino acid content. Associated with the polymeric amino acids were some phenols. The exact nature of the polymeric fraction is still in question, particularly since repolymerization was observed to occur spontaneously on allowing the hydrolyzed samples to stand at pH 8.0. The recombination may be highly selective and may distort the observed amino acid distribution. Spontaneous aggregation and precipitation have been reported also with humic acids and seaweed natural products (33).

Further uses of ligand exchange chromatography in studies of amino acid distribution in natural waters have been the work of Webb and Wood (29), Hobbie et al. in the York River Estuary (30), and Siegel in the Gulf of Maine (34). As had been reported previously (28), the predominant FAA were found to be glycine, serine, and ornithine. Attempts to examine larger samples have been moderately successful, particularly when the copper-Chelex resin was used in a batchwise process (34). Two treatments of a sea water sample with resin, i.e., fresh resin for each step, have recovered up to 95% of spiked glycine at the micromole level.

Ligand exchange chromatography has been utilized to isolate "humic" matter from near shore sea water (35). The infrared spectrum of fractions of this material was similar to published spectra of terrestrial and fresh-water humates. Of interest is the metal-associating ability of this "humic" matter as demonstrated by its interaction with the copper ion on the resin column. Barsdate and Matson (36) isolated fresh water humates which associated with copper and lead; an apparent stability constant for lead was 10^{20}. Basu et al. (37) had reported similar results; the transition metals were picked up by humic acids from soils. Copper was the most strongly complexed and some of the adsorbed metals were apparently fixed and not released even on acid extraction.

In a further work, Barsdate (38) established that much of the cobalt, manganese, and zinc in highly colored lake waters was non-dialyzable and associated with large organic species. These organics were not found in Bering sea water although in a near shore sample 60% of the exchangeable copper (labile to exchange with ^{64}Cu) was in the form of large complexes. The dialysis technique is of some interest as two containers were used for the "organics," to one of which was added the labeled copper ion. Equilibrium was established through the external medium, thus ensuring the approach to equilibrium from both directions.

The earlier work on the nature of these yellow-colored (Gelbstoff), "humic-like" organic materials in sea water has been reviewed by Kalle (39). More recently, Sieburth and Jensen (33,40) found that _Fucus vericulosus_ exudated into sea water solution up to 42 mg C/100 g dry weight per hour. Within a short time this material, together with carbohydrate and proteinaceous material, formed a yellow solution, probably by polyphenol condensation. The slow process of aggregation continued, until actual particles form and precipitate. These particles appear indistinguishable from the organic particles found at sea by Riley (41). Khailov (42) also reported the aggregation and precipitation of yellow organic compounds produced by brown algae. Of some significance, these exudate aggregates have a high ash content indicating possible metal ion interactions. Copper ion interaction has been found with these external exudates of brown algae and also with nondialyzable organic products of single celled algae (43). Similar results with other algae have been reported by Fogg (44) and by Fogg and Westlake (45).

Other natural products produced by the biosphere exhibited protective effects versus toxic metal ions along the lines suggested by Saunders (15). Button (46), working with a marine yeast chosen because it grew well in continuous culture without added chelators (presumably because it produced natural organic chelators), found that the addition of copper ion at the micromolar level reduced the population somewhat but that the population approached its original level after a time. A second addition of copper had almost no effect on the population level. Fitzgerald (47), using silver as an algistatic agent, determined that a reinoculation of an algal culture with silver required sixteen times the amount of silver which originally was found to limit growth. The algae were able to detoxify the silver in some manner. Small amounts of added EDTA (3×10^{-6}M) were sufficient to protect algae from equal amounts of copper (48), demonstrating, with a known chelating agent, the protective effects of chelation. More recently, Jones (49) studied the toxic effects of nickel and copper on a variety of marine and terrestrial bacteria. Some species released large amounts of organic material which interacted with the added metal ions. The conditions of the experiment were somewhat removed from those in sea water.

III. STATE OF THE TRACE METALS AND ORGANIC INTERACTIONS

In addition to the previously cited examples of organic-metal ion interaction in natural waters (35-38, 43-49), there is a body of information which indicates the extent to which this phenomenon occurs in the environment. Unfortunately, some of this information is of the one of a kind variety. Only rarely has a body of water been examined with some hope of seeing the microstructure of the trace metals present. The chief reason for this state of affairs has been the lack of a simple, convenient, and widely applicable tool which could distinguish between the various possible species of a metal ion. This situation, hopefully, is improved considerably now, with the increasing use of anodic stripping voltammetry. The next several years will see a series of papers in which the detailed speciation of some of the trace metals, in particular, copper, lead, and zinc, will be worked out.

In the literature there are some significant works that merit serious consideration. Trace element analysis of sea water by neutron activation (50) showed an increase in concentrations of silver, cobalt, and nickel with depth in areas of high organic productivity. In near shore environments (Long Island Sound) removal mechanisms can lower the concentrations of these elements to values below that of open ocean areas. A possible mechanism could be precipitation of terrestrial humic material at the fresh water - sea water interface, with the consequent removal of associated trace metals. Humic materials are known to salt-out at interfaces and convert trace metals to insoluble species (40,51,52).

Slowey et al. (53) extracted copper from Gulf of Mexico samples with chloroform. From 8 to 50% of the total copper in five samples could be extracted, presumably in nonlabile organic combination since less than 1% of divalent ^{64}Cu added to identical samples could be extracted by chloroform. Other reports of the existence of nonlabile, organically associated copper have been Corcoran and Alexander (54), in which dithiozone was used to extract copper from waters of the Florida current, both before and after perchloric oxidation of the organics present, Slowey and Hood (55), who used diethyldithiocarbamate as a chelating agent to extract copper, both before and after persulphuric acid oxidation, and Alexander and Corcoran (56), who used a colorimetric method for the determination of copper, both before and after perchloric acid oxidation. The latter report indicated that up to 90% of the total copper present in nearshore waters off

Florida is associated with soluble organics. Corcoran (57) found copper in a soluble organic complex form and iron in two principal forms - particulate and soluble complex. The high percentage of copper as an organic complex in nearshore waters is not surprising. Hodgson et al. (58, 59) reported up to 99% of the copper present in soil solution as organic complexes. Other authors (60, 61, 62) have found high uptake of transition metal ions on soil fulvic acids.

More recently Williams (63) analyzed 20 samples of Southern California sea water for copper content. The procedure was a colorimetric one and was also applied to aliquots of the samples which had been irradiated with ultraviolet light to destroy the organics (64). In 19 of the samples, the irradiation treatment produced higher values of copper which, in some cases, amounted to an increase of up to 28% of the total copper present. Williams suggests that the increase in reactive copper may represent a copper-organic complex with a log stability constant greater than 18 or a copper colloid surrounded by an organic sheath. The latter possibility is similar to one proposed by Shapiro (65-68) for iron in highly colored fresh waters. Yellow organic acids present in lake waters were found to hold iron in solution (defined as the iron which could pass through a 0.45 micron pore diameter membrane filter). However, this iron was retained on a 0.10 micron pore diameter filter. It appears that the mechanism of "solubilization" is the formation of a protected sol of colloidal ferric iron. Similar effects were observed with copper (67). High molecular weight organic components were more effective than low weights in peptizing iron (69).

Ghassemi and Christman (70) and Christman (71) also reported iron to be associated with colored organics in fresh waters. Their results indicated strong bonding in the complex as some of the iron could not be removed by EDTA.

Lai and Goya (72) and Callahan et al. (73) found that a small fraction of the cobalt in sea water samples was not absorbed in passage through a Chelex-100 ion exchange column. The uptake on the resin of cobalt (II) and cobalt (III) freshly added to the sample was quantitative, leading them to believe that some of the natural cobalt (III) in the samples was associated with some unknown species in sea water. Whether this is organic or inorganic is unknown. However, activated carbon retained the natural cobalt (III), indicating a possible organic nature.

Duursma and Sevenhuysen (19) attempted to detect chelating activity under sea water conditions by means of solubility determinations of Fe^{3+}, Zn^{2+}, Co^{2+}, Cu^{2+}, and Ni^{2+} ions in the presence of added known organic compounds and natural organic material found in lake waters. The quantities of added metal ions were such that massive amounts of the metal ions were precipitated from solution as hydroxides or as carbonates. These hydroxides, at similar concentration levels, have been used to concentrate and remove organics from sea water. Ferric hydroxide can coprecipitate many organics of various functional classes (74) and amino acids (75, 76); copper hydroxide has coprecipitated organic nitrogen compounds (77). In the latter work, Chapman and Rae were successful in bringing down with the copper hydroxide precipitate such chelating species as citric acid, aspartic acid, and succinic acid, although not at the 100% recovery level. Similar coprecipitation must have occurred in the Duursma work, making any interpretation of the data hazardous. The point to be noted is that chelation may not always lead to solubilization; interaction with three dimensional gels may lead to coprecipitation of the organic species.

By a line of reasoning similar to that which indicates the existence of copper-organic complexes in sea water, it has been suggested that boron-boric acid complexes with organic polyhydroxy compounds exist in the oceans. The amount of measurable boric acid in sea water samples was increased upon oxidation of the organic matter (78, 79). Williams and Strack (80) calculated that "such complexes would not be significant in sea water at low concentrations of diols that have equivalent complexing power to mannitol." They found that adding mannitol to sea water produced no significant reduction of the titratable boric acid nor did oxidation with ultraviolet light produce an increase in boric acid. The existence of "mannitol" - boric acid complexes in sea water is, therefore, still open to question. Ghassemi and Christman (70) found that the presence of boric acid caused an increase in the molecular size of "yellow" colored organics from fresh water sources. This was attributed to the ability of the borate ion to form chelates with the polyhydroxy-polycarboxylic organics. It is, therefore, possible that in sea water, the borate ion may be associated with humic material, perhaps in colloidal form and thus be unavailable to titration.

IV. INTERACTIONS WITH SOLID PHASES

Studies of organic-inorganic interactions with solid phases under sea water conditions have involved clays, carbonates, and organic aggregates. The adsorption of zinc-glycine complexes to synthetic cation exchangers and various clays indicated that the zinc monoglycinate sorbed almost as strongly as did the divalent zinc ion, whereas the diglycinate (zero charged species) did not appear to sorb at all (81). At low levels of organic concentrations, the effect of univalent ligands on the uptake of trace metals on solid phases may be minimal, therefore, as the cationic complex may still be sorbed, ligand and all.

Chave and others (82, 83, 84) demonstrated the protective effect of organic matter in sea water on natural and synthetic calcium carbonates. The lack of reactivity of these aggregates to reduction of the pH value was due to a coating of brown, amorphous organics. This protective coating could be destroyed by hydrogen peroxide with restoration of the reactivity of the solid phase. In a further work, Chave (85) showed that the dissolved organic compounds in sea water help regulate the precipitation of carbonate minerals, allow for the protection of non-equilibrium minerals, and prevent the formation of calcium carbonate. This maintains the concentration of $CaCO_3$ at supersaturated levels for most surface and near surface sea waters.

Kitano and others (86, 87) have studied the effect of organics at high concentrations (1-10g/l) on the precipitation of calcium carbonate. Those organics with strong affinity for calcium (citrate, malate, succinate, etc.) slowed down the rate of formation of calcium carbonate, favoring formation of the more stable calcite rather than aragonite, even in the presence of magnesium ion. Calcium carbonate precipitation in the form of concretions has been attributed to the decomposition of organic matter (88) or the presence of an organic matrix in naturally-occurring oolites (89).

Organic aggregates, presumably formed as a result of bubbling processes, have been suggested as a concentration and transport mechanism for trace metals. MacKenzie (90) reported the presence of a strontium/chlorinity maximum at depth in the Sargasso Sea, possibly due to the uptake of strontium on bubble produced aggregates at the sea surface, their transport down to this depth, and their release on the oxidation of the organic aggregates. Siegel and Burke (91), in laboratory experiments utilizing labeled tracers and nearshore waters, found only low uptake of calcium, zinc, and strontium on bubble produced aggregates, too low to account for the reported strontium/chlorinity maximum. Of perhaps some significance, manganese ion showed some uptake in this work. Wangersky found that organic aggregates in sea water contained significant levels of calcium and manganese (4).

V. ANODIC VOLTAMMETRY

One of the new and potentially very useful methods of examining the question of the nature of the state of the trace metals in natural waters is anodic stripping voltammetry (ASV) (92, 93, 94). Of particular relevance to sea water work is the use of a composite mercury-graphite electrode (CMGE), first reported by Matson et al. (95) and more fully described in Ref. (51) which examined the use of the CMGE in the investigation of the state of the trace metals in various fresh waters, the Charles River, and Boston Harbor.

Matson used a thin film of mercury (electroplated on graphite) as an electrode in which to form various metal amalgams by plating from natural waters at a reducing potential for a relatively long time and then anodically stripping the metal from the amalgam with a linearly varying oxidizing potential over a short time. The current at various specific voltage values is a sensitive measure of the amount of a metal previously plated out.

The method, while not perfect (for example, there are interferences in the determination of zinc in the presence of high levels of nickel), is of great value since it provides a method for determination of the concentration of "free" metal ion. The plating process can distinguish the free metal ions from those bound in complexes, even with such labile chelators as aspartic acid or EDTA, at the 10^{-7}M level added to sea water. Fitzgerald (96) examined the effect of additions of aspartic acid on the determination of free copper ion in sea water. The sample was irradiated (UV) for four hours to destroy the background level of organics before the aspartic acid was added and the plating started. For the near-shore waters off Woods Hole, Massachusetts, the half-life for organics, as determined by the method of Menzel and Vaccaro (97), was 94 minutes. Even at levels as low as 8×10^{-7}M aspartic acid, the association of the added organic with the naturally present copper was evident, as noted by a decrease in the area of the free copper peak. Calculations, based on the decrease in the free copper ion, indicate that the actual amount of the copper bound by aspartic acid is less than expected, even when taking into account the competition of the hydrogen and magnesium ions (Table 2). Calcium ion can be ignored due to its relatively low stability constant with aspartic acid. For Point 1 the concentration of the copper-organic chelate is 20% of that calculated using stability constants, ignoring all other trace metal ions; for Point 2 the value equals 44%. Note that even at the 7×10^{-8}M level the chelation effects of the added aspartic acid are obvious. As Duursma indicated (18), the competition between trace metals for organic chelators will establish the actual equilibrium point of the system. However, the nature of the chelator and its selective affinities will determine, for any one metal ion, the importance of the other ions. As the example of aspartic acid has shown, copper ion may be well able to compete with iron and chromium.

In Fitzgerald's work, only 0.3 to 3% of the total ion is removed from solution during the plating step (10 minutes to one hour). Therefore, there isn't any significant mass action effect which would cause

TABLE 2

Chelation of Copper (II) by Aspartic Acid Added to Sea Water[a]

Point 1

Total Copper (II) = 7×10^{-8}M
Total Aspartic Acid added = 8.4×10^{-7}M

Calculated Aspartic Ligand Concentration = 1.7×10^{-8}M
Calculated Magnesium-Ligand Complex Concentration = 2.2×10^{-7}M
Calculated Copper-Ligand Complex Concentration = 5.6×10^{-8}M

Experimentally Determined $(Cu^{2+})_{Free} = 5.85 \times 10^{-8}$M

$[Cu^{2+}]_{Total} - [Cu^{2+}]_{Free} = [\text{Copper-Ligand}]_{Experimental} = 1.15 \times 10^{-8}$M

$[\text{Cu-Ligand}]_{Exp.} / [\text{Cu-Ligand}]_{Cal.} = 0.20$

Point 2

Total Copper (II) = 7×10^{-8}M
Total Aspartic Acid added = 5.9×10^{-6}M

Calculated Aspartic Ligand Concentration = 1.17×10^{-7}M
Calculated Magnesium-Ligand Complex Concentration = 1.7×10^{-6}M
Calculated Copper-Ligand Complex Concentration = 6.77×10^{-8}M

Experimentally Determined $(Cu^{2+})_{Free} = 4 \times 10^{-8}$M

$[\text{Copper-Ligand}]_{Exp.} = 3 \times 10^{-8}$M

$[\text{Cu-Ligand}]_{Exp.} / [\text{Cu-Ligand}]_{Cal.} = 0.44$

---

[a]Experimental values for $(Cu)_{Free}$ are from Fitzgerald (96).

the release of metal from labile complexes unless virtually all of the metal was complexed.

Anodic stripping voltammetry on standard solutions and natural waters, which have been acidified or have had their organic matter oxidized, show increased levels of metals. This increase apparently is due to metal ions released from complexes. The increase of plated metal produced by acidification agrees with the increase obtained upon oxidation of the organic matter with persulfate or replacement of the exchangeable metal by addition of nonelectroactive metals (10^{-4} to 10^{-5}M Mn^{2+} or Fe^{3+}) at 90° C. and pH 3-4, both described by Matson (51).

Fitzgerald (96) has found that the use of high energy ultraviolet light to oxidize organic matter in sea water or acidification (pH 3.0) previous to ASV releases relatively large amounts of metal ions, particularly in the case of copper. Lead, masked by EDTA, was released by acidification and upon readjustment of the pH value back to 8.0, was again masked, indicating, for this system, the reversible nature of the process.

ASV of natural water samples has indicated many features of interest. Matson (51), working in lakes and the Charles River, has found association of copper and lead with high molecular weight organic fractions. However, there was little evidence of the existence of high molecular weight complexes of copper or lead in Boston Harbor. In one experiment, filtered Charles River water containing dissolved "bound" lead, was mixed with equal amounts of filtered Boston Harbor sea water under aseptic conditions. After several days, filtration through a 0.45 micron pore diameter membrane filter indicated that a particulate form of lead had been produced in the mixed flasks, whereas unmixed Charles River water showed no production of particulate forms of lead. Sieburth and Jensen (40) reported that a considerable fraction of the humic material of terrestrial origin precipitates on contact with sea water. This represents a mechanism for the removal of the river-transported trace metals at the sea water interface.

Examination of some of the data collected by Fitzgerald indicates the great potential of this technique to answer some of the questions raised in this and previous papers. For example, the copper content of a sample of Buzzards Bay Sea Water is presented in Table 3A. In this sample of nearshore waters 70% of the total copper is apparently held in organic complexes. Woods Hole dock water (Table 3B) has 64% of the total copper, and Sargasso Sea Water (Table 3C) has 84% of the total copper in organic combination, respectively. In the Woods Hole sample, some of the copper is in organic combination in a form other than organic acids whose chelates release the metal ion at pH 3.0 or the copper is associated with non-labile species (UV treatment release greater amounts of copper than did the pH drop). In open sea water (Table 3C) the situation is reversed, but the values presented in the UV treatment may represent incomplete release since, for these open

TABLE 3

Metal Speciation in Natural Waters[a]

A. Filtered Buzzards Bay Sea Water: Nearshore Waters

Treatment	Total Determined	Concentration
Determined at pH 8.0	Free Copper Ion	0.70 μg/l
Determined at pH 8.0	Free Copper Ion	0.69 μg/l
Determined at pH 3.0	Free Ion and "Acid" Bound	2.3 ± 0.3 μg/l
UV Treatment (12 hrs) and Determined at pH 8.0	Free Ion and "Organic" Bound	2.3 μg/l
UV Treatment (12 hrs) and Determined at pH 3.0	Free Ion, "Organic" Bound and "Acid" Bound	2.3 μg/l

Conclusions: Free Cu^{2+} at pH 8.0 = 0.70 μg/l
"Organic"-bound Copper = 1.60 μg/l
Percent of Total Copper Bound in Organic Complex (virtually all as organic acids) = 70%

B. Filtered Woods Hole Sea Water: "Polluted" Waters

Treatment	Total Determined	Concentration
Determined at pH 8.0	Free Copper Ion	1.6 μg/l
UV Treatment at pH 8.0	Free and Organic Bound	4.4 μg/l
Determined at pH 2.5	Free and Acid Bound	3.2 μg/l

Conclusions: Free Cu^{2+} at pH 8.0 = 1.6 μg/l
"Organic"-bound Copper = 2.8 μg/l
"Organic Acid"-bound Copper = 1.6 μg/l

C. Unfiltered Sargasso Sea Surface Sea Water

Treatment	Concentrations (μg/l)			
	Zn	Cu	Pb	Cd
Determined at pH 8.0 - "Free"	0.30	0.053	0.29	0.026
UV Treatment (4 hours) - "Free" and "Organic"-bound	0.58	0.38	0.19(?)	0.053
Determined at pH 3.0 - "Free" and "acid" bound	---	0.75	0.35	---

[a] Fitzgerald (96).

ocean samples, the UV half-life may be different than that of near-shore water and irradiation for four hours may not be sufficient to destroy all the organics.

Fitzgerald has found that butanol extraction of some nearshore sea water samples reduced the level of copper remaining in the aqueous phase. The presence of copper in sea water in a form which could be extracted into organic systems had been previously reported by Slowey et al. (53). Spencer (98), examining mid-ocean waters with the use of atomic absorption spectrophotometry, found no increase in the levels of zinc, copper, and nickel when his samples were treated with ultraviolet light. However, since the extraction procedure previous to use of spectrophotometry is conducted at pH 2.5, acidification in both untreated and UV treated samples may have released all the bound metals. Fitzgerald's open ocean samples (Table 3C) indicate that acidification to pH 3.0 released the maximum amount of metal ion.

VI. PRODUCTIVITY STUDIES

The last area of inquiry into the possible importance of metal-organic interactions in the environment may be the most vital -- the control and initiation of phytoplankton productivity in sea water. It has long been obvious that apparently similar bodies of water, containing equal amounts of inorganic nutrients and exposed to equal amounts of light, may differ greatly in their productivity, i.e., the production of particulate organic matter, as determined by the incorporation of labeled carbon (added as $NaH^{14}CO_3$), into a nonfilterable, particulate form. Further, there may be a lag period when a nutrient rich water rises into the euphotic zone, during which time little or no growth occurs.

Provasoli (99) and Johnston (16,17) helped establish the concept of "good" and "bad" waters based, apparently, on the presence or absence of "available" trace metals. The availability of these elements depended on the presence of natural organics with some chelating ability. Johnston found that addition of trace quantities of chelating agents to "bad" waters often improved their productivity (17). In laboratory cultures it is now common practice to incorporate trace levels of EDTA in the medium if growth is desired. Indeed, the growth enhancing effect of chelated metals is not restricted to sea water or culture media. Schelske (100) increased the primary productivity of lake waters by adding chelated iron (in the form of the N'-(2-hydroxyethyl) ethylene-NNN'-triacetic acid, with a log K_1 = 19.8 or approximately five orders of magnitude less than that for EDTA).

The state of the trace metals in sea water is considerably in question -- particularly that of iron. What determines availability? Johnston (17) evidently believed that iron in sea water exists in a "weathered" and "aged" particulate form and would not be as available as recently added $FeCl_3$ in laboratory experiments. The slow conversion of freshly precipitated amorphous ferric hydroxide to a more crystalline form, α- FeOOH, was cited which is from 100 to 1000 times less soluble. Organic chelators could make iron more available.

On the other hand, excess amounts of chelators could swing the system over again, making the iron again unavailable. Provasoli et al. (101) reported that with sufficient excess of EDTA, in a mixture of Fe-EDTA, the iron will be unavailable to algae. Prakash and Rashid (102) produced a growth response in artificial sea water cultures of marine dinoflagellates (increased yield, growth rate,and ^{14}C uptake) by the addition of small amounts of humic substances (from "yellow" humic fresh waters). Controls which received charcoal treated river water did not grow at all. However, concentrations of humic material greater than 35 μg/ml produced a decrease in growth rate, and yield.

Iron exists in the environment in various complexed forms. Shapiro (69) added equal amounts of total iron from different lakes to iron starved algal cultures and found that the culture densities varied at the end of 7-11 days. The iron from the various sources was evidently not equally available, since all other factors (trace metals, nutrients, etc.) were constant. The rank order of culture densities was found to correspond to those lake waters whose iron was proportionately more reactive with thiocyanate at higher pH values, i.e., there were indications that the various lake waters varied in their proportions of complexed and uncomplexed iron. Those with high amounts of complexed iron required lower pH conditions before releasing iron. In these particular samples, strong associations of iron with unknown species present in varying amounts evidently made a fraction of the total iron unavailable for biological use.

This precarious balance of "free" and "unavailable" trace metal has now been vividly demonstrated by recent work by Barber and Ryther (103) and Barber (20). The data to be presented comes entirely from the latter work, and is not in the literature yet.

Table 4 presents the results of productivity studies in the Gulf of Mexico. All the samples, with the sole exception of No. 1, were 1:1 mixtures of 1000m water and 10m water, the former bringing up inorganic nutrients and the latter serving as an inoculum of phytoplankton. It is evident that this water would be classified as bad water. Even after 7 days, samples without any additives (No. 2), with additional trace metals (No. 3), and with ammonium ion (No. 5) have not produced

TABLE 4

Productivity Studies: Gulf of Mexico[a]

Composition[b]	Counts Per Minute/100ml after 5 days	7 days
1. 1000m water alone	6×10^1	$6x10^1$
1:1 Mixtures of 1000m and 10m water		
2. No addition	$1.6x10^2$	$1.7x10^2$
3. Trace metals	"	"
4. Desferriferrioxamine B	"	"
5. Ammonium ion	"	"
6. Histidine	$3.7x10^2$	$7.2x10^2$
7. Arginine	"	"
8. Glyclglycine	$4.6x10^2$	$9.5x10^3$
9. $FeC1_3$	"	"
10. Fe-Histidine	"	"
11. Fe-Glyclglycine	"	"
12. Fe-EDTA	"	"
13. EDTA	$1x10^3$	$1.9x10^4$
14. EDTA and Trace metals	$3.7x10^3$	$2.3x10^4$

[a]Barber (20).

[b]All additives are at the micromole/l level. The trace metals consist of copper, zinc, cobalt, manganese, and molybdenum.

any significant growth. Of even more importance, addition of desferriferrioxamine B (DES), a natural product occurring in soils with six oxygen functional groups and a very high specific affinity for ferric ion (log K for $Fe^{3+} = 30.6$ whereas log K for $Zn^{2+} = 10$ and for $Mg^{2+} = 4.3$), does not aid in growth at all. Some amino acids (No. 6 and 7) do increase productivity but not as much as does the addition of fresh ferric ion (No. 9) or iron in various chelated forms (No. 10, 11, and 12). Samples with added EDTA (No. 13) and EDTA and trace metals (No. 14) showed the greatest increase in productivity.

If productivity enhancement was solely a function of log K of the chelation step of the organic species, the expected increase in productivity would have been DES > EDTA > AMINO ACIDS (log K for Fe^{3+} is DES = 30.6, EDTA = 25, AMINO ACIDS = 11). However, the DES is unproductive, probably due to its extremely high affinity for iron, making it unavailable to phytoplankton. Indeed the presence of DES can negate the advantages of adding EDTA to cultures. Table 5 presents another series of studies with Gulf of Mexico water (20). Samples without added organic chelator (No. 13,14,15, and 16) are generally low in productivity, in contrast to those samples which have had EDTA added (No. 5,6,7, and 8). However, addition of DES to EDTA samples (No. 1,2,3, and 4) reduces the productivity to levels even lower than set No. 13-16. Note that addition of DES to any sample, at the 3.4 micromole/l level, prevents any significant development of productivity.

This relationship of log K of the organic chelator and its effect on productivity can be presented in more quantitative terms. Table 6 presents data from productivity studies on upwelling waters off Peru (20). Again, the data are presented with the least productive samples first and the most productive at the bottom. The expected order of productivity increase based on the values of the log K for Fe (DPTA log K = 28.6, EDTA log K = 25, Cysteine log K = 10, and NTA log K = 8) would be:

$$\text{DPTA} > \text{EDTA} > \text{CYSTEINE} > \text{NTA.}$$

The experimental results indicate the following order:

$$\begin{array}{l} 1\text{x}10^{-6}\text{M DPTA} \\ 1\text{x}10^{-6}\text{M EDTA} \\ 10\text{x}10^{-6}\text{M EDTA} \end{array} > \begin{array}{c} 10\text{x}10^{-6}\text{M} \\ \text{NTA} \end{array} > \begin{array}{c} 1\text{x}10^{-6}\text{M} \\ \text{CYSTEINE} \end{array} > \begin{array}{c} 1\text{x}10^{-6}\text{M} \\ \text{NTA} \end{array} > \text{NONE} > \begin{array}{c} 10\text{x}10^{-6}\text{M} \\ \text{DPTA} \end{array}$$

Note the relationship between log K, concentration, and effect on productivity. Whereas EDTA (log K_{Fe} = 25) is effective at both the 1 and 10 micromole/l level, DPTA (log K_{Fe} = 28.6) is effective only at the 1 micromole/l level. At the 10 micromole/l level it depresses the productivity below the level of no addition at all. A chelator, which has a log stability constant for Fe^{3+} = 28.6, approaches, at the 10 micromole/l level, the effect of DES at the 1 micromole/l level (log K_{Fe} for DES = 30.6). The product of the chelator concentration times

TABLE 5

Productivity Studies: Gulf of Mexico[a]
1:1 mixture of 1000m and 10 m water

	Composition				Counts Per Minute/100ml After 7 Days		
					Exp. #39-5	Exp. #37-3	Exp. #37-4
1.	EDTA[b]	DES[c]	TM[d]	Fe[e]	2390	1170	1720
2.	EDTA	DES	TM		970	1030	1330
3.	EDTA	DES		Fe	630	1250	3090
4.	EDTA	DES			1120	1570	2110
5.	EDTA		TM	Fe	62,500	40,000	90,000
6.	EDTA		TM		59,700	80,000	111,000
7.	EDTA			Fe	50,630	66,670	99,000
8.	EDTA				43,500	57,140	100,000
9.		DES	TM	Fe	480	3520	15,500
10.		DES	TM		480	1760	14,800
11.		DES		Fe	480	3100	6500
12.		DES			1300	2960	7800
13.			TM	Fe	97,500	780	5150
14.			TM		7310	830	3390
15.				Fe	17,470	4650	21,280
16.					940	2830	5710

[a] Barber (20).
[b] EDTA = 3.4 micromole/l Ethylenediamine tetraacetic acid.
[c] DES = 3.4 micromole/l Desferriferrioxamine B.
[d] TM = Trace levels of Cu^{2+}, Zn^{2+}, Co^{2+}, Mn^{2+}, Mo^{2+}. The concentrations of the trace metals in experiments #37-3 and 37-4 are tenfold greater than that of experiment #39-5.
[e] $[Fe]$ = 1.0 micromole/l $FeCl_3$.

TABLE 6

Productivity Studies: Peru Upwelling[a]
9:1 mixture of 1000m and 10m water

Additives	Counts Per Minute/100ml After		
	1 day	3 days	5 days
1. 10 micromole/l Diethylenetriamine-pentacetic acid (DPTA)	1.8×10^3	3.0×10^3	6.6×10^3
2. No addition	2.7×10^3	4.7×10^3	7.8×10^3
3. 1 micromole/l Nitrilotriacetic acid (NTA)	3.0×10^3	7.2×10^3	1.7×10^4
4. 1 micromole/l Cysteine	4.0×10^3	1.1×10^4	3.1×10^4
5. 1 micromole/l $FeCl_3$	5.2×10^3	2.5×10^4	3.3×10^4
6. 10 micromole/l NTA	5.8×10^3	2.8×10^4	3.8×10^4
7. 1 micromole/l Ethylenediaminetetra-acetic acid (EDTA)	6.6×10^3	3.4×10^4	4.8×10^4
8. 10 micromole/l EDTA	"	"	"
9. 1 micromole/l DPTA	"	"	"

[a]Barber (20).

TABLE 7

Productivity Studies -- Peru Upwelling:
Effect of Charcoal Filtration on "Good" and "Bad" Waters[a]

Table 7A: "Good" Water - Station 053: 9:1 mixture of glass wool filtered 75 m water and unfiltered 75m water.

	Counts Per minute/100ml After		
	1 day	3.5 days	7 days
1. No addition	1.8×10^3	2.4×10^4	6.2×10^4
2. Charcoal filtered - No addition	1.7×10^3	6.2×10^3	2.5×10^4
3. 1 micromole/1 EDTA	2.1×10^3	3.7×10^4	6.4×10^4
4. Charcoal filtered + 1 micromole/1 EDTA	3.1×10^3	2.7×10^4	5.1×10^4
5. 1 micromole/1 Fe-EDTA	2.7×10^3	4.1×10^4	6.4×10^4
6. Charcoal filtered + 1 micromole/1 Fe-EDTA	3.4×10^3	3.8×10^4	7.2×10^4
7. 1 micromole/1 $FeCl_3$ + Trace metals	3.7×10^3	4.3×10^4	6.4×10^4
8. Charcoal filtered + 1 micromole/1 $FeCl_3$ + Trace metals	3.6×10^3	4.2×10^4	7.6×10^4

[a]Barber (20).

TABLE 7 (continued)

Productivity Studies -- Peru Upwelling:
Effect of Charcoal Filtration on "Good" and "Bad" Waters[a]

Table 7B: "Bad" Water - Station 062: 9:1 mixture of glass wool filtered 0m water and unfiltered 0m water.

	Counts Per Minute/100ml After		
	1 day	4 days	6 days
1. No Addition	$1.2x10^3$	$3.8x10^3$	$8.9x10^3$
2. Charcoal filtered - No addition	$1.2x10^3$	$3.5x10^3$	$8.1x10^3$
3. 1 micromole/1 $FeCl_3$	$1.4x10^3$	$4.4x10^3$	$1.0x10^4$
4. Charcoal filtered + 1 micromole/1 $FeCl_3$	$1.3x10^3$	$4.7x10^3$	$1.5x10^4$
5. 1 micromole/1 EDTA	$1.7x10^3$	$1.4x10^4$	$2.0x10^4$
6. Charcoal filtered + 1 micromole/1 EDTA	$1.4x10^3$	$1.3x10^4$	$1.8x10^4$
7. 1 micromole/1 Fe-EDTA	$2.4x10^3$	$1.7x10^4$	$2.3x10^4$
8. Charcoal filtered + 1 micromole/1 Fe-EDTA	$1.7x10^3$	$1.6x10^4$	$2.1x10^4$

[a] Barber (20).

the equilibrium constant is a measure of the ratio of the concentration of the iron-organic complex to the concentration of the free iron, neglecting any consideration of the specific ionization process which produces the free organic ligand. At the 10 micromole/l level DPTA evidently makes iron unavailable, while at the 1 micromole/l level it increases productivity. NTA, on the other hand, at the 10 micromole/l level becomes even more effective than at the 1 micromole/1 level. The NTA log stability constant for Fe^{3+} is just below that of typical amino acids and only at the 10 micromole/l level will NTA surpass cysteine in productivity enhancement. A reasonable conclusion for this series of experiments is that the availability of iron and other trace metals to phytoplankton in sea water is determined by a combination of factors -- chelator species and chelator quantity. Excessive amounts of some chelators (1 micromole/l DES or 10 micromole/l DPTA) can be inhibiting to productivity.

These results have identified the effects of added chelator to sea waters. What are the natural organics in sea water which serve the same function? This we do not as yet know. Some of Barber's results have demonstrated the presence in sea water of such species (20). Table 7 shows the effect of charcoal filtration on productivity of upwelling waters off Peru. These results were obtained with 9:1 mixtures of glass wool filtered 75m water and an inoculum of unfiltered 75m water (Station 053). Every pair of samples was treated identically except for the addition of a charcoal filtration step for the even numbered members of each pair. Pairs 3-4, 5-6, and 7-8 are alike in productivity, and, within each pair, the charcoal treatment has not made much of a difference. However, in pair 1-2, there is a considerable effect of the charcoal filtration. Sample No. 1, without any addition, while it was lower in growth at the first day, realized the same productivity after seven days as did samples 3-8. However, sample No. 2, which had been filtered through charcoal, never achieved as high a level of growth as did the others. The sea water at this station would be characterized as "good" since sample No. 1 (no additions) produced as high a level of productivity as did the ones to which additional chelators had been added. The naturally present, unknown species had evidently been retained by the charcoal filter, causing the poor productivity of sample No. 2.

Table 7B whose samples were composed of water from an adjacent area (Station 062) shows the behavior of a typically "bad" water. Sample 1 (no additions) has low productivity even though it has not been filtered through charcoal. Sample 2 (no addition), filtered through charcoal, is no worse than sample No. 1, i.e., the charcoal treatment has not decreased productivity (as it did for the good water in Table 7A, sample No. 2) since there were no natural organic chelators originally present. Note further that even the addition of $FeCl_3$ does not bring samples 3 and 4 up to that of No. 5-8, to which chelators had been added.

VII. CONCLUSION

It is evident that the interaction of organic chelators and trace metals in sea water is of great importance to such processes as productivity rates, adsorption to other phases, oxidation states of the metal ions, and transport to the sediments. Through the increased application of studies applying anodic stripping voltammetry, atomic absorption spectrophotometry, ligand exchange chromatography, activation analysis, and biological parameters such as productivity and metal uptake, we will help open these areas to quantitative treatment and remove them from the realm of speculation.

REFERENCES

1. J. R. Vallentyne, J. Fish Res. Board Can., 14, 33 (1957).

2. E. K. Duursma, Neth. J. Sea Res., 1, 1 (1961).

3. E. K. Duursma, in Chemical Oceanography (J. P. Riley and G. Skirrow, eds.), Vol. 1, Academic Press, London, 1965, p. 443.

4. P. J. Wangersky, Amer. Scientist, 53, 358 (1965).

5. E. T. Degens, Symposium on Organic Matter in Natural Waters, College, Alaska, September, 1968.

6. N. Handa, Symposium on Organic Matter in Natural Waters, College, Alaska, September, 1968.

7. L. M. Jeffrey, Symposium on Organic Matter in Natural Waters, College, Alaska, September, 1968.

8. D. W. Menzel and J. H. Ryther, Symposium on Organic Matter in Natural Waters, College, Alaska, September, 1968.

9. J. F. Corwin, Symposium on Organic Matter in Natural Waters, College, Alaska, September, 1968.

10. W. D. Garrett, Symposium on Organic Matter in Natural Waters, College, Alaska, September, 1968.

11. E. J. Kuenzler and J. P. Perras, Woods Hole Oceanographic Institution Ref. No. 65-59, Unpublished Manuscript , 1965.

12. O. T. Hogdahl, The Trace Elements in the Ocean: A Bibliographic Compilation, Publ. of Central Inst. Industr. Res., Oslo, 1963.

13. D. F. Martin, in Equilibrium Concepts in Natural Water Systems (W. Stumm, ed.), American Chemical Society, Washington, D.C., 1967, p. 255.

14. R. T. Barber, Nature, 220, 274 (1968).

15. G. W. Saunders, Bot. Rev., 23, 389 (1957).

16. R. Johnston, J. Mar. Biol. Assoc. U.K., 43, 427 (1963).

17. R. Johnston, J. Mar. Biol. Assoc. U.K., 44, 87 (1964).

18. E. K. Duursma, Symposium on Organic Matter in Natural Waters, College, Alaska, September, 1968.

19. E. K. Duursma and W. Sevenhuysen, Neth. J. Sea Res., 3, 95 (1966).

20. R. T. Barber, 1969 (In press).

21. A. E. Martell and L. G. Sillen, Stability Constants of Metal-ion Complexes. Special Publication No. 17, The Chemical Society, Burlington House, London, 1964.

22. M. Blumer, in Equilibrium Concepts in Natural Water Systems (W. Stumm, ed.), American Chemical Society, Washington, D.C., 1967, p. 312.

23. S. Y. Tyree, Jr., in Equilibrium Concepts in Natural Water Systems (W. Stumm, ed.), American Chemical Society, Washington, D.C., 1967, p. 194.

24. J. A. Hellebust, Symposium on Organic Matter in Natural Waters, College, Alaska, September, 1968.

25. G. E. Fogg, in Oceanogr. Mar. Biol. Ann. Rev. (H. Barnes, ed.), Vol. 4, George Allen and Unwin, Ltd., London, 1966, p. 195.

26. R. E. Johannes and K. L. Webb, Symposium on Organic Matter in Natural Waters, College, Alaska, September, 1968.

27. J.A.C. Nicol, The Biology of Marine Animals, 2nd ed., Interscience, New York, 1967.

28. A. Siegel and E. T. Degens, Science, 151, 1098 (1966).

29. K. L. Webb and L. Wood, in Automation in Analytical Chemistry, Technicon Symposium, 1966, Vol. 1, Mediad, White Plains, New York, 1967, p. 440.

30. J. E. Hobbie, C. C. Crawford, and K. L. Webb, Science, 159, 1463 (1968).

31. K. L. Webb and R. E. Johannes, Limnol. Oceanogr., 12, 376 (1967).

32. A. Siegel, in Pollution and Marine Ecology (T. A. Olsen and F. J. Burgess, eds.), Interscience, New York, 1967, p. 235.

33. J. McN. Sieburth and A. Jensen, Symposium on Organic Matter in Natural Waters, College, Alaska, September, 1968.

34. A. Siegel, unpublished work, 1967.

35. A. Siegel and J. Parson, 1969 (In preparation).

36. R. J. Barsdate and W. R. Matson, in Radioecological Concentration Processes (B. Aberg and F. Hungate, eds.), Pergamon Press, Oxford, 1966.

37. A. N. Basu, D. C. Mukherjee, and S. K. Mukherjee, J. Ind. Soc. Soil Sci., 12, 311 (1964).

38. R. J. Barsdate, Symposium on Organic Matter in Natural Waters, College, Alaska, September, 1968.

39. K. Kalle, in Oceanogr. Mar. Biol. Ann. Rev. (H. Barnes, ed.), Vol. 4, George Allen and Unwin, Ltd., London, 1966, p. 91.

40. J. McN. Sieburth and A. Jensen, J. Exp. Mar. Biol. Ecol., 2, 174 (1968).

41. G. A. Riley, Limnol. Oceanogr., 8, 372 (1963).

42. K. M. Khailov, Dokl. Akad. Nauk, USSR, 147, 1355 (1963).

43. K. M. Khailov, Dokl. Akad. Nauk, USSR, 155, 933 (1964).

44. G. E. Fogg, Proc. Roy. Soc. London, 139, 372 (1952).

45. G. E. Fogg and D. F. Westlake, Verh. Int. Ver. Limnol., 12, 219 (1955).

46. D. K. Button, Symposium on Organic Matter in Natural Waters, College, Alaska, September, 1968.

47. G. P. Fitzgerald, Water and Sewage Works, 114, 185 (1967).

48. G. P. Fitzgerald and S. L. Faust, Appl. Microbiol., 11, 345 (1963).

49. G. E. Jones, Symposium on Organic Matter in Natural Waters, College, Alaska, September, 1968.

50. D. F. Schutz and K. K. Turekian, Geochim. Cosmochim. Acta., 29, 259 (1965).

51. W. R. Matson, Ph.D. Thesis, Massachusetts Inst. of Tech., Cambridge, Mass., 1968.

52. A. Szalay and M. Szilagyi, Geochim. Cosmochim. Acta, 31, 1 1967.

53. J. F. Slowey, L. M. Jeffrey, and D. W. Hood, Nature, 214, 377 (1967).

54. E. F. Corcoran and J. E. Alexander, Bull. Mar. Sci. Gulf and Carib., 14, 594 (1964).

55. J. F. Slowey and D. W. Hood, Ann. Rep. AEC-Con. No. AT-(40-1)-2799, Texas A and M Rep. 66-2F (1966).

56. J. E. Alexander and E. F. Corcoran, Limnol. Oceanogr., 12, 236 (1967).

57. E. F. Corcoran, Studies Trop. Oceanogr. Miami, 5, 290 (1967).

58. J. F. Hodgson, J. R. Geering, and W. A. Norvell, Soil Sci. Soc. Amer. Proc., 29, 665 (1965).

59. J. F. Hodgson, W. L. Linsay, and J. F. Trierweiler, Soil Sci. Soc. Amer. Proc., 30, 723 (1966).

60. M. Schnitzer and S.I.M. Skinner, Soil Science, 102, 361 (1966).

61. M. Levesque and M. Schnitzer, Soil Science, 103, 183 (1967).

62. M. Schnitzer and S.I.M. Skinner, Soil Science, 103, 247 (1967).

63. P. M. Williams, Limnol. Oceanogr., 14, 156 (1969).

64. F. A. J. Armstrong, P. M. Williams, and J. D. H. Strickland, Nature, 211, 481 (1966).

65. J. Shapiro, Limnol. Oceanogr., 2, 161 (1957).

66. J. Shapiro, Science, 127, 702 (1958).

67. J. Shapiro, J. Amer. Water Works Assoc., 56, 1062 (1964).

68. J. Shapiro, Verh. Int. Verein. Limnol., 16, 477 (1966).

69. J. Shapiro, I.B. P. Symposium, Amsterdam and Nieuwerslius, October, 1966.

70. M. Ghassemi and R. F. Christman, Limnol. Oceanogr., 13, 583 (1968).

71. R. F. Christman, Symposium on Organic Matter in Natural Waters, College, Alaska, September, 1968.

72. M. G. Lai and H. A. Goya, USNRDL-TR-67-11, November, 1966.

73. C. M. Callahan, J. M. Pascual, and M. G. Lai, USNRDL-TR-67-10, December, 1966.

74. R. G. Bader, D. W. Hood, and J. B. Smith, Geochim. Cosmochim. Acta, 19, 236 (1960).

75. K. Park, W. T. Williams, J. M. Prescott, and D. W. Hood, Science, 138, 531 (1962).

76. M. Tatsumoto, W. T. Williams, J. M. Prescott, and D. W. Hood, J. Mar. Res., 19, 89 (1961).

77. G. Chapman and A. C. Rae, Nature, 214, 627 (1967).

78. J. E. Noakes and D. W. Hood, Deep-Sea Res. 8, 121 (1961).

79. J. A. Gast and T. G. Thompson, Anal. Chem., 30, 1549 (1958).

80. P. M. Williams and P. M. Strack, Limnol. Oceanogr., 11, 401 (1966).

81. A. Siegel, Geochim. Cosmochim. Acta, 30, 757 (1966).

82. K. E. Chave, Science, 148, 1723 (1965).

83. K. E. Chave and R. F. Schmalz, Geochim. Cosmochim, Acta, 30, 1037 (1966).

84. K. E. Chave and E. Suess, Trans. N. Y. Acad. Sci., 29, 991 (1967).

85. K. E. Chave, Symposium on Organic Matter in Natural Waters, College, Alaska, September, 1968.

86. Y. Kitano and D. W. Hood, Geochim. Cosmochim. Acta, 29, 29 (1965).

87. Y. Kitano, N. Kanamori, and A. Tokuyama, Symposium on Organic Matter in Natural Waters, College, Alaska, September, 1968.

88. R. A. Berner, Science, 159, 195 (1968).

89. R. M. Mitterer, Science, 162, 1498 (1968).

90. F. T. MacKenzie, Science, 146, 517 (1964).

91. A. Siegel and B. Burke, Deep-Sea Res., 12, 789 (1965).

92. E. Barendrecht, Chem. Weekblad, 60, 345 (1964).

93. L. Meites, Polarographic Techniques, 2nd ed., Interscience, New York, 1965.

94. D. N. Hume, Anal. Chem., 38, 261R (1966).

95. W. R. Matson, D. K. Roe, and D. Carritt, Anal. Chem., 37, 1594 (1965).

96. W. Fitzgerald, private communication, 1969.

97. D. W. Menzel and R. F. Vaccaro, Limnol. Oceanogr., 9, 138 (1964).

98. D. W. Spencer, private communication, 1969.

99. L. Provasoli, in The Sea (M.N. Hill, ed.), Vol. 2, Interscience, New York, 1963, p. 165.

100. C. L. Schelske, Science, 136, 45 (1962).

101. L. Provasoli, J. J. A. McLaughlin, and M. R. Droop, Arch. für Microbiol., 25, 392 (1957).

102. A. Prakash and M. A. Rashid, Limnol. Oceanogr., 13, 598 (1968).

103. R. T. Barber and J. H. Ryther, J. Exp. Mar. Biol. Ecol. (In press).

CHAPTER 13

METAL-ORGANIC MATTER INTERACTIONS IN SOILS AND WATERS

M. Schnitzer

Soil Research Institute
Canada Department of Agriculture

I. INTRODUCTION

Humic compounds are among the most widely occurring natural products on the earth's surface. They occur in soils, lakes (1), rivers (2), and in the sea (3). It is not inconceivable that they will be found on Mars when man lands there. In spite of their wide distribution, little is known about their chemical structure, properties, synthesis, and degradation. We have been studying the chemical structure and reactions of humic compounds for many years, attempting to relate structure to reactivity in order to comprehend the role of these substances in nature. We were interested especially in their ability to form stable water-soluble and insoluble complexes with metal ions and hydrous oxides and to interact with clay minerals. While our work has dealt exclusively with soil humic compounds, there is increasing evidence that humic compounds in water are, in many respects, similar to those occurring in soils.

II. HUMIC COMPOUNDS

Soils contain a large variety of organic compounds that can be grouped conveniently into non-humic and humic substances. Non-humic compounds include those with definite physical and chemical characteristics such as carbohydrates, proteins, peptides, amino acids, fats, waxes, and low molecular weight organic acids. Most of these compounds are attacked relatively easily by microorganisms and have, therefore, only a short life span in the soil.

Humic compounds are acidic, dark coloured, partially aromatic, chemically complex substances of molecular weights ranging from a few hundred to several thousand, that lack the specific chemical and physical characteristics (such as melting point, refractive index, exact elementary composition, etc.) usually associated with well-defined organic compounds. Humic substances have an appreciable exchange capacity, due primarily to carboxyl and phenolic hydroxyl groups, and can complex metal ions and react with clay minerals. They constitute the bulk of the organic matter in most soils, and possibly also in waters, and exhibit considerable resistance to microbial decomposition.

Based on their solubilities, humic compounds are partitioned usually into three main fractions: (a) fulvic acid, which is soluble in both alkali and acid, and is considered to have the lowest molecular weight; (b) humic acid, soluble in alkali but insoluble in acid and of intermediate molecular weight; and (c) humin, insoluble in both alkali and acid, and apparently of the highest molecular weight.

III. FULVIC ACID

Fulvic acid is a water-soluble humic compound that occurs in all soils, accounting for 25 to 75% of the total organic matter. Recently, analysis of a soil leachate, collected over a period of six months 2.5 cm below the mineral soil surface of a Humic Podzol, showed that 85% of the dry ash-free weight was fulvic acid (4). There is increasing evidence that fulvic acid is the most important organic component of the "soil solution", and that it likely affects practically all reactions that occur in soils. According to recent Russian work (5), 85% of the organic matter in swamp water is fulvic acid. Most of our experiments were done with a fulvic acid extracted from a podzol Bh horizon. This fulvic acid was practically free of carbohydrates and contained only little N, so that it was a more or less "pure" humic compound.

A. Chemical Structure.

The chemical structure of this fulvic acid has been investigated in our laboratory by chemical (6,7,8), spectroscopic (9), thermogravimetric (10), and, more recently, by X-ray methods (11). A number of its analytical characteristics is shown in Table 1. Note that 61% of the weight of the fulvic acid molecule is in functional groups, and that the latter contains all of the oxygen. The ratio of COOH to total OH groups is approximately 1, which is favorable for

TABLE 1

Analytical Characteristics of Fulvic Acid

Elementary composition (dry, ash-free basis)	
C	50.90%
H	3.35%
N	0.75%
S	0.25%
O	44.75%
Oxygen-containing functional groups (meq/g dry, ash-free)	
Total Acidity	12.4
Carboxyl	9.1
Total Hydroxyl	6.9
Phenolic Hydroxyl	3.3
Alcoholic Hydroxyl	3.6
Carbonyl	3.1
Number-average molecular weight	670
Molecular formula	$C_{20}H_{12}(COOH)_6(OH)_5(CO)_2$
Solubility	100% in water, 96% in methanol, 60% in acetone, insoluble in benzene, chloroform and carbon tetrachloride.
Aromaticity	61% of nuclear C
Physical Properties	Optically inactive; $d(20^o) = 1.61$ g/cm^3 Age by radiocarbon dating: 600 ± 50 years Stable free radicals: 0.58×10^{18} spins/g
UV Spectrum (methanol)	λ max at 215 mn (log ϵ = 4.61) Shoulder at 260 mn (log ϵ = 4.45)

metal-fulvic acid interactions, while the ratio of COOH to phenolic OH groups is about 3.0. While the molecular formula tells little about the structure, it is nonetheless useful in illustrating the relation between functional groups and the "nucleus". Radiocarbon dating indicates that fulvic acid has considerable resistance to biological and/or

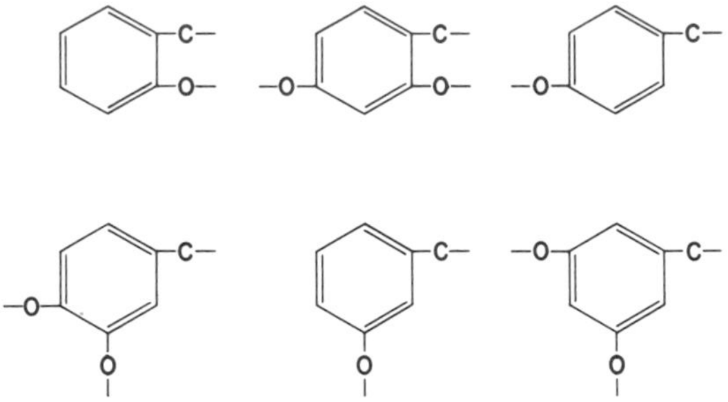

FIG. 1

Chemical structures of hydroxycarboxylic acids produced by the oxidation of fulvic acid with 2 N HNO_3. Reprinted from Ref. (13), p. 79, by courtesy of the Soil Science Society of America.

chemical degradation in nature. The fulvic acid contains one stable free radical per approximately 1,500 number-average molecular weight units or per 42,000 carbon and 29,000 oxygen atoms. Little is known about the nature of the free radicals and about their role in metal-fulvic acid reactions.

The oxidative degradation of the fulvic acid with alkaline permanganate solution and with nitric acid produced aliphatic mono- and di-carboxylic acids, benzenecarboxylic acids, and nitrophenols (8,12, 13). Reductive degradation such as zinc-dust distillation and fusion, yielded polycyclic aromatic compounds, ranging from naphthalene to perylene (14). Chemical structures of the major degradation products are shown in Figures 1, 2, and 3. X-ray studies indicated that the carbon skeleton of fulvic acid consisted of a broken network of poorly condensed aromatic rings with appreciable numbers of disordered aliphatic chains or alicyclic structures around the edges of the aromatic layers (11). Carbon skeletons of four possible molecular models for fulvic acid which are consistent with X-ray, chemical, and spectroscopic data are presented in Figure 4. In structure (a) the two isolated rings are joined together via an alicyclic ring; in structure (b) they are linked by two aliphatic chains; in structure (c) one long aliphatic chain joins two aromatic rings, whereas in structure (d) the two aromatic rings are condensed and the aliphatic portion of the molecule extends from one of the aromatic rings.

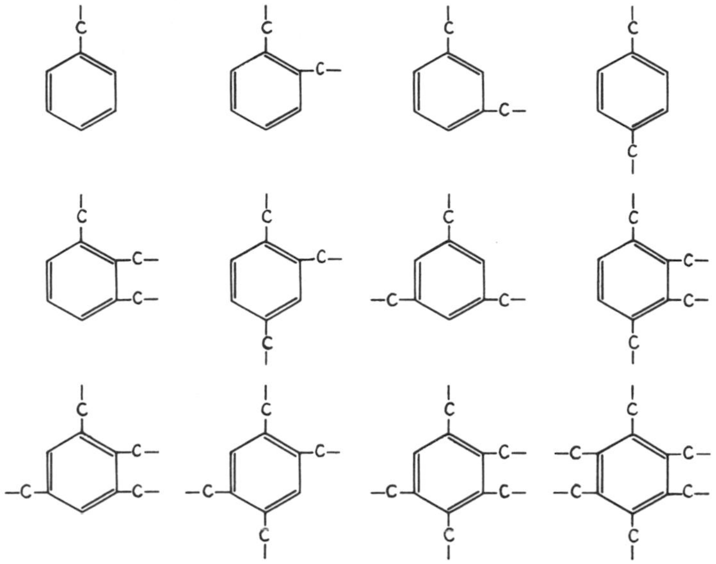

FIG. 2

Carbon skeletons of benzene carboxylic acids produced by the oxidation of fulvic acid with either alkaline potassium permanganate, 2 N, or 7.5 N HNO_3. Reprinted from Ref. (13), p. 79, by courtesy of the Soil Science Society of America.

B. IR Spectrum.

The IR spectrum of fulvic acid is shown in Figure 5. The main absorption bands are in the regions of 3,400 (hydrogen-bonded OH), 2,900 (aliphatic C-H stretch), 1,725 (C=O of COOH, C=O stretch of ketonic carbonyl), 1,630 (aromatic C=C (?) hydrogen-bonded C=O of carbonyl, double bond conjugated with carbonyl, COO^-), 1,400 (COO^-, aliphatic C-H), 1,200 (C-O stretch or OH deformation of COOH), 1,050 cm^{-1} (Si-O of silicate impurity). The absorption bands are broad, likely, because of extensive overlapping of individual absorptions. The IR spectrum reflects the preponderance of oxygen-containing functional groups, that is, COOH, OH, and C=O, but provides little

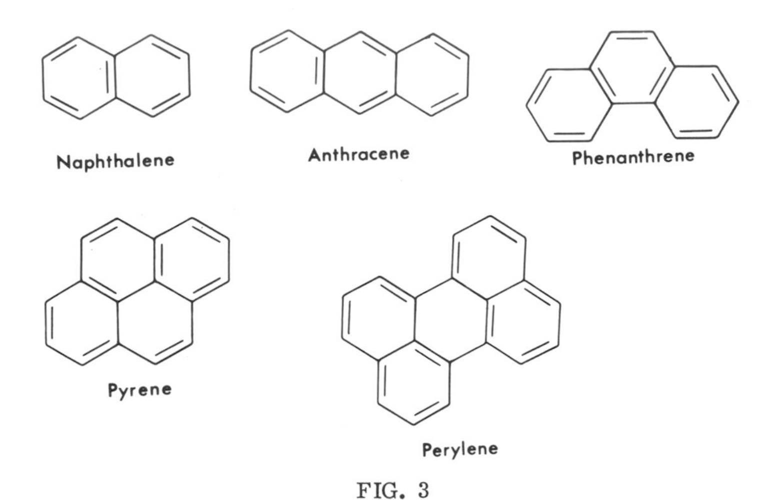

FIG. 3

Major reaction products resulting from the Zn-dust distillation, and Zn-dust fusion of fulvic acid.

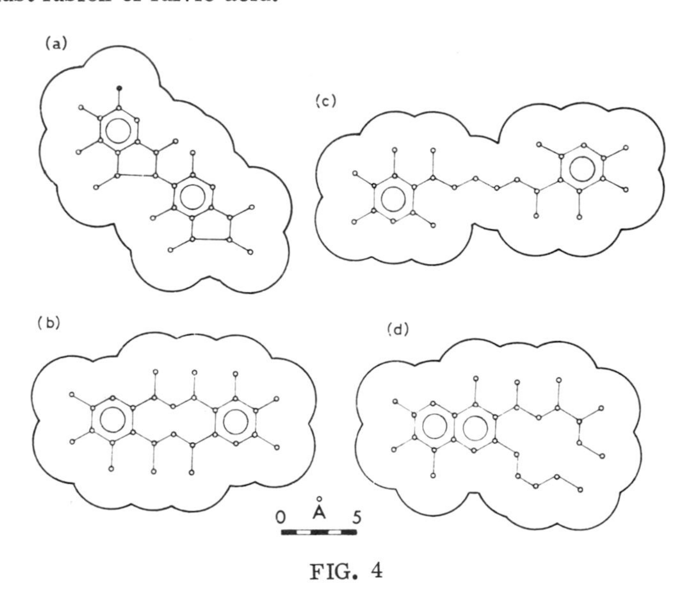

FIG. 4

Four possible carbon skeletons for the planar fulvic acid molecule. Reprinted from Ref. (11), p. 87, by courtesy of Butterworths Scientific Publications.

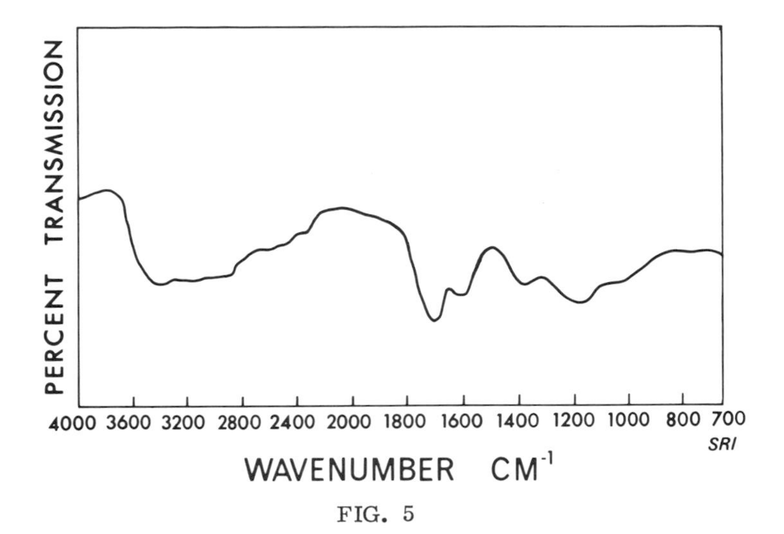

FIG. 5

IR spectrum of fulvic acid.

information on the "nucleus" of the fulvic acid molecule. It is noteworthy that the IR spectra of complex polymeric hydroxycarboxylic acids recently extracted by Lamar (2) from surface waters from widely differing geographical locations in the U.S.A., were similar to that of fulvic acid.

IV. DETERMINATION OF STABILITY CONSTANTS OF METAL-FULVIC ACID COMPLEXES

Stability constants of a number of metal-fulvic acid complexes were measured by two independent methods: the ion-exchange equilibrium method (15) and by the method of continuous variations (16).

A. The Ion-Exchange Equilibrium Method.

According to Martell and Calvin (15), the equilibrium reaction for chelate or complex formation can be written as:

$$M + x\,Ke = MKe_x \qquad (13\text{-}1)$$

where M is the metal ion, Ke is the complexing agent, and x is the number of moles of complexing agent. The chelate formation constant is then:

$$K = \frac{(MKe_x)}{(M)(Ke)^x} \qquad (13\text{-}2)$$

If x is an integer, that is, if the complex is mononuclear with respect to M, Equation (13-2) can be rewritten, as shown in detail by Martell and Calvin (15), in the following form:

$$K = \frac{\left(\frac{\lambda_o}{\lambda} - 1\right)}{(Ke)^x} \qquad (13\text{-}3)$$

or $\log\left(\frac{\lambda_o}{\lambda} - 1\right) = \log K + x \log (Ke)$. Log K is the intercept and x is the slope of a plot of $\log\left(\frac{\lambda_o}{\lambda} - 1\right)$ vs. log (Ke).

B. The Method of Continuous Variations.

This method, usually ascribed to Job (16), is based on variations of optical densities of solutions containing different ratios of metal ion and complexing agent, while simultaneously maintaining a constant total concentration of reactants.

Job assumed that only one complex was present in solution, and considered the systems formed by mixing 1-V volumes of solution B (concentration B_o) with V volumes of solution A, (concentration $A_o = rB_o$), where V varies from 0 to 1. If the mononuclear species, BAn, is the only complex formed, its concentration will be maximal (16) when:

$$\beta_n B_o r^{n-1} \left[(n + r)\,Vmax - n\right]^{n+1} = (r-1)^n \left[n-(n + 1)\,Vmax\right] \qquad (13\text{-}4)$$

If n = 1 and r = 2, Equation (13-4) simplifies to:

$$\beta = \frac{1 - 2\,Vmax}{B_o \left[3\,Vmax - 1\right]^2} \qquad (13\text{-}5)$$

where β is the stability constant of complex BA. Vmax is the volume of a solution of A at the maximum. When non-equimolar solutions are employed, the value of Vmax depends on B_0 and r, and the value of β can be calculated from Equation (13-4) if n, B_0, and r are known and Vmax has been measured experimentally.

C. Ratio of Moles of Fulvic Acid to Moles of Metal.

Prior to measuring stability constants, it was necessary to determine how many moles of metal combined with one mole of fulvic acid, that is, whether the complexes were mono- or poly-nuclear with respect to the metal. We found that at ionic strength (I) = 0.1, molar metal-fulvic acid ratios at pH values of 3 and 5 were 1 for Cu^{+2}, Ni^{+2} (see Fig. 6), Mn^{+2}, and Zn^{+2}. The ratio was also 1 at about pH 2 and I = 0.1 for Fe^{+3} and Al^{+3}. At pH 3, molar divalent metal-fulvic acid ratios were not affected by lowering the ionic strength from 0.1 to near 0. At pH 5, however, molar divalent metal-fulvic acid ratios tended to increase from 1 to 2 as I was lowered from 0.1 to near 0.

Since at I = 0.1, the complexes were mononuclear at both pH levels, and also since the ion exchange equilibrium method required a constant ionic medium (16), we felt justified in using Equations (13-3) and (13-4) for determination of stoichiometric stability constants. In connection with this, it is noteworthy that Geering and Hodgson (17) recently reported that at pH < 4.5 both Zn^{+2} and Cu^{+2} formed 1:1 molar complexes with what they referred to as the "natural ligand" of the soil solution.

D. Stability Constants.

Stoichiometric stability constants determined by two independent methods (Table 2) were in good agreement. Log K values increased linearly as the ionic strength decreased. Thus, at I = 0 and pH 3, log K values for the four metal-fulvic acid complexes were as follows: Cu-complex: 4.7, Ni-complex: 4.5, Mn-complex: 2.9, and Zn-complex: 3.2.

Complexes of fulvic acid with trivalent metals (Table 3) were investigated near pH 2 in order to ensure that the complexes were mononuclear with respect to the metals and to minimize the formation of hydroxides. Effects of different ionic strengths on the stability constants of Fe^{+3} - and Al^{+3} - fulvic acid complexes at pH 2 are illustrated in Table 3. The constants at I = 0 are thermodynamic stability constants. What is the ionic strength of a natural soil solution still remains a matter of conjecture. A value of $I \approx 0.03$ has

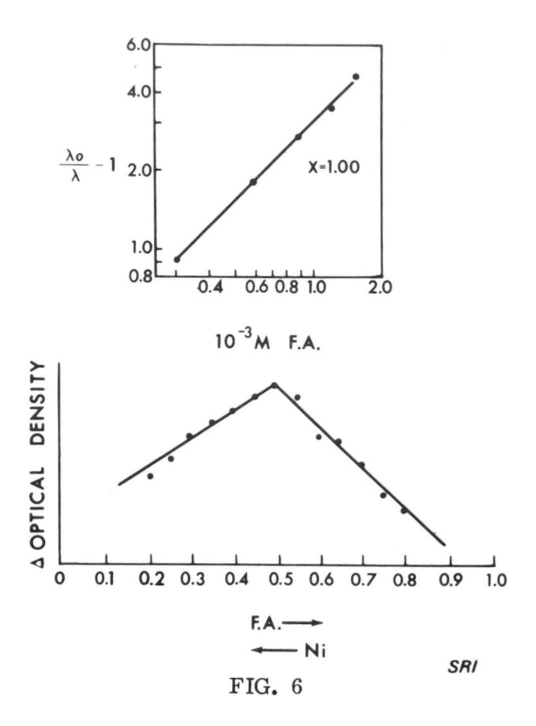

FIG. 6

Ratio of nickel to fulvic acid in complex formed at pH 3. Plot of log $\left\{\frac{\lambda_o}{\lambda} - 1\right\}$ vs. fulvic acid concentration (upper figure); Job plot (lower figure).

been suggested (17). If we accept this value, log K values determined at a very low ionic strength are likely to be more relevant to soils than those measured at I = 0.1.

V. MECHANISM OF METAL-FULVIC ACID INTERACTIONS

From the relative abundance of functional groups in fulvic acid, it seemed reasonable to assume that these groups were involved in reactions with metals. To obtain more specific information, measurements were made of the amounts of metals taken up by fulvic acid preparations in which the alcoholic OH, phenolic OH, and COOH groups were blocked selectively by methylation, acetylation, and esterification (18). Alcoholic OH groups played no part in metal-fulvic acid

TABLE 2

Stability Constants of Metal-Fulvic Acid Complexes at I = 0.1

Metal	Log K			
	pH 3		pH 5	
	CV[a]	IE[b]	CV[a]	IE[b]
Cu^{+2}	3.3	3.4	4.0	4.0
Ni^{+2}	3.1	3.2	4.2	4.2
Mn^{+2}	2.5	2.4	3.7	3.7
Zn^{+2}	2.4	2.3	3.7	3.6

[a]CV = method of continuous variations.
[b]IE = ion exchange equilibrium method.

TABLE 3

Effect of Ionic Strength on Log K Values of Fe^{+3} - and Al^{+3} - Fulvic Acid Complexes at pH 2.0

Metal	Log K			
	Ionic Strength (I)			
	0	0.05	0.10	0.15
Fe^{+3}	7.6	6.9	6.1	5.4
Al^{+3}	5.3	4.6	3.7	2.9

reactions. Essentially, two types of reactions occurred with metal ions and hydrous oxides: a major one, involving both acidic COOH and phenolic OH groups simultaneously, and a minor one, in which only less acidic COOH groups participated. A possible partial structure for 1:1 Fe^{+3} - fulvic acid complex is shown in Figure 7. Supporting evidence for this structure comes also from thermogravimetric experiments which showed that, in the 1:1 molar Fe^{+3} - fulvic acid complex, the iron occurred as $Fe(OH)^{+2}$(19).

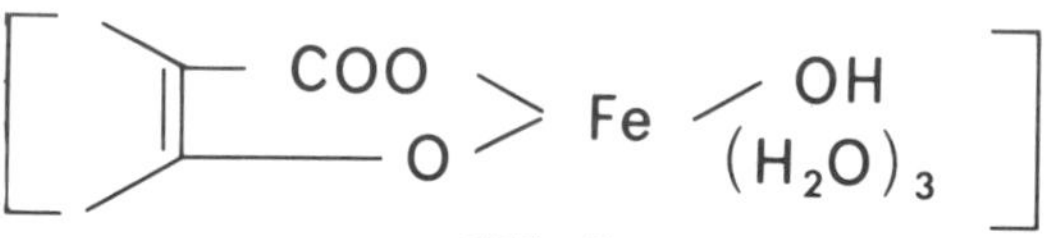

$$\left[\ \text{—COO},\ \text{—O} > Fe \begin{matrix} \diagup OH \\ (H_2O)_3 \end{matrix}\ \right]$$

FIG. 7

Possible partial structure for a 1:1 molar Fe^{+3} - fulvic acid complex.

VI. FORMATION OF WATER-INSOLUBLE METAL-FULVIC ACID COMPLEXES

As iron and aluminum are among the most abundant elements in soils, we were interested especially in their reactions with fulvic acid. As increasing amounts of iron and aluminum were added to fulvic acid solutions maintained at pH values of 2.5 and 4.0, respectively, the complexes became more and more insoluble in water. While 1:1 molar Fe^{+3} - and Al^{+3} - fulvic acid complexes were completely water-soluble, 2:1 complexes were less so, and 6:1 complexes were water-insoluble. Chemical, spectroscopic, and thermogravimetric characteristics of water-insoluble metal-fulvic acid complexes have been described (19, 20). Thus, fulvic acid forms two types of complexes with iron and aluminum: water-soluble ones if the molar metal-fulvic acid ratio is 1:1 or less, and water-insoluble complexes if the ratio is greater than 1:1. These observations have practical implications in soils and possibly also in waters. For example, they offer a reasonable explanation for the genesis of the B horizon (rich in Fe, Al, and fulvic acid) in Podzol soils. Thus, after being produced in the 0 (surface) horizon, fulvic acid, on its path down the soil profile, forms, at first, water-soluble complexes with metal ions and hydroxylated metal compounds. The water-solubility of these compounds decreases as more and more metal reacts with the fulvic acid until the complexes become completely insoluble and precipitate in the B horizon. The amounts of metal that can be dissolved and transported by fulvic acid within soils are quite substantial. Under acidic conditions, 670 g of fulvic acid can dissolve and maintain in solution up to 56 g of iron or 27 g of aluminum, which is equivalent to 84 mg of iron or 40 mg of aluminum per g of fulvic acid.

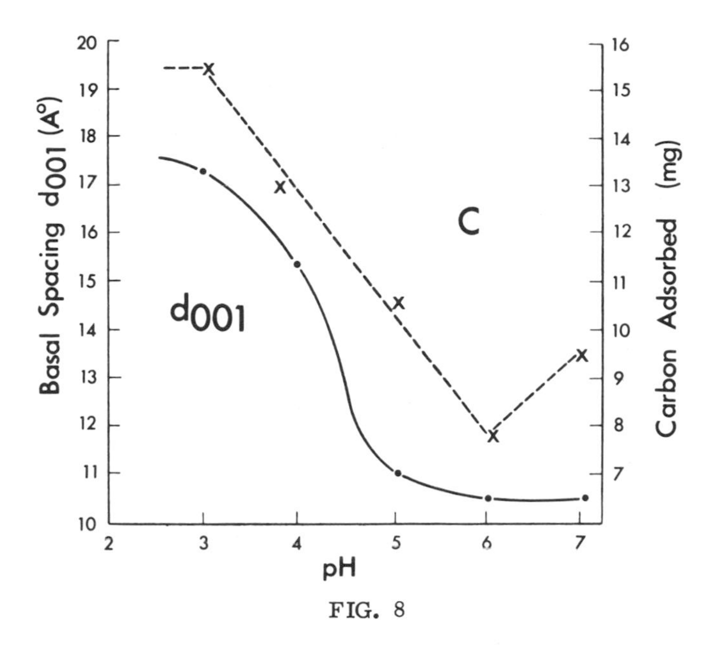

FIG. 8

Effect of pH on d_{001} and adsorption of carbon by Na-montmorillonite. Reprinted from Ref. (21), p. 70, by courtesy of the Amer. Assoc. Advan. Sci.

VII. REACTIONS BETWEEN FULVIC ACID AND CLAYS

A. Effect of pH.

Other reactions that are likely to occur in soils and waters are those between humic compounds and clay minerals. The effect of pH on the adsorption, both interlamellar and on external surfaces, of fulvic acid on Na-montmorillonite is shown in Figure 8. The d_{001} of untreated Na-montmorillonite was 9.87Å; after interaction with fulvic acid at pH 2.5 it was increased to 17.80Å (21). Interplanar spacings were pH dependent, and decreased with an increase in pH. The steepest decrease occurred between pH values of 4 and 5. It is noteworthy that the apparent pK value of the COOH groups in the fulvic acid was 4.5, which suggests that the magnitude of the d_{001} was related to the degree of ionization of the COOH groups in the fulvic acid. At pH $<$ 4.0, relatively few of these groups had ionized, so that the fulvic acid behaved essentially like an uncharged molecule that could penetrate the interlamellar spaces and displace water from between the silicate layers. As the pH rose, more and more functional groups ionized to result in an increased negative charge; thus, at pH $>$ 5.0, the d_{001} was less than 11.0Å, indicating repulsion of negatively charged fulvic acid by the negatively charged clay. The curve showing adsorption of C (% C x 2 = fulvic acid) resembled the one showing d_{001} values. At pH 2.5, 33.2 mg of fulvic acid were adsorbed on 40 mg of Na-montmorillonite.

B. Effect of Interlayer Cations.

In addition to pH, the fulvic acid concentration and, especially, interlayer cations had a marked effect on the adsorption of fulvic acid by montmorillonite. As illustrated in Figure 9, the magnitude of the d_{001} spacings was proportional to the amounts of fulvic acid adsorbed (22). Variations in spacings appeared, however, not to be related to the ability of the different cations to form stable complexes with fulvic acid, nor to the ionization potentials of the metal ions. The high interlamellar adsorption of fulvic acid by clays saturated with Pb^{+2}, Cu^{+2}, and Na^{+} is due, likely, to the relatively low energy with which these ions hold on to water, and which, in turn, is more easily displaced by the fulvic acid.

C. Adsorption of Fulvic Acid on and in the Clay.

DTA analysis was found to be suitable for differentiating between adsorption of fulvic acid on external clay surfaces and in interlamellar spaces (23). DTA curves for untreated fulvic acid, sodium montmorillonite, and for two fulvic acid-montmorillonite complexes are shown in Figure 10. Peaks in the DTA curve of untreated fulvic acid (curve A) may be interpreted in the following manner (10): the shallow and broad endotherm near 100°C. is due to dehydration, the shoulder-like exotherm at about 330°C., and the prominent exotherm at 450°C. arise from decarboxylation and oxidation of the fulvic acid "nucleus", respectively. The DTA curve for the physical mixture of fulvic acid and sodium montmorillonite (curve C) is essentially a composite of the two constituents, except that the main exotherm is lowered from 450 to 400°C. The DTA curve for the fulvic acid-unheated clay complex (curve D) differs from that of the physical mixture in a number of respects: the exotherm in the 400 to 500°C. region is much broader, indicating that the decomposition occurs over a wider temperature

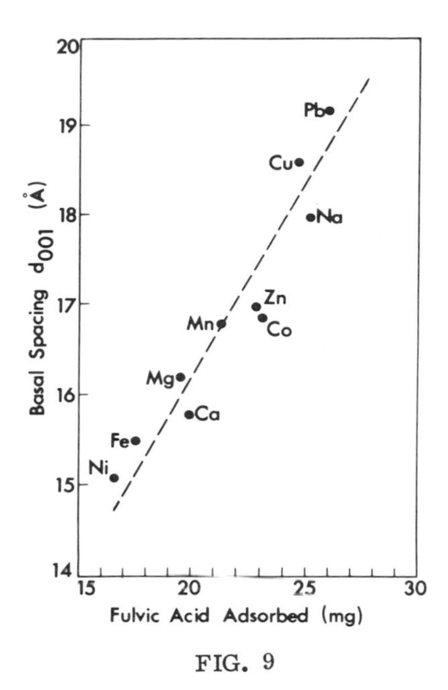

FIG. 9

Relationship between d_{001}-spacings and amounts of fulvic acid adsorbed by 40 mg of montmorillonite saturated with different cations. Reprinted from Ref. (22), p. 73, by courtesy of Soil Science.

range. Near 670°C., an exotherm appears which is symptomatic of the presence of an interlayer complex and is similar to exotherms in this temperature range exhibited by clay-protein and other clay-organic complexes (24, 25, 26). By contrast, the DTA pattern of the fulvic acid-heated clay complex (curve E) shows no exotherm near 670°C. Thus, the two principal adsorption reactions are well separated. The reaction extending from 350 to 550°C. is associated most likely with the combustion of fulvic acid adsorbed by montmorillonite on its external surfaces, while the reaction occurring between 550 and about 800°C. is due to fulvic acid adsorbed in the interlamellar spaces. Curve (D) also shows that approximately one half of the adsorbed fulvic acid is in the interlamellar spaces, with the remainder on external clay surfaces.

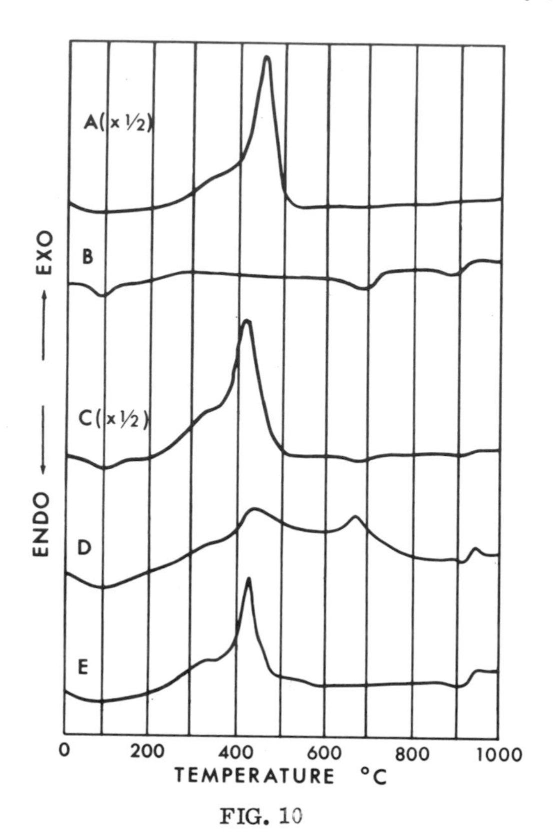

FIG. 10

DTA curves of: (A) untreated fulvic acid, (B) Na-montmorillonite, (C) physical mixture of fulvic acid and Na-montmorillonite, (D) fulvic acid-unheated montmorillonite complex, (E) fulvic acid-heated montmorillonite complex.

VIII. PHYSIOLOGICAL EFFECTS OF FULVIC ACID

Since little is known about the more direct effects of humic compounds, such as those caused when they are taken up by plants and how plant metabolism is influenced, we investigated the effect of fulvic acid on root initiation in bean stem segments. Root formation was stimulated by fulvic acid concentrations $>$ 500 ppm and was more than 300% greater than that of the control when between 3,000 and 6,000 ppm of fulvic acid was administered (27). Fulvic acid concentrations up to 4,000 ppm inhibited indoleacetic acid stimulated cell enlargement in Alaska pea stems regardless of whether the plant hormone was endogenous or externally applied (28). The physiological effects of fulvic appear to be related to its metal complexing ability and possibly also to its relatively high content of stable free radicals.

IX. SUMMARY

Fulvic acid is a highly oxidized, and chemically and biologically stable water-soluble humic compound that is found in soils and waters. As much as 60% of the weight of the fulvic acid is composed of functional groups such as carboxyls, hydroxyls, and carbonyls which are attached to a predominantly aromatic "nucleus". Fulvic acid is a naturally-occurring metal complexing agent that can complex di- and trivalent metal ions, and bring into and maintain in stable solution metals from practically insoluble hydroxides and oxides. It can mobilize and transport within soils and waters substantial amounts of metals at pH levels at which these metals would be insoluble otherwise.

Fulvic acid can also interact with clay minerals such as montmorillonite by adsorbing on external clay surfaces regardless of pH and in interlayer spaces at pH values < 4.5.

In view of its physical and chemical properties, it is likely that fulvic acid plays a prominent role in controlling the supply and availability of metals in soils and waters. It may so affect practically all reactions that occur in these media. Of special interest are those reactions that involve living systems, requiring metals for their survival.

REFERENCES

1. R. Ishiwatari, Soil Sci., 107, 53 (1969).

2. W. L. Lamar, Geol. Survey Prof. Paper, 600-D, D24 (1968).

3. M. A. Rashid and L. H. King, Geochim. Cosmochim. Acta, 33, 147 (1969).

4. M. Schnitzer and J. G. Desjardins, Can. J. Soil Sci., 49, 151 (1969).

5. A. V. Fotiyev, Doklady Akad. Nauk. (USSR), 179, 183 (1967).

6. D.H.R. Barton and M. Schnitzer, Nature, 198, 417 (1963).

7. M. Schnitzer and J. G. Desjardins, Soil Sci. Soc. Amer. Proc., 26, 362 (1962).

8. M. Schnitzer and J. R. Wright, Transact. 7th Intl. Congr. Soil Sci. Madison, II, 112 (1960).

9. M. Schnitzer, Can. Spectrosc., 10, 121 (1965).

10. M. Schnitzer and I. Hoffman, Geochim. Cosmochim. Acta., 29, 278 (1965).

11. H. Kodama and M. Schnitzer, Fuel, 46, 87 (1967).

12. M. Schnitzer and J. G. Desjardins, Can. J. Soil Sci., 44, 272 (1964).

13. E. H. Hansen and M. Schnitzer, Soil Sci. Soc. Amer. Proc., 31, 79 (1967).

14. E. H. Hansen and M. Schnitzer, Soil Sci. Soc. Amer. Proc., 33, 28 (1969).

15. A. E. Martell and M. Calvin, Chemistry of Metal Chelate Compounds, 1st. ed., Prentice Hall, Inc., New York, 1952, p. 94.

16. P. Job, in The Determination of Stability Constants, (F. J. C. Rossotti and H. Rossotti), 1st ed., McGraw-Hill Book Co., Inc., New York, 1961, p. 47.

17. J. R. Geering and J. F. Hodgson, Soil Sci. Soc. Amer. Proc., 33, 54 (1969).

18. M. Schnitzer and S.I.M. Skinner, Soil Sci., 99, 278 (1965).

19. M. Schnitzer and S.I.M. Skinner, Soil Sci., 98, 197 (1964).

20. M. Schnitzer and S.I.M. Skinner, Soil Sci., 96, 86 (1963).

21. M. Schnitzer and H. Kodama, Science, 153, 70 (1966).

22. H. Kodama and M. Schnitzer, Soil Sci., 106, 73 (1968).

23. H. Kodama and M. Schnitzer, Transact. Intl. Clay Conference (Tokyo), I, 765 (1969).

24. W. F. Bradley and R. Grim, J. Phys. Chem., 52, 1404 (1948).

25. W. H. Allaway, Soil Sci. Soc. Amer. Proc., 13, 183 (1949).

26. O. Talibudeen, J. Soil Sci., 3, 251 (1952).

27. M. Schnitzer and P. A. Poapst, Nature, 213, 598 (1967).

28. P. A. Poapst, C. Genier, and M. Schnitzer, Plant and Soil, in press.

CHAPTER 14

THE ESTIMATION OF THE FREE ENTHALPY OF FORMATION OF GASEOUS ORGANIC COMPOUNDS BY MEANS OF GROUP CONTRIBUTIONS

H.A.G. Chermin

Central Laboratory, Staatsmijnen/DSM,
Geleen, The Netherlands

I. INTRODUCTION

In the theory of thermodynamics, the condition of equilibrium at constant temperature and pressure is expressed by the relation:

$$dG = 0 \tag{14-1}$$

i.e., the free enthalpy of a system at equilibrium is minimum. Writing this condition in terms of equilibrium constants for, say, the reaction:

$$\sum \nu_A A \rightleftharpoons \sum \nu_B B \tag{14-2}$$

we get:

$$\Delta G_r = - RT \ln K = - RT \ln \left\{ \frac{\pi_{(B)}^{\nu_B}}{\pi_{(A)}^{\nu_A}} \right\} \tag{14-3}$$

where K is the equilibrium constant and ΔG_r is the change in free enthalpy caused by the reaction. This difference in free enthalpy may be written as:

$$\Delta G_r = \sum \nu_B \, \Delta G_{fB} - \sum \nu_A \, \Delta G_{fA} \tag{14-4}$$

where ΔG_{fA} and ΔG_{fB} are the free enthalpies of formation of components A and components B, respectively.

The equilibrium constant K is a measure of the maximum yield of products obtainable from given reactants. Obviously, if one wants to produce a large quantity of reaction products, the equilibrium constant should be at least unity. In terms of free enthalpies, the change of free enthalpy caused by the reaction, ΔG_r, should be zero or negative. It should be stressed here that the equilibrium constants, and hence the free enthalpies, enable us to predict the possible, so not the actual, composition of the mixture after the reaction. The actual composition of the reacted mixture depends on the relative velocities with which the various products are formed.

Equation (14-4) shows how the free enthalpy change caused by the reaction, ΔG_r, can be calculated from the free enthalpies of formation of products and reactants. It is important, therefore, to know these free enthalpies of formation.

The free enthalpies of formation of various compounds have been compiled in tabular form (1, 2, 3). These compilations, however, cover only a very small range of the immense variety of known or possible chemical compounds. Measuring the thermodynamic properties of all these compounds is an unfeasible job. This is why attempts have been, and still are being undertaken to establish correlations between thermodynamic properties and structure of a chemical compound.

II. GROUP CONTRIBUTIONS

If attempts are made to calculate the thermodynamic properties of a compound with a minimum of data, and if one demands a reasonable accuracy, one soon comes to depend on calculation methods involving the use of group contributions. In these methods it is assumed that, within certain limits, the molecular structural groups all contribute equally to the thermodynamic property, whatever the total molecular structure may be. For example, the thermodynamic properties of nC_6H_{14}:

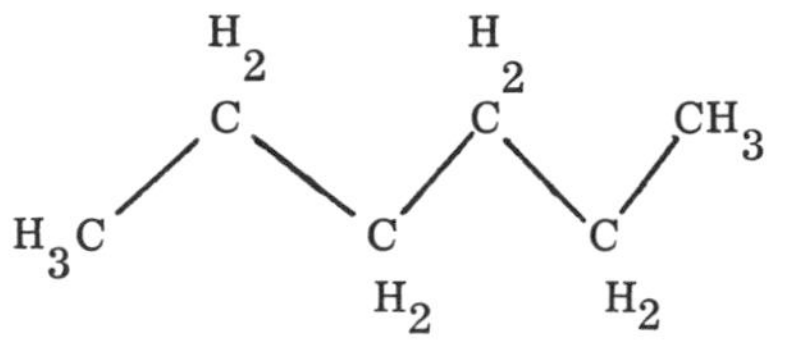

H_2 H_2
C C CH_3
H_3C C C
H_2 H_2

are supposed to be the sum of two contributions of a CH_3 group and four contributions of a CH_2 group. A statistical-mechanical calculation, however, (4, 5, 6) shows that the thermodynamic property is built up not only of group contributions, but also of structural contributions. These structural contributions comprise corrections for external rotation of the molecule as a whole, and corrections for what might roughly be called steric hindrance in the molecule.

The correction for external rotation can be calculated easily. From statistical thermodynamics, the symmetry contribution to the free enthalpy function, $-\left(\frac{G^o - H^o_o}{T}\right)$, is given by $-R \ln \sigma$ where σ is the symmetry number for rigid rotation of the molecule as a whole. The symmetry number is defined as the number of identical positions the molecule can assume by rigid rotation. Calculation of the correction for steric hindrance is more difficult. For normal and branched hydrocarbons the steric factor can be found by inspection of molecular models, as was demonstrated by Pitzer (4). The "steric influence" in other molecules is less easily found. In the method of group contributions, therefore, the "steric influence" is accounted by empirically established correction factors. Some 18 years ago, van Krevelen and Chermin (7) developed a correlation for calculating the free enthalpy of formation of gaseous organic compounds. Supposing:

$$\Delta G^o_{f\ \text{component}} = \sum \Delta G^o_{f\ \text{composing groups}} + \sum \Delta G^o_{f\ \text{correction for deviation}} + RT \ln \sigma \qquad (14\text{-}5)$$

where:

$$\Delta G^o_{f\ \text{group}} = A + B \cdot 10^{-2} \cdot T \qquad (14\text{-}6)$$

and:

$$\Delta G^o_{f\ \text{correction for deviation}} = A + B \cdot 10^{-2} \cdot T \qquad (14\text{-}7)$$

A and B are constants. A linear temperature relationship was found not to hold throughout the temperature range considered (300 °K. . . . 1000°K.), but that the group contributions and corrections could be

TABLE 1
Alkanes and Cyclo-Alkanes

Group	300 . . . . 600 °K.	600 . . . . 1500 °K.
$-CH_3$	$-10.833 + 2.176 \cdot 10^{-2}\ T$	$-12.393 + 2.436 \cdot 10^{-2}\ T$
$-CH_2-$	$-5.283 + 2.443 \cdot 10^{-2}\ T$	$-5.913 + 2.548 \cdot 10^{-2}\ T$
$-\overset{\vert}{C}\ H$	$-0.756 + 2.942 \cdot 10^{-2}\ T$	$-0.756 + 2.942 \cdot 10^{-2}\ T$
$-\underset{\vert}{\overset{\vert}{C}}-$	$+3.060 + 3.636 \cdot 10^{-2}\ T$	$+3.840 + 3.506 \cdot 10^{-2}\ T$
Correction	**300 . . . . 600°K.**	**600 . . . . 1500°K.**
Branching:		
3 adjacent $-\overset{H}{C}-$ groups	+ 2.312	+ 2.312
adjacent $-\overset{H}{C}-$ and $-\underset{\vert}{\overset{\vert}{C}}-$	+ 1.625	+ 1.625
2 adjacent $-\underset{\vert}{\overset{\vert}{C}}-$ groups	+ 2.543	+ 2.543
Ring formation:		
3 ring	$+27.215 - 3.147 \cdot 10^{-2}\ T$	$+26.495 - 3.027 \cdot 10^{-2}\ T$
4 ring	$+25.689 - 2.901 \cdot 10^{-2}\ T$	$+24.459 - 2.696 \cdot 10^{-2}\ T$
5 ring	$+5.511 - 2.583 \cdot 10^{-2}\ T$	$+3.417 - 2.234 \cdot 10^{-2}\ T$
6 ring	$-0.707 - 1.623 \cdot 10^{-2}\ T$	$+1.145 - 1.550 \cdot 10^{-2}\ T$
Ring branching:		
5 ring		
single branching	$-0.665 - 0.065 \cdot 10^{-2}\ T$	$-0.443 - 0.102 \cdot 10^{-2}\ T$
double branching:		
1.1 position	$-1.880 - 0.138 \cdot 10^{-2}\ T$	$-1.112 - 0.266 \cdot 10^{-2}\ T$
cis 1.2 position	$-1.485 + 0.245 \cdot 10^{-2}\ T$	$-0.939 - 0.159 \cdot 10^{-2}\ T$

TABLE 1 (continued)

Correction	300 . . . . 600 °K.	600 . . . . 1500 °K.
trans 1.2 position	$-2.163 - 0.138 \cdot 10^{-2}$ T	$-1.377 - 0.269 \cdot 10^{-2}$ T
cis 1.3 position	-1.423	$-0.637 - 0.131 \cdot 10^{-2}$ T
trans 1.3 position	$-1.838 - 0.138 \cdot 10^{-2}$ T	$-1.048 - 0.278 \cdot 10^{-2}$ T
6 ring		
single branching	$-0.370 - 0.106 \cdot 10^{-2}$ T	$+0.452 - 0.243 \cdot 10^{-2}$ T
double branching:		
1.1 position	-1.722	$-0.432 - 0.215 \cdot 10^{-2}$ T
cis 1.2 position	-0.500	$+1.432 - 0.322 \cdot 10^{-2}$ T
trans 1.2 position	$-2.003 - 0.138 \cdot 10^{-2}$ T	$-0.173 - 0.443 \cdot 10^{-2}$ T
cis 1.3 position	-3.055	$-1.687 - 0.228 \cdot 10^{-2}$ T
trans 1.3 position	-1.718	$+0.262 - 0.330 \cdot 10^{-2}$ T
cis 1.4 position	-1.152	$-0.192 \cdot 10^{-2}$ T
trans 1.4 position	$-3.125 + 0.034 \cdot 10^{-2}$ T	$-1.205 - 0.286 \cdot 10^{-2}$ T

TABLE 2
Alkenes, Cycloalkenes, and Alkynes

Group	300 . . . . 600 °K.	600 . . . . 1500 °K.
$H_2C=CH-$	$+14.281+1.642 \cdot 10^{-2}\,T$	$+13.513+1.770 \cdot 10^{-2}\,T$
$H_2C=C<$	$+16.823+1.864 \cdot 10^{-2}\,T$	$+15.785+2.037 \cdot 10^{-2}\,T$
$-CH=CH-$ (cis)	$+18.407+1.834 \cdot 10^{-2}\,T$	$+16.781+2.114 \cdot 10^{-2}\,T$
$-CH=CH-$ (trans)	$+17.019+2.007 \cdot 10^{-2}\,T$	$+16.755+2.051 \cdot 10^{-2}\,T$
$-CH=C<$	$+20.273+2.306 \cdot 10^{-2}\,T$	$+19.913+2.366 \cdot 10^{-2}\,T$
$>C=C<$	$+23.955+2.839 \cdot 10^{-2}\,T$	$+25.731+2.543 \cdot 10^{-2}\,T$
$H_2C=C=CH-$	$+48.871+1.063 \cdot 10^{-2}\,T$	$+48.133+1.186 \cdot 10^{-2}\,T$
$H_2C=C=C<$	$+51.159+1.481 \cdot 10^{-2}\,T$	$+51.159+1.481 \cdot 10^{-2}\,T$
$-CH=C=CH-$	$+53.176+1.528 \cdot 10^{-2}\,T$	$+52.690+1.609 \cdot 10^{-2}\,T$
$H_2C\leftrightarrow$	$+9.634+1.088 \cdot 10^{-2}\,T$	$+8.980+1.197+10^{-2}\,T$
$HC\leftrightarrow$ (ring)	$+3.100+0.610 \cdot 10^{-2}\,T$	$+2.536+0.704 \cdot 10^{-2}\,T$
$HC\leftrightarrow$ (ring, substituted)	$+12.302+1.438 \cdot 10^{-2}\,T$	$+12.408+1.420 \cdot 10^{-2}\,T$
$-C\leftrightarrow$ (ring)	$+5.280+0.994 \cdot 10^{-2}\,T$	$+5.634+0.935 \cdot 10^{-2}\,T$
$HC\equiv$	$+27.104-0.775 \cdot 10^{-2}\,T$	$+26.678-0.704 \cdot 10^{-2}\,T$
$-C\equiv$	$+27.478-0.617 \cdot 10^{-2}\,T$	$+27.346-0.595 \cdot 10^{-2}\,T$

TABLE 2 (continued)

Correction	300 . . . . 600 °K.	600 . . . . 1500 °K.
Ring formation:		
Pentene ring	$+3.455 - 2.448 \cdot 10^{-2}$ T	$+2.189 - 2.237 \cdot 10^{-2}$ T
Hexene ring	$-1.043 - 2.070 \cdot 10^{-2}$ T	$-0.737 - 2.121 \cdot 10^{-2}$ T

TABLE 3

Aromatics

Group	300 . . . . 600 °K.	600 . . . . 1500 °K.
HC⟷⟷	+ 3.100 + 0.610 . 10^{-2} T	+ 2.536 + 0.704 . 10^{-2} T
- C⟷⟷	+ 5.280 + 0.994 . 10^{-2} T	+ 5.634 + 0.935 . 10^{-2} T
⟷ C⟷⟷	+ 2.260 + 0.553 . 10^{-2} T	+ 2.566 + 0.502 . 10^{-2} T
Correction		
Branching		
Double branching		
1.2 position	+ 0.955 + 0.055 . 10^{-2} T	+ 1.687 - 0.067 . 10^{-2} T
1.3 position	+ 0.352 - 0.057 . 10^{-2} T	+ 0.574 - 0.094 . 10^{-2} T
1.4 position	- 0.183 + 0.105 . 10^{-2} T	+ 0.615 - 0.028 . 10^{-2} T
Triple branching		
1.2.3 position	+ 1.453 - 0.112 . 10^{-2} T	+ 1.039 - 0.043 . 10^{-2} T
1.2.4 position	+ 0.297 - 0.070 . 10^{-2} T	+ 0.243 - 0.061 . 10^{-2} T
1.3.5 position	- 0.320 - 0.137 . 10^{-2} T	- 0.436 - 0.116 . 10^{-2} T
Four fold branching		
1.2.3.4 position	+ 3.664 - 0.007 . 10^{-2} T	+ 4.456 - 0.139 . 10^{-2} T
1.2.3.5 position	+ 2.861 + 0.025 . 10^{-2} T	+ 3.359 - 0.058 . 10^{-2} T
1.2.4.5 position	+ 2.736 - 0.150 . 10^{-2} T	+ 3.072 - 0.206 . 10^{-2} T

TABLE 3 (continued)

Correction	300 . . . . 600 °K.	600 . . . . 1500 °K.
Five fold branching	$+4.400 + 0.091 \cdot 10^{-2}$ T	$+5.468 - 0.087 \cdot 10^{-2}$ T
Six fold branching	$+8.254 + 0.260 \cdot 10^{-2}$ T	$+10.006 - 0.032 \cdot 10^{-2}$ T

TABLE 4

Sulphur Compounds

Group	300 . . . . 600 °K.	600 . . . . 1000 °K.
$- SH_{prim.}$	$-11.789 + 1.069 \cdot 10^{-2}$ T	$-11.234 + 0.975 \cdot 10^{-2}$ T
$- SH_{sec.}$	$-12.935 + 1.082 \cdot 10^{-2}$ T	$-12.131 + 0.948 \cdot 10^{-2}$ T
$- SH_{tert.}$	$-14.173 + 1.279 \cdot 10^{-2}$ T	$-12.955 + 1.076 \cdot 10^{-2}$ T
$- SH_{phen.}$	$-10.678 + 1.189 \cdot 10^{-2}$ T	$-10.186 + 1.107 \cdot 10^{-2}$ T
- S -	$- 5.408 + 1.288 \cdot 10^{-2}$ T	$- 4.766 + 1.181 \cdot 10^{-2}$ T
- S - S -	$-18.532 + 2.806 \cdot 10^{-2}$ T	$-17.362 + 2.611 \cdot 10^{-2}$ T
S (aromatic)	$- 0.974 + 0.561 \cdot 10^{-2}$ T	$- 0.248 + 0.440 \cdot 10^{-2}$ T
=SO	$-30.634 + 3.523 \cdot 10^{-2}$ T	$-29.014 + 3.253 \cdot 10^{-2}$ T
$=SO_2$	$-82.784 + 5.620 \cdot 10^{-2}$ T	$-80.354 + 5.215 \cdot 10^{-2}$ T
Correction	**300 . . . . 600 °K.**	**600 . . . . 1000 °K.**
Ring formation:		
C, C, S (three-membered ring)	$+18.328 - 3.011 \cdot 10^{-2}$ T	$+16.828 - 2.761 \cdot 10^{-2}$ T
C, C, S, C (four-membered ring)	$+18.343 - 2.707 \cdot 10^{-2}$ T	$+16.687 - 2.431 \cdot 10^{-2}$ T

TABLE 4 (continued)

Correction	300 . . . . 600 °K.	600 . . . . 1000 °K.
C — C | S C — C (5-membered ring: 4 C, 1 S)	- 0.808 - 2.263 . 10^{-2} T	- 1.294 - 2.082 . 10^{-2} T
C — C C S C — C (6-membered ring: 5 C, 1 S)	- 1.879 - 1.691 . 10^{-2} T	- 2.113 - 1.652 . 10^{-2} T
Reference state S_2g		

TABLE 5

Nitrogen Compounds

Group	300 . . . . 600 °K.	600 . . . . 1000 °K.
$-NH_2$	$- 3.654 + 2.615 \cdot 10^{-2}\ T$	$- 4.564 + 2.750 \cdot 10^{-2}\ T$
$-\overset{H}{N}-$	$+ 2.262 + 2.937 \cdot 10^{-2}\ T$	$- 2.634 + 2.875 \cdot 10^{-2}\ T$
$-N<$	$+ 4.960 + 3.721 \cdot 10^{-2}\ T$	$+ 6.430 + 3.476 \cdot 10^{-2}\ T$
$-NH_{2\,phen.}$	$- 1.752 + 2.745 \cdot 10^{-2}\ T$	$- 1.164 + 2.647 \cdot 10^{-2}\ T$
N (aromatic)	$+ 16.746 + 1.064 \cdot 10^{-2}\ T$	$+ 16.746 + 1.064 \cdot 10^{-2}\ T$
$- NO_2$	$- 8.048 + 3.562 \cdot 10^{-2}\ T$	$- 8.048 + 3.562 \cdot 10^{-2}\ T$
$- C \equiv N$	$+ 28.668 - 1.001 \cdot 10^{-2}\ T$	$+ 30.344 + 1.280 \cdot 10^{-2}\ T$
$- N \equiv C$	$+ 43.952 - 1.093 \cdot 10^{-2}\ T$	$+ 45.638 - 1.374 \cdot 10^{-2}\ T$

described by two linear temperature functions, one for the range $300^{o}K$. . . . $600^{o}K$. and one for temperatures above $600^{o}K$.

It became necessary to rework the old correlation in the light of new data. Revised values of group contributions and corrections are listed in Tables 1-7. The group contributions and corrections given in these tables have been calculated from the following data:

Table 1 - Alkanes and cycloalkanes: group contributions and corrections calculated from data by Rossini (1). Table 2 - Alkenes, cycloalkenes, and alkynes: group contributions and corrections calculated from data by Rossini (1). Table 3 - Aromatics: group contributions and corrections calculated from data by Rossini (1). Table 4 - Sulphur compounds: group contributions and corrections calculated from data supplied by members of the thermodynamics group at the Bureau of Mines, Bartlesville, Oklahoma (8, 9, 10, 11, 12, 13, 14, 15, 16, 17, 18, 19). Contributions for $>SO$ and $>SO_2$ taken from data by Pitzer and Barrow (20). Table 5 - Nitrogen compounds: - NH_2, $>NH$, and $>N$-group contributions calculated from data by Seha (21) for the methyl amines. - C≡N and - N≡C group contributions calculated from thermodynamic properties published by Pillai and Cleveland (22). The value for the - NO_2 group was calculated from data by McCullough (23). The phenolic amine contribution has been derived from data for aniline (24). The N ⇆ contribution was calculated from the thermodynamic properties of the methyl pyridenes (25). Table 6 - Oxygen compounds: group contributions for the - OH groups, - C(=O)H, - C(=O) -, - O - , and - C(=O) - OH groups were taken from unpublished calculations of thermodynamic properties by the author (26). The calculated thermodynamic properties agree well with literature values on alcohols (27, 28, 29, 30, 31), aldehydes (32, 33), ketones (34, 35), acetic acid (36), and ethers (37). The correction for the formation of a C–O–C ring was derived from thermodynamic properties published by Green (38). Table 7 - Halogen compounds: the contribution for a - F group was calculated from thermodynamic properties listed in the IANAF tables (39). The - Cl and - Br group contributions were calculated from values by Green and Holden (40). Finally, the contribution for the - I group was taken from data by Pitzer and Gelles (41).

III. ACCURACY OF THE CORRELATION

It can be shown that on the average, the free enthalpies of formation of hydrocarbons calculated with the above correlations agree

TABLE 6

Oxygen Compounds

Group	300 . . . . 600 °K.	600 . . . . 1000 °K.
$-OH_{prim.}$	$-42.472 + 1.126 \cdot 10^{-2}$ T	$-43.216 + 1.250 \cdot 10^{-2}$ T
$-OH_{sec.}$	$-44.913 + 1.212 \cdot 10^{-2}$ T	$-44.500 + 1.143 \cdot 10^{-2}$ T
$-OH_{tert.}$	$-49.913 + 1.121 \cdot 10^{-2}$ T	$-49.190 + 1.004 \cdot 10^{-2}$ T
$-OH_{phen.}$	$-44.498 + 1.419 \cdot 10^{-2}$ T	$-44.522 + 1.348 \cdot 10^{-2}$ T
$-C(=O)H$	$-29.160 + 0.663 \cdot 10^{-2}$ T	$-29.994 + 0.802 \cdot 10^{-2}$ T
$-C(=O)-$	$-33.280 + 1.000 \cdot 10^{-2}$ T	$-33.280 + 1.000 \cdot 10^{-2}$ T
$-C(=O)OH$	$-94.755 + 2.764 \cdot 10^{-2}$ T	$-94.755 + 2.764 \cdot 10^{-2}$ T
$-O-$	$-49.178 + 1.783 \cdot 10^{-2}$ T	$-49.178 + 1.783 \cdot 10^{-2}$ T
Correction	**300 . . . .600 °K.**	**600 . . . .1000 °K.**
Ring formation		
C–C–O (three-membered ring)	$+45.934 - 3.207 \cdot 10^{-2}$ T	$+45.934 - 3.207 \cdot 10^{-2}$ T

TABLE 7

Halogen Compounds

Group	300 . . . . 600 °K	600 . . . . 1000 °K
-F	$-45.953 - 0.218 \cdot 10^{-2}\ T$	$-45.953 - 0.218 \cdot 10^{-2}\ T$
- Cl	$-11.702 - 0.225 \cdot 10^{-2}\ T$	$-11.906 - 0.191 \cdot 10^{-2}\ T$
- Br	$-4.665 - 0.218 \cdot 10^{-2}\ T$	$-4.017 - 0.321 \cdot 10^{-2}\ T$
- I	$-6.401 + 1.693 \cdot 10^{-2}\ T$	$-8.801 + 2.093 \cdot 10^{-2}\ T$

Reference state

Br_2 g
I_2 g

with literature values within 0.8 kcal/mole. The accuracy of the free enthalpies of formation of non-hydrocarbons thus calculated is not so good. Deviations of up to 3-5 kcal/mole may occur. This is due to the fact that most of the group contributions have been calculated from the first few members of a homologous series, which deviate from "normal behavior". Lack of data prevented calculation of corrections for compensating these effects. It is recommended, therefore, to use this correlation only as a means for investigating whether or not a product can be formed in a certain reaction. If it is desired to know the equilibrium concentration, other methods for predicting the thermodynamic properties must be employed.

IV. HOW TO USE THE CORRELATION

Some examples:

(a) Naphthalene

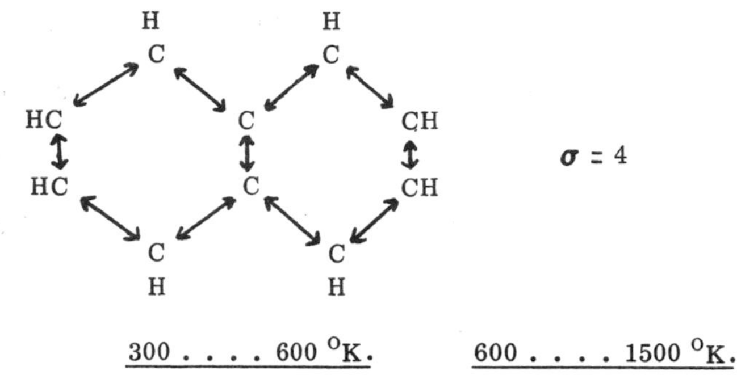

	300 . . . . 600 °K.	600 . . . . 1500 °K.
8 HC ⇆	$24.800 + 4.880 \cdot 10^{-2} \cdot T$	$20.288 + 5.632 \cdot 10^{-2} \cdot T$
2 ↔ C ⇆	$4.520 + 1.106 \cdot 10^{-2} \cdot T$	$5.132 + 1.004 \cdot 10^{-2} \cdot T$
$RT\ln\sigma$	$0.276 \cdot 10^{-2} \cdot T$	$0.276 \cdot 10^{-2} \cdot T$
+		
ΔG_f^o =	$29.320 + 6.262 \cdot 10^{-2} \cdot T$	$25.420 + 6.912 \cdot 10^{-2} \cdot T$

Literature value (1):

$$\Delta G_f^o = 29.284 + 6.288 \cdot 10^{-2} \cdot T \qquad 25.639 + 6.896 \cdot 10^{-2} \cdot T$$

(b) Fluorobenzene

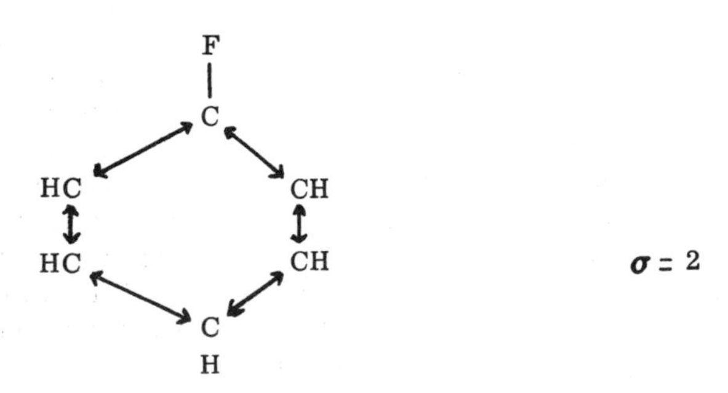

	300 . . . . 600 °K.	600 . . . . 1000 °K.
5 $HC\leftrightarrows$	$15.500 + 3.050 \cdot 10^{-2} \cdot T$	$12.680 + 3.520 \cdot 10^{-2} \cdot T$
1 — $C\leftrightarrows$	$5.280 + 0.994 \cdot 10^{-2} \cdot T$	$5.634 + 0.935 \cdot 10^{-2} \cdot T$
1 — F	$-45.953 - 0.218 \cdot 10^{-2} \cdot T$	$-45.953 - 0.218 \cdot 10^{-2} \cdot T$
$RT\ln\sigma$ +	$0.138 \cdot 10^{-2} \cdot T$	$0.138 \cdot 10^{-2} \cdot T$
$\Delta G_f^o =$	$-25.173 + 3.964 \cdot 10^{-2} \cdot T$	$-27.639 + 4.375 \cdot 10^{-2} \cdot T$

Literature value (42):

$\Delta G_f^o =$	$-27.508 + 4.153 \cdot 10^{-2} \cdot T$	$-29.380 + 4.465 \cdot 10^{-2} \cdot T$

(c) 4-Fluorotoluene

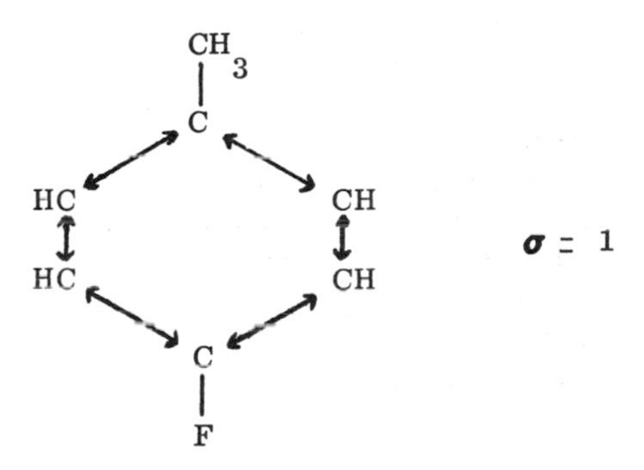

	300 . . . . 600 °K.	600 . . . . 1000 °K.
4 $HC\leftrightarrows$	$12.400 + 2.440 \cdot 10^{-2} \cdot T$	$10.144 + 2.816 \cdot 10^{-2} \cdot T$
2 — $C\leftrightarrows$	$10.560 + 1.988 \cdot 10^{-2} \cdot T$	$11.268 + 1.870 \cdot 10^{-2} \cdot T$
1 — CH_3	$-10.833 + 2.176 \cdot 10^{-2} \cdot T$	$-12.393 + 2.436 \cdot 10^{-2} \cdot T$
1 — F	$-45.953 - 0.218 \cdot 10^{-2} \cdot T$	$-45.953 - 0.218 \cdot 10^{-2} \cdot T$

1 (1.4 Corr.)	$-0.183 + 0.105 \cdot 10^{-2} \cdot T$	$0.615 - 0.028 \cdot 10^{-2} \cdot T$
$RT\ln\sigma$	$+ 0.0$	$+ 0.0$
$\Delta G_f^o =$	$-34.009 + 6.491 \cdot 10^{-2} \cdot T$	$-36.319 + 6.876 \cdot 10^{-2} \cdot T$

Literature value (43):

$\Delta G_f^o =$	$-35.448 + 6.663 \cdot 10^{-2} \cdot T$	$-35.982 + 6.752 \cdot 10^{-2} \cdot T$

V. DISCUSSION

The group contributions in the foregoing tables are given in more figures than is warranted by the accuracy of the contributions. This has been done to preserve the internal consistency of the correlation.

Calculation of thermodynamic properties by means of the correlation based on the group contribution concept is very simple. However, the accuracy of such a group contribution method leaves much to be desired, especially in those cases where the thermodynamic properties of the first few compounds of a homologous series have to be calculated. Unfortunately, it is exactly these compounds that are of great importance for the chemical industry. The inaccuracy can be remedied by a lavish introduction of corrections for structural effects. We have done so for the hydrocarbon groups so as to achieve an average deviation of not more than 0.8 kcal/mole. At the moment, corrections for non-hydrocarbon groups, cannot be calculated due to lack of data. However, even if sufficient data had been available, we would have refrained from calculating these corrections because this would have detracted from the simplicity of the correlation. In addition, the number of corrections would have been extremely large.

APPENDIX

A MORE ACCURATE ESTIMATION OF THERMODYNAMIC PROPERTIES

Separate the thermodynamic property as follows:

$$\phi = \phi_{transl.} + \phi_{ext.rot.} + \phi_{vibr.} + \phi_{int.rot.} + \phi_{el.} \qquad (14\text{-}8)$$

Assume for the moment that only the lowest electronic level of the molecule is populated. Then $\phi_{el.}$ is zero. Once the composition of the molecule and its structure is known, $\phi_{transl.}$ and $\phi_{ext.rot.}$ can be calculated "exactly" (in rigid rotator approximation).

The sum of $\phi_{vibr.} + \phi_{int.\,rot.}$ can be found by the method of group equation. To give an example, the thermodynamic properties of gaseous cyclohexanol were calculated:

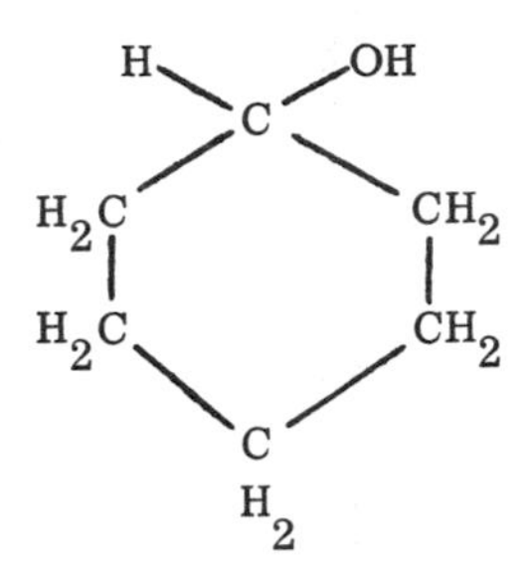

The molecular weight is 100.16.

With: $r_{C-C} = 1.54$ Å $\qquad r_{C-H} = 1.07$ Å

$r_{C-O} = 1.43$ Å $\qquad r_{O-H} = 0.96$ Å

$\Delta COH = 111^o$ and with all other angles seeing tetrahedral, calculation of the product of the moments of inertia (along the main axes) yields:

$$I_A I_B I_C = 3.038 \,.\, 10^{-113}\ g^3\ cm^6$$

For the sum of the contributions from translation and external rotations, we find in the usual way (44):

$$C_p^o = 7.9488$$

$$\frac{H^o - H_o^o}{T} = 7.9488$$

$$-\left(\frac{G^o - H_o^o}{T}\right) = 18.3027\ {}^{10}\log T + 13.0475$$

We assume that the sum of the contributions for vibration and internal rotation plus contributions for boat-chair equilibrium and polar-equatorial equilibrium is given by the group equation.

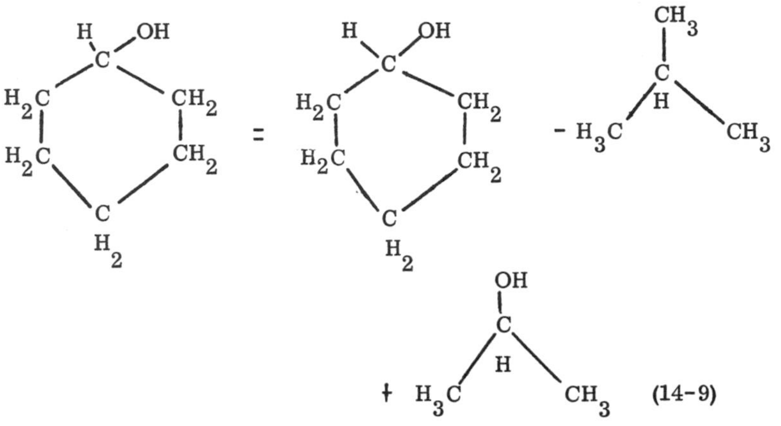

H OH
C
H_2C CH_2 = H OH C H_2C CH_2 − CH_3 C H H_3C CH_3
H_2C CH_2 H_2C CH_2
C C
H_2 H_2

OH
C
H
+ H_3C CH_3 (14-9)

This sum of the contributions is found in Table 8.

TABLE 8

Temp. °K.	C_p^{o*}	$\frac{H^o - H_o^{o*}}{T}$	$-\left(\frac{G^o - H_o^{o*}}{T}\right)$
298.15	22.42	9.28	5.72
300	22.62	9.36	5.77
400	32.99	13.98	9.09
500	42.43	18.72	12.73
600	50.32	23.47	15.56
700	56.92	27.71	20.62
800	64.41	31.75	24.46
900	67.12	35.40	28.42
1000	71.10	38.78	31.32

The total thermodynamics properties of cyclohexanol in this approximation are given in Table 9.

TABLE 9

Temp. °K.	C_p^o *	$\frac{H^o - H_o^o}{T}$ *	$-\left(\frac{G^o - H_o^o}{T}\right)$ *	S^o *
298.15	30.37	17.22	65.43	82.65
300	30.56	17.31	65.53	82.84
400	40.93	21.91	71.14	93.05
500	50.36	26.58	76.53	103.11
600	58.27	31.32	81.84	113.16
700	64.86	35.67	87.01	122.68
800	70.34	39.69	92.00	131.69
900	75.04	43.35	96.90	140.25
1000	79.07	46.73	101.64	148.37

*cal/mole °K.

Kelley (45) measured the entropy of cyclohexanol. From the entropy, using the vapor pressures as a function of the temperature (46), we calculated for the "measured" entropy of gaseous cyclohexanol:

$$S^o_{298.15\ ^oK.} = 82.8 \pm 0.5 \text{ cal/mol } ^oK.$$

As can be seen, there is excellent agreement between the estimated and measured entropies. Analogous calculations have been done on a number of other molecules, and good to excellent agreement has been found always.

REFERENCES

1. F. D. Rossini, Selected Values of Physical and Thermodynamic Properties of Hydrocarbons and Related Compounds., Carnegie Press, Pittsburgh, 1953.

2. F. D. Rossini, Selected Values of Chemical Thermodynamic Properties., National Bureau of Standards, Circular 500, U. S. Government Printing Office, Washington, 1950.

3. Landolt-Börnstein , Zahlenwerte and Funktionen aus Physik-Chemie-Astronomie-Geophysik-Technik 4 Teil Kalorische Zustandsgrossen, Springer, Berlin, 1961.

4. K. S. Pitzer, J. Chem. Phys., 8, 711 (1940).

5. W. B. Person and G. C. Pimentel , J. Amer. Chem. Soc., 75, 532 (1953).

6. D. R. Herschbach, H. S. Johnston, and D. Rapp, J. Chem. Phys., 31, 1661 (1959).

7. D. W. van Krevelen and H.A.G. Chermin , Chem. Eng. Sci., 1, 66 (1951).

8. D. W. Scott and J. P. McCullough , The Chemical Thermodynamic Properties of Hydrocarbons and Related Substances., Contribution No. 93, Thermodynamics Laboratory, Petroleum Research Center, Bureau of Mines., Region IV, Bartlesville, Oklahoma.

9. D. W. Scott, J. Amer. Chem. Soc., 78, 5463 (1956).

10. G. B. Guthrie, D. W. Scott, and G. Waddington , J. Amer. Chem. Soc., 74, 2795 (1952).

11. D. W. Scott, J. Amer. Chem. Soc., 75, 2795 (1953).

12. W. N. Hubbard, J. Amer. Chem. Soc., 80, 3547 (1958).

13. D. W. Scott, J. Chem. Eng. Data, 4, 246 (1959).

14. R. E. Pennington, J. Amer. Chem. Soc., 78, 2055 (1956).

15. J. P. McCullough, J. Amer. Chem. Soc., 79, 561 (1957).

16. D. W. Scott, J. Amer. Chem. Soc., 79, 1062 (1957).

17. R. E. Pennington, J. Amer. Chem. Soc., 78, 3266 (1956).

18. J. P. McCullough, J. Amer. Chem. Soc., 76, 4796 (1954).

19. J. P. McCullough, J. Amer. Chem. Soc., 80, 4786 (1958).

20. K. S. Pitzer and G. M. Barrow, Ind. Eng. Chem., 41, 2737 (1949).

21. Z. Seha , Collect. Czech. Chem. Commun., 26, 2435 (1961).

22. M. G. Krishna Pillai and F. F. Cleveland, J. Molec. Spectrosc., 5, 212 (1960).

23. J. P. McCullough, J. Amer. Chem. Soc., 76, 4791 (1954).

24. W. E. Hatton, J. Chem. Eng. Data, 7, 2229 (1962).

25. D. W. Scott, J. Phys. Chem., 67, 680 (1963), 67, 685 (1963).

26. H.A.G. Chermin., Unpublished results.

27. J.H.S. Green., J. Appl. Chem., 11, 397 (1961).

28. H.A.G. Chermin, Petrol. Refiner., 40, 127 (1961).

29. N. S. Berman and J.J. McKetta, J. Phys. Chem., 66, 1444 (1962).

30. J.H.S. Green., Trans. Faraday Soc., 59, 1559 (1963).

31. H.A.G. Chermin., Petrol. Refiner., 40, 234 (1961).

32. H.A.G. Chermin., Petrol. Refiner.,40, 181 (1961).

33. I.A. Vasilev and A.A. Vvedenskii., Russian J. Phys. Chem., 31, 1092 (1965), 40, 459 (1966).

34. J. K. Nickerson, K. A. Kobe, and J.J. McKetta, J. Phys. Chem., 65, 1037 (1961).

35. R. E. Pennington and K.A. Kobe., J. Amer. Chem. Soc., 79, 300 (1957).

36. W. Weltner., J. Amer. Chem. Soc., 77, 3427 (1955), 83, 5045 (1961).

37. S. C. Banerjee and K.L. Doraiswamy, British Chem. Eng., 9, 311 (1964).

38. J.H.S. Green, Chem. Ind., 369 (1961).

39. D. Stull, IANAF Thermochemical Tables, (loose leaf issue).

40. J.H.S. Green and D.J. Holden , J. Chem. Soc., London, 1794 (1962).

41. K.S. Pitzer and E. Gelles, J. Amer. Chem. Soc., 75, 5259 (1953).

42. D.W. Scott, J. Amer. Chem. Soc., 78, 5457 (1956).

43. D. W. Scott, J. Phys. Chem., 37, 867 (1962).

44. G. J. Janz , Estimation of Thermodynamic Properties of Organic Compounds, Academic Press, New York, 1958.

45. K.K. Kelley , J. Amer. Chem. Soc., 51, 1400 (1929).

46. I. Nitta and S. Seki, J. Chem. Soc. Japan, Pure Chem. Section, 69, 141 (1948).

CHAPTER 15

THERMODYNAMIC STABILITY OF SELECTED ORGANIC PESTICIDES IN AQUATIC ENVIRONMENTS

Hassan M. Gomaa and Samuel D. Faust

Rutgers - The State University

I. INTRODUCTION

Modern pesticides are an important example of many synthetic organic chemicals which have helped man in the continuous struggle for improvement of his environment. Since 1940, there have been concerted efforts to prepare compounds that would be effective against various pests. The success of these efforts can be judged by the numbers of compounds currently used to control such diverse pests as insects, fungi, nematodes, rodents, and weeds.

The production and sale of organic pesticides, sometimes called economic poisons, have increased in almost every year since 1954. Over 750 million pounds of synthetic organic pesticides were used in the United States in 1964. The usage versus time plot is exponential with one billion pounds per year predicted for application by 1970 (1).

Since 1945, organic pesticides and herbicides were reported in drinking, recreational, irrigational, fish, and shellfish waters, and sediments and bottom muds wherefrom the evidence was obtained, in the main, from physiological responses of aquatic organisms. More recent (1961 to date) evidence from direct analysis by gas-liquid chromatography and other techniques indicates that most natural waters and their equilibrium solid phases contain trace amounts of organic pesticides (2).

Organic pesticides enter natural waters from direct application for control of aquatic weeds, trash fish, and aquatic insects; percolation and runoff from agricultural lands; drift from aerial and land applications; discharge of industrial waste waters; and discharge of waste waters from cleanup of equipment used for pesticidal formulations and application. Instances of pollution from all of these sources have been reviewed (3,4,5,6,7,8,9,10). These papers indicated also those specific organic pesticides that have been detected in water, sediments, and bottom muds from the "500 compounds in more than 54,000 formulations" (11).

Man's priority use of water for ingestion always raises the health issue. No concentration limits were established by the U. S. Public Health Service in 1962 when they considered drinking water standards for pesticides (12). Insufficient information, at that time, was cited as the reason not to establish specific limits. Where information was available, however, the concentrations of organic pesticides were below levels which would constitute a known health hazard (13). Nevertheless, the committee recommended continued surveillance of the problem.

The National Technical Advisory Subcommittee on Public Water Supplies found it necessary to make some rather arbitrary decisions in order to develop water quality criteria for organic pesticides (14). Table 1 shows the permissible levels of three groups of pesticides: the chlorinated hydrocarbons, herbicides, and the cholinesterase inhibiting group which include the organic phosphorus types and the carbamates. These values were derived by a group of toxicologists as those residual levels which, if ingested over extensive periods, could not cause harmful or adverse physiological changes in man. In the case of aldrin, heptachlor, chlordane, and parathion, the committee adopted even lower than physiologically safe levels; namely, amounts which, if present, can be detected by their taste and odor. The committee attempted to set limits at the lowest practical level in order to minimize the amount of a toxicant contributed by water, particularly when such other sources as milk, food, or air are known to represent the major exposures to man (12). If pesticides do gain entry into water resources in spite of all reasonable precautions in their use, public health and water treatment officials must have the knowledge and the plant facilities to reduce their concentrations to the permissible levels. Some organic pesticides are resistant to conventional water and waste water treatments (15) and many resist biologically-mediated degradation in aquatic environments where they persist for considerable periods of time (5,16,17,18). Numerous studies have been conducted on processes for removal of organic compounds of all types from aqueous solutions (19,20,21). One possibility is the use of such chemical oxidants as ozone, chlorine dioxide, chlorine, the peroxides, and potassium permanganate. These have been used in water treatment plants to reduce the concentration of organic contaminants (22,23,24,25).

TABLE 1

Surface Water Criteria for Public Water Supplies

Pesticide	Permissible Level mg/l
Pesticides:	
Aldrin	0.017
Chlordane	0.003
DDT	0.042
Dieldrin	0.017
Endrin	0.001
Heptachlor	0.018
Heptachlor epoxide	0.018
Lindane	0.056
Methoxychlor	0.035
Organic phosphate plus carbamates[a]	0.100
Toxaphene	0.005
Herbicides:	
2,4-D plus 2,4,5-T plus 2,4,5-TP	0.100
Phenols	0.001

[a]As 0.1 mg/l parathion equivalent.

Occasionally water chemists are faced with the question of whether or not a chemical oxidative treatment actually will remove trace organic pollutants from water. Thermodynamics permits the calculation of chemical equilibria which, in turn, yields information concerning the feasibility of the occurrence of a reaction. This is valid provided that sufficient free energy of formation data are available for reactants and products.

Whether or not a reaction proceeds to equilibrium depends on kinetic factors. A certain amount of activation energy is necessary to initiate any reaction. Whereas thermodynamic data provide an answer to the feasibility of an oxidative reaction, kinetics determine the overall rate of the process. A combined knowledge of the driving forces and of the kinetic factors provides the information from which the water chemist may evaluate the optimum conditions for the chemical degradation of specific pollutants.

The objectives, therefore, of this paper are to explore the feasibility and the kinetics of chemical oxidation of selected organic pesticides by potassium permanganate, chlorine, and chlorine dioxide. Three aquatic herbicides, 2,4-dichlorophenoxyacetic acid (2,4-D), diquat, and paraquat, and the most widely used organophosphorus insecticide, parathion, and its intermediate oxidation products paraoxon and p-nitrophenol were studied (Table 2). Also the kinetics of hydrolysis of parathion and paraoxon were examined in order to determine their persistence in aqueous systems.

II. THERMODYNAMIC STABILITY AND EQUILIBRIUM

Thermodynamics gives the relation between an equilibrium constant and the free energy change in a reaction:

$$\Delta G^{o}_{Reaction} = -RT \ln K_{eq} \qquad (15\text{-}1)$$

In this case, $\Delta G^{o}_{Reaction}$ is calculated from the arithmetic difference of the sum of the free energies of formation of the products and reactants:

$$\Delta G^{o}_{Reaction} = \sum \Delta G^{o}_{Products} - \sum \Delta G^{o}_{Reactants} \qquad (15\text{-}2)$$

If the free energies of formation of reactants and products in an equilibrium are known, it is possible to calculate the net free energy change of a hypothetical reaction.

Occasionally the application of thermodynamics to organic reactions is handicapped by a lack of the free energy of formation of compounds. Only a very small fraction of organic compounds have been examined for their thermodynamic behaviour.

A. Calculation of Free Energy of Formation of Organic Pesticides.

Starting from a statistical - mechanical calculation, Franklin (26) proved that a thermodynamic function of an arbitrary paraffin

TABLE 2

Structures of Organic Pesticides Examined in This Study

Diquat (1,1'-ethylene-2,2'-dipyridylium dibromide monohydrate)

$2\,Br^- \cdot H_2O$

Paraquat (1,1'-dimethyl-4,4'-dipyridylium dimethyl sulfate)

$2\,CH_3 \cdot SO_4^-$

2,4-D (2,4-dichlorophenoxyacetic acid)

Parathion (O,O-diethyl-O-p-nitrophenyl phosphorothioate)

Paraoxon (O,O-diethyl-O-p-nitrophenyl phosphate)

p-Nitrophenol

---

hydrocarbon may be the sum of contributions from the various functional groups. If required, there may be a correction for the symmetry of the molecule:

$$\Delta G_f = \sum \begin{matrix}\text{Contributions of} \\ \text{composing groups}\end{matrix} + RT \ln \sigma \qquad (15\text{-}3)$$

in which $RT\ln \sigma$ represents the corrective factor. (The symmetry number, σ, is the number of indistinguishable positions in space that a molecule may occupy by simple rigid rotation. It may be found by considering a space model of the molecule concerned.)

Van Krevelen and Chermin (27) modified Franklin's method by considering the group contributions as a linear function of temperature:

$$\Delta G_{f\,group} = A + \frac{B}{100} T \qquad (15\text{-}4)$$

This equation is based on an argument by Scheffer (28) and shows a similarity to the general equation:

$$\Delta G = \Delta H - T\Delta S \qquad (15\text{-}5)$$

If Equations (15-4) and (15-5) can be compared, it follows that A has the dimension of heat of formation and $\frac{B}{100}$ has the dimension of entropy of formation. According to Ulich (29) A is, by approximation, equal to the heat of formation at 298.1°K. and $\frac{B}{100}$ is approximately the entropy of formation at 298.1°K. Separate classes of group contributions as well as their free energies of formation were calculated and tabulated by Chermin (30) for several organic compounds. This method was used for the calculation of free energy of formation of organic pesticides in this study. Table 3 shows an example for the calculation of ΔG^o_f for 2,4-D. Table 4 shows the calculated free energy of formations of the selected organic pesticides in this study.

B. Thermodynamic Oxidation-Reduction Models.

A water that is in equilibrium with the atmosphere has a well-defined redox potential (P_{O_2} = 0.21 atm., pH = 7.0, 25°C., pE = 13.75 or $E_h \simeq 800$ mv). A simple equilibrium calculation shows that, at this pE, all organic carbon should be present as C(+IV) or as CO_2, HCO_3^-, CO_3^{-2}; S and N should occur in the form of SO_4^{-2} and NO_3^-.

TABLE 3

Calculation of Free Energy of Formation of 2,4-Dichlorophenoxyacetic Acid

Groups Present	A	B
$-\overset{\displaystyle O}{\overset{\|\|}{C}}-OH$	-94.755	+2.764
$-CH_2-$	- 5.283	+2.443
-O -	-49.178	+1.783
3 $\overset{\|}{C}$ (aromatic, ↙↘)	3(+5.100)	3(+0.994)
3 $\overset{H}{\overset{\|}{C}}$ (aromatic, ↙↘)	3(+3.100)	3(+0.610)
2 -Cl	2(-11.702)	2(-0.225)
1,2,4-Branching	+0.297	-0.070

$$\Delta G^{o}_{f_{2,4-D}} = \sum \begin{matrix}\text{Contributions of} \\ \text{Composing Groups}\end{matrix} + RT \ln \sigma$$

$$= \sum \left(A + \frac{B}{100} T\right)_{group} + RT \ln \sigma$$

At T = 293°K., $\sigma = 1$

$$\Delta G^{o}_{f_{2,4-D}} = -114.127 \quad \text{K. Cal. Mole}^{-1}$$

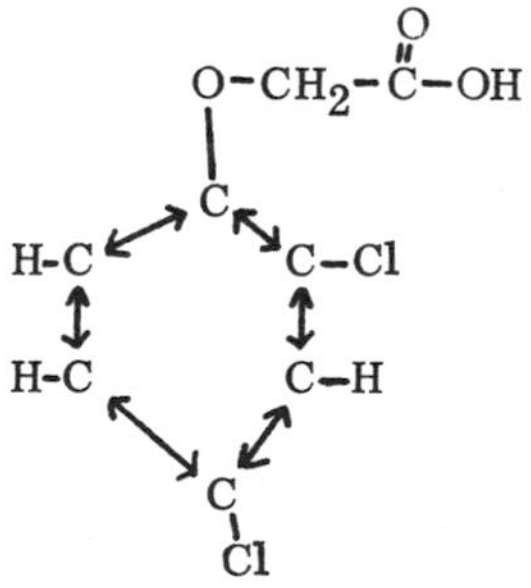

O-CH₂-C(=O)-OH attached to aromatic ring C; ring: C, C-Cl, C-H, C(Cl), H-C, H-C

TABLE 4

Calculated Free Energies of Formation of Selected Organic Pesticides

Compound	ΔG_f^o K. Cal. Mole^{-1}
Diquat	+112.47
Paraquat	+108.84
2,4-D	-114.13
Parathion	- 36.77
Paraoxon	-211.94
p-Nitrophenol	- 2.02

That organic carbon compounds are unstable in aquatic environments can be shown from a thermodynamic model:

$$CH_2O + H_2O = CO_{2(g)} + 4\,H^+ + 4\bar{e} \qquad (15\text{-}6)$$

$$O_{2(g)} + 4H^+ + 4\bar{e} = 2\,H_2O \qquad (15\text{-}7)$$

$$CH_2O + O_{2(g)} = CO_2 + H_2O \qquad (15\text{-}8)$$

$$\Delta G^o_{Reaction} = \Delta G^o_{f\,(CO_2)} + \Delta G^o_{f\,(H_2O)} - \Delta G^o_{f\,(CH_2O)} - \Delta G^o_{f\,(O_2)}$$

$$= -\,92.30 - 56.69 + 31.00 = -117.99 \text{ K. Cal. Mole}^{-1} \qquad (15\text{-}9)$$

$$\Delta G^o_{Reaction} = -\,n\,FE^o_{Reaction} \qquad (15\text{-}10)$$

$$\text{or } E^o_{Reaction} = -\,\frac{\Delta G^o\text{Reaction}}{nF} = +\,1.28 \text{ volts}$$

From the free energy value and $E^o_{Reaction}$ it appears that the reaction is thermodynamically feasible. A similar model can be calculated for organic pesticides for their thermodynamic stability in aquatic environments, i.e., in the presence of dissolved oxygen and also to predict whether or not oxidation of an organic pesticide would be feasible with a specific oxidant. Thus, the reaction between diquat and $KMnO_4$ in an acid medium can be calculated as follows:

$$3(C_{12}H_{12}N_2)^{2+} + 72\ H_2O = 36\ CO_2 + 6\ NH_4^+ + 156\ H^+ + 156\ \bar{e} \quad (15\text{-}11)$$

$$52\ MnO_4^- + 208\ H^+ + 156\ \bar{e} = 52\ MnO_2 + 104\ H_2O \quad (15\text{-}12)$$

$$3(C_{12}H_{12}N_2)^{2+} + 52\ MnO_4^- + 52\ H^+ = 36\ CO_2 + 6\ NH_4^+ + 52\ MnO_2 + 32\ H_2O \quad (15\text{-}13)$$

$$\Delta G^o_{Reaction} = 36\ \Delta G^o_{f_{CO_2}} + 6\ \Delta G^o_{f_{NH_4^+}} + 52\ \Delta G^o_{f_{MnO_2}} + 32\ \Delta G^o_{f_{H_2O}} - 3\ \Delta G^o_{f_{Diquat}} - 52\ \Delta G^o_{f_{MnO_4^-}} - 52\ \Delta G^o_{f_{H^+}} \quad (15\text{-}14)$$

$$\Delta G^o_{Reaction} = -\ 5781.04\ \text{K. Cal.}$$

$$\text{or } E^o_{Reaction} = +1.61 \text{ volts}$$

The assumptions involved in the above redox model are: (a) MnO_4^- is reduced to MnO_2 and not to Mn^{2+} (31), (b) carbon is oxidized to CO_2(31), (c) nitrogen is oxidized to NH_4^+ (31), and (d) equilibrium is obtained between the reactants and the products. With these assumptions in mind, the redox reaction models suggest that oxidation of diquat with MnO_4^- is feasible. Table 5 shows similar calculations for the organic pesticides and oxidants selected for this study.

Stewart (31) reported that many organic compounds are degraded to CO_2 by $KMnO_4$, although in basic solution oxalate is isolated frequently as a major reaction product. This is because oxalate suffers further rapid oxidation only in acid solution. When aromatic nitro

TABLE 5
Reaction Models for Chemical Oxidation of Selected Organic Pesticides

	Redox Reaction Model	$\Delta G^o_{Reaction}$ K. Cal.
	2,4-D	
(15-15)	$2\ C_8H_6O_3Cl_2 + 15O_2 = 16\ CO_2 + 4H^+ + 4\ Cl^- + 4\ H_2O$	-1600.78
(15-16)	$C_8H_6O_3Cl_2 + 15\ Cl_2 + 13H_2O = 8CO_2 + 32\ H^+ + 32\ Cl^-$	- 915.33
(15-17)	$C_8H_6O_3Cl_2 + 11\ ClO^- + 10\ OH^- = 4\ C_2O_4^{-2} + 13\ Cl^- + 8\ H_2O$	- 910.69
(15-18)	$C_8H_6O_3Cl_2 + 6\ MnO_4^- + 16\ H^+ = 8\ CO_2 + 11\ H_2O + 6\ Mn^{+2} + 2\ Cl^-$	- 992.64
(15-19)	$3\ C_8H_6O_3Cl_2 + 22\ MnO_4^- + 8\ OH^- = 12\ C_2O_4^{-2} + 6\ Cl^- + 13\ H_2O + 22\ MnO_2$	-2504.38
(15-20)	$C_8H_6O_3Cl_2 + 6\ ClO_2 + H_2O = 8\ CO_2 + 8\ Cl^- + 8\ H^+$	- 936.66
	Diquat	
(15-21)	$(C_{12}H_{12}N_2)^{2+} + 26\ Cl_2 + 24\ H_2O = 12\ CO_2 + 2\ NH_4^+ + 52\ Cl^- + 52\ H^+$	-1571.51
(15-22)	$(C_{12}H_{12}N_2)^{2+} + 20\ ClO^- + 12\ OH^- = 6\ C_2O_4^{-2} + 2\ NH_4^+ + 20\ Cl^- + 8\ H_2O$	-1573.11
(15-23)	$3(C_{12}H_{12}N_2)^{2+} + 52\ MnO_4^- + 52\ H^+ = 36\ CO_2 + 6\ NH_4^+ + 52\ MnO_2 + 32\ H_2O$	-5781.04
(15-24)	$3(C_{12}H_{12}N_2)^{2+} + 40\ MnO_4^- + 2\ OH^- = 18\ C_2O_4^{-2} + 6\ NH_3 + 40\ MnO_2 + 10\ H_2O$	-3884.47
(15-25)	$5\ (C_{12}H_{12}N_2)^{2+} + 52\ ClO_2 + 16\ H_2O = 60\ CO_2 + 10\ NH_4^+ + 52\ Cl^- + 52\ H^+$	-8038.49

TABLE 5 (continued)

	Redox Reaction Model	$\Delta G^o_{Reaction}$ K. Cal.
	Paraquat	
(15-26)	$(C_{12}H_{14}N_2)^{2+} + 27\ Cl_2 + 24\ H_2O = 12\ CO_2 + 2\ NH_4^+ + 54\ Cl^- + 54\ H^+$	-1631.45
(15-27)	$(C_{12}H_{14}N_2)^{2+} + 21\ ClO^- + 12\ OH^- = 6\ C_2O_4^{-2} + 2\ NH_4^+ + 21\ Cl^- + 9\ H_2O$	-1633.76
(15-28)	$(C_{12}H_{14}N_2)^{2+} + 18\ MnO_4^- + 18\ H^+ = 12\ CO_2 + 2\ NH_4^+ + 18\ MnO_2 + 12\ H_2O$	-2001.43
(15-29)	$(C_{12}H_{14}N_2)^{2+} + 14\ MnO_4^- = 6\ C_2O_4^{-2} + 2\ NH_3 + 14\ MnO_2 + 4\ H_2O$	-1356.52
(15-30)	$5\ (C_{12}H_{14}N_2)^{2+} + 54\ ClO_2 + 8\ H_2O = 60\ CO_2 + 10\ NH_4^+ + 54\ Cl^- + 54\ H^+$	-8575.96
	Parathion	
(15-31)	$C_{10}H_{14}O_5\ NSP + 2O_2 + H_2O = C_{10}H_{14}O_6NP + SO_4^{-2} + 2\ H^+$	- 295.92
(15-32)	$C_{10}H_{14}O_5\ NSP + 30\ Cl_2 + 26\ H_2O = 10\ CO_2 + SO_4^{-2} + NO_3^- + PO_4^{-3} + 60\ Cl^- + 66\ H^+$	-1791.76
(15-33)	$C_{10}H_{14}O_5\ NSP + 25\ ClO^- + 16\ OH^- = 5\ C_2O_4^{-2} + NO_3^- + SO_4^{-2} + PO_4^{-3} + 25\ Cl^- + 15\ H_2O$	-2164.78
(15-34)	$C_{10}H_{14}O_5NSP + 20\ MnO_4^- + 14\ H^+ = 20\ MnO_2 + 10\ CO_2 + SO_4^{-2} + NO_3^- + PO_4^{-3} + 14\ H_2O$	-6608.58
(15-35)	$3\ C_{10}H_{14}O_5\ NSP + 50\ MnO_4^- = 50\ MnO_2 + 15\ C_2O_4^{-2} + 3\ NO_3^- + 3\ SO_4^{-2} + 3\ PO_4^{-3} +$ $20\ H_2O + 2\ OH^-$	-5021.29
(15-36)	$C_{10}H_{14}O_5NSP + 12\ ClO_2 + 2\ H_2O = 10\ CO_2 + SO_4^{-2} + PO_4^{-3} + NO_3^- + 12\ Cl^- + 18\ H^+$	-1834.42

TABLE 5 (continued)

	Redox Reaction Models	$\Delta G^{o}_{Reaction}$ K. Cal.
	Paraoxon	
(15-37)	$C_{10}H_{14}O_6NP + 26\ Cl_2 + 21\ H_2O = 10\ CO_2 + NO_3^- + PO_4^{-3} + 52\ Cl^- + 56\ H^+$	-1465.30
(15-38)	$C_{10}H_{14}O_6NP + 21\ ClO^- + 14OH^- = 5\ C_2O_4^{-2} + NO_3^- + PO_4^{-3} + 21\ Cl^- + 14\ H_2O$	-1595.37
(15-39)	$3\ C_{10}H_{14}O_6NP + 52\ MnO_4^- + 40\ H^+ = 52\ MnO_2 + 30\ CO_2 + 3NO_3^- + 3\ PO_4^{-3} + 41\ H_2O$	-5464.75
(15-40)	$C_{10}H_{14}O_6NP + 14\ MnO_4^- = 14\ MnO_2 + 5C_2O_4^{-2} + NO_3^- + PO_4^{-3} + 7\ H_2O$	-3915.65
(15-41)	$5\ C_{10}H_{14}O_6NP + 52\ ClO_2 + H_2O = 50\ CO_2 + 5\ NO_3^- + 5\ PO_4^{-3} + 52\ Cl^- + 72\ H^+$	-7511.35
	p-Nitrophenol	
(15-42)	$C_6H_5NO_3 + 14\ Cl_2 + 12\ H_2O = 6\ CO_2 + NO_3^- + 28\ Cl^- + 29\ H^+$	- 798.89
(15-43)	$C_6H_5NO_3 + 11\ ClO^- + 7\ OH^- = 3\ C_2O_4^{-2} + NO_3^- + 11\ Cl^- + 6\ H_2O$	- 826.54
(15-44)	$3\ C_6H_5NO_3 + 28\ MnO_4^- + 25\ H^+ = 18\ CO_2 + 3\ NO_3^- + 28\ MnO_2 + 20\ H_2O$	-2972.22
(15-45)	$3\ C_6H_5NO_3 + 22\ MnO_4^- = 9\ C_2O_4^{-2} + 3\ NO_3^- + 22\ MnO_2 + 7\ H_2O + OH^-$	-2023.66
(15-46)	$5\ C_6H_5NO_3 + 28\ ClO_2 + 4\ H_2O = 30\ CO_2 + 5\ NO_3^- + 28\ Cl^- + 33\ H^+$	-3293.20

compounds are degraded by permanganate, the nitro unit becomes a nitrate ion in acid solution. Making the assumption that organic phosphorus and sulfur groups are oxidized to PO_4^{-3} and SO_4^{-2}, the reaction models for parathion, paraoxon, and p-nitrophenol are given in Table 5.

III. KINETICS OF CHEMICAL OXIDATION OF DIQUAT AND PARAQUAT

A. Potassium Permanganate.

Before proceeding to the kinetic studies, experiments were conducted to determine the stoichiometry of the reactions. Also, it was necessary to verify the reduced state of manganese as MnO_2(s). The oxidation of diquat (4.16×10^{-6}M) and paraquat (3.65×10^{-6}M) was conducted by the addition of an excess of $KMnO_4$ (10^{-3}M). The reactions were conducted under acidic and alkaline conditions (pH 5.1 and 9.1). After the complete disappearance of either diquat or paraquat, as indicated by a spectrophotometric measurement at 306 and 256 mμ, respectively, the amount of MnO_4^- consumed in the oxidation was determined. Table 6 shows the agreement between the experimental data and the proposed redox model. Thus, an application of MnO_4^- at a molar concentration 25 times the herbicide's concentration should be sufficient for stoichiometric oxidation.

The rates of oxidation of diquat and paraquat by potassium permanganate were found to be dependent upon the hydronium ion concentration. The rate of oxidation is also proportional to the product of the residual concentrations of permanganate and herbicide in accord with an integrated form of a second-order kinetic equation:

$$\frac{2.303}{C^o_{KMnO_4} - C^o_{HERB}} \log \frac{C^o_{HERB} C_{KMnO_4}}{C^o_{KMnO_4} C_{HERB}} = K_{ob} t \qquad (15\text{-}47)$$

where t is the time in minutes, C_{KMnO_4} is the residual molar concentration of $KMnO_4$ in solution at time t, C_{HERB} is the residual herbicide's concentration at time t, and K_{ob} is the observed rate constant in Liter Mole^{-1} Min.$^{-1}$. That the rates of $KMnO_4$ oxidation of diquat and paraquat are pH dependent is seen in Table 7.

TABLE 6

Potassium Permanganate Consumed by the Oxidation of Diquat and Paraquat

Compound	pH	Contact Time hours	Mole MnO_4^-/Mole Compound Calculated	Experimental
Diquat	5.1	120	17.33	18.40
	9.1	72	13.33	12.95
Paraquat	5.1	120	18.00	17.58
	9.1	72	14.00	13.28

TABLE 7

Effect of pH Upon the Rate Constants of Oxidation of Diquat and Paraquat by $KMnO_4$[a]

pH	K_{ob} Liter $Mole^{-1}$ $Min.^{-1}$ Diquat-$KMnO_4$	Paraquat-$KMnO_4$[b]
5.1	.123	.112
6.0	.087	.085
6.9	.186	.058
8.0	1.18	.035
9.1	14.8	9.89
10.0	54.6	15.6

[a] I = 0.02M; Temp. = $20^o \pm 0.2^oC.$; Average values of three experimental runs at the indicated pH.

[b] Initial molar concentration of diquat is 4.16×10^{-5}; that for paraquat is 3.65×10^{-5}, and $KMnO_4$ is 1×10^{-3}.

B. Chlorine.

From the preliminary experiment conducted to test for the feasibility of chlorine in the oxidation of diquat and paraquat, it was found that pH was a very important factor. Chlorine had no oxidative effect on diquat and paraquat under acidic conditions. On the other hand, as the pH value of the system was increased, the velocity of the oxidative reaction was increased. Thus, it was concluded tentatively that the oxidation proceeded via a ClO^- mechanism rather than by Cl_2 or HOCl.

As in the case of the permanganate oxidation, the same assumptions were made for chlorine about the products of oxidation. A molar ratio of 5:1 chlorine to herbicide concentration was used because of the analytical convenience for the determination of residual Cl_2 and also for the simulation of dosages used in water treatment plants.

In a similar manner to the permanganate oxidations, the reactions of diquat and paraquat with chlorine were fitted to the integrated form of a second-order kinetic equation. The reactions were studied in the pH range of 5 to 10. Chlorine had no oxidative effect on diquat at pH 5.0. An increase of the pH value of the system from 6.1 to 10.1 resulted in an increase of the velocity of the oxidation. Paraquat, in the pH range 5.0 to 7.1, was not oxidized within 24-hours. In the pH range 8.1 to 10.1, the rates of the paraquat reaction are pH dependent as shown in Table 8.

C. Chlorine Dioxide.

In the oxidation of diquat and paraquat with chlorine dioxide at pH values of 8.1, 9.0, and 10.1, the rates were extremely fast and could not be measured with the experimental techniques. These reactions were almost complete in less than one minute. On the other hand, chlorine dioxide had no effect on diquat and paraquat at pH values of 5.0, 6.1, and 7.1. Table 9 summarizes these oxidative reactions at 20^oC. Lowering the temperature from 20^oC. to 10^oC. had no effect on the oxidative kinetics.

D. Activation Energies.

Another portion of this study examined the effect of temperature on the rate constants which, in turn, led to the calculation of the activation energies of the oxidations. The rate constant, K_{ob}, was evaluated at 10°, 20°, 30°, and 40^oC. That the rate constants of the reaction vary greatly with temperature is shown in Table 10. Slight

TABLE 8

Effect of pH Upon the Rate Constants of Oxidation of Diquat and Paraquat by Chlorine[a]

pH	K_{ob} Liter Mole^{-1} Min.$^{-1}$	
	Diquat-Cl_2[b]	Paraquat-Cl_2[b]
5.1	No reaction	No reaction
6.2	.074	No reaction
7.1	1.87	No reaction
8.1	4.14	.019
9.0	12.4	.056
10.0	27.0	.099

[a]Temp. = 20° ± 0.2°C.; I = 0.02M; Average of three experimental runs at the indicated pH.

[b]Initial molar concentration of diquat is 4.16×10^{-5}; paraquat is 3.65×10^{-5}, and chlorine is 2.0×10^{-4}.

deviations appear at 30° and 40°C. The activation energies were evaluated with the Arrhenius equation as shown in Table 11.

The observed activation energy for the diquat-$KMnO_4$ reaction at pH 9.1 is approximately 1 K. Cal. smaller than that for the paraquat-$KMnO_4$ oxidation under the same conditions. Activation energy was increased by about 4 K. Cal. as a result of decreasing the pH value from 9.1 to 5.1. This was observed also for the paraquat-$KMnO_4$ system. Chlorine oxidations are characterized, generally, by lower activation energies under alkaline conditions when compared with neutral or slightly acidic conditions.

The rate of reaction of diquat - $KMnO_4$ at pH 9.1 is about 100 times faster than the rate at pH 5.1. The same is true with respect to paraquat oxidation. This means that potassium permanganate is a more effective oxidant of dipyridylium cations in an alkaline medium than in an acid medium. This may be due to the decrease of activation energy of the oxidation as the pH value is increased.

TABLE 9

Diquat and Paraquat - Chlorine Dioxide Oxidation[a]

pH	Herbicide, mg/l Initial	Residual	Residual ClO_2 mg/l	Reaction Time
Diquat - ClO_2				
10.1	15.0	0.00	2.61	1 Min.
	30.0	8.93	0.00	1 Min.
9.0	15.0	0.00	2.60	1 Min.
	30.0	8.83	0.00	1 Min.
8.1	15.0	0.00	2.59	1 Min.
	30.0	8.83	0.00	1 Min.
7.1	15.0	15.0	6.74	1 Min.
	15.0	15.0	6.72	3 Hrs.
	15.0	15.0	6.66	24 Hrs.
Paraquat - ClO_2				
10.1	15.0	0.00	2.84	1 Min.
	30.0	9.23	0.00	1 Min.
9.0	15.0	0.00	2.81	1 Min.
	30.0	9.23	0.00	1 Min.
8.1	15.0	0.00	2.85	1 Min.
	30.0	9.33	0.00	1 Min.
7.1	15.0	15.0	6.73	1 Min.
	15.0	15.0	6.72	3 Hrs.
	15.0	15.0	6.72	24 Hrs.

[a]Initial chlorine dioxide concentration = 6.75 mg/l; I = 0.02 M, and Temp. = $20^{o} \pm 0.2^{o}C$.

TABLE 10

Potassium Permanganate and Chlorine Oxidations - Effect of Temperature[a]

Temp., °C.	K_{ob} Liter Mole^{-1} Min.$^{-1}$			
	Diquat - $KMnO_4$		Paraquat - $KMnO_4$	
	pH 5.1	pH 9.1	pH 5.1	pH 9.1
10	0.053	6.71	0.040	4.73
20	0.126	14.54	0.111	9.64
30	0.275	23.46	0.232	17.99
40	0.501	41.28	0.425	33.61
	Diquat - Cl_2		Paraquat - Cl_2	
	pH 6.2	pH 9.0	pH 8.1	pH 9.0
10	0.024	6.72	0.004	0.025
20	0.074	12.42	0.018	0.056
30	0.133	19.11	0.048	0.130
40	0.190	28.19	0.126	0.294

[a]I = 0.02M; Initial concentrations were: diquat = 4.16×10^{-5}M, paraquat = 3.65×10^{-5}M, chlorine = 2.0×10^{-4}M, and $KMnO_4$ = 10^{-3}M.

TABLE 11

Activation Energies of the Diquat and Paraquat Oxidation Reactions

Reaction	pH	E_{ob} K. Cal. Mole^{-1}
Diquat - $KMnO_4$	5.1	14.25
	9.1	10.12
Paraquat - $KMnO_4$	5.1	15.12
	9.1	11.27
Diquat - Cl_2	6.2	14.82
	9.0	8.87
Paraquat - Cl_2	8.1	17.74
	9.0	14.46

The rates of reaction of aqueous chlorine with diquat and paraquat may vary with pH because of the effect of (H_3^+O) on the hydrolysis and protolysis of chlorine species in water. The relative concentrations of molecular chlorine, hypochlorous acid, and hypochlorite ion in aqueous solutions are controlled by two equilibria. When chlorine is dissolved in water it hydrolyzes:

$$Cl_2 + 2\ H_2O = HO\ Cl + H_3^+O + Cl^- \quad (15\text{-}48)$$

$$K_h = \frac{(HOCl)\ (H_3^+O)\ (Cl^-)}{(Cl_2)} \quad (15\text{-}49)$$

The equilibrium constant of the above reaction has been determined as 3.88×10^{-4} at 20^oC. (32). Hypochlorous acid protolyzes in water:

$$HOCl + H_2O = H_3^+O + OCl^- \quad (15\text{-}50)$$

$$K_a = \frac{(H_3^+O)\ (OCl^-)}{(HOCl)} \quad (15\text{-}51)$$

The acidic dissociation constant was determined by Morris (33) as 2.62×10^{-8} at 20^oC. Thus, molecular chlorine in water will immediately distribute itself between HOCl and OCl^- with the molar ratio of the two species controlled entirely by pH.

Apparently, diquat can be oxidized by hypochlorite ion, but not by hypochlorous acid. At pH 5.1 chlorine had no effect on diquat. But, at pH 7.1 chlorine is present as hypochlorite in approximately 25%, the oxidation proceeds with an appreciable velocity (K_{ob} = 1.87 Liter Mole^{-1} Min.$^{-1}$). At pH 9.0 the percentage of hypochlorite ion in solution is 97.1. Consequently, the rate constant increases to 12.4 Liter Mole^{-1} Min.$^{-1}$. Likewise, paraquat, in the pH range 5.1 to 7.1, was not oxidized during a 24-hour period. Whereas, in the pH range of 8.1 to 10.1, the rate constant value was increased. Concurrently, the activation energies of the chlorine oxidation reactions (Table 11) were decreased as the pH value was increased.

IV. HYDROLYSIS AND CHEMICAL OXIDATION OF PARATHION

A. Stability of Parathion and Paraoxon in Aqueous Systems.

During the last decade, parathion has been a most widely used organophosphorus insecticide for crop protection and as an aquatic insecticide (34). However, the wide spread use of this compound has resulted also in a great number of accidental intoxications many of which have been lethal to aquatic species.

Many organophosphorus compounds yield toxic metabolites by oxidation and isomerization processes (35). The conversion of thiophosphate esters to their oxygen-analogs may be accomplished by chemical or enzymic oxidation which, in turn, increases their cholinesterase inhibitive activity and hydrolyzability. Natural waters may provide the enzymic oxidation opportunity by aquatic insects, fish, and microorganisms. Chemical oxidation may be afforded by dissolved oxygen in natural waters or by chlorine and potassium permanganate that are used in conventional water treatment plants. Should oxygen analogs appear in natural or treated waters then their persistence and effect on potable water quality must be considered.

Since most organophosphorus pesticides hydrolyze, their persistence in natural waters may be controlled by chemical forces rather than by biological activities. An order of magnitude of persistence and/or appearance of hydrolysis products of organophosphorus pesticides in natural waters may be obtained from kinetic studies. Reference should be made to the researches of Peck (36), Ketelaar (37), Muhlman and Schrader (38), Weiss and Gakstatter (39), and Ruzicka et al. (40) who investigated the hydrolysis of several organophosphorus compounds in a variety of liquid media and under a variety of temperatures and pH values.

B. Hydrolysis of Parathion and Paraoxon.

This study reexamined the hydrolysis of parathion and paraoxon because of various deficiencies in the above cited references and because it was necessary to evaluate the extent of hydrolysis during the chemical oxidation studies. The rates of hydrolysis of parathion or paraoxon were found to be pH dependent, i.e., the reaction was catalyzed by hydronium or hydroxide ions:

$$\text{O}_2\text{N-C}_6\text{H}_4\text{-OP(S(O))(OC}_2\text{H}_5)_2 + H_2O \xrightarrow[\bar{O}H]{H_3\overset{+}{O}} \text{O}_2\text{N-C}_6\text{H}_4\text{-OH} + \text{HOP(S(O))(OC}_2\text{H}_5)_2 \qquad (15\text{-}52)$$

According to the mass action, the velocity should be dependent on the concentration of the two reactants:

$$-\frac{dC_{Ins.}}{dt} = KC_{Ins.}\ C_{Cat.} \qquad (15\text{-}53)$$

where $C_{Ins.}$ is the residual molar concentration of parathion at time t, and $C_{Cat.}$ is the concentration of catalyst, i.e., either $H_3\overset{+}{O}$ or OH^- in solution. As the amount of catalyst consumed in the course of the reaction is negligible, its concentration may be regarded as constant. The velocity equation then takes the form:

$$-\frac{dC_{Ins.}}{dt} = K_{ob}\ C_{Ins.} \qquad (15\text{-}54)$$

Thus, the process should observe first order kinetics that was verified by experimental observations. Integration of Equation (15-54) yields:

$$\log C_{Ins.} = \log C^{o}_{Ins.} - \frac{K_{ob}}{2.303}\ t \qquad (15\text{-}55)$$

where $C^{o}_{Ins.}$ is the initial molar concentration of parathion or paraoxon, and K_{ob} is the observed rate constant (t^{-1}). Table 12 shows the hydrolysis rate constants and half-lives of parathion and paraoxon at different $H_3\overset{+}{O}$ concentrations and at a temperature of 20°C.

An order of magnitude of persistence and/or appearance of hydrolysis products of parathion and paraoxon in natural water is obtained from the kinetic data. In general, at 20°C., parathion hydrolysis proceeds slightly faster than paraoxon under acidic or

TABLE 12

Hydrolysis of Parathion and Paraoxon at 20°C.

pH	Parathion[b] K_{ob} hr^{-1}	$t_{1/2}$ hrs	Paraoxon[c] K_{ob} hr^{-1}	$t_{1/2}$ hrs
3.1	1.65×10^{-4}	4182	1.46×10^{-4}	4726
5.0	1.88×10^{-4}	3670	1.66×10^{-4}	4156
7.4	2.66×10^{-4}	2594	2.00×10^{-4}	3450
9.0	1.32×10^{-3}	523	9.87×10^{-3}	69.9
10.4	2.08×10^{-2}	33.2	1.15×10^{-1}	6.0

[a]I = 0.02M.
[b]Initial parathion concentration = 3.948×10^{-5}M.
[c]Initial paraoxon concentration = 4.812×10^{-5}M.

neutral conditions. The reverse is true under alkaline conditions, where paraoxon hydrolysis is approximately seven times faster than parathion at pH 9.0 and five times faster at pH 10.4.

Hydrolysis products of organophosphorus pesticides should not be overlooked since their ultimate fate and effect on potable water quality are largely unknown at this date. One clue may be provided by p-nitrophenol, the hydrolysis product of parathion and paraoxon. Substituted phenols, especially chlorophenols, affect the taste and odor qualities of potable water (42). Chlorination of p-nitrophenol at a water treatment plant may produce an odorous product. Similar effects may be observed with other hydrolysis products. Otherwise, these impurities presently should be considered as contributing to the total organic content of natural waters.

C. Oxidation of Parathion, Paraoxon, and p-Nitrophenol by Potassium Permanganate.

There have not been any publications of significance concerning the kinetics of chemical oxidation of organophosphorus insecticides. Two nonkinetic studies are those of Robeck et al. (25), and Shiavone and Torrado (41) who investigated the effectiveness of water treatment processes for pesticide removal. These researchers found that conventional treatment with chemical coagulation, followed by sand filtration gave a maximum removal of 20% for parathion.

In this study, the oxidations of parathion (3.95×10^{-6}M), paraoxon (4.81×10^{-6}M), and p-nitrophenol (7.18×10^{-5}M) were conducted by adding an excess of $KMnO_4$ (2.0×10^{-3}M) under acidic and alkaline conditions (pH 3.1 and 9.0). After contact times of 2 to 5 days, the number of moles of $KMnO_4$ required for complete oxidation of one mole of compound was calculated. The data are shown in Table 13. It is clear that the observed experimental data are in good agreement with the hypothetical reactions proposed in Table 5.

Using gas-liquid and thin-layer chromatographic techniques, in conjunction with IR and UV spectrophotometric analyses, the only intermediate oxidation product detected during the oxidation of parathion under acidic and neutral conditions was paraoxon. However, under alkaline conditions, p-nitrophenol accumulates in the medium as well as trace amounts of paraoxon. Table 14 shows the differences in rates of build-up of paraoxon and p-nitrophenol during the oxidation of parathion. It is clear now that paraoxon is not the only oxidation product of parathion oxidation by $KMnO_4$ as was suggested by Robeck et al. (25) and Shiavone and Torrado (41).

The data obtained for parathion, paraoxon, and p-nitrophenol oxidations conform with the second-order rate Equation (15-47). A deviation from the second-order equation occurs as the reaction proceeds over a long period of time. Such a deviation can only mean that some intermediate product in the reaction is produced that is characterized by a slower rate of degradation and, as a consequence, will affect the observed rate constant of oxidation. In order to compare the efficiency of potassium permanganate for the oxidation of parathion and its intermediate oxidation products, paraoxon and p-nitrophenol, at different pH values, the observed rate constants determined after 30 minutes of reaction were used. This will avoid complications due to interference by intermediate products formed during the oxidation of the parent compound. By comparison of the rate constants of oxidation of paraoxon and p-nitrophenol with those for parathion, it will be possible to explain the differences in their rates of accumulation under acid, neutral, and alkaline conditions (Table 14).

Table 15 summarizes the rate constants and shows that the rates of oxidation of parathion are faster than those of paraoxon under acid or neutral conditions. The parathion - $KMnO_4$ reaction proceeds with a rate five times faster than the paraoxon - $KMnO_4$ reaction at pH 3.1 and about twenty-two times faster at pH 7.4. This may explain why paraoxon accumulates in the system as the parathion - $KMnO_4$ oxidation

TABLE 13

Potassium Permanganate Consumed in the Oxidation of Parathion, Paraoxon, and p-Nitrophenol[a]

Compound	pH	Contact Time, hours	Mole $KMnO_4$/Mole Compound Experimental	Calculated
Parathion	3.1	120	19.48	20.00
	9.0	48	17.08	16.66
Paraoxon	3.1	120	16.69	17.33
	9.0	48	14.17	14.00
p-Nitrophenol	3.1	96	9.10	9.33
	9.0	96	7.62	7.33

[a]Temp. = $20^{o} \pm 0.2^{o}C$.

reaction proceeds either under acid or neutral conditions. The reverse is true under an alkaline condition (pH 9.0) where trace concentrations of the paraoxon are detected in the system.

p-Nitrophenol may accumulate in the system from the hydrolysis or oxidation of parathion and paraoxon. However, the rates of hydrolysis of parathion and paraoxon are very slow under acid or neutral conditions, and, as a result, most of the p-nitrophenol that is formed comes from the paraoxon - $KMnO_4$ oxidation reaction. At a pH value of 3.1, the rate of p-nitrophenol - $KMnO_4$ is approximately forty-five times faster than parathion and 225 times faster than paraoxon. Consequently, any p-nitrophenol formed either from the hydrolysis or oxidation will be oxidized without any appreciable accumulation in the system. As the pH value of the system was increased from 3.1 to 9.0 the overall rate of oxidation of p-nitrophenol decreased appreciably, while the overall rates of oxidation of parathion and paraoxon increase. Thus, at pH 9.0 the overall rate of oxidation of parathion by $KMnO_4$ is about twice as fast as that determined for p-nitrophenol. Also, the paraoxon - $KMnO_4$ reaction proceeds with a rate approximately four times faster than the p-nitrophenol reaction. Consequently, any p-nitrophenol from the hydrolysis or oxidation of parathion or paraoxon will accumulate in the system. The oxidation of parathion and paraoxon at pH 9.0 nearly proceeds to completion within two hours. This is a

TABLE 14

Intermediate Oxidation Products of Parathion - $KMnO_4$ Oxidations[a]

Sampling Time		pH 3.1			pH 5.0			pH 7.4			pH 9.0	
	P_t[b]	$(PO)_t$[c]	$(PNP)_t$[d]	$(P)_t$	$(PO)_t$	$(PNP)_t$	$(P)_t$	$(PO)_t$	$(PNP)_t$	$(P)_t$	$(PO)_t$	$(PNP)_t$
Min.	mg/l	mg/l	mg/l	mg/l	mg/l	mg/l	mg/l	mg/l	mg/l	mg/l	mg/l	mg/l
0	11.5	None	None	11.5	None	None	11.5	None	None	11.5	None	None
15	--	--	--	--	--	--	--	--	--	9.66	0.003	0.005
30	10.9	0.02	None	11.2	0.006	None	11.3	0.013	None	8.27	0.007	0.014
60	10.5	0.07	None	10.95	0.050	None	--	--	--	6.15	0.008	0.027
120	9.90	0.17	None	10.6	0.074	None	11.0	0.064	None	3.36	0.011	0.089
180	--	--	--	--	--	--	--	--	--	1.88	0.012	0.168
240	8.92	0.40	None	9.93	0.189	None	10.8	0.132	None	1.03	0.008	0.176
600	6.75	0.89	None	8.59	0.435	None	10.2	0.345	None	---	---	---
1200	3.84	1.49	Trace	7.08	0.809	None	9.31	0.90	None	---	---	---
2640	1.95	2.11	Trace	5.41	1.422	Trace	7.89	1.47	None	---	---	---
4080	--	--	--	--	--	--	6.91	2.31	Trace	---	---	---

[a] Temp. = 20° ± 0.2°C; Average values of three experimental runs; I = 0.02 M.
[b] $(P)_t$ = parathion concentration.
[c] $(PO)_t$ = concentration of paraoxon.
[d] $(PNP)_t$ = concentration of p-nitrophenol.

TABLE 15

Rate Constants for the Oxidation of Parathion, Paraoxon, and p-Nitrophenol by $KMnO_4$

System	Rate Constant Liter $Mole^{-1}$ $Min.^{-1}$ [a]			
	pH 3.1	pH 5.0	pH 7.4	pH 9.0
Parathion[b]	.415	.205	.088	7.84
Paraoxon[c]	.083	.032	.004	17.0
p-Nitrophenol[d]	18.6	9.24	8.64	4.14

[a]Temp. = 20° ± 0.2°C. I = 0.02 M, and the rate constants are average values of three experimental runs.
[b]Initial concentration of parathion is 3.95×10^{-5}M, and $KMnO_4$ is 8.0×10^{-4}M.
[c]Initial concentration of paraoxon is 4.81×10^{-5}M, and $KMnO_4$ is 8.0×10^{-4}M.
[d]Initial concentration of p-nitrophenol is 7.18×10^{-4}M, and $KMnO_4$ is 2.0×10^{-3}M.

very short period of time compared to the hydrolytic half-lives of 522 hours for parathion and 69 hours for paraoxon at pH 9.0. Therefore, most of the p-nitrophenol accumulating in the system comes from the oxidative step and not from the hydrolysis step.

Figure 1 represents the suggested mechanism of oxidation of parathion by $KMnO_4$. Gas-liquid and thin-layer chromatographic techniques and infrared, ultraviolet, and mass spectral analyses were used for the separation and identification of the intermediate products from the oxidation of p-nitrophenol. 2,4-Dinitrophenol was identified as the major intermediate oxidation product coming from

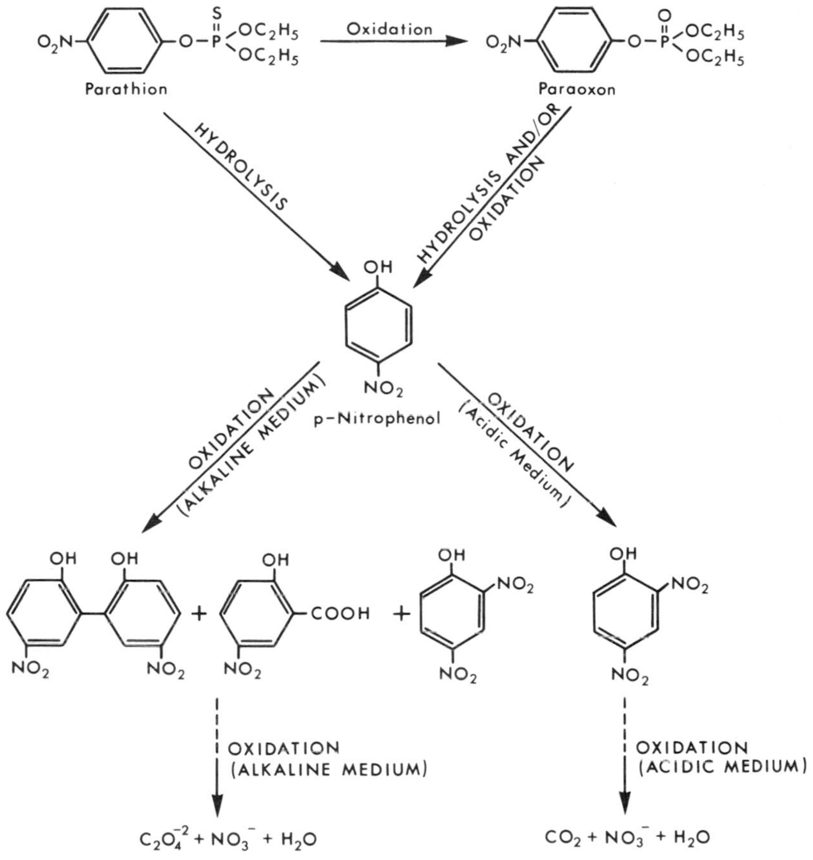

Parathion — Oxidation → Paraoxon

HYDROLYSIS

HYDROLYSIS AND/OR OXIDATION

p-Nitrophenol

OXIDATION (ALKALINE MEDIUM)

OXIDATION (Acidic Medium)

OXIDATION (ALKALINE MEDIUM)

$C_2O_4^{-2} + NO_3^- + H_2O$

OXIDATION (ACIDIC MEDIUM)

$CO_2 + NO_3^- + H_2O$

FIG. 1

Proposed pathways for the potassium permanganate oxidation of parathion.

either an acid or alkaline oxidation. However, two additional intermediate products were identified from an alkaline oxidation of p-nitrophenol. One of which is 2-hydroxy - 5- nitrobenzoic acid, that is formed from the further oxidation of the dimer 2,2'-dihydroxy - 5,5'-dinitrodiphenyl (43):

OH, NO_2 + OH, NO_2 $\xrightarrow[^{-}OH]{KMnO_4}$ OH OH, NO_2 NO_2 $\xrightarrow[^{-}OH]{KMnO_4}$

OH, —COOH, NO_2

(15-56)

V. OXIDATION OF 2,4-DICHLOROPHENOXY ACETIC ACID

Chlorine and potassium permanganate were administered in an unsuccessful attempt to oxidize the sodium salt of 2,4-D and the isopropyl, isooctyl, and butyl esters of 2,4-D at a pH value of 7.0 and a temperature of 25°C. (44). Oxidant concentrations of 1.0 to 10.0 mg/l were applied to 1.0 mg/l of the 2,4-D compounds. Contact times of 30 and 60 minutes were used. This study reexamined the failure of $KMnO_4$ and Cl_2 to oxidize 2,4-D. It was hypothesized that the contact time should be increased above those used in the previous experimental systems. Another factor investigated is the pH of the medium. Chlorine dioxide, which proved to be the most efficient oxidant for degradation of the dipyridylium herbicides, was tested also for the oxidation of 2,4-D.

A. Potassium Permanganate.

The effect of potassium permanganate on the oxidation of 2,4-D was investigated in the pH range 3.1 to 10.1 and temperature of 20°C. Twenty-five mg/l of 2,4-D were treated with 15.8 and 158 mg/l $KMnO_4$ under acid, neutral, and alkaline conditions, pH 3.1, 7.4, and 10.1, respectively, and after contact times of 3 and 24 hours. Again, no oxidation of 2,4-D was observed under these conditions. Consequently, it was decided to examine conditions of high temperature (100°C.) and lower pH values of 5.0, 3.1, and 2.0. That some oxidation of 2,4-D was observed at 100°C. and under acidic conditions is seen in Table 16. Over a contact time of one hour, 23% of oxidation of 2,4-D occurs at a pH value of 3.1 whereas complete oxidation was observed at pH 2.0.

B. Chlorine and Chlorine Dioxide.

Chlorine and chlorine dioxide were examined also for their oxidative effects on 2,4-D at pH values of 3.1, 7.4, and 10.1 at 20°C. No oxidation was observed in contact times up to 24 hours and under 1:1 weight ratios of oxidant to 2,4-D.

VI. GENERAL DISCUSSION

A kinetic approach is taken for the chemical oxidation of pure organic compounds in order to determine rates whereby the reactions proceed and the effect of various environmental factors on the overall process. Also, kinetic data can be used to compare and evaluate the efficiencies of various oxidants towards the degradation of specific organic contaminants, and to explain, in part, the mechanism of the oxidative reaction.

This study offers examples of how thermodynamic and kinetic data may be applied by the water chemist for the removal of trace concentrations of organic pollutants from water by chemical oxidation. First, the thermodynamic feasibility of a redox reaction should be calculated as seen in Table 5. This involves some assumptions about the end products of the reactions. The next step is to test for the experimental validity of the redox reaction model by comparing the calculated stoichiometry given by the reaction model with that determined experimentally.

The data collected from this study substantiates the notion that, under given environmental conditions, various organic compounds will react differently to a particular chemical oxidant. No general statement can be made concerning the oxidation of organic materials by a particular chemical.

Chlorine dioxide seems to be the oxidant of choice for the removal of dipyridylium quaternary herbicides from water. Its application would require preadjustment of pH of water to slightly alkaline values (above pH 8.0). Temperature has little or no effect on the velocity of reaction which requires a contact time of less than one minute. Also, the required dosages (approximately 0.3 mgClO_2/mg

TABLE 16

Effect of pH on Oxidation of 2,4-D by $KMnO_4$ at 100°C.

pH	Residual 2,4-D mg/l	Residual $KMnO_4$ mg/l Test	Blank	Contact Time hours
7.4	25.0	158.2	158.2	0
	25.0	157.1	157.1	1
	25.0	156.8	156.8	3
5.0	25.0	158.2	158.2	0
	25.0	157.1	157.1	1
	25.0	156.8	156.9	3
3.1	25.0	158.2	158.2	0
	19.25	135.0	157.1	1
	17.0	125.5	156.9	3
2.0	25.0	158.2	158.2	0
	00.0	57.6	157.1	1

dipyridylium herbicide) of chlorine dioxide are within normal operation at a water treatment plant. Chlorine also requires adjustment of the pH value to the slightly alkaline side of 8.0 where it will be more active as an oxidant for the removal of dipyridylium herbicides. It is, however, not as efficient as chlorine dioxide. Potassium permanganate, on the other hand, would be the oxidant of choice under neutral or slightly acidic conditions. Similar to Cl_2 and ClO_2, the efficiency of $KMnO_4$ towards the oxidation of the dipyridylium herbicides increases under alkaline conditions when compared to acid or neutral conditions.

Potassium permanganate was found to be an efficient oxidant for parathion and its major intermediate by-products of oxidations. The parathion-$KMnO_4$ reaction at pH 9.0 proceeds with a rate 90 times faster than that determined at pH 7.4. However, no paraoxon is accumulated in the system during the oxidation of parathion at pH 9.0 when compared to neutral or acid conditions. The last point is important because of the fact that paraoxon is at least five times more toxic to fish and animal than parathion (44). Therefore, unless the pH of the water is carefully controlled, the application of $KMnO_4$ may appreciably aggravate the situation instead of improving it. However, the only major intermediate oxidation product of the parathion - $KMnO_4$ reaction at pH 9.0 that accumulates in the system is p-nitrophenol. A kinetic study indicated that p-nitrophenol is oxidized further by $KMnO_4$ with a rate less than that determined for the parathion - $KMnO_4$ reaction. Thus, a careful adjustment of the pH value of the water and the contact time of the reaction is essential in order to oxidize any residual phenol that is formed either by hydrolysis or oxidation of the parent compound, parathion.

2,4-Dichlorophenoxy acetic acid resists oxidation by $KMnO_4$, Cl_2, and ClO_2 under environmental conditions (pH, temperature, and ionic strength) comparable to those normally present in a water treatment plant. Using $KMnO_4$ under highly acidic conditions (pH less than 3.0) and a temperature of $100^oC.$, the oxidation proceeds with an appreciable velocity. However, these conditions are impractical for large scale treatment of water.

Water chemists will face the problem that trace concentrations of organic pesticides are usually present in natural waters together with different kinds of organic constituents that will compete for the amount of oxidant added to the water. Therefore, separation and identification of those constituents are essential steps in order to obtain an idea about their rates of degradation using a particular oxidant. However, chemical kinetic data collected in the laboratory will help the water chemist in consideration of the optimum conditions under which the rate of a specific reaction can be enhanced while some other oxidation reactions of secondary importance will be retarded or proceed at slower rates at the same time.

The following calculations show how the rate constants of the oxidation reactions can be used to calculate the contact times necessary for the degradation of residual organic pesticides present in water.

Take the case of a water supply contaminated with diquat at a concentration of 1 mg/l. The plant is practicing $KMnO_4$ treatment for the oxidation of organic contaminants. Since the oxidation of diquat by $KMnO_4$ follows second-order kinetics, i.e., the rate of oxidation reaction is proportional to the oxidant and herbicide concentration in solution:

$$-\frac{dC_{Diq.}}{dt} = K_2 C_{Oxd.} \cdot C_{Diq.} \qquad (15\text{-}57)$$

Where K_2 is the second-order rate constant (Liter $Mole^{-1}$ $Min.^{-1}$), $C_{Oxd.}$ is the $KMnO_4$ concentration in solution (Mole/Liter), and $C_{Diq.}$

is the diquat concentration in solution (Mole/Liter). Assume that the oxidant is present in excess concentration compared to the herbicide in solution. Therefore, Equation (15-57) can be modified to a pseudo first-order expression with respect to the herbicide:

$$- \frac{dC_{Diq.}}{dt} = K_1 \quad C_{Diq.} \tag{15-58}$$

Where, $K_1 = K_2 \cdot C_{Oxd.}$ $(Min.^{-1})$. Integration of Equation (15-58) gives:

$$K_1 \, t = 2.303 \log \frac{C^o_{Diq.}}{C_{Diq.}} \tag{15-59}$$

from which the t_{50}, t_{90}, and t_{99} values or the times at which 50%, 90%, and 99% of the diquat concentration will disappear from the solution can be calculated:

$$t_{50} = \frac{2.303}{K_1} \log \frac{C^o_{Diq.}}{C^o_{Diq.} \times 50/100}$$

$$t_{50} = \frac{2.303}{K_1} \log \frac{100}{50} = \frac{0.691}{K_1} \tag{15-60}$$

$$t_{90} = \frac{2.303}{K_1} \log \frac{C^o_{Diq.}}{C^o_{Diq.} \times 10/100}$$

$$t_{90} = \frac{2.303}{K_1} \log \frac{100}{10} = \frac{2.303}{K_1} \tag{15-61}$$

$$t_{99} = \frac{2.303}{K_1} \log \frac{C^o_{Diq.}}{C^o_{Diq.} \times 1/100}$$

$$t_{99} = \frac{2.303}{K_1} \log 100 = \frac{4.606}{K_1} \tag{15-62}$$

TABLE 17

Contact Times Required for $KMnO_4$-Diquat Reaction at Different pH Values[a]

pH	K_1[b] $Min.^{-1}$	t_{50}	t_{90}	t_{99}
5.1	1.23×10^{-4}	5635.2	18723.5	37447.0
6.0	8.69×10^{-5}	7950.5	26501.7	53003.4
6.9	1.86×10^{-4}	3714.5	12381.7	24763.4
8.0	1.18×10^{-3}	585.5	1951.7	3903.4
9.1	1.48×10^{-2}	46.7	155.6	311.2
10.0	5.46×10^{-2}	12.7	42.2	84.4

[a]Contact times reported in minutes.
[b]Calculated by multiplying K_2 (the second-order rate constant) by $KMnO_4$ concentration (10^{-3}M).

Table 17 shows the contact times calculated for the diquat-$KMnO_4$ reaction at different pH values.

ACKNOWLEDGMENTS

Paper of the Journal Series, New Jersey Agricultural Experiment Station, Rutgers - The State University of New Jersey, Department of Environmental Sciences, New Brunswick, New Jersey. This research was sponsored by Research Grant ES-00016, Office of Resource Development, Bureau of State Services, U. S. Public Health Service, Washington, D. C.

REFERENCES

1. L. E. Mitchell, *Pesticide: Properties and Prognosis - Organic Pesticide in the Environments.*, *Advances in Chemistry Series*, No. 60, ACS, Washington, D. C., 1966, p. 1.

2. S. D. Faust and I. H. Suffet, in *Microorganic Matter in Water*, ASTM-STP, 448, Amer. Soc. for Testing and Materials, 1969, p. 24.

3. E. Brown and Y. A. Nishioka, *Pestic. Monit. J.*, *1*, 38 (1967).

4. S. D. Faust and O. M. Aly, *J. Amer. Water Works Assoc.*, *56*, 267 (1964).

5. J. M. Ginsburg, *J. Agr. Food Chem.*, *3*, 322 (1955).

6. A. R. Grzenda, H. P. Nicholson, J. I. Teasley, and J. H. Patric, *J. Econ. Entomol.*, *57*, 615 (1964).

7. E. Hindin, D. S. May and G. H. Dunstan, *Residue Rev.*, *7*, 130 (1964).

8. F. M. Middleton and J. J. Lichtenberg, *Ind. Eng. Chem.*, *52*, 99A (1960).

9. A. A. Rosen and F. M. Middleton, *Anal. Chem.*, *31*, 1729 (1959).

10. L. Weaver, C. G. Gunnerson, A. W. Breidenbach, and J. J. Lichtenberg, *Public Health Reports*, *80*, 481 (1965).

11. S. D. Faust and O. M. Aly, *J. Amer. Water Works Assoc.*, *56*, 267 (1964).

12. Anonymous, *Public Health Service Drinking Water Standards*, Publication 956, U. S. Government Printing Office, Washington, D. C., 1962.

13. H. P. Nicholson, *Science*, *108*, 871 (1967).

14. Anonymous, *Report on the Committee on Water Quality Criteria*, Federal Water Pollution Control Admin., U. S. Dept. of the Interior, 1968.

15. E. Edgerley, R. T. Skrinde, and D. W. Ryckman, in Proceeding of the Fourth Rudolfs Research Conference, (S. D. Faust and J. V. Hunter, eds.), John Wiley and Sons, Inc., New York, 1967, p. 405.

16. E. P. Lichtenstein, J. Econ. Entomol., 50, 545 (1957).

17. E. P. Lichtenstein, L. J. DePew, E. L. Eshbaugh, and J. P. Sleesman, J. Econ. Entomol., 53, 136 (1960).

18. W. E. Westlake and J. P. San Antonio, Agr. Res. Serv., ARS 20-9, 105 (1960).

19. R. G. Spicher and R. T. Skrinde, J. Amer. Water Works Assoc., 55, 1200 (1963).

20. H. G. Schwartz, Removal of Selected Organic Pesticides by Specific Water Treatment Unit Process. Master of Science Thesis: Sewer Institute of Technology, Washington Univ., 1962.

21. R. G. Spicher and R. T. Skrinde, J. Amer. Water Works Assoc., 57, 472 (1965).

22. C. A. Buescher, J. H. Dougherty, and R. T. Skrinde, J. Water Pollut. Contr. Fed., 36, 1005 (1964).

23. N. S. Chamberin and A. E. Griffin, Sewage Ind. Wastes, 24, 750 (1952).

24. G. F. Lee and J. C. Morris, Int. J. Air Water Pollut., 6, 419 (1962).

25. G. G. Robeck, K. A. Dostal, J. M. Cohen, and J. F. Kreissel, J. Amer. Water Works Assoc., 57, 181 (1965).

26. J. L. Franklin, Ind. Eng. Chem., 41, 1070 (1949).

27. D. W. VanKrevelen and H.A.G. Chermin, Chem. Eng. Sci., 1, 66 (1951).

28. F.E.C. Scheffer, De toepassing van de thermodynamic op chemische processen., Waltman, Delft, 1945.

29. H. Ulich, Chemische Thermodynamik, Dresden und Leipzig, 1930.

30. H. A. G. Chermin, Chap. 14, this volume.

31. R. Stewart, in Oxidation in Organic Chemistry, (K. Wiberg, ed.), Academic Press, New York, 1965, p. 36.

32. R. E. Connick and Yuan-tsan Chia, J. Amer. Chem. Soc., 81, 1280 (1959).

33. J. C. Morris, J. Phys. Chem., 70, 3798 (1966).

34. T. Fredrikson, W. L. Farrior, and R. F. Witter, Acta Dermato-Venereologica, 41, 335 (1961).

35. R. D. O'Brien, Toxic Phosphorus Esters - Chemistry, Metabolism, and Biological Effects, Academic Press, New York, 1960.

36. D. R. Peck, Chem. Ind., 67, 526 (1948).

37. J. A. A. Ketelaar, Rec. Trav. Chim., 69, 649 (1950).

38. R. Muhlman and G. Schrader, Z. Naturforsch, 12b, 196 (1957).

39. C. M. Weiss and J. H. Gakstatter, The Decay of Anticholinesterase Activity of Organic Phosphorus Insecticides on Storage in Waters of Different pH, The Proc. of the 2nd International Water Pollution Research Conf., Pergamon Press, 1965, p. 83.

40. J. H. Ruzicka, J. Thomson, and B. B. Wheals, J. Chromatogr., 31, 37 (1967).

41. E. L. Shiavone and O. A. Torrado, Sem. Med. (Buenos Aires), 129, 449 (1966).

42. R. A. Baker, J. Amer. Water Works Assoc., 55, 913 (1963).

43. H. M. Gomaa, J. D. Rosen, and S. D. Faust, (In press).

44. R. T. Skrinde, J. W. Caskey, and C. K. Gillespie, J. Amer. Water Works Assoc., 54, 1407 (1962).

CHAPTER 16

CHEMICAL OXIDATIONS OF ORGANIC COMPOUNDS

Donald B. Denney

Rutgers - The State University

I. INTRODUCTION

Chemical oxidation can be thought of as proceeding by three basic mechanisms. These reaction types are quite common and occur in many systems. The first involves transfer of hydrogen. Representative examples are the oxidation of an alcohol to an aldehyde or the dehydrogenation of a hydrocarbon. These reactions are

$$RCH_2OH \longrightarrow R\text{-}\overset{O}{\overset{\|}{C}}\text{-}H$$

$$\text{cyclohexane} \longrightarrow \text{benzene}$$

clearly oxidation processes, although oxygen itself is not introduced into the organic substrate. The second process does involve the introduction of oxygen. Examples of oxygen introduction include conversion of an aldehyde to an acid and oxidation of an

$$R\text{-}\overset{O}{\overset{\|}{C}}\text{-}H \longrightarrow R\text{-}\overset{O}{\overset{\|}{C}}\text{-}OH$$

$$R\text{-}NH_2 \longrightarrow R\text{-}NO_2$$

amine to a nitro compound. The last mode of oxidation involves an electron transfer from the substrate to the oxidizing agent. Such reactions are often called one electron oxidation reactions. Examples include the oxidation of organic radicals by transition metal ions such as cupric ion to give olefins, cuprous ion, and a proton. Electron transfer reactions between metal ions also fall into this class.

$$-\dot{\mathrm{C}}-\underset{|}{\overset{\mathrm{H}}{\overset{|}{\mathrm{C}}}}- + \mathrm{Cu(II)} \longrightarrow \;\rangle\mathrm{C}=\mathrm{C}\langle\; + \mathrm{Cu(I)} + \mathrm{H}^{+}$$

II. AIR OXIDATIONS

One of the most important areas of oxidation chemistry is that of air oxidations. These reactions have been studied in great detail and some provide the basis for the production of several valuable chemicals. The same air oxidation reactions also contribute highly to the weathering and deterioration of organic based materials.

The reactions involved in air oxidation are a set of simple reactions each one of which contributes to the free radical chain process. The first reaction in the set involves initiation, i.e., the production of a free radical. There are many ways in which initiation can occur.

$$-\overset{|}{\underset{|}{\mathrm{C}}}-\mathrm{H} + \mathrm{Z}\cdot \longrightarrow -\overset{|}{\underset{|}{\mathrm{C}}}\cdot + \mathrm{ZH}$$

In controlled processes, unstable substances, i.e., initiators are introduced into the reaction medium and they then decompose to give radicals Z· which then attack the material to be oxidized. Other methods of initiation include photo and thermal dissociation of bonds. Because of the chain nature of air oxidation reactions very little initiation is required to achieve extensive conversion to product. The second step in the process involves reaction of the radical with oxygen to give a peroxy radical which reacts with more starting material to

$$-\overset{|}{\underset{|}{\mathrm{C}}}\cdot + \mathrm{O_2} \longrightarrow -\overset{|}{\underset{|}{\mathrm{C}}}-\mathrm{O}-\mathrm{O}\cdot$$

$$-\overset{|}{\underset{|}{\mathrm{C}}}-\mathrm{H} + -\overset{|}{\underset{|}{\mathrm{C}}}-\mathrm{O}-\mathrm{O}\cdot \longrightarrow -\overset{|}{\underset{|}{\mathrm{C}}}\cdot + -\overset{|}{\underset{|}{\mathrm{C}}}-\mathrm{O}-\mathrm{O}-\mathrm{H}$$

regenerate the radical and give hydroperoxide. This sequence of steps explains the chain character of the air oxidation process. One of the interesting features of chain reactions is that once a chain has been started it cannot be stopped unless two radicals react to give nonradical products. Such a combination reaction is not very likely and as a consequence very long chains can often be observed with reactive substrates.

Those substrates which are oxidized most readily contain allylic, benzylic, and tertiary hydrogens. Some representative oxidation reactions follow:

$$+ \; O_2 \longrightarrow \text{OOH} \qquad (16\text{-}1)$$

$$+ \; O_2 \longrightarrow \text{OOH} \qquad (16\text{-}2)$$

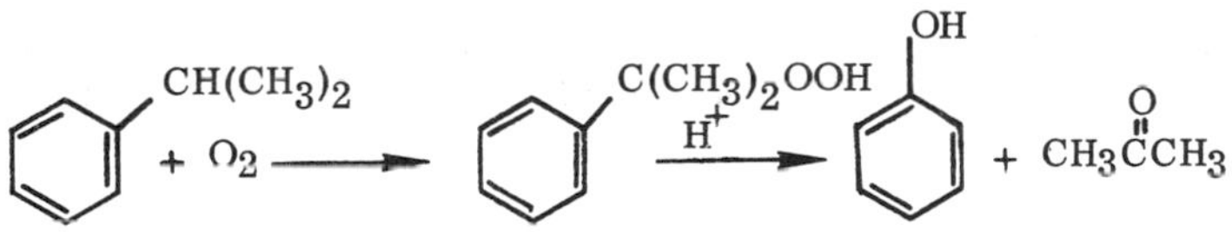

$$\text{CH(CH}_3)_2 + O_2 \longrightarrow \text{C(CH}_3)_2\text{OOH} \xrightarrow{H^+} \text{OH} + CH_3\overset{O}{\overset{\|}{C}}CH_3 \qquad (16\text{-}3)$$

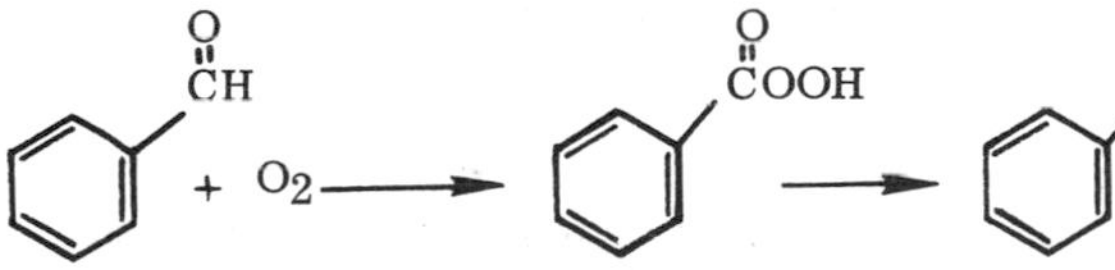

$$\overset{O}{\overset{\|}{C}}H + O_2 \longrightarrow \overset{O}{\overset{\|}{C}}OOH \longrightarrow \overset{O}{\overset{\|}{C}}OH \qquad (16\text{-}4)$$

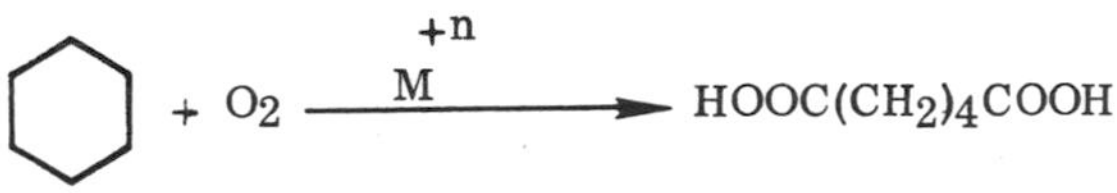

$$+ \; O_2 \xrightarrow{M^{+n}} HOOC(CH_2)_4COOH \qquad (16\text{-}5)$$

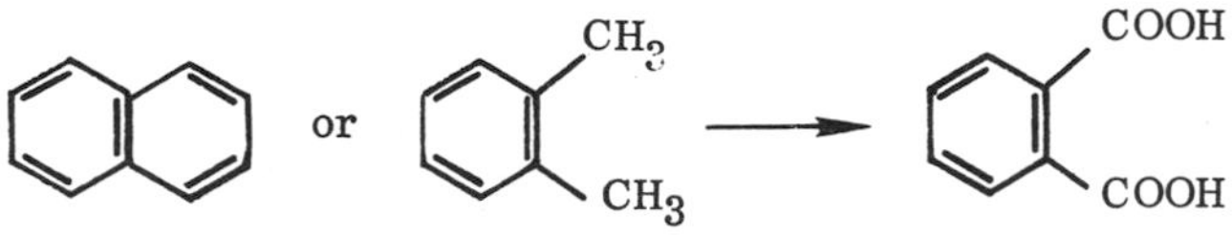

$$\text{or} \quad \begin{matrix} CH_3 \\ CH_3 \end{matrix} \longrightarrow \begin{matrix} COOH \\ COOH \end{matrix} \qquad (16\text{-}6)$$

The first four illustrate the high reactivity of allylic, tertiary, benzylic, and aldehyde hydrogens. Oxidation of cumene, Reaction (16-3), is used commercially as a means of producing phenol and acetone by acid catalyzed decomposition of the intermediate cumene hydroperoxide. Reactions (16-5) and (16-6) are also commercially important. Production of adipic acid, Reaction (16-5), by air oxidation of cyclohexane in the presence of metal ion catalysts has become a very important route to this chemical. Air oxidation of napthalene or o-xylene at high temperatures is the only important commercial method for the production of phthalic anhydride.

III. HYDRIDE TRANSFERS

Another important oxidation reaction involves hydride, H^-, transfer. In general, H^- is not generated but rather a hydrogen is transferred to an electrophilic site with its bonding pair of electrons. The general equation is:

$$-\overset{|}{\underset{|}{C}}-H + E^+ \longrightarrow -\overset{|}{\underset{|}{C}}^+ + HE \qquad (16\text{-}7)$$

In this case an electrophile attacks on a carbon-hydrogen bond to give a carbonium ion and HE. An example of this kind of reaction follows:

$$C_7H_8 \text{ (cycloheptatriene)} + (C_6H_5)_3C^+ \longrightarrow C_7H_7^+ \text{ (tropylium)} + (C_6H_5)_3C\text{-}H \qquad (16\text{-}8)$$

where triphenylmethylcarbonium ion attacks on cycloheptatriene to give triphenylmethane and the tropylium cation. Hydride transfers between carbonium ions are a commercially important method for the isomerization of petroleum hydrocarbons.

Another hydride transfer reaction is the classic Cannizzaro reaction in which an aldehyde without α-hydrogens undergoes oxidation-reduction in basic media to give the corresponding acid and alcohol. This transformation can be illustrated by considering the disproportionation of benzaldehyde:

$$C_6H_5\overset{O}{\overset{\|}{C}}H + OH^- \rightleftharpoons C_6H_5\overset{O^-}{\underset{H}{C}}OH \qquad (16\text{-}9)$$

$$C_6H_5-\overset{\overset{\delta-}{O}}{\underset{H}{C}}\cdots H\cdots\overset{\overset{\delta-}{O}}{\underset{OH}{C}}-C_6H_5 \longrightarrow C_6H_5-CH_2O^- + C_6H_5-COOH \quad (16\text{-}10)$$

Addition of hydroxide to the carbonyl carbon yields an anion which transfers a hydride to the carbonyl carbon of another benzaldehyde molecule to give ultimately benzyl alcohol and benzoate ion.

Other hydride transfer reactions involve the oxidation of Grignard reagents by hindered ketones:

$$>C=O + RCH_2CH_2MgX \longrightarrow H-\overset{|}{\underset{|}{C}}-\overset{-}{O}\ \overset{+}{Mg} + RCH=CH_2 \quad (16\text{-}11)$$

and the ammoniative reduction of ketones by ammonia and formate:

$$-\overset{O}{\overset{\|}{C}}- + NH_4^+ + HCO_2^- \longrightarrow -\overset{NH_2}{\overset{|}{\underset{H}{\underset{|}{C}}}}- + CO_2 + H_2O \quad (16\text{-}12)$$

In the latter reaction, formate ion acts as the hydride transfer agent and it is oxidized to carbon dioxide.

IV. OXIDATION OF CARBON-CARBON BONDS

One of the most important classes of oxidation reactions are those which lead to the cleavage of carbon-carbon bonds. In general carbon-carbon single bonds are quite inert towards direct oxidation and those oxidation reactions which appear to have resulted by direct cleavage of a carbon-carbon single bond are more complex and have involved formation of a double bond as an intermediate.

Carbon-carbon double bonds are quite susceptible to oxidation either with complete cleavage of the double bond or by reaction with the π-bond. For example, under appropriate conditions, permanganate will oxidize double bonds by cis addition to give the glycols:

$$>C=C< + MnO_4^- \longrightarrow -\overset{O}{\overset{|}{\underset{|}{C}}}-\overset{O}{\overset{|}{\underset{|}{C}}}- \ (\text{O–}MnO_2^-\text{–O bridge}) \longrightarrow >\overset{OH}{C}-\overset{OH}{C}< \quad (16\text{-}13)$$

Under more strenuous conditions, permanganate will cleave the glycols to carbonyl compounds:

$$\begin{matrix} OH & OH \\ | & | \\ >C & - & C< \end{matrix} + MnO_4^- \longrightarrow >C=O + O=C< \qquad (16\text{-}14)$$

Peracids, such as peracetic, attack on carbon-carbon double bonds to give epoxides which can be hydrolyzed to glycols:

$$>C=C< + CH_3\overset{O}{\overset{\|}{C}}OOH \longrightarrow -\overset{O}{C{-}C}- \xrightarrow{H_3O^+} \begin{matrix} OH & OH \\ >C & -C< \end{matrix} \qquad (16\text{-}15)$$

Ozone reacts with double bonds extremely rapidly to give highly unstable ozonides which decompose quite readily to give carbonyl compounds:

$$>C=C< + O_3 \longrightarrow \text{ozonide} \longrightarrow >C=O + O=C< \qquad (16\text{-}16)$$

Atmospheric ozone can act as an important initiator for free radical oxidation reactions.

Interestingly carbon-carbon triple bonds are quite resistant to chemical oxidation and usually require considerably more stringent conditions than are required for the oxidation of carbon-carbon double bonds.

Aldehydes are easily oxidized to carboxylic acids which are quite resistant to further oxidation. Ketones are not readily oxidized and when oxidation occurs it is usually the enol which is oxidized:

$$\text{ketone (H)} \rightleftharpoons \text{enol (OH)} \xrightarrow{[O]} -\overset{O}{\overset{\|}{C}}-OH + HO\overset{O}{\overset{\|}{C}}- \qquad (16\text{-}17)$$

V. OXIDATIONS BY PERMANGANATE AND DICHROMATE

Perhaps the two most important oxidizing agents that are used in organic chemistry are permanganate and acidified dichromate. Permanganate can act by electron abstraction, (16-18), hydrogen atom abstraction, (16-19), hydride transfer, (16-20), and oxygen transfer, (16-21):

$$MnO_4^- + e \longrightarrow MnO_4^{=} \quad (16\text{-}18)$$

$$MnO_4^- + RH \longrightarrow HMnO_4^- + R \quad (16\text{-}19)$$

$$MnO_4^- + ZH \longrightarrow HMnO_4^{=} + Z^+ \quad (16\text{-}20)$$

$$MnO_4^- + Z \longrightarrow MnO_3^- + ZO \quad (16\text{-}21)$$

Permanganate can also act as an initiator of free radical air-oxidation reactions, e.g., the oxidation of tetralin to tetralin hydroperoxide:

$$\text{tetralin} + O_2 \xrightarrow{MnO_4^-} \text{tetralin hydroperoxide (OOH)} \quad (16\text{-}22)$$

Oxidation of acetic anhydride in the presence of napthalene yields the substitution product napthylacetic acid:

$$\text{naphthalene} + (CH_3\overset{O}{\overset{\|}{C}})_2O + MnO_4^- \longrightarrow \text{naphthalene-}CH_2COOH \quad (16\text{-}23)$$

This sequence involves initial abstraction of hydrogen from acetic anhydride to give a free radical which then adds to the napthalene. Another example of hydrogen transfer involves the oxidation of alcohols by permanganate in basic solution. The reaction involves several steps. The first is formation of an alkoxide ion by hydrogen abstraction from the alcohol. Permanganate abstracts an α-hydrogen atom from the alkoxide ion to give $HMnO_4^-$ and the radical anion, $R\dot{C}H\text{-}O^-$,

$$RCH_2OH + OH^- \rightleftharpoons RCH_2O^- + H_2O \quad (16\text{-}24)$$

$$RCH_2O^- + MnO_4^- \longrightarrow R\dot{C}H\text{-}O^- + HMnO_4^- \quad (16\text{-}25)$$

$$R\dot{C}HO^- + Mn(VII) \text{ or } Mn(VI) \longrightarrow R\text{-}\overset{O}{\overset{\|}{C}}\text{-}H \quad (16\text{-}26)$$

which is rapidly oxidized by Mn(VII) or (VI) to the aldehyde or ketone.

Aldehyde oxidations under acidic conditions follow a different course; permanganate adds across the carbonyl double bond to give an intermediate which decomposes to the carboxylic acid:

$$\mathrm{Ar\overset{\overset{O}{\|}}{C}\text{-}H + MnO_4^- + \overset{+}{H} \longrightarrow Ar\text{-}\underset{H}{\overset{\overset{OH}{|}}{C}}\text{—}O\text{-}MnO_3} \quad \text{Acid} \qquad (16\text{-}27)$$

$$\mathrm{Ar\text{-}\underset{\underset{\underset{H\text{-}O\text{-}H}{\uparrow}}{H}}{\overset{\overset{OH}{|}}{C}}\text{-}OMnO_3 \longrightarrow ArCO_2H + H_3O^+ + MnO_3^-} \qquad (16\text{-}28)$$

Under basic conditions the oxidations are similar to those of alcohols. Hydroxide ion adds to the carbonyl group to give an intermediate anion which reacts with permanganate by hydrogen atom transfer. Conversion to acid is rapidly effected by an electron transfer reaction.

$$\mathrm{Ar\text{-}\underset{H}{\overset{\overset{O^-}{|}}{C}}\text{-}OH + MnO_4^- \longrightarrow Ar\text{-}\underset{\cdot}{\overset{\overset{O^-}{|}}{C}}\text{-}OH + HMnO_4^-} \quad \text{Base} \qquad (16\text{-}29)$$

Oxidation by chromate, Cr(VI), under acid condtions proceeds ultimately to give Cr (III); intermediate oxidation states, Cr(V) and Cr(IV), are usually formed and these ions act as potent oxidzing agents. Chromous ion, Cr(II), is a reducing agent and it is readily converted to Cr(III) under a variety of conditions.

The oxidations of alcohols by Cr (VI) has received extensive study. The first step is esterification of the alcohol to give a chromate ester which decomposes to an aldehyde or ketone and Cr (IV), Cr (IV) reacts rapidly with Cr (VI) to give Cr (V) which is a very powerful oxidizing agent. These reactions are summarized by the following equations:

$$\mathrm{(CH_3)_2\,CHOH + HCrO_4^- + 2H^+ \longrightarrow (CH_3)_2CHOCrO_3H_2^+ + H_2O} \qquad (16\text{-}30)$$

$$\mathrm{(CH_3)_2CHOCrO_3H_2^+ + B \longrightarrow (CH_3)_2CO + BH^+ + H_2CrO_3\,(Cr\ IV)} \qquad (16\text{-}31)$$

$$Cr\ (IV) + Cr\ (VI) \longrightarrow 2\ Cr\ (V) \qquad (16\text{-}32)$$

$$2\ Cr\ (V) + 2\ (CH_3)_2CHOH \longrightarrow 2\ Cr\ (III) + 2(CH_3)_2CO + 4H^+ \qquad (16\text{-}33)$$

During the course of some of these reactions fragmentations and rearrangements of the carbon skeleton occur. Such reactions are believed to be due to the formation of carbonium ions as intermediates.

Aldehydes are oxidized by Cr (VI) to give acids and esters. The mechanism involves the formation of an intermediate ester much like that found with alcohol oxidations:

$$R\text{-}\overset{\overset{\displaystyle O}{\|}}{C}\text{-}H + H^+ + HCrO_4^- \longrightarrow R\text{-}\underset{\underset{\displaystyle H}{|}}{\overset{\overset{\displaystyle OH}{|}}{C}}\text{-}OCrO_3H \longrightarrow RCO_2H + Cr\ (IV) \qquad (16\text{-}34)$$

Some hydrocarbons are oxidized by Cr (VI) to give alcohols, and olefins are often oxidized to give $\alpha\beta$- unsaturated ketones or epoxides as the primary products.

During the past few years considerable attention has been devoted to the study of oxidations of organic compounds by transition metal ions. Perhaps one of the most interesting is the conversion of ethylene to acetaldehyde by Pd (II). This process is now used commercially. Pd (II) is reduced to Pd metal which is converted back to Pd (II) by oxygen and copper ion. Under these circumstances only oxygen and ethylene are consumed and thus the process is economically feasible. The development of other processes of a similar nature seems to be only a matter of time.

CHAPTER 17

PRINCIPLES OF PHOTOCHEMICAL REACTIONS IN AQUEOUS SOLUTION

E. D. Owen

University of California

I. INTRODUCTION

Photochemical reactions in aqueous solution undoubtedly played a significant role in the formation of the first organic molecules to appear on the Earth. The flux of electromagnetic radiation from the Sun represented a large proportion of the energy source required to convert the inorganic material which composed the bulk of the Earth's atmosphere and its crust, into organic molecules. The amino acids, purines, pyrimidines, and carbohydrates so formed were the precursors of the proteins, nucleic acids, and polysaccharides which in turn led to the appearance of life itself. The papers that follow will show that photochemical reactions are involved also in the fate of a variety of organic molecules. It is appropriate, therefore, that the subject of photochemical reactions in aqueous solution should be included. This introduction to the subject will be a review of some of the aspects which are most appropriate to the papers which follow.

II. INTERACTION BETWEEN MATTER AND RADIATION

A. General Principles.

The visible and ultraviolet regions of the electromagnetic spectrum together constitute the radiation which has a wavelength lying between about 2,000Å and about 7,000Å (1Å = 10^{-8}cm.). Any boundaries which are defined necessarily must be purely arbitrary, but it is often convenient to regard the region between 2,000Å and 4,000Å as the ultraviolet and that between 4,000Å and 7,000Å as the visible region. The main justification for this choice is the fact that the visible region as so defined is the one to which the retina of the human eye is sensitive. Radiation whose wavelength lies in either of these regions may interact with molecules and in so doing cause them to undergo photochemical reactions, and, hence, to exhibit a chemistry which is quite different to that with which they would be associated in the absence of radiation. To understand how these reactions occur requires some knowledge of: (a) how and why the light is absorbed and the nature of the molecule so formed, and (b) the various fates which may befall the excited molecule in the several different environments in which it may find itself. The main concern in this treatment will be with molecules in aqueous solution, but the same principles apply to molecules in non-aqueous solvents and to the gaseous and solid states.

Before proceeding, it is necessary to recall the simple relationships which exist between the wavelength λ (cms.), the frequency ν (time^{-1}), and the velocity c (cms. time^{-1}) of the radiation, namely:

$$\lambda \nu = c \tag{17-1}$$

and the dependence of the energy E of one photon of the radiation on its frequency, thus:

$$E \text{ (ergs)} = h\nu \tag{17-2}$$

where h = Planck's constant (h = 6.6 x 10^{-27} erg. sec.). An Avogadro's number N of photons or one mole of photons is known as one Einstein. Table 1 clarifies these relationships for various frequencies of interest.

B. Absorption of Radiation.

When radiation of a particular frequency falls on a molecule it may or may not be absorbed, if it is not absorbed it may be transmitted or reflected. Radiation such as sunlight, which consists of a mixture or range of frequencies in the visible region, may interact with molecules in such a way that some selected frequencies only are absorbed. The rest may be transmitted or reflected in which case the system will appear colored. The fact that water is colorless means that the molecules have no absorption in the visible region and, in fact, water is highly transparent in the ultraviolet region also. Dyes likes crystal violet or methylene blue, on the other hand, appear colored because they transmit only the blue portion of the light (i.e., about 4,000Å) into our eyes while absorbing strongly the red frequencies

TABLE 1

Relationships between the Wavelengths, Frequencies, and Energies for Radiation in the Visible and Ultraviolet Range of the Electromagnetic Spectrum

Wavelength Å	Description	Frequency cycles/sec $\times 10^{-14}$	Energy	
			Ergs/molecule $\times 10^{12}$	Kcal/mole
2,537	Ultraviolet	11.82	7.83	112.8
3,130	Ultraviolet	9.58	6.35	99.3
3,660	Black light	8.19	5.43	78.2
4,050	Blue	7.40	4.91	70.7
5,460	Green	5.49	3.64	52.4
7,000	Red	4.28	2.84	40.9
10,000	Infra Red	2.99	1.99	28.6

around 7,000Å. Figure 1 shows the absorption spectrum of crystal violet which is a measure of how strongly the radiation of each particular frequency is absorbed.

Experimentally the amount of light absorbed by a molecule at any wavelength is a function of its concentration (c moles/1) and of the optical path length (l cms.). The relation between these two quantities and the intensities of incident and transmitted light (I_0 and I, respectively) is the Beer-Lambert Law which may be written in various forms, one of the most useful being:

$$\log (I_0/I) = \epsilon cl \qquad (17\text{-}3)$$

where the quantity $\log (I_0/I)$ is called the optical density. The proportionality constant ϵ is called the molar decadic extinction coefficient and is a measure of the intensity of the absorption of any wavelength. Most spectrophotometers operate in a way which allows the optical density or percentage transmission to be read directly as the wavelength is scanned and, if c and l are known, then the optical density can be converted (Equation 17-3) to ϵ at each wavelength and the absorption spectrum presented in a much more satisfactory way by showing ϵ as a function of the frequency.

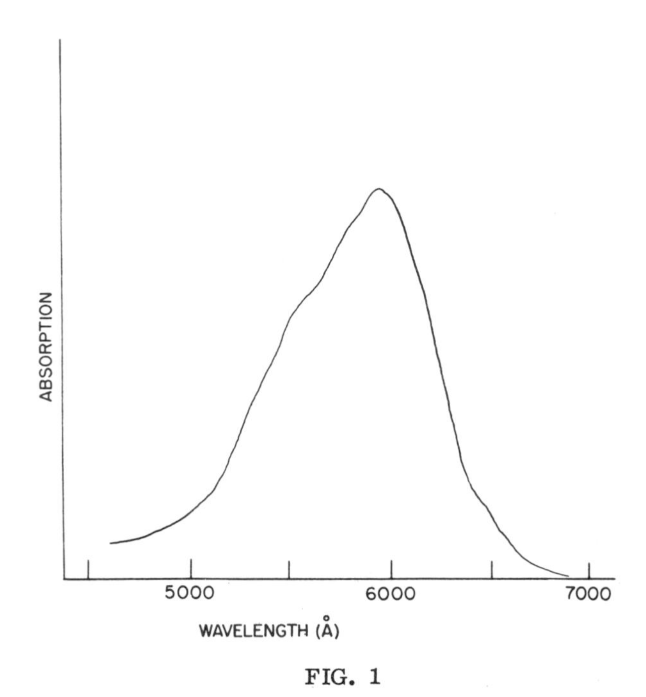

FIG. 1

Visible absorption spectrum of crystal violet.

The most important factor in determining whether or not a particular molecule will absorb radiation of a particular frequency is the detailed structure of the molecule. The quantum theory insists that every molecule has a certain minimum energy content at any temperature and in this condition the molecule is said to be in its ground state. Further, its energy may not increase by any arbitrary amount but only by certain allowed increments. Such an increase converts the molecule into a higher quantum state which is no longer the ground state and is called an excited state. When the increase in energy results from the absorption of a photon of radiation of frequency ν , the condition is that the energy difference between ground and excited states must be $E = h\nu$.

Molecules in the ground state have their electrons arranged in molecular orbitals, each orbital having a characteristic energy which contributes to the total energy of the molecule. Each orbital contains two electrons spinning about their axes in opposite directions. Absorption of radiation in the visible or ultraviolet region promotes an

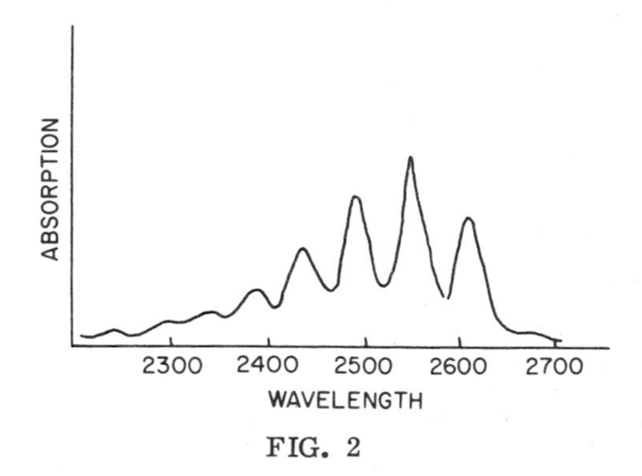

FIG. 2

Ultraviolet absorption spectrum of benzene.

electron to a higher (i.e., more energetic) molecular orbital and the energy of the molecule is increased accordingly. The molecule is now said to be in an excited electronic state. The absorption of crystal violet (Fig. 1) corresponds to such a transition, the energy involved in this case is about 40 kcal/mole. Molecules such as benzene which have no absorption in the visible region and, hence, are colorless, may have several strong absorptions in the ultraviolet (Fig. 2), each corresponds to a transition which promotes the molecule to some higher quantum state.

The excitation energy possessed by a molecule makes it behave in several ways in an attempt to lose the energy and so revert to the ground state. One way is for the molecule to take part in a photochemical reaction, although, as seen below, there are usually several alternative modes of deactivation available so that the absorption of a photon does not necessarily lead to a photochemical reaction.

The existence of a suitable higher energy state is the first but not the only factor in deciding whether a particular photon will be absorbed by a particular molecule. The probability of the transition occurring depends on the so-called selection rules. These are a set of rules which determine the mutual compatability of the quantum mechanical wave functions of the ground and excited states. The details of the selection rules and how they are applied need not concern us here, we need only remember that they determine the probability of a particular transition by examining the two wave functions concerned. If the probability is high then that particular transition will result in a strong absorption and will have a high extinction coefficient.

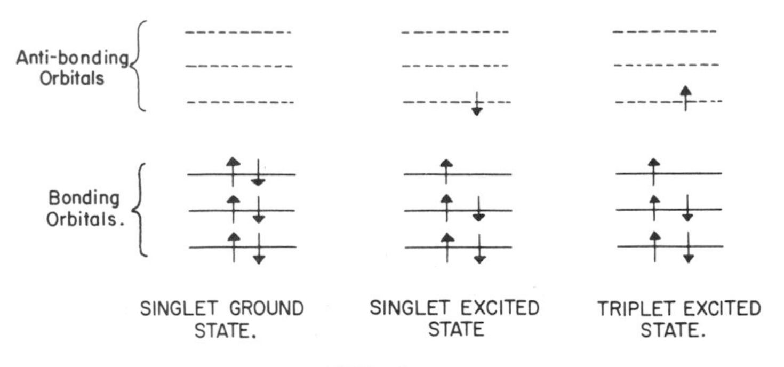

FIG. 3

Singlet and triplet excited electronic states.

Although most molecules possess several possible different excited states it is only the first excited state which need concern us when considering photochemical reactions since molecules which are excited to states higher than the first lose energy very rapidly ($\sim 10^{-12}$sec.) and revert to the first excited state. All photochemical reactions which occur do so from lowest excited states. When the transition occurs, the spin of the electron is unchanged normally (Fig. 3), the total spin angular momentum is zero, and the excited state is called a singlet state. It is possible, however, for the direction of the spin to be reversed during the transition, in which case the promoted electron and the one remaining will have their spins in the same sense. The total spin angular momentum will no longer be zero, and such a state is called a triplet state (Fig. 3). The energy of the triplet state is always lower than that of the corresponding singlet state (i.e., the one involving the same pair of orbitals). Transitions which involve a change in the direction of the spin disobey one of the selection rules and are known as forbidden transitions. This means that their probability is low and they result in a very weak absorption. Nevertheless we shall see later that it is precisely the very low probability of singlet $\rightarrow$ triplet transitions which makes the triplet state an extremely important intermediate in photochemistry.

It is only rarely that molecules in their ground state possess unpaired electrons and exist in a triplet state which makes them paramagnetic. One familiar example of such a molecule is molecular oxygen (O_2) which has a triplet ground state and two low lying excited singlet states (Fig. 4).

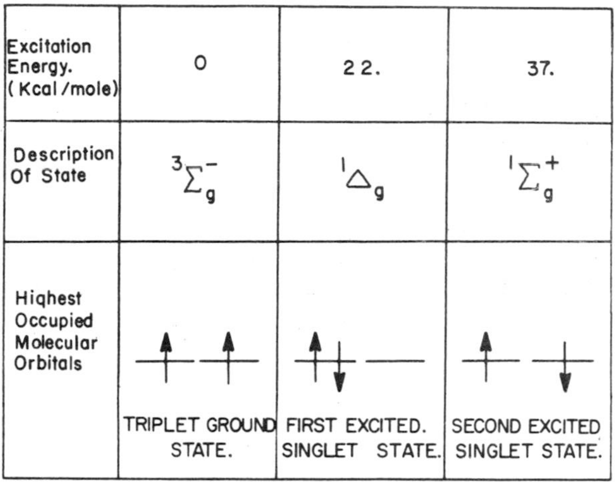

Excitation Energy. (Kcal/mole)	0	22.	37.
Description Of State	$^3\Sigma_g^-$	$^1\Delta_g$	$^1\Sigma_g^+$
Highest Occupied Molecular Orbitals	↑ ↑ TRIPLET GROUND STATE.	↑↓ — FIRST EXCITED. SINGLET STATE.	↑ ↓ SECOND EXCITED SINGLET STATE.

FIG. 4

Ground and lower excited states of molecular oxygen.

C. Molecules in Excited States.

A molecule in an excited singlet state (S_n) is a highly energetic species and has a lifetime of the order of 10^{-9} to 10^{-8} sec. after which time it will lose its excitation energy in some way and revert to the ground state (S_0). One possible mode of deactivation of the singlet state is by the emission of a photon, a process known as fluorescence. The absorption process which populates the excited singlet state usually occurs between the lowest vibrational level of the ground state and some higher vibrational level of the excited state. This excess vibrational energy is lost very quickly (10^{-13} sec.) by collision with solvent molecules, a process known as vibrational relaxation and the fluorescence emission occurs from the lowest vibrational level of the lowest excited singlet state. The photon emitted, therefore, will be of lower energy, i.e., longer wavelength than the one absorbed. Figure 5 makes this point clear. A second possibility is that the transition from $S_1 \rightarrow S_0$ may occur without a photon being emitted. This radiationless process is called internal conversion, and the excitation energy in this case is converted into translational and vibrational energy of the surrounding matrix, that is, the matrix warms up. Internal conversion may be defined as a radiationless transition between different electronic states of the same multiplicity.

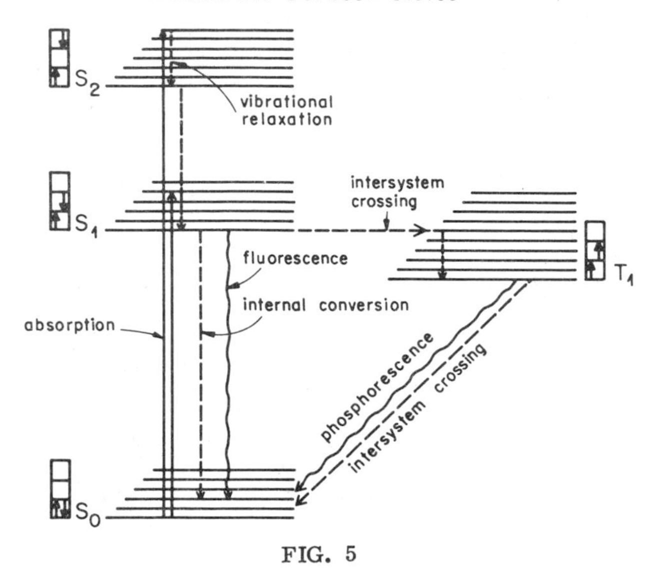

FIG. 5

Modes of formation and deactivation of molecules in excited states.

An important third possibility is that the molecule may undergo a spin change and convert from the singlet state S_1 to the triplet state T_1 of slightly lower energy. This radiationless intersystem crossing process, which is strongly forbidden by the selection rules, normally has a low probability since it is a singlet-triplet transition, but certain perturbing influences such as the presence of a paramagnetic substance or a heavy atom may increase its probability very substantially. This process of absorption to the singlet state followed by intersystem crossing is the most important means of populating the triplet state since the direct singlet$\longrightarrow$triplet absorption process is strongly forbidden. A fourth possibility is that S_1 may transfer its energy in some way to a suitable acceptor molecule and, in so doing, raise the acceptor into an excited state. Equation (17-4) describes the process:

$$\text{Donor } (S_1) + \text{Acceptor } (S_0) \longrightarrow \text{Donor } (S_0) + \text{Acceptor } (S_1) \tag{17-4}$$

A necessary condition, of course, is that the acceptor molecule has an excited state which is of the correct energy and that some interaction is possible which allows the energy transfer to occur. Finally, the molecule may undergo a photochemical reaction such as a dimerization, elimination, dissociation, isomerization, hydration, etc., in fact, as many different types of reactions as occur from the ground state.

The various modes of deactivation listed for the singlet state are also applicable to the triplet state. The radiative $T_1 \rightarrow S_0$ process is known as phosphorescence and, since T_1 is lower in energy than S_1, it will occur at a longer wavelength than fluorescence. Another important difference in the deactivation modes is that any $T_1 \rightarrow S_0$ process, whatever the detailed mechanism, is a forbidden process and, therefore, of low probability. As a result, the lifetime of the triplet state is much longer than that of the singlet, anything between 10^{-6} sec. and several seconds. This longer lifetime is the most significant factor in making T_1 a very important intermediate in photochemistry.

D. Quantum Yields.

A direct measure of the rate of a photochemical process in terms of moles/1 of products formed or reactants consumed in unit time is seldom a useful quantity to quote since it depends on the intensity of the absorbed light. However, when investigating the mechanism of photochemical, photophysical, or photobiological processes, it is essential to have some means of deciding how efficient a particular process is and what are the effects of variables on it. The process may be fluorescence, intersystem crossing, photochemical reaction, and so on. A most useful parameter in this respect is the quantum yield of the process of interest. The quantum yield is defined as the rate of the process of interest divided by the rate of absorption of photons by the system. The quantum yield of fluorescence, for example, is the fraction of the singlet excited state molecules which fluoresce and clearly the sum of the quantum yields for all the various fates which may befall molecules in the singlet excited state is unity,

$$\text{i.e.,} \sum \phi_i = 1.$$

The first requirement in measuring quantum yields is a source of monochromatic light which may be produced by a suitable light source and filter system. Calibration of the number of photons emitted by the lamp for some particular system geometry is conducted with a suitable actinometer system. The ferrioxalate actinometer described by Parker (1) is one of the most widely used in the visible and ultraviolet regions. Quantum yields of fluorescence can be measured absolutely but more often use is made of a comparative method and some standard substance the fluorescence quantum yield of which is accurately known.

The quantum yield idea is illustrated best by a simplified reaction sequence such as that described by Equations (17-5) to (17-12) below. The superscripts 1 and 3 refer to singlet and triplet excited states, respectively, and k values are the unimolecular specific velocity constants for the various processes described.

$M + h\nu \longrightarrow {}^1M$	(absorption)	(17-5)
${}^1M \xrightarrow{k_f} M + h\nu'$	(fluorescence)	(17-6)
${}^1M \xrightarrow{k_7} M + \text{heat}$	(internal conversion)	(17-7)
${}^1M \xrightarrow{k_8} {}^3M$	(intersystem crossing)	(17-8)
${}^1M \xrightarrow{k_9} \text{products}$	(photochemical reaction)	(17-9)
${}^3M \xrightarrow{k_p} M + h\nu''$	(phosphorescence)	(17-10)
${}^3M \xrightarrow{k_{11}} M + \text{heat}$	(intersystem crossing)	(17-11)
${}^3M \xrightarrow{k_{12}} \text{products}$	(photochemical reaction)	(17-12)

$$\phi_f = \frac{k_f}{k_f + k_7 + k_8 + k_9}$$

$$\phi_p = \phi_{isc.} \cdot \frac{k_p}{k_p + k_{11} + k_{12}}$$

where $\phi_{isc.}$ is the quantum yield of intersystem crossing in Equation (17-8).

E. Lifetimes of Molecules in Excited States.

Molecules in singlet excited states may emit a photon as fluorescence and this process is governed by a unimolecular reaction rate law which is characterized by an exponential decay and a mean radiative lifetime, τ_0, which is defined as the time for the concentration of excited states to be reduced to 1/e of their original value. As has been stated already, the same selection rules which govern the probability of an electronic transition due to absorption of a photon also govern the reverse radiative transition which is fluorescence. A transition which is strongly allowed and results in a strong absorption is also strongly allowed in the reverse sense and results in a short radiative lifetime. The degree of allowedness of an absorption is related to the area under the absorption band and so the mean radiative lifetime is proportional to the reciprocal of this area. The relation of Equation (17-13), which strictly is only applicable to atomic line emissions, is often a useful approximation when applied to molecular systems:

$$\frac{1}{\tau_0} = 3 \times 10^{-9} n^2 \nu_0^2 \int \epsilon \, d\nu \qquad (17\text{-}13)$$

where n is the refractive index of the solvent, ν_0 the wave number (cm^{-1}) of the band center, and the quantity in the integral is the area under the absorption band. The value of τ_0 calculated in this way is what the lifetime would be if fluorescence was the only mode of deactivation, i.e., if $\phi_f = 1$. Usually ϕ_f is less than one, and, in many cases, is considerably less. This is because the molecule can deactivate in ways other than fluorescence and, for this reason, the actual measured lifetime τ is less than τ_0 by an amount which depends on how much less ϕ_f is than unity. The relation between τ_0, τ and ϕ_f is verified easily by reference to Equations (17-6) to (17-9) thus:

$$\tau_0 = 1/k_f \text{ and } \tau = 1/(k_f + k_7 + k_8 + k_9)$$

and since: $$\phi_f = k_f/(k_f + k_7 + k_8 + k_9)$$

then: $$\phi_f = \tau/\tau_0$$

or: $$\tau = \phi_f \cdot \tau_0$$

Lifetimes of triplet states are considerably longer as a consequence of the fact that the singlet-triplet processes in absorption or phosphorescence are forbidden, and, therefore, the areas under the absorption spectra for singlet-triplet absorptions may be vanishingly small.

III. FLUORESCENCE AND THE SINGLET STATE

A. Requirements for Fluorescence.

As has been stated already, fluorescence is the result of a radiative transition between the lowest excited singlet state and the ground state. Since comparatively few organic molecules fluoresce strongly in solution, those which do so clearly possess some particularly suitable molecular structure. Experimental measurements of the fluorescence quantum yields of molecules and how they are affected by experimental conditions have shed a good deal of light on the structure of molecules in the singlet excited state and on the mechanism of photochemical reactions which occur from that state.

It is difficult to make generalizations since small changes in environment can have very large effects on the fluorescence but in general there are some factors which favor a high fluorescence quantum yield. The molecule must have an intense longest wavelength absorption band in the visible or ultraviolet region of the spectrum which is due to the promotion of a π electron into an antibonding(π*) level. This is known as a $\pi \rightarrow \pi^*$ transition. The lifetime of the excited state so formed should not be so long that the singlet $\longrightarrow$ triplet intersystem crossing process is facilitated nor should it be too short since this implies some very efficient alternative mode of deactivation of the singlet state. The competing singlet $\longrightarrow$ triplet process is facilitated also when the singlet and triplet states lie close together. Beyond stating these very broad requirements, the fluorescence properties of molecules cannot be predicted with accuracy since many of the factors mentioned are very much affected by solvent polarity, viscosity, temperature, and so on.

It appears that a great many molecules owe their fluorescence to a rigid, planar structure presumably because the increased number of vibrational modes associated with greater flexibility can more easily convert the electronic energy into heat. This is well illustrated by comparing the fluorescence quantum yields of the structurally similar molecules biphenyl (I) and fluorene (II).

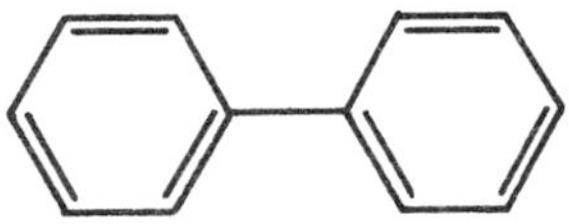

C
H_2

(I) 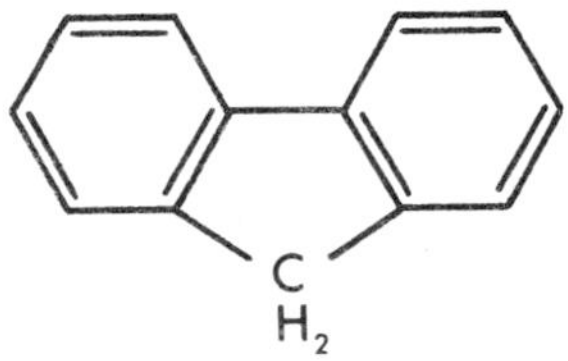 (II)

The former has a quantum yield of fluorescence of 0.23 while for the more rigid latter molecule, $\phi_f = 0.54$. Similarly, the highly fluorescent molecule fluorescein (III) in its basic form differs only in its greater rigidity from the non-fluorescent molecule phenolphthalein (IV).

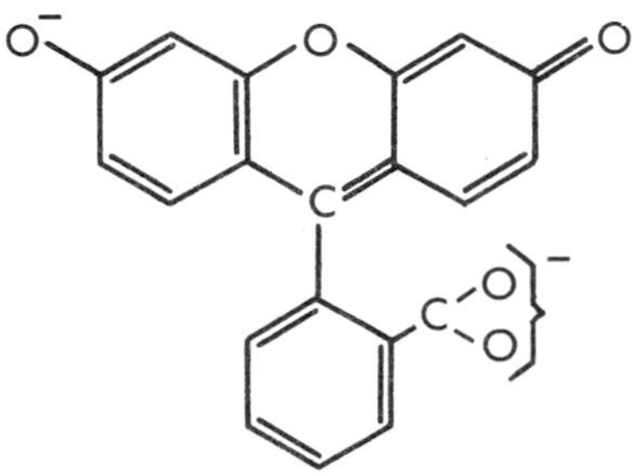

(III) 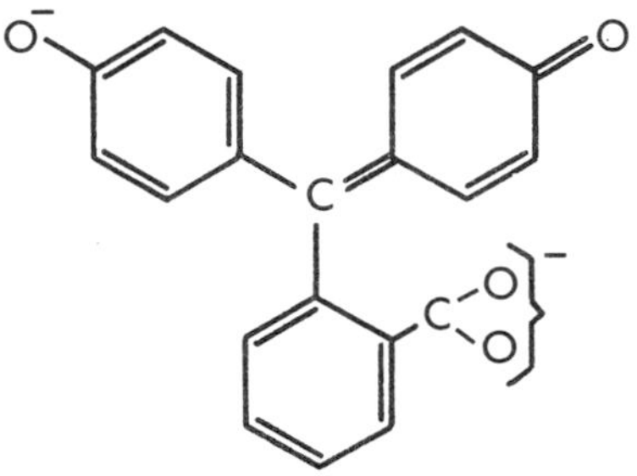 (IV)

The effect of substituents on a π electron system is generally to increase the fluorescence if the substituent increases the π electron density (i.e., if it is an electron donor) and vice versa. Care must be taken, however, in predicting such effects since the substituent may have an important steric effect. It may, for example, distort an otherwise planar steric configuration. Alternatively, the substituents, iodine or bromine in particular, which have large magnetic fields associated with their nuclei, may provide a perturbation which increases the probability of a singlet $\longrightarrow$ triplet transition and so provides an alternative mode of de-excitation and a decrease in the fluorescence. The fluorescence quantum yield of fluorescein is 0.70, that of eosin, its tetra bromo derivative, is 0.15, and erythrosin, its tetra iodo derivative, is 0.03.

The detailed and characteristic fluorescence spectra which many molecules possess together with the development of very sophisticated equipment has made fluorometry a very sensitive and versatile analytical technique in a great many fields.

B. Quenching of Fluorescence.

Certain substances when added to a fluorescent solution are able to reduce drastically the fluorescence intensity of molecules by a process known as fluorescence quenching. Since the fluorescence lifetime is of the order of 10^{-8} sec., any process which can compete in the sense that it can react with molecules in the excited singlet state before they can fluoresce must be a very fast process indeed. Very fast reactions of this kind can be investigated by measuring the decrease in the fluorescence as a function of the concentration of added quencher. A simple reaction scheme which describes the quenching process is shown below where M represents the fluorescent molecule and Q the added quencher:

$$M + h\nu \longrightarrow {}^1M \quad (17\text{-}5)$$

$${}^1M \longrightarrow M + h\nu' \quad (17\text{-}6)$$

$${}^1M \longrightarrow M \quad (17\text{-}7)$$

$${}^1M \longrightarrow {}^3M \quad (17\text{-}8)$$

$${}^1M + Q \xrightarrow{k_q} M + Q \quad (17\text{-}14)$$

Application of simple steady-state kinetics to this system leads to the relationship of Equation (17-15) originally due to Stern and Volmer (2):

$$(F^o/F) - 1 = K_q [Q] \quad (17\text{-}15)$$

where F^o and F are the fluorescent intensities in the absence and presence of Q. K_q, the so-called Stern-Volmer quenching constant, is a measure of the efficiency of the quenching process and is obtained from the slope of the graph of $(F^o/F)-1$ versus $[Q]$. It is a simple matter to show that $K_q = k_q \tau$, where k_q is the bimolecular rate constant for Equation (17-14) and τ the actual mean lifetime of the singlet state. Since τ is typically about 10^{-8} sec., the values of K_q obtained for strong quenchers indicate that k_q is usually close to the value characteristic of diffusion controlled reactions.

The detailed quenching mechanism is usually thought to invoke some kind of "encounter complex" ${}^1MQ$. The overall process is written:

$${}^1M + Q \rightleftharpoons {}^1MQ \longrightarrow M + Q \quad (17\text{-}16)$$

This kind of strong quenching is known as collisional quenching and may be compared with the static quenching described by Equation (17-17):

$$M + Q \rightleftharpoons MQ + h\nu \longrightarrow {}^1MQ \longrightarrow M + Q \qquad (17\text{-}17)$$

in which the complex is formed in the ground state and becomes important when the quencher is weak but present in high concentration, e.g., when Q is the solvent. The detailed nature of ^1MQ and the way in which the actual energy transfer occurs within the complex has been and is being investigated extensively. Iodo and bromo compounds, which are strong quenchers, probably facilitate the intersystem crossing of ^1M to the triplet state within the complex and the same may be true when the paramagnetic O_2 is the quencher. Another possibility is that the quencher may react with the singlet state by donating or withdrawing an electron thus forming a redox system with the details and direction of electron transfer depending on the redox potential of the molecule in its excited state.

The system is often more complicated than that described above and departs from simple Stern-Volmer kinetics with the result that the graph of $(F^0/F) - 1$ versus $[Q]$ is not linear. One reason may be the formation of the non-fluorescent complexes already mentioned, and fluorescence quenching can be a valuable technique in their study. Scheraga et al. (3), for example, have made a detailed study of the thermodynamics of the various types of interactions which occur between phenolic compounds and various sodium monocarboxylic acid salts.

C. Fluorescence of Biomolecules.

1. Flavins and flavoproteins.

Fluorescence is a powerful tool for the study of the important group of compounds, the flavin co-enzymes and flavo-proteins, which participate in such processes as photosynthesis, respiration, bioluminescence, and so on. Riboflavin (V), a molecule which contains the planar isoalloxazine skeleton, possesses an intense yellow-green fluorescence in aqueous solutions which has a quantum yield of 0.25 and which is even higher (0.71) in dioxane. Flavin adenine dinucleotide (VI) is also fluorescent but considerably weaker due to the fact that the flexibility of the sugar-phosphate portion of the molecule allows the flavin and adenine portions to come into close proximity making possible the transfer of electronic energy from the flavin to the adenine portion. The details of this folding process are of great importance since the conformation of these molecules is a major

(V)

(VI)

(VII)

factor in their performance as enzymes or co-enzymes, a function which requires a very precise stereochemistry. In this sense, fluorescence is a powerful tool in the study of enzyme conformation and mechanism. This is a rather special case of the phenomenon of static quenching of fluorescence since many aromatic molecules are able to form the nonfluorescent complexes already mentioned. Although adenine is the energy acceptor in this case it can also play an opposite role, e.g., when present in the reduced form of nicotinamide-adenine dinucleotide, NADH (VII), in which case the energy is absorbed by the adenine moiety, transferred with high efficiency to the nicotinamide portion, and emitted as fluorescence.

In the type of quenching in which a ground state complex is involved the fluorescence lifetime is unchanged. This distinguishes it from collisional quenching when the lifetime is reduced considerably. The importance of the photochemistry of flavins themselves as well as their role as sensitisers of the photochemical reactions of other molecules has made them among the most intensively studied group of photobiological molecules.

2. Fluorescence of chlorophylls.

The unique position held by the various chlorophylls in the process of photosynthesis is the main reason why they have been studied so much in <u>in vivo</u> and <u>in vitro</u> systems. Chlorophyll a (VIII) is strongly fluorescent only when solvated by bases (such as water) and in a highly

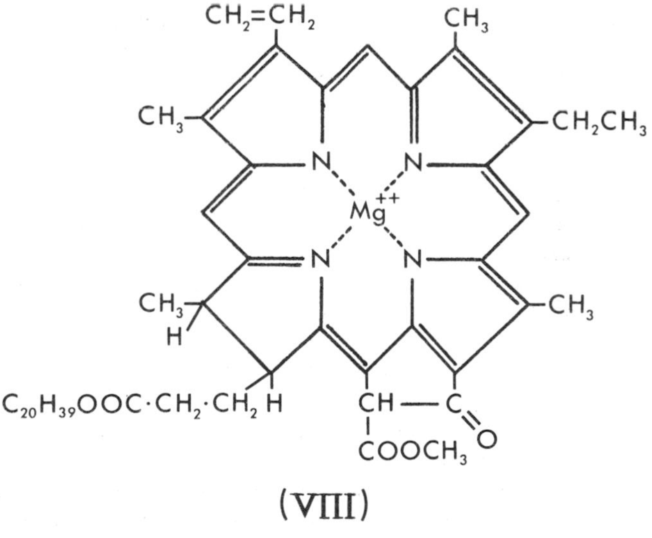

(VIII)

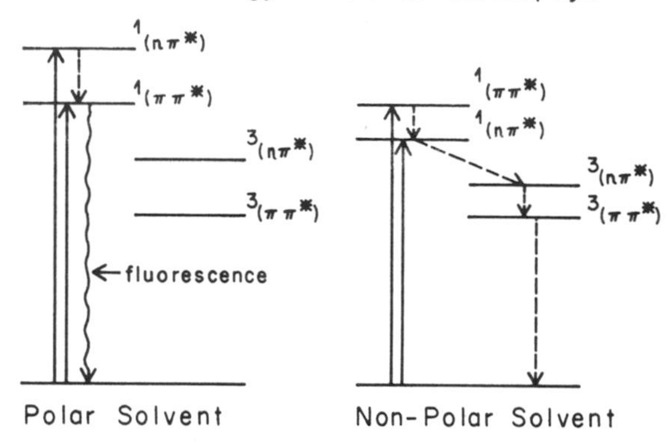

FIG. 6

Lowest $\pi\pi^*$ and $n\pi^*$ excited states of chlorophyll.

purified non-polar solvent or in the absence of the central Mg^{2+} ions the fluorescence is reduced drastically. The most probable explanation of this behavior appears to be that the presence of bases lowers the $\pi\pi^*$ excited state to a level where it is lower than the $n\pi^*$ state (Fig. 6). Under these conditions, the deactivation of the $\pi\pi^*$ state by fluorescence successfully competes with other non-radiative modes. The most obvious interest, however, is not in the fluorescence and photochemistry of isolated and solvated chlorophyll molecules but in molecular aggregates such as those which are involved in the photosynthetic apparatus and in the transfer of energy among the various components of the aggregates. In this context, the formation and photochemistry of thin layers and films of chlorophyll spread on water have been studied extensively. Experiments such as these have led to great advances in the understanding of the primary photophysical processes in photosynthesis. Many aspects of this study have involved the photochemistry of the quasi solid state and are, therefore, beyond the scope of this discussion.

3. Fluorescence of proteins and amino acids.

Proteins contain up to about twenty different amino acids but of this number only three, namely, tyrosine, tryptophan, and, to a lesser extent, phenylalanine have a measurable fluorescence in the visible region and so the fluorescence of proteins depends on how much of these three amino acids they contain. Their fluorescence quantum yield, however, is reduced considerably when they become incorporated into proteins and the various quenching processes which bring about this reduction have been studied since this kind of interaction is associated with the detailed secondary and tertiary structure of the protein. The fluorescence of both tryptophan and tyrosine is quenched in both acid and alkaline solutions, in the former case by the -COOH and NH_3^+ groups present, and this, together with the high sensitivity of tryptophan fluorescence to a polar environment, makes the fluorescence of both molecules a sensitive probe for detailed protein configuration.

The fluorescence spectra of the homopolymers of these amino acids, namely, poly-L-tyrosine and poly-L-tryptophan, exhibit new emission bands the intensities of which are solvent dependent and which are most intense when the solvent is dimethylformamide when the polymers exist in the α-helical form. When this happens it appears that the excitation may not be located on one residue but delocalized over much larger areas of the polymer. The intensity of the new peak depends on whether the polymer is in the helical or uncoiled form which may depend very critically on such factors as pH. It is often of interest to determine the point at which this helical coil transition occurs and to investigate the factors which influence it. Fluorescence is a valuable tool for this purpose particularly when used in conjunction with other techniques such as optical rotatory dispersion, spectroscopy, and chemical studies.

4. Fluorescence of dyes and pigments.

Dyes or pigments when bound to polymeric substrates whether natural or synthetic often undergo a profound modification of their photochemical and photophysical behavior including the intensity of their fluorescence emission. The direction of the change is not always the same, however, for whereas the fluorescence of a dye such as eosin is decreased when bound to the water soluble polyvinylpyrrolidone, aqueous solutions of the triphenylmethane dye crystal violet exhibit a measurable fluorescence only on binding to a polymer such as polymethacrylic acid (4). The changes in the photochemistry reflect the different environments of the dye resulting in some cases in dye-dye interactions the probability of which increases when the dye molecules are bound onto adjacent polymer sites but which are negligible when the dye is free in solution. Interactions of this kind may result in the dye acquiring some of the properties of an aggregate with the consequent increased probability of energy transfer and intersystem crossing leading to the formation of the longer lived triplet state. Binding may also protect a dye from destructive photo-oxidation. A dilute aqueous solution of rose bengal, for example, is photo-oxidized and becomes colorless in a few hours in sunlight but is several hundred times more stable when bound to polyvinylpyrrolidone. Similarly the sensitivity of chlorophyll to atmospheric photo-oxidation is reduced considerably when it is bound in the form of a protein-chlorophyll complex. Whether a dye, in general, becomes more or less photo labile on binding depends on the mechanism of its photochemistry.

The binding of acridine dyes such as proflavine or acridine orange to nucleic acids has been explained by postulating several types of interaction. One, for which a good deal of evidence exists, is the so-called intercalation complex in which the planar dye is thought to assume a position between two adjacent pairs of stacked bases in the nucleic acid. This type of binding is a consequence of the more regular structure of these polymers compared with proteins. These systems have served as excellent models in the study of energy transfer between dye and polymer.

In the cases of dye binding mentioned above, the binding, in most cases has been electrostatic in nature and, therefore, reversible in the sense that the dye could be removed leaving both dye and polymer intact. In recent years a related technique which makes use of "fluorescent protein conjugates" has become important. In this technique, a fluorescent dye containing a reactive group - a familiar example is fluorescein isothiocyanate (FITC) - is allowed to form a conjugate in which the dye is attached to the protein by a chemical bond. The course of the now fluorescent labelled protein can be followed easily in the animal body using fluorescence microscopy. This has become a very powerful technique in the field of immunochemistry and in the study of antibody-antigen interactions.

IV. PHOSPHORESCENCE AND THE TRIPLET STATE

A. Historical Development.

It has been known for about one hundred years that some molecules on illumination with ultraviolet radiation while in a very viscous or rigid environment exhibit a luminescence which has a much longer lifetime than that of the normal fluorescence and appears at longer wavelength. In 1895, for example, Weidemann and Schmidt (5) observed the afterglow of certain dyes by adding gelatin to the aqueous solution to produce the required increase in viscosity. Jablonski and others, around 1935, made an attempt to rationalize the different kinds of luminescence by means of a three level diagram, and associated the phosphorescence with a metastable state of the molecule. The credit for the first definite assignment of phosphorescence to the triplet state with any certainty seems to belong to Terenin (6), although Lewis, Lipkin, and Magel (7) had mentioned the triplet state as a possibility two years earlier. Lewis and Kasha (8) identified and recorded the triplet levels of a large number of compounds by measuring their phosphorescence spectra. The light induced reversible paramagnetism of the triplet state and the involvement of two unpaired electrons, using fluorescein for the study, was confirmed by Lewis and Calvin (9) and later by Lewis, Calvin, and Kasha (10). The electrical dipole nature of the phosphorescence transition was verified later by Weissman and Lipkin (11). The experiments of Evans (12) on fluorescein showed that the paramagnetism decayed at the same rate as the phosphorescence. Since that time many other experiments have put the matter beyond doubt.

B. Spin-Orbit Coupling.

The selection rules which, to a first approximation, makes the probability of singlet $\longrightarrow$ triplet transitions zero, do so because they assume that there is no coupling between the resultant of all the spin angular momenta of all the electrons in the molecule and the resultant of all the orbital angular momenta of all the electrons. This is only true for small atoms but not for larger atoms or molecules for which the extent of the spin-orbital coupling may be appreciable. The overall result of this is that the wave function for a pure singlet state must be modified to include a spin-orbit coupling contribution and, when this is done, a small amount of triplet character is introduced. Similarly the triplet wave function contains a term which introduces a small amount of singlet character. This perturbation which the spin-orbit coupling introduces is responsible for the observed finite nature of singlet $\longrightarrow$ triplet transitions.

The effect of a heavy atom, most commonly the halogens iodine or bromine, whether internal in the sense of being a part of the molecule under consideration or external meaning present in the solvent, is a factor which may, by nature of the intense magnetic field associated with these nuclei, increase the amount of spin-orbit coupling and, hence, the singlet $\longrightarrow$ triplet transition probability. The effect of such an increase is to enhance the $S_0 \longrightarrow T_1$ absorption processes, decrease the radiative lifetime of the triplet state, increase the rate of intersystem crossing ($S_1 \longrightarrow T_1$) and, hence, decrease the fluorescence quantum yield, and increase the intersystem crossing process $T_1 \longrightarrow S_0$. Extensive studies of the roles played by heavy atoms in electronic transitions have been conducted by McGlynn et al. and are detailed in their excellent book (13).

C. Phosphorescence Spectra.

Aromatic hydrocarbons represent, probably, the most studied class of compounds from the point of view of their phosphorescence spectra. Although sometimes difficult to measure, they are a source of important information about the electronic structure of the molecule in its lowest triplet state. As the size of the conjugated system increases, the $T_1 \longrightarrow S_0$ energy separation decreases and so the phosphorescence emission moves to longer wavelength. This is illustrated in Table 2. The spacings between the vibrational components of the emission band are the same as those observed in the fluorescence spectrum indicating that, like the singlet state, the emission occurs from the lowest vibrational level and the spacings observed are those between the vibrational levels of the ground state. The energy gap between S_1 and T_1 the so-called singlet-triplet splitting is fairly large for this class of compounds (e.g., 18,580 cm^{-1} for benzene), but generally decreases as the size of the molecule increases. Substituents on the whole have little effect on the position of the phosphorescence emission although, as mentioned previously, a heavy

TABLE 2

Lowest Triplet Levels and Lifetimes of Some Compounds of Interest Measured from Their Phosphorescence Spectra

Compound	λ phos, A°	Lifetimes sec.
Benzene	3401	7.0
Naphthalene	4695	2.6
Anthracene	6803	--
Toluene	3472	8.8
1-Methylnapthalene	4796	2.5
1-Fluoronaphthalene	4768	1.5
1-Chloronaphthalene	4831	0.30
1-Bromonaphthalene	4831	0.018
1-Iodonaphthalene	5263	0.0025
Triphenylene	4202	15.9
Biphenyl	4348	3.1
Benzoic Acid	3676	2.5
2-Naphthoic Acid	4785	2.5

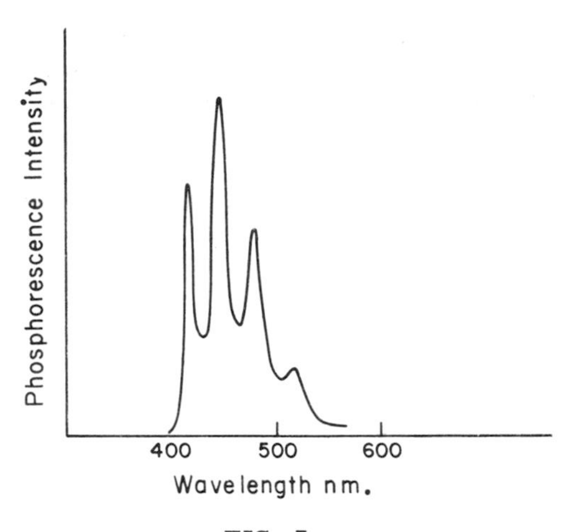

FIG. 7

Phosphorescence spectrum of benzophenone.

atom substituent may have a drastic effect on the transition probability and hence on the phosphorescence lifetime (Table 2).

Molecules which contain heteroatoms and, therefore, unlike aromatic hydrocarbons possess non-bonding electrons, usually have a lowest singlet state which is called an **nπ* state. This is because** non-bonding electrons are held less strongly and on absorption of a photon it is the non-bonding electron which is promoted into a π* state. Non-bonding electrons are not conjugated with the π system and the low overlap between n and π* orbitals accounts for the low intensity of n π* absorptions. The importance of these transitions is that they are characterized by a small S-T splitting (~ 2,000-5,000 cm^{-1}) which means that the $^3n\pi^*$ state lies close to the $^1n\pi^*$ state which will facilitate transitions between the two. There are, however, several possible ways in which the relative energies of the n π and $\pi\pi$ singlet and triplet states can be arranged and so the n π state is not necessarily the lowest energy. In some dyestuff molecules, for example, the lowest singlet and triplet states are $\pi\pi^*$. A useful generalization is that the larger the amount of charge transfer character in the transition, the smaller will be the S-T splitting. Molecules, in which the n π* state is the lowest, usually have a low quantum yield of fluorescence and high quantum yield of phosphorescence the position of which is fairly close to that of the n $\longrightarrow\pi^*$ absorption. Benzophenone (Fig. 7) is a typical example. It is not always easy to determine whether a particular emission originates from a $^3n\pi^*$ or a $^3\pi\pi^*$ state but measurement of the phosphorescence lifetime can be useful since the lifetimes of $^3n\pi^*$ states are generally much shorter than those of $^3\pi\pi^*$ states. Examination of the vibronic spacing which should be around 1730 cm^{-1} for $>C = O$, for example, may also be helpful.

The effect of changing solvent may affect the relative separations of $n\pi^*$ and $\pi\pi^*$ states since each has its own electronic distribution which will be affected differently by solvent polarity. In changing from hydrocarbon to more polar solvents, for example, $n \longrightarrow \pi^*$ transitions show a blue shift whereas $\pi \longrightarrow \pi^*$ transitions show a small red shift. This new electronic configuration of the excited molecule means that its chemistry may be quite different. Many acids, for example, exhibit new pK_a values when in the triplet state. Mention has already been made of McGlynn's fine book which deals with many aspects of the triplet state (13).

D. Triplet States in Photochemical Reactions.

Backstrom (14) was probably the first to realize the chemical involvement of the triplet state although he described it as a biradical when investigating the hydrogen abstracting species in the photoreduction of alcohols by benzophenone. Since that time its importance as an intermediate has been demonstrated clearly in a very wide variety of systems. This particular system subsequently has been studied extensively. In a carefully deoxygenated solution, the overall reaction may be described:

$$2\ \phi_2CO + CH_3 \cdot CHOH \cdot CH_3 \xrightarrow{h\nu} \phi \cdot CHOH \cdot CH \cdot OH \cdot \phi\ + CH_3 \cdot COCH_3 \qquad (17\text{-}18)$$

The mechanism of the reaction is now well established as one in which absorption of a photon converts benzophenone into its $^1n\pi^*$ state which undergoes intersystem crossing with almost 100% efficiency to the $^3n\pi^*$ state. The triplet benzophenone then extracts a hydrogen atom from the isopropanol to form the ketyl radical, ϕ_2COH, which may dimerize to pinacol, one of the observed products. The other products may be rationalized on the basis of the reactions of the ketyl radical with the various possible radicals derived from the solvent. The overall quantum yield for the disappearance of benzophenone in isopropanol as the solvent is around 0.8, but falls to 0.02 when water is the solvent presumably due to the much less favorable hydrogen abstraction process. The participation of the triplet state has been confirmed in this and a great many other reactions by a variety of techniques including flash spectroscopy, physical quenching, and emission spectroscopy.

The triplet state is formed usually in the way described, namely, by intersystem crossing from the singlet state or by energy transfer from a triplet sensitizer which itself has been formed in that way. The efficiency with which the intersystem crossing proceeds is, therefore, of considerable interest, but it was a quantity which was not readily accessible until Lamola and Hammond (15) developed a simple kinetic method in which the substance for which $\phi_{isc.}$ was to be measured was made the sensitizer in a cis-trans isomerization reaction. The sequence of events may be summarized:

$$S + h\nu \longrightarrow {}^1S \tag{17-19}$$

$${}^1S \longrightarrow S + \text{heat}/h\nu' \tag{17-20}$$

$${}^1S \longrightarrow {}^3S \tag{17-21}$$

$${}^3S + \text{trans olefin} \longrightarrow S + {}^3\text{trans olefin} \tag{17-22}$$

$${}^3\text{trans olefin} \longrightarrow \text{cis-olefin} \tag{17-23}$$

$${}^3\text{trans olefin} \longrightarrow \text{trans-olefin} \tag{17-24}$$

When conditions are arranged correctly with a moderate concentration of olefin and with the triplet level of S higher than that of the trans-olefin, Equation (17-22) is diffusion controlled and, consequently, all the ${}^3S$ formed reacts in this way. Under these conditions the easily measurable quantum yield of isomerization gives a direct measure of the yield of intersystem crossing in S. Table 3 lists some values obtained for some compounds of interest.

V. ELECTRONIC ENERGY TRANSFER

A. Transfer from the Singlet State.

This treatment will be concerned only with the modes of energy transfer which apply to aqueous solutions, omitting the solid and gaseous phases, each of which requires a special treatment with a different emphasis.

TABLE 3

Values of Quantum Yield of Intersystem Crossing for Some Molecules of Interest

Compounds	$\phi_{isc.}$
Acetophenone	0.99
Benzophenone	0.99
Triphenylamine	0.95
Fluorenone	0.93
Benzil	0.92
Triphenylamine	0.88
9,10-Anthraquinone	0.88
Phenanthracene	0.76
1-Fluoronaphthalene	0.63
2-Methylnaphthalene	0.51
Napthalene	0.39
Diphenylamine	0.38
Carbazole	0.36
Fluorene	0.31
1-Napthoic acid	0.20
1-Naphthylamine	0.15

Energy transfer processes between donor (D) and acceptor (A) molecules may be described in principle by an equation of the type:

$$D^* + A \longrightarrow D + A^* \qquad (17\text{-}25)$$

where the asterisk denotes the site of the excitation energy. The transfer clearly requires that there is some type of interaction between D* and A and the most obvious kind of interaction is simply a collision. Such collisional transfer processes have been studied extensively by Dubois and his co-workers (16) who found that so long as the transfer process is exothermic (i.e., the excitation energy of A is less than that of D) the process is very efficient, with transfer occurring at every collision. Rates of such processes then depend entirely on how quickly the molecules are able to reach each other in solution and are described as diffusion controlled. Bimolecular rate constants for such processes in water at 20°C. would be around

6.5×10^9 1. mole^{-1}sec.$^{-1}$. These may be calculated after measuring the decrease in the fluorescence of D as the concentration of A is increased and the accompanying increase in the sensitized fluorescence of A.

It is also possible for the transfer of energy described by Equation (17-25) to occur without a collision occurring between the molecules concerned. This can only happen when there is a close resonance between the initial and final states of the molecules concerned and is usually due to a Coulombic interaction of the dipole-dipole type. This means that the electric dipole induced in the donor as a result of the absorption of a photon may induce a similar dipole in a suitable acceptor and energy may be transferred as a result. Förster (17), in a quantum mechanical treatment of this type of dipole-dipole interaction, predicted that the rate of such a transfer would be proportional to the strengths of the electric dipole transitions in donor and acceptor which means that the transfers are governed by the same selection rules which apply to ordinary radiative transitions. A necessary condition is that the energy transferred must correspond to a frequency which is common to the emission spectrum of the donor and the absorption spectrum of the acceptor when the coupling is weak, i.e., it must lie in a region where their spectra overlap. The greater the overlap, the greater will be the number of transitions of the type shown in Figure 8 and, hence, the greater is the probability of a transfer occurring. Making the appropriate substitutions Förster's treatment results in the expression given by Equation (17-26) for the bimolecular rate constant $k_{D^* \to A}$ for the transfer process:

$$k_{D^* \to A} = \frac{8.8 \times 10^{-25} K^2 \phi_D}{\eta^4 \tau_D R^6} \int_0^\infty F_D(\nu) \epsilon_A(\nu) \frac{d\nu}{\nu^4} \tag{17-26}$$

where K^2 is an orientation factor which is 2/3 for a random arrangement, ϕ_D is the quantum yield of the donor fluorescence, η is the refractive index of the solvent, τ_D is the mean lifetime of D^*, R is the distance between D and A, $F_D(\nu)$ is the spectral distribution of the donor emission, and ϵ_A the molar extinction coefficient of the acceptor. One important aspect of this result is the fact that the rate depends on R^{-6} and so falls off very rapidly with distance. As the value of R increases, the probability of a transfer decreases and that of spontaneous deactivation of D^* increases. For this reason it has been convenient to define a critical distance R_0 at which both transfer and spontaneous decay are equally probable. Transfers which

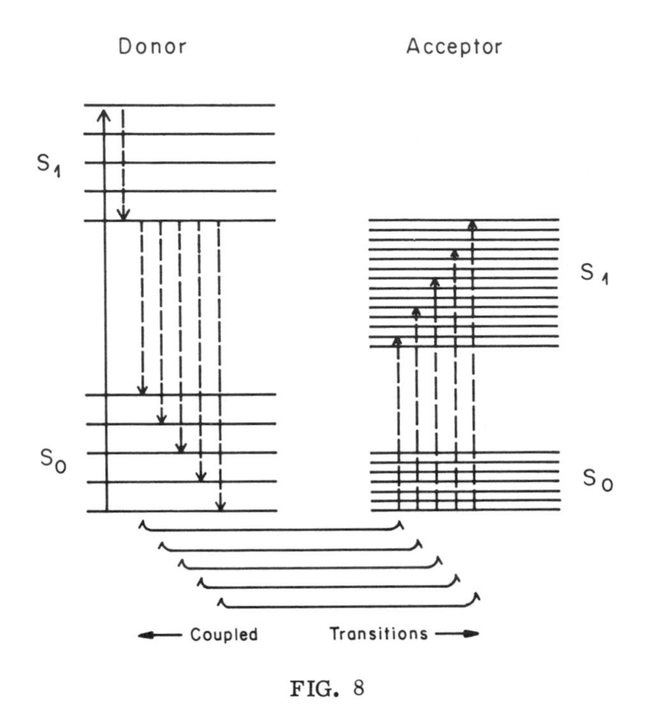

FIG. 8

Resonance transitions in molecules leading to long range energy transfer.

are fully allowed by the selection rules and for which the spectral overlap is large predict values of R_0 of up to 100Å and biomolecular rate constants which are appreciably greater than those for diffusion controlled reactions.

The first experimental confirmation of Förster's theory was made by Bowen and Brocklehurst (18) in a classical experiment designed to exclude all the other possible kinds of energy transfer. Using 1-chloroanthracene as the donor and perylene as the acceptor, bimolecular rate constants as high as 2.0×10^{11} l. mole^{-1}sec.$^{-1}$ were obtained which corresponded to R_0 values around 40Å. Since that time many suitable donor acceptor pairs have been studied (19,20,21, 22) most of which gave values which agreed closely with those predicted by the Förster theory. More recently, Stryer (23) has obtained a very elegant confirmation of the validity of the Förster equation in a system where the groups of the donor, α-naphthyl, and the acceptor, 1-dimethylamino naphthalene-5-sulphonyl, were attached to the ends of various oligomers of poly-L-proline of varying known lengths and in which donor and acceptor were at known distances apart. The efficiency of transfer varied from 100% at 12Å to 16% at 46Å which extrapolated to 50% (the R_0 value) at 34.6Å. This agreed well with the value of 27.2Å obtained from the calculation using the spectral overlap and Equation (17-26). In a similar experiment, Latt et al. (24) used a donor-acceptor pair attached to opposite sides of a bisteroid molecule which held them at a distance of 20Å apart. The values of R_0 of 21.3Å for one pair and 16.7Å for another again showed excellent agreement.

B. Transfer from the Triplet State.

The triplet-triplet transfer process described by Equation (17-27) is strongly forbidden by the selection rules:

$$D^* \text{ (triplet)} + A \text{ (singlet)} \longrightarrow D \text{ (singlet)} + A^* \text{ (triplet)} \quad (17\text{-}27)$$

since both transitions D^* (triplet) $\longrightarrow$ D (singlet) and A (singlet) $\longrightarrow$ A^* (triplet) involve spin changes even though the increased lifetime of the donor would make the energy transfer more favorable. It may occur, however, by an exchange mechanism, the details of which need not concern us here. To ensure an unambiguous demonstration of this type of transfer requires that the acceptor possess a singlet level higher in energy than that of the donor and inaccessible by direct absorption, but a triplet level lower than that of the donor (Fig. 9). A filter system can be used which ensures that only the donor singlet is populated. Terenin and Ermolaev (25) found that benzophenone and naphthalene were such a donor-acceptor pair. They showed that, in a frozen solution at 77°K., radiation of 3660Å which is not absorbed by the napthalene nevertheless is emitted as naphthalene phosphorescence. The concurrent quenching of the phosphorescence of benzophenone showed conclusively that energy transfer was occurring from triplet benzophenone to form triplet naphthalene, a process described by Equation (17-27). These workers showed that each excited molecule could be regarded as being surrounded by a sphere of radius R_C. An acceptor molecule lying within this sphere can accept the energy whereas one lying outside cannot. The values of R_C were found to be of the order of 11-15Å for the various pairs studied. Such short distances are predicted by the exchange mechanism which, quite unlike the Förster long range mechanism, requires that there is an overlap of the wave functions of donor and acceptor. Similar results were obtained by

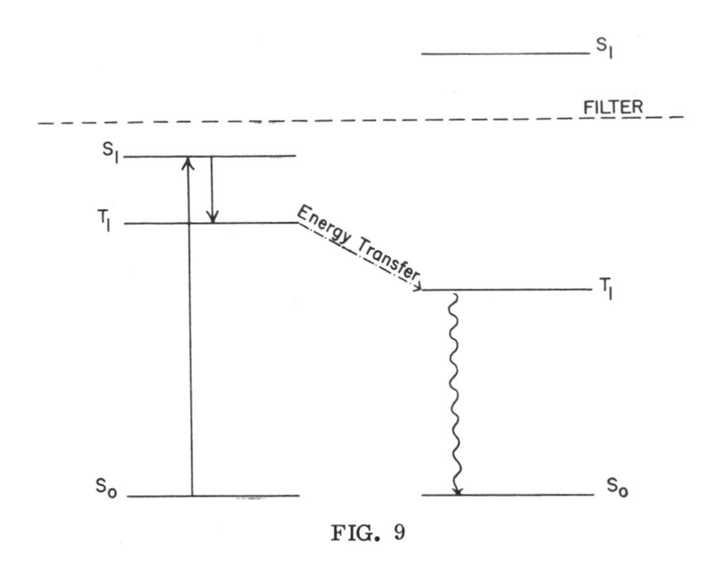

FIG. 9

Triplet-triplet energy transfer.

Backstrom and Sandros (26) and also by Porter and Wilkinson (27) using flash photolysis in fluid solutions. Their results indicate that so long as the triplet energy level of the donor exceeds that of the acceptor by about 3 to 4 kcal/mole then transfer is, to all intents, diffusion controlled. As the difference in energies is reduced, the rate of transfer decreases also but does not fall to zero even when the acceptor level is slightly higher than that of the donor.

Resonance transfer over long distances of the type described for singlet-singlet transitions can also be applied to triplet-triplet transitions. The forbidden nature of the $T_1 \longrightarrow S_0$ phosphorescence means that transfer can often compete with phosphorescence. Ermolaev and Sveshnikova (28) detected a transfer process from triplet triphenylamine to a variety of acceptors including chlorophylls a and b in an ethanol glass at 77°K, with the resultant sensitization of their fluorescence. The transfer which is of the type:

$$D^* \text{ (triplet)} + A \text{ (singlet)} \longrightarrow D \text{ (singlet)} + A^* \text{ (singlet)} \quad (17\text{-}28)$$

leads to R_0 values of 40-50Å. In this case, the calculated value of R_0 is obtained from the overlap of the phosphorescence spectrum of the donor and the absorption spectrum of the acceptor.

C. Intramolecular Energy Transfer.

The transfer of excitation energy within molecules as distinct from that between different molecules is also of great interest particularly in the study of the photochemistry of large molecules. There is considerable evidence that the energy absorbed at one part of the molecule may eventually manifest itself at some other part. For fully conjugated aromatic and heterocyclic systems this process, of course, is well understood and is a direct consequence of the delocalization of the π electron system. It also appears, however, that energy transfer can occur even between portions of a molecule which are themselves conjugated but "insulated" from each other by non-conjugated groups.

This phenomenon has been demonstrated very clearly for many systems, one of the simplest being 4-(1-naphthyl alkyl)-benzophenone (IX). The absorption spectrum of this compound is identical with that of an equimolar mixture of benzophenone and 1-alkyl naphthalene which shows that it behaves as though it contained two independently absorbing systems one due to the benzophenone portion and one due to the naphthalene. Any interaction between these two systems would be

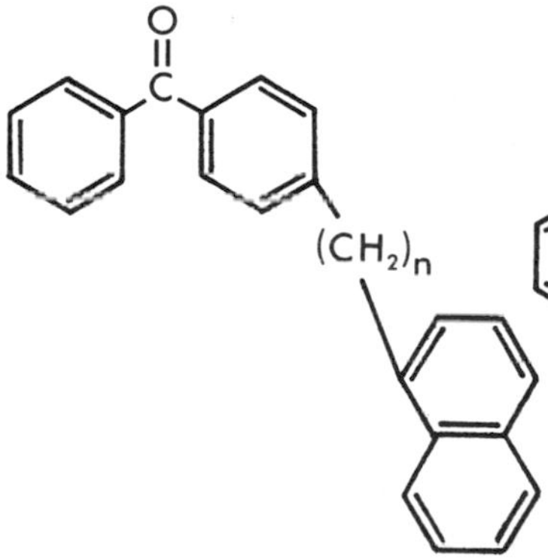

$(CH_2)_n$ $(CH_2)_3$

(IX) 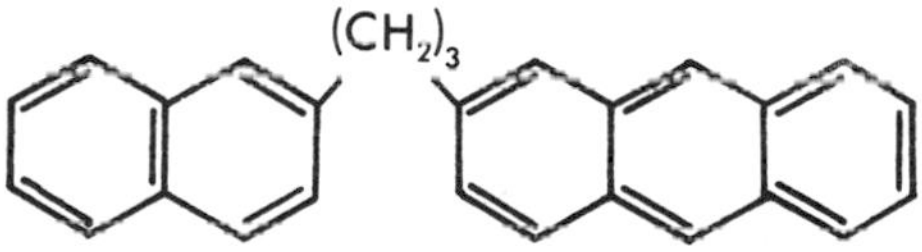 (X)

reflected in the absorption spectrum. By irradiating the molecule with radiation of a frequency absorbed only by the benzophenone (which undergoes intersystem crossing with almost 100% efficiency) Leermakers et al. (29) demonstrated that triplet excitation energy can be transferred from the benzophenone to the naphthalene moiety and detected as the phosphorescence of the naphthalene. When the frequency of the radiation is such that it is absorbed only by the singlet naphthalene portion it is efficiently transferred to the lower singlet level of the benzophenone. Varying the length of the methylene chain appears to indicate that the singlet-singlet transfer is affected much more than is the triplet-triplet transfer. The singlet-singlet transfer of energy from the naphthalene to the anthracene part of molecule (X) on the other hand appears to be highly efficient even when the insulating chain contains up to three methylene groups.

D. Energy Transfer and Reaction Mechanisms.

A dilute aqueous solution of uracil (XI) when photolysed (2537Å) undergoes photochemical reactions and gives two major products, one a cyclobutane dimer (XII) and one a hydrate (XIII). In each case, the site of reactivity is the C_5-C_6 carbon-carbon double bond.

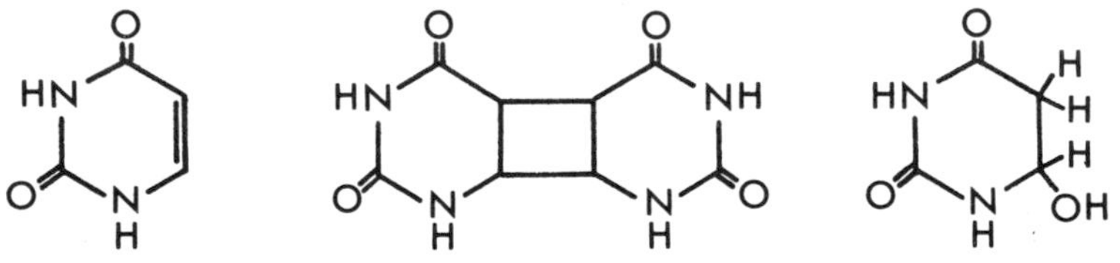

(XI) (XII) (XIII)

When a substance is added which is able to quench the triplet state, examples are 2,4-hexadien-1-ol or the paramagnetic molecular O_2, the dimerization is quenched and the photohydrate is the only product. On the other hand, when acetone is added to the solution and the wavelength of irradiation changed to 3130Å, which is absorbed only by the acetone, the photo dimer is the only product. The conclusion from these observations is that the hydrate is formed from the singlet state and the dimer from the triplet state. Reference to the simple energy level diagram of Figure 10 will make this clear. Since the triplet state of 2,4-hexadien-1-ol is lower than that of uracil it effectively drains off the uracil triplets by energy transfer and so dimerization, the reaction occurring via the triplet state is quenched.

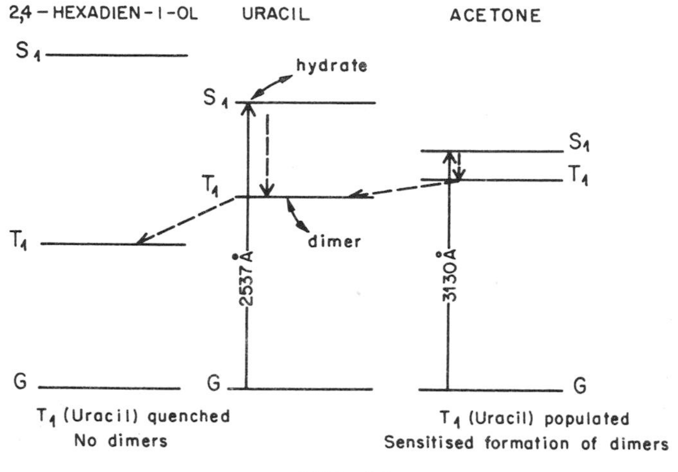

FIG. 10

Energy transfer properties of uracil.

The triplet state of acetone, on the other hand, populated by intersystem crossing from the singlet state, lies above that of uracil and so the energy transfer in this case is from acetone to uracil thus enhancing the dimer formation. Since the singlet state of acetone is lower than that of uracil, the formation of the hydrate is prevented. Oxygen, because of its paramagnetic triplet ground state, strongly quenches triplet states by facilitating the $T_1 \rightarrow S_0$ process.

The ability of uracil and the structurally similar bases cytosine and thymine to form such dimers and hydrates assumes considerable importance when these bases occur in DNA and RNA. This accounts for a good deal of the photobiology of viruses and bacteria which contain these genetic materials (30, 31, 32). The way in which these reactions may be reversed in the process known as photorepair or photorecovery may give a valuable clue to the mechanism which enabled molecules of this type to survive in the early stages of the history of life on Earth. There is good reason to believe that molecules of this type were a constituent of the early forms of life at a time when the ultraviolet flux from the Sun contained a considerably greater range of frequencies than it does at present.

When investigating the mechanism of photochemical reactions involving the triplet state it is often desirable to know the quantum yield of intersystem crossing $\phi_{isc.}$ for a particular molecule in solution.

TABLE 4

Triplet Sensitizers and Quenchers Which Have Been Used to Study Reactions in Aqueous Solution

Sensitizers	
Acetone	Biacetyl
Acetophenone	Proflavin
Tyrosine	Riboflavin
4,4'-benzophenone disulfonic acid sodium salt	Acridine Orange
	Eosin
2,6-naphthalene disulfonic acid sodium salt	Fluorescein
	Rose Bengale
Quenchers	
2,6-naphthalene disulfonic acid sodium salt	
2,4-hexadien-1-ol	
trans-cinnamic acid	
trans-stilbene-4-carboxylic acid	

This may be accomplished by using the particular compound to be studied as a sensitizer for a reaction which is known to proceed via the triplet state. The quantum yield of the sensitized reaction $\phi_{sens.}$ is measured and $\phi_{isc.}$ obtained from the relation:

$$\phi_{sens.} = C \cdot \phi_{isc.} \tag{17-29}$$

where C is a constant which can be determined using a sensitizer for which $\phi_{isc.}$ is known. We have already seen how this technique has been used employing a cis-trans isomerization reaction of certain olefins to determine $\phi_{isc.}$ for a large number of compounds in benzene solution. The method may be useful also in aqueous solution since several rare earth ions such as Eu^{3+} or Tb^{3+} have states which luminesce in aqueous solution and which may be populated by energy transfer from donors with higher triplet states. Using this kind of technique, several groups of workers have obtained parameters of interest for such important compounds as thymidylic acid (33) and orotic acid (34).

The triplet sensitizers must be chosen with care having regard for such things as their optical properties (i.e., correctly positioned levels), photochemical stability, possible dark reactions, solubility, and so on. Table 4 contains a few of the compounds which have been used in aqueous solutions.

ACKNOWLEDGMENT

At the time of preparation of this chapter, Dr. Owen was on leave of absence from University College, Cathays Park, Cardiff, Great Britain.

REFERENCES

1. (a) C. A. Parker, Proc. Roy. Soc. (London), A220, 104 (1953).
   (b) C. G. Hatchard and C. A. Parker, Proc. Roy. Soc. (London), A235, 518 (1965).

2. D. Stern and M. Volmer, Phys., Z., 20, 183 (1919).

3. (a) A. Y. Moon, D. C. Poland, and H. A. Scheraga, J. Phys. Chem., 69, 2960 (1965).
   (b) D. K. Kunimitsu, A. Y. Woody, E. R. Stimson, and H. A. Scheraga, J. Phys. Chem., 72, 856 (1968).

4. J. S. Bellin, Photochem. Photobiol., 8, 383 (1968).

5. E. Wiedemann and G. C. Schmidt, Ann. Physik., 56, 201 (1895).

6. (a) A. Terenin, Acta Physicochim., U.R.S.S., 18, 210 (1943).
   (b) A. Terenin, Zhur. Fiz. Khim., 18, 1 (1944).

7. G. N. Lewis, D. Lipkin, and T. T. Magel, J. Amer. Chem. Soc., 63, 3005 (1941).

8. G. N. Lewis and M. Kasha, J. Amer. Chem. Soc., 66, 2100 (1944) 67, 994 (1945).

9. G. N. Lewis and M. Calvin, J. Amer. Chem. Soc., 67, 1232 (1945).

10. G. N. Lewis, M. Calvin, and M. Kasha, J. Chem. Phys., 17, 804 (1949).

11. S. I. Weissman and D. Lipkin, J. Amer. Chem. Soc., 64, 1916 (1942).

12. D. F. Evans, Nature, 176, 777 (1955).

13. S. P. McGlynn, T. Azumi, and M. Kinoshita, Molecular Spectroscopy of the Triplet State, Prentice Hall, Engelwood Cliffs, N. J., 1969.

14. H. L. J. Backstrom, Z. Physik. Chem., B25, 99 (1934), (b) H. L. J. Backstrom, The Svedberg, (Memorial Volume), (A. Tiselius and K. O. Pedersen, eds.), Almqvist e Wiksells, Uppsala, Sweden, 1944, p. 45.

15. A. A. Lamola and G. S. Hammond, J. Chem. Phys., 43, 2129 (1965).

16. (a) J. T. Dubois and M. Cox, J. Chem. Phys., 38, 2536 (1963).
(b) F. Wilkinson and J. T. Dubois, J. Chem. Phys., 39, 377 (1963).

17. (a) Th. Förster, Disc. Faraday Soc., 27, 7, (1959).
(b) Th. Förster, in Modern Quantum Chemistry, (O. Sinanoglu, ed.), Vol. III, Academic Press, New York, 1965.

18. E. J. Bowen and B. Brocklehurst, Trans. Faraday Soc., 49, 1131 (1953).

19. W. R. Ware, J. Amer. Chem. Soc., 83, 4374 (1961).

20, G. Weber, Trans. Faraday Soc., 50, 552 (1964).

21. A. G. Tweet, W. D. Bellamy, and G. L. Gaines, J. Chem. Phys., 41, 2068 (1964).

22. R. G. Bennett, J. Chem. Phys., 41, 3037 (1964).

23. L. Stryer, Proc. Nat. Acad. Sci., 58, 719 (1967).

24. S. A. Latt, H. T. Cheung, and E. R. Blout, J. Amer. Chem. Soc., 87, 995 (1965).

25. A. N. Terenin and V. L. Ermolaev, (a) Dokl. ANSSR, 85, 547 (1952); (b) Trans. Faraday Soc., 52, 1042, (1956).

26. H. L. J. Backstrom and K. Sandros, Acta. Chem. Scand., 12, 823 (1958); 14, 48 (1960).

27. G. Porter and F. Wilkinson, Proc. Roy. Soc., (London), A264, 1 (1961).

28. V. L. Ermolaev and E. B. Sveshnikova, Izo. Akad. Nauk. SSSR, ser. Fiz., 26, 29 (1962).

29. P. A. Leermakers, G. W. Byers, A. A. Lamola, and G. S. Hammond, J. Amer. Chem. Soc., 85, 2670, (1963).

30. A. D. McLaren, Photochem. Photobiol., 8, 521 (1968).

31. G. D. Small, M. Tao, and M. P. Gordon, J. Mol. Biol., 38, 75 (1968).

32. R. B. Setlow, Photochem. Photobiol., 7, 643 (1968).

33. A. A. Lamola and J. Eisinger, International Conference on Molecul-A Luminescence, Chicago, 1968.

34. D. W. Whillans and H. E. Johns, Biophys. J. (Society Abstracts), 8, A 83 (1968).

CHAPTER 18

PHOTODECOMPOSITION OF ORGANIC PESTICIDES

Joseph D. Rosen

Rutgers - The State University

I. INTRODUCTION

Sunlight does not necessarily cause the breakup of pesticides into their constituent parts or into simpler compounds. In many cases, sunlight transforms pesticides into materials of similar or even greater structural complexity. The toxicology and persistence of these materials are, for the most part, unknown.

After a pesticide is placed into the environment it does not remain stationary nor does it retain its original molecular structure. Pesticides are dispersed through our environment by wind, water, insects, birds, fish, and animals. At the same time, the molecular structures of pesticides are being altered by hydrolysis, biological metabolism, and photolysis. In many cases, a combination of all these forces is at work. Although we will concern ourselves, in the main, with photolysis, it should be kept in mind that the photoproducts themselves might then undergo biological metabolism and/or hydrolysis. The new metabolites, in turn, could be further susceptible to photolysis and so on. It should also be pointed out that, under actual environmental conditions, one process may predominate over another. Thus, if a pesticide, dissolved in distilled water, is converted to a photoproduct in good yield when exposed to sunlight, it does not necessarily mean that the photoproduct will be found in the environment where a competing biological metabolism might be occurring at a much faster rate. Furthermore, the water in which the pesticide is dissolved or dispersed may contain other materials which may absorb solar energy and fail to transfer it to the pesticide. On the other hand, some materials may absorb solar energy and transfer it to a pesticide that would not ordinarily undergo solar transformation.

II. PHOTOCHEMISTRY OF PESTICIDES

In spite of the difficulties involved in extending laboratory findings to the environment, much useful information can be obtained from studying the photochemistry of pesticides. A great number of such studies has been conducted but it will be impossible to discuss them all. For those interested in further pursual of this topic, two excellent reviews, one by Crosby and Li (1) and the other by Plimmer (2) are recommended. In this discussion we will concern ourselves mainly with photoreactions involving isomerization, aromatic substitution, and oxidation.

The chlorinated cyclodiene insecticides, in general, undergo photoisomerization. The first member of this family to be photoisomerized was isodrin (1) (See Appendix). Photoisodrin (2) was obtained after photolysis with a 2537 A^{o} lamp for seven days in a carbon dioxide atmosphere (3,4). Films of endrin (3), a widely-used insecticide, were converted by 2537 A^{o} light to the ketone (4) and the aldehyde (5) (5). When the reaction was conducted in sunlight, however, the aldehyde (5) was not found among the reaction products (6). Dieldrin (6), a geometric isomer of endrin, was photoisomerized in sunlight (5) and under laboratory conditions (7,8) to photodieldrin (7). Similarly, aldrin (8) was converted to photoaldrin (9) in sunlight (9). The reaction mixtures obtained from a 1-month exposure of an aldrin film contained 2.6% unchanged aldrin and 9.6% photoaldrin. The remainder consisted of 4.1% dieldrin, 24.1% photodieldrin, and 59.7% of the major product, an unidentified material which had an average molecular weight of 482. Since photoaldrin could not be photoepoxidized to photodieldrin, the combined yields of dieldrin and photodieldrin indicated that a significant portion of aldrin was photoepoxidized to dieldrin. Of particular interest to those concerned with pesticide residues is that the major product of aldrin photolysis was subsequently found to be approximately five times more toxic to flies than DDT. Yet, this material would not pass through gas chromatographic columns normally used by residue chemists. Thus, should this material be an actual environmental product, it would be undetected.

The photoreactions of endrin, aldrin, and dieldrin were conducted in the absence of solvent. However, since the intramolecular photoisomerizations almost certainly proceed via a free radical mechanism, it is most probable that these reactions will also occur in water. This is so because the high bond energy (116 kcal/mol) of the H-O bond in water makes hydrogen abstraction from water extremely unlikely. Thus, Henderson and Crosby (10) achieved a 63% yield of photodieldrin after exposing an aqueous solution of dieldrin to sunlight for three months. By contrast, irradiation of dieldrin (11) and aldrin (11,12) in solvents from which hydrogen atoms could be more easily abstracted resulted in the formation of the pentachloro derivatives (10) and (11), respectively, in addition to the photoisomers.

Irradiation of aldrin, dieldrin, and isodrin in the presence of the photosensitizer, benzophenone, afforded the respective photoisomers in yields of 58-90% (6,13,14). These results indicated that the photoisomers were formed via the triplet state. By the same method, heptachlor (12) was converted to photoheptachlor (13) (14). Although the latter reaction was not performed in sunlight, one can predict, on the basis of the chemistry so far discussed, that both compound (13) and heptachlor epoxide (14) would probably be formed.

Photoisodrin is less toxic to flies (Musca domestica) than isodrin (15), but photodieldrin, photoaldrin, and photoheptachlor are, in general, 2-3 times more toxic than their respective precursors to flies and mosquito larvae (Aedes egypti) (5,9,14). Photoaldrin exhibited exceptional toxicity to mosquito larvae, being as much as 11 times more toxic than aldrin (16). To a wide variety of vertebrates, photodieldrin was 2 to 10 times more toxic than dieldrin (9,17).

The enhanced toxicities of photoaldrin and photodieldrin to flies and mosquito larvae were found (18) to be due to rapid metabolism to the ketone (15), a material which expresses its toxicity more rapidly than either dieldrin or photoaldrin (18,19). The results suggest that compound (15) might be an environmental photoproduct of aldrin and dieldrin arising by a combination of photolysis and insect metabolism. The ultimate fate of compound (15) has not been studied.

The first rule of photochemistry is that a material must absorb light energy (directly or indirectly) if it is to undergo a photoreaction. Because of a thin band of ozone in the earth's atmosphere, practically all of the sun's emitted radiation in the ultraviolet region below 285mμ is absorbed and does not reach the earth's surface. It is, therefore, believed that any material which absorbs light below 285mμ will not be significantly photolyzed by sunlight. The chlorinated cyclodiene insecticides, which absorb in the 215mμ region, fall into this category. In fact, research on the photolysis of dieldrin was initiated only because of a published report (20) that indicated dieldrin was indeed altered in sunlight. Recently, it has been calculated that the solar flux between 200-285mμ reaching the earth's surface is of the order of 10^{16} photons/cm^2/month (21). Because the absorption coefficient of ozone drops off sharply at 220mμ, most of this solar flux is concentrated in the 200-220 mμ region. Since the cyclodiene insecticides absorb light in this region, it is not surprising that they undergo appreciable photolysis in sunlight.

Two other types of photochemical reactions by a large number of pesticides are aromatic substitution and oxidation of N - methyl and

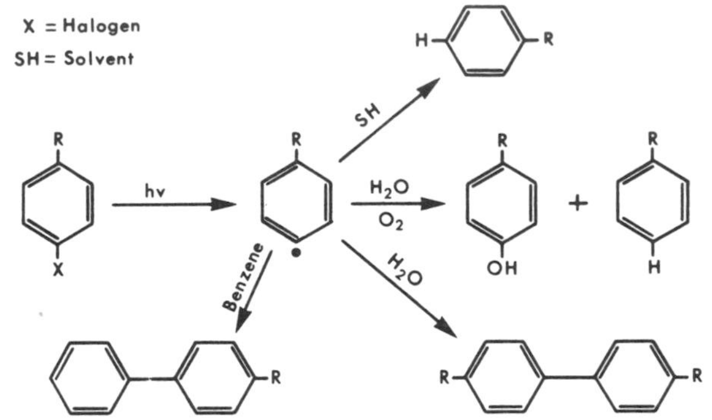

X = Halogen
SH = Solvent

FIG. 1

Solvent effects on photolysis of aryl halides.

N-methoxy substituents. The photochemical aromatic substitution of a large number of halogenated aromatic pesticides has been reviewed by Plimmer (2). In general, light energy causes rupture of the aryl-halogen bond to form a substituted aryl radical (Fig. 1). The fate of this radical is dependent upon its surroundings. In a solvent which has easily abstractable hydrogen atoms, such as methanol, the radical will abstract a hydrogen atom (22). In benzene, where hydrogen abstraction is not as easy, the preferred route is condensation of the aryl radical with benzene (23). In water, phenols are the major products, and, in some cases, hydrogen substitution (24, 25) and dimerization (26) have been observed.

The detailed mechanism of these reactions has not been elucidated. Joscheck and Miller (26) have proposed cleavage of the aryl-halogen bond, followed by reaction of the resulting aryl radicals with water to form phenols. Kharasch and Sharma (23) have speculated on the involvement of hydroperoxy radicals as intermediates in the formation of phenols. A general mechanism illustrating the latter is proposed in Figure 2. Fission of the aryl-halogen bond is followed by reaction of the aryl radical with oxygen to form a hydroperoxy radical, two of which can then interact "head to head" to form phenoxy radicals and oxygen. The latter reaction is well documented (27, 28). The phenoxy radicals may then abstract a hydrogen atom from a ring substituent to give phenols. Hydrogen abstraction by aryl radicals from ring substituents (rather than water) can account for the several observed substitutions of hydrogen for halogen.

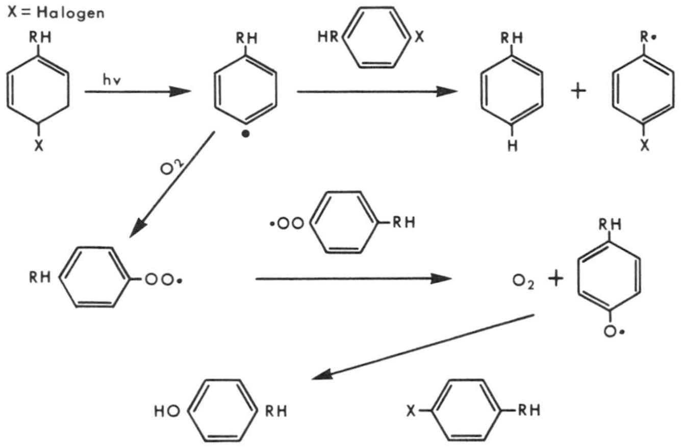

FIG. 2

Possible mechanism for photoconversion of aryl halides to phenols and arenes.

Several examples of these photoreactions follow: in aqueous solution, the widely-used herbicide, 2,4-dichlorophenoxyacetic acid (16), in addition to fission of its ether bond, underwent stepwise substitution of hydroxyl for chlorine until 1,2,4-trihydroxybenzene was formed (29). This material was rapidly air-oxidized to the major product, a polymer whose infrared spectrum was similar to a sample of "humic acid" prepared from 2-hydroxybenzoquinone.

The herbicide, metobromuron (17), was converted in a 20% yield to the phenol (18) after exposure of a 225 mg/l aqueous solution to sunlight for 17 days (30). In addition, oxidation to compounds (19) and (20) occurred. There was also isolated a material whose mass spectrum indicated that it was a condensation product of compounds (17) and (18).

The photolysis of linuron (21) in aqueous solution proceeded in a manner similar to metobromuron (31). The major phenol obtained, (22), had the hydroxyl group in the 4-, not the 3- position. Again, the expected oxidation products (23) and (24) were found.

On the basis of the foregoing results, one would predict that monuron (25) would be converted to the phenol (26) by sunlight in aqueous solution. This was indeed the case (31). Tang and Crosby (32), however, under similar experimental conditions, did not observe formation of compound (26). Instead, they isolated the phenolic compound (27). Several monuron oxidative products were also isolated

(28 - 33). These products suggest that demethylation proceeds via oxidation of the N-methyl groups to formamides. Similar oxidative products were isolated after the photolyses of zectran (33), matacil (33), and diphenamid (34).

Of further interest is the isolation of the symetrical urea (34), which probably was formed by the action of water on the isocyanate (35), which, in turn, was formed by fission of the carbonyl-dimethylaniline bond.

In closing, it should be pointed out that almost all studies of the photolyses of pesticides in sunlight have been conducted by exposing pesticides to direct irradiation. Under actual environmental conditions, however, pesticides may be in intimate contact with materials which act as photosensitizers. These photosensitizers may absorb light energy and transfer this energy to the pesticide, thus altering the products of direct irradiation or causing pesticides which do not absorb light to photolyze. For example, 3, 4-dichloroaniline, a plant and soil metabolite of linuron (35), diuron (36), and propanil (37, 38), is fairly stable to sunlight in aqueous solution. However, in the presence of riboflavin-5'-phosphate, a significant amount of photolysis occurs in a few hours (39). A number of photoproducts of the sensitized reactions has been isolated, two of which have been identified as the azo compounds (36) and (37). These materials are of interest in that they belong to a family of compounds which includes several members with carcinogenic properties (40).

APPENDIX

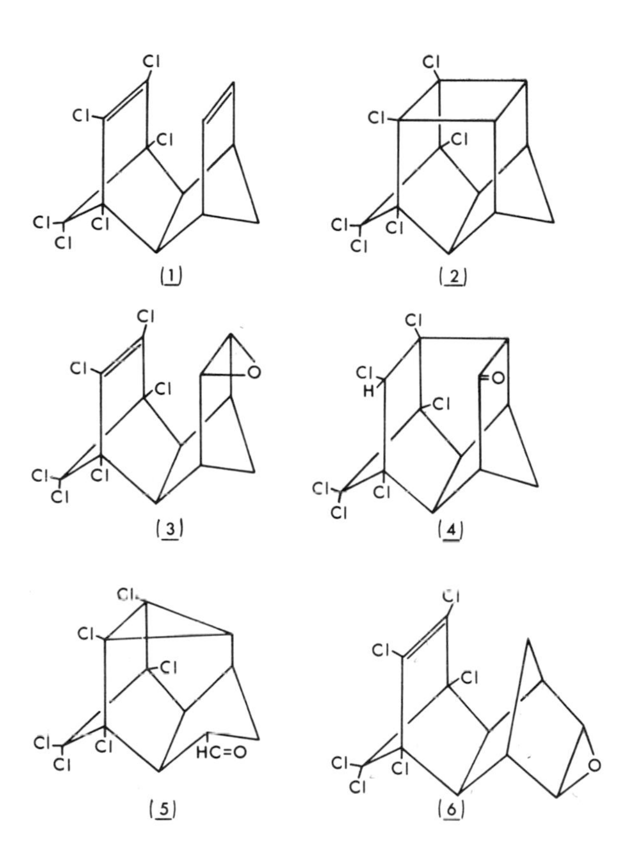
Cl
Cl
Cl
Cl
Cl
Cl
(1)
Cl
Cl
Cl
Cl
Cl
Cl
(2)
Cl
Cl
Cl
O
Cl
Cl
Cl
(3)
Cl
Cl
H
=O
Cl
Cl
Cl
Cl
(4)
Cl
Cl
Cl
Cl
Cl
Cl
HC=O
(5)
Cl
Cl
Cl
Cl
Cl
Cl
O
(6)

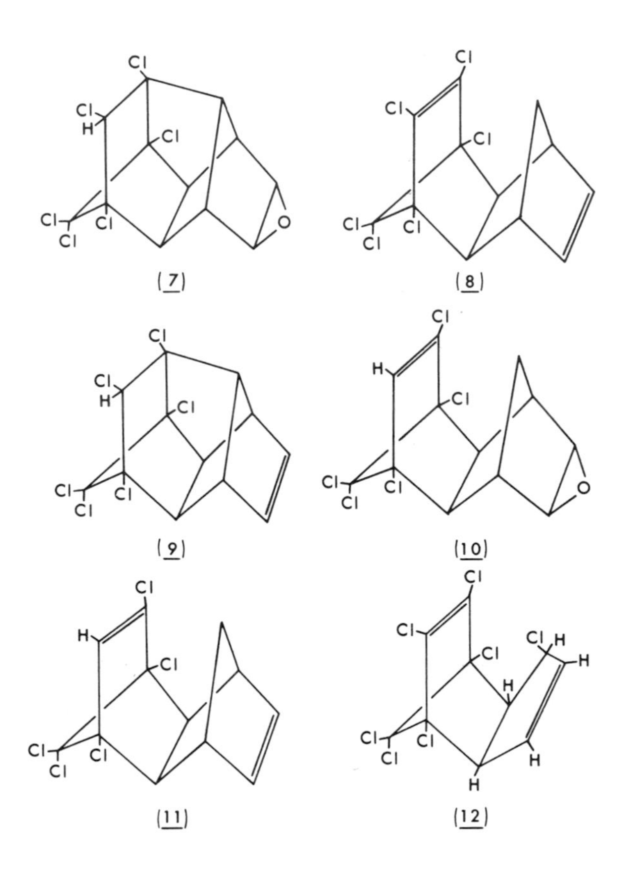
(7)
(8)
(9)
(10)
(11)
(12)

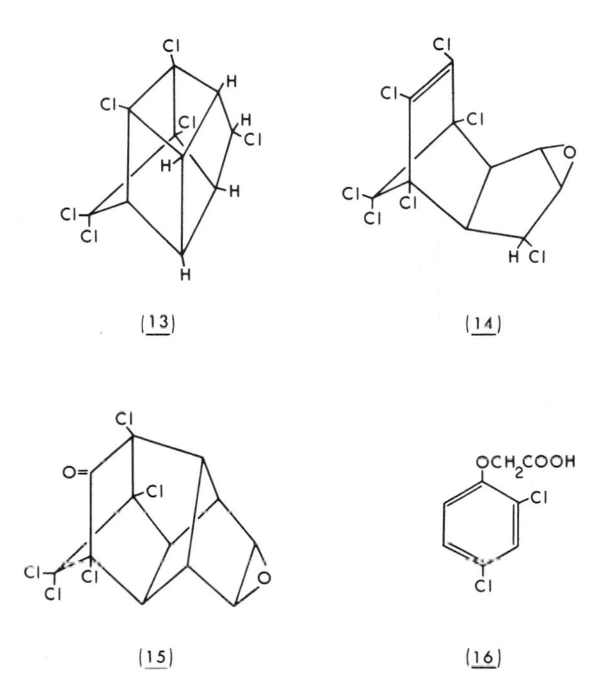
Cl
H
Cl
Cl
H
Cl
H
H
Cl
Cl
H
(13)
Cl
Cl
Cl
O
Cl
Cl
Cl
H Cl
(14)
Cl
O=
Cl
Cl
Cl
Cl
O
(15)
OCH2COOH
Cl
Cl
(16)

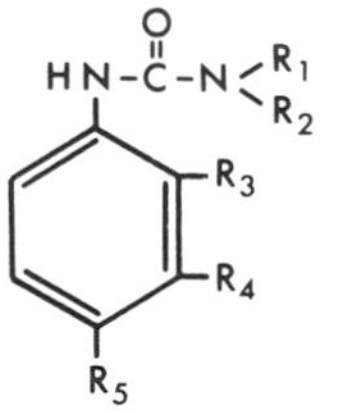

	R_1	R_2	R_3	R_4	R_5
(17)	CH_3	OCH_3	H	H	Br
(18)	CH_3	OCH_3	H	H	OH
(19)	CH_3	H	H	H	Br
(20)	H	H	H	H	Br
(21)	CH_3	OCH_3	H	Cl	Cl
(22)	CH_3	OCH_3	H	Cl	OH
(23)	CH_3	H	H	Cl	Cl
(24)	H	H	H	Cl	Cl
(25)	CH_3	CH_3	H	H	Cl
(26)	CH_3	CH_3	H	H	OH
(27)	CH_3	CH_3	OH	H	Cl
(28)	HCO	CH_3	H	H	Cl
(29)	H	CH_3	H	H	Cl
(30)	H	HCO	H	H	Cl
(31)	H	H	H	H	Cl

(32) (33)

(34) (35)

(36)

(37)

REFERENCES

1. D. G. Crosby and M.-Y. Li, in *Degradation of Herbicides* (P. C. Kearney and D. D. Kaufman, eds.), Marcel Dekker, New York, 1969, p. 321.

2. J. R. Plimmer, in *Residue Reviews* (F. A. Gunther, ed.), Springer-Verlag, New York, in press, Vol. 31.

3. R. C. Cookson and E. Crundwell, *Chem. Ind.*, 1004 (1958).

4. C. W. Bird, R. C. Cookson, and E. Crundwell, *J. Chem. Soc.*, 4809 (1961).

5. J. D. Rosen, D. J. Sutherland, and G. R. Lipton, Bull. Environ. Contam. Toxicol., 1, 133 (1966).

6. J. D. Rosen, unpublished information.

7. J. Robinson, A. Richardson, B. Bush, and K. E. Elgar, Bull. Environ. Contam. Toxicol., 1, 127 (1966).

8. A. M. Parsons and D. J. Moore, J. Chem. Soc., 2026 (1966).

9. J. D. Rosen and D. J. Sutherland, Bull. Environ. Contam. Toxicol, 2, 1 (1967).

10. G. L. Henderson and D. G. Crosby, Bull. Environ. Contam. Toxicol. 3, 131 (1968).

11. G. L. Henderson and D. G. Crosby, J. Agr. Food Chem., 15 888 (1967).

12. J. D. Rosen, Chem. Commun., 189 (1967).

13. J. D. Rosen and W. F. Carey, J. Agr, Food Chem., 16, 536 (1968).

14. J. D. Rosen, D. J. Sutherland, and M.A. Q. Khan, J. Agr. Food Chem., 17, 404 (1969).

15. S. B. Soloway, Advan. Pest. Control Res., 6, 85 (1965).

16. D. J. Sutherland and J. D. Rosen, Mosquito News, 28, 155 (1968).

17. V.K.H. Brown, J. Robinson, and A. Richardson, Food Cosmet. Toxicol., 5, 771 (1967).

18. M.A.Q. Khan, J. D. Rosen, and D. J. Sutherland, Science, 164, 318 (1969).

19. A. K. Klein, J. D. Link, and N.F. Ives, J. Assoc. Offic. Anal. Chem., 51, 805 (1968).

20. J. Roburn, Chem. Ind., 38, 1555 (1963).

21. R. E. Barker, Photochem. Photobiol., 7, 275 (1969).

22. J. R. Plimmer and B. E. Hummer, Division of Agricultural and Food Chemistry, 156th Meeting, American Chemical Society, San Francisco, Calif., April 1968.

23. N. Kharasch and R. K. Sharma, Chem. Commun., 106 (1966).

24. D. G. Crosby, Division of Agricultural and Food Chemistry, 152nd Meeting, American Chemical Society, New York, Sept. 1966.

25. J. R. Plimmer and B. E. Hummer, J. Agr. Food Chem., 17, 83 (1969).

26. H.-I. Joscheck and S.I. Miller, J. Amer. Chem. Soc., 88, 3269 (1966).

27. T. G. Traylor and P. D. Bartlett, Tetrahedron Letters, 24, 30 (1960).

28. D. B. Denney and J. D. Rosen, Tetrahedron, 20, 271 (1964).

29. D. G. Crosby and H. O. Tutass, J. Agr. Food Chem., 14, 596 (1966).

30. J. D. Rosen and R. F. Strusz, J. Agr. Food Chem., 16, 568 (1968).

31. J. D. Rosen, R. F. Strusz, and C. C. Still, J. Agr. Food Chem., 17, 206, (1969).

32. C. S. Tang and D. G. Crosby, Division of Agricultural and Food Chemistry, 156th Meeting, American Chemical Society, Atlantic City, N. J., September 1968.

33. A. M. Abdel-Wahab and J. E. Casida, J. Agr. Food Chem., 15, 479 (1967).

34. J. D. Rosen, Bull. Environ. Contam. Toxicol., 2, 349 (1967).

35. R. B. Nashed and R. D. Ilnicki, Proc. Northeast. Weed Contr. Conf. 21, 564 (1967).

36. R. L. Dalton, A. W. Evans, and R. C. Rhodes, Weeds, 14, 31 (1966).

37. R. Bartha and D. Pramer, Science, 156, 1617 (1967).

38. C. C. Still and O. Kuzirian, Nature, 216, 799 (1967).

39. J. D. Rosen, M. Siewierski, and G. Winnett, Division of Pesticide Chemistry (Probationary), 158th Meeting, American Chemical Society, New York, September, 1969.

40. J. H. Weisburger and E. K. Weisburger, Chem. Eng. News, p. 124, February 7, 1966.

CHAPTER 19

THE PHOTODECOMPOSITION OF 3, 4-BENZPYRENE SORBED ON CALCIUM CARBONATE (1)

Julian B. Andelman and Michael J. Suess

Graduate School of Public Health
University of Pittsburgh

I. INTRODUCTION

Polynuclear aromatic hydrocarbons (PAH) are widely dispersed in man's environment. Many of these compounds are carcinogenic to animals, and probably to man as well. 3, 4-Benzpyrene (BP) is one of the most potent of the carcinogenic PAH (2), and its presence in the environment has been widely studied. Although PAH are known to be formed as combustion products at high temperature (3), there is evidence of their endogenous formation in plants and microorganisms (4, 5, 6, 7, 8, 9).

Carcinogenic PAH have been found in many locations in fresh water. One study by Borneff indicated that their total concentration varies from 0.001 to 0.01 μg/l in ground water, from 0.01 to 0.025 μg/l in treated river and lake water, from 0.025 to 0.1 μg/l in a typical surface water, and greater than 0.1 μg/l in a strongly polluted surface water (10). Both BP and other carcinogenic PAH have been found in treated and untreated drinking water, and it was concluded that, in most cases, the carcinogenic PAH did not exceed 0.025 μg/l (11). These compounds have also been found in the marine environment, including sediments, plankton, algae, and a variety of fish. Many of these studies have been conducted by Mallet and co-workers (12, 13).

As would be expected from their chemical structure, the solubilities of PAH in pure water are very low. For example, after equilibrating distilled water for two years with crystals of 1, 2, 5, 6-dibenzanthracene, another highly carcinogenic PAH, none of this material was detected in the water by an analytical technique sensitive to 0.01 μg/l (14). However, their solubilities may be increased by the presence of a variety of water soluble organics, as has been shown for BP by the addition of lactic acid, purines, acetone, and ethyl alcohol (15, 16, 17, 18). They may also be brought into colloidal solution through the process of "solubilization" due to their association with micelles (19). Synthetic detergents serve for this purpose, but it is necessary that they be at least at the critical micelle concentration (CMC). In general, the CMC for synthetic detergents in natural waters is higher than is likely to be achieved as a result of pollution. One study indicated a measured CMC of 40 mg/l for a linear alkylbenzene sulfonate in drinking water (20). An additional mechanism accounting for the presence of PAH in environmental waters is their ability to concentrate by sorption onto such surfaces as calcareous material and silica (18, 21, 22). Thus BP and other carcinogenic PAH have been found associated with suspended solids in a variety of surface water samples (5, 23).

The stability of PAH, especially as affected by light and oxygen, is an important characteristic in relation to their incidence in water and elsewhere in man's environment. One study of the photodecomposition of BP and several other PAH in cyclohexane and dichloromethane, utilizing sunlight, fluorescent lamps, and ultraviolet sources, found that naphthacene and 9, 10-dimethyl-1, 2-benzanthracene were the most light sensitive, with anthracene and BP less sensitive (24). The other PAH generally did not decompose. The extent of decomposition was similar under exposure to daylight type of fluorescent lamps and sunlight for reasonably comparable light intensities and exposure times. With the ultraviolet source, it was found that BP decomposition was comparable for a solution purged by nitrogen and that exposed to air; however, BP in an oxygen saturated solution decomposed at a much higher rate. Another study of BP decomposition in cyclohexane under ultraviolet irradiation in air and argon atmospheres also found no differences (25).

Another investigation measured the photodecomposition of BP in benzene solutions exposed to solar radiation at various times of the year, indicating seasonal and other variations in rate, presumably correlating with light intensity (26). One investigation of the photodecomposition of BP under ultraviolet radiation of 0.1 and 0.66 mw/cm^2 showed that the rate was higher in acetone than in benzene; also the decomposition rate appeared to be first order in BP, at least in acetone (27).

There have also been studies of the photodecomposition of BP in aqueous solutions with detergents and other organics added as solubilizing agents. In one such study it was found that, in a 0.16 percent aqueous solution of beta-lactoglobulin exposed to air and ultraviolet irradiation at 366 mμ, BP decomposed according to a first order rate, the comparable decomposition for 90 minutes under nitrogen being much smaller (28). In contrast, using caffeine as the solubilizing agent, the decomposition rate was much smaller, indicating that the nature of the solubilizing agent can have a marked effect. Another study found that the photodecomposition of BP was greatest in aqueous detergent solutions, less in vegetable oil solutions, and smallest when the BP was in crystalline form (29). Short wave length ultraviolet light was found to be more effective than daylight.

The question of the nature of the photodecomposition products is of interest. Following ultraviolet irradiation from a mercury lamp of BP in oxygenated benzene solutions, three quinones were isolated after chromatographic separation; namely, 1,6-BP-quinone, 3,6-BP-quinone, and 6,12-BP-quinone (30). Their ultraviolet absorption spectra are quite different and easily distinquished from that of BP. Another study has indicated that the relative amounts of these quinones produced depend on such variables as the nature of the solvent, light intensity and wavelength, and oxygen and BP concentration (31). In one instance where a BP solution in benzene was exposed to sunlight for 9 hours, approximately 1/3 of the BP decomposed; of that which had decomposed approximately 10 percent consisted of the three quinones indicated above.

In the present paper, a study is made of the photodecomposition rate of BP sorbed onto surfaces of calcium carbonate in aqueous suspensions exposed to cool white fluorescent light. Such a system is likely to simulate BP behavior in natural water environments, particularly because of the demonstrated ability of BP to sorb onto mineral surfaces. The sorption is effected from acetone-water solutions and the sorption isotherm is briefly studied. The effects of oxygen concentration, temperature, light intensity, pH value, and ionic strength on the photodecomposition rate are determined, and some comparisons are made with rates in acetone. The mechanism of the photodecomposition process is discussed.

II. EXPERIMENTAL

A. Materials and Equipment.

Crystalline 3, 4-benzpyrene (BP) was used as obtained from K and K Laboratories, Inc., Plainview, N. Y. The reagent grade calcium carbonate used to sorb BP was from the Fisher Scientific Co. (Cat. No. C-65). Under microscopic observation it was found to be crystalline, with the particles for the most part varying from 2 to 20 microns in size. Other chemicals were also reagent grade from Fisher. The nitrogen, helium, and oxygen gases used in the degradation studies were a "prepurified" grade supplied by Airco. The oxygen concentrations in the nitrogen and helium were stated to be 0.002 and 0.0008 percent, respectively. The water was double-distilled with a conductivity of approximately 10^{-6} mho/cm.

For determining the concentration of BP, a Beckman Model DB spectrophotometer was used. Aqueous calcium concentrations were measured by a Perkin-Elmer Model 303 atomic absorption spectrophotometer. Illumination was determined with a YSI-Kettering Model 65 radiometer and a General Electric direct-reading pocket-size light meter, color and angle corrected.

B. BP Analysis.

BP has several absorption peaks in the ultraviolet (UV). An example of its spectrum in acetone is given in Figure 1. In this work there was a need to measure BP concentration in acetone-water mixtures, as well as in acetone alone. For this purpose, using absorption at 384 mμ, it was found that the Beer-Lambert law was followed quite well over BP concentrations ranging from 0.2 to 4 mg/l, and the absorbance at a given concentration was relatively independent of the acetone-water ratio of the solvent, as shown in Figure 2. It has been noted that for aromatic hydrocarbons there may be small shifts in wave length for UV absorption peaks with a change in solvent (32). Some preliminary work on the stability of BP under laboratory light conditions in benzene, dioxane, methanol, and ethanol also utilized spectrophotometric analysis and it was found that there were shifts in the peak wave lengths as high as 6 mμ among these solvents.

In addition to the direct analysis by absorption at 384 mμ in organic and mixed aqueous solvents, BP was also extracted from calcium carbonate surfaces with acetone and its concentration

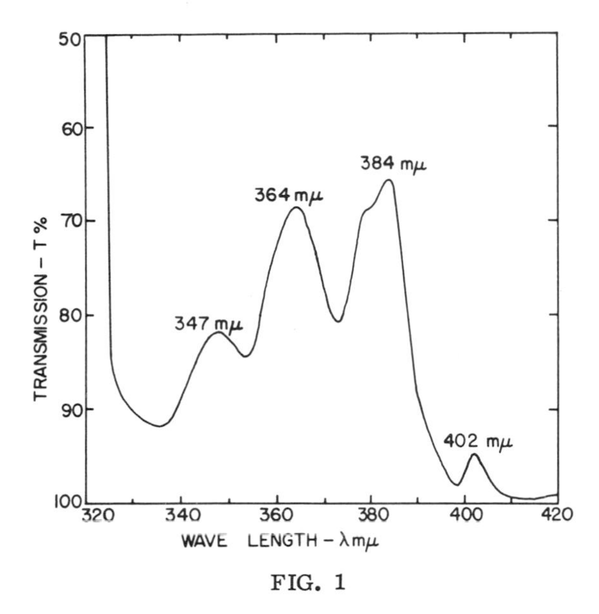

FIG. 1

Absorption spectrum of 1.7 mg/l BP in acetone.

similarly determined. This technique was utilized both in the sorption equilibrium measurements and in the experiments on photochemical degradation.

C. Adsorption Experiments.

In some previous unrelated studies in this laboratory in 20 percent acetone (by volume in water), it was observed that BP adsorbs onto a variety of materials. With initial BP concentrations of 0.5 mg/l in such acetone-water solutions it was found that the surface concentration of BP after 3 hours in darkness was 0.009 $\mu g/cm^2$ for aluminum, 0.006 for stainless steel, 0.05 for Lucite, 0.038 for Teflon, and 0.008 for borosilicate glass. In each case the BP solution volume was 100 ml and the exposed material surface was approximately 70 cm^2. These results indicate the need for care in the choice of materials exposed to BP solutions in order to minimize material losses.

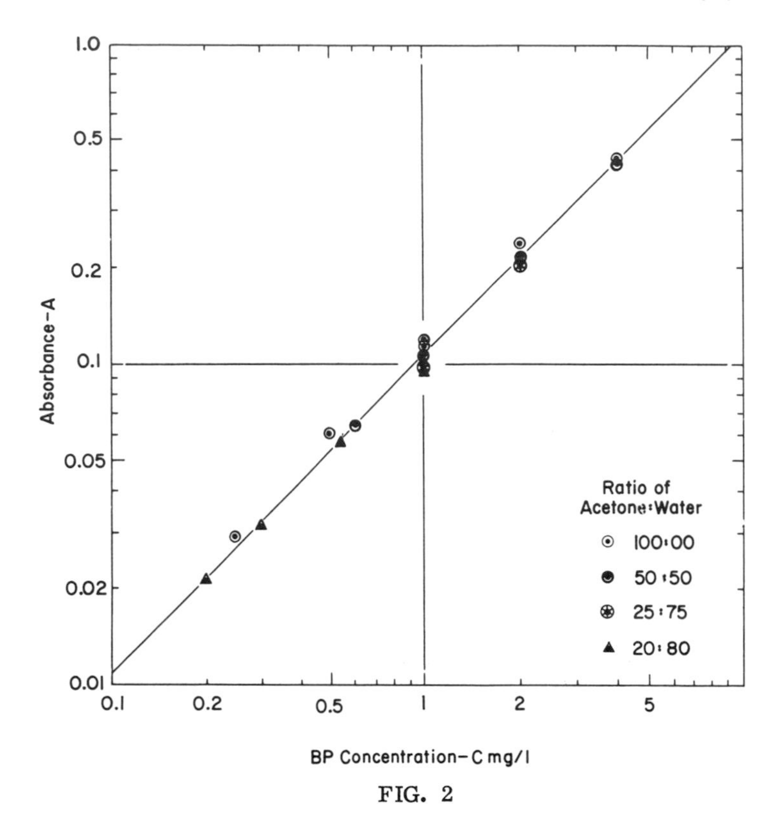

FIG. 2

3,4-Benzpyrene absorption spectra calibration curve (at 383 mμ).

In order to sorb the BP onto the calcium carbonate, typically, 30 g of the latter was shaken for 30 minutes in 100 ml of a 20 percent acetone solution of BP whose initial concentration varied from 0.3 to 1 mg/1. The suspension was then vacuum filtered on filter paper, followed by washing with pH 9 water in order to remove adhering films of acetone BP solution. The filter cake was then dried for several hours at room temperature or in an oven at 35°C. In order to analyze the BP on the dry calcium carbonate powder, approximately 5 g of the latter were shaken in 7 ml of acetone in a glass stoppered centrifuge tube for 2 minutes, followed by centrifugation. The clear acetone supernatant was collected and the BP concentration measured spectrophotometrically.

It was found that 30 minutes of shaking in the sorption process was sufficient to attain equilibrium. Shaking and contact times up to 24 hours did not increase the amount of BP sorbed. To the contrary, there was a decrease of sorption by as much as 20 percent in 5 hours of contact compared to 30 minutes, with a relatively small change subsequently up to 24 hours. This decrease in sorption after the first 30 minutes was probably due to an increase in temperature of the ambient BP acetone solution from the heat generated by the reciprocal shaker. It was observed that this increase was as high as 10°C. above room temperature. Routinely 30 minutes of shaking time was used for the sorption studies and to prepare BP sorbed onto calcium carbonate for the photodecomposition experiments.

The reproducibility of the extraction procedure for BP was within about 4 percent. Neither oven drying the calcium carbonate, nor delaying the acetone extraction up to 9 days had any significant effect on the recoverability of BP, provided the material was kept in the dark. Similarly, suspending the powder in water and shaking in the dark had no effect up to 9 days. One acetone extraction was sufficient to remove essentially all the sorbed BP.

The isotherm for the adsorption of BP onto calcium carbonate from a 20 percent acetone solution is shown in Figure 3. It is reasonably linear in the range studied. The material balance between the BP reduction in the ambient acetone solution and that recovered from the calcium carbonate surfaces was within the reproducibility of the extraction method.

D. Photodecomposition Studies.

Preliminary experiments on the stability of BP were performed in several organic and mixed aqueous solvents in order to select one in which BP was relatively stable. This involved maintaining these solutions that initially contained from 0.5 to 10 mg/l of BP in sealed borosilicate glass containers up to 2 months in darkness or exposing them to typical laboratory white fluorescent light. Absorbance measured by the spectrophotometer in the vicinity of 384 mμ was used to determine BP concentration in these and subsequent photodecomposition studies. It should be emphasized that no attempt was made to isolate and measure decomposition products in either these preliminary or the comprehensive studies. However, the spectrum from 320 to 420 mμ was scanned frequently and no deviation from the typical BP spectrum was ever noted.

These preliminary decomposition experiments indicated that, in dioxane-water solutions ranging from 50 percent (by volume) to 15 percent dioxane, the BP was unstable under laboratory illumination with as much as 80 to 90 percent lost in as little as 17 hours. A

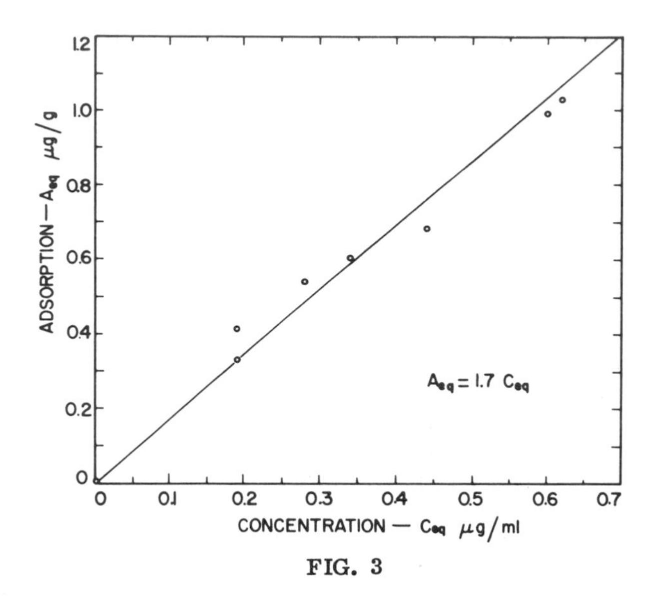

FIG. 3

The adsorption isotherm on calcium carbonate for BP in 20 percent acetone.

100 percent dioxane solution was, however, relatively light stable for 70 hours, as was a 50 percent methanol solution for 17 hours. A 48 percent ethanol solution was light stable for 70 hours, while 24 percent solutions were unstable in the light and dark. In the dark both 100 and 25 percent acetone solutions were stable for at least 57 days and 50 hours, respectively. In all acetone solutions the BP decomposed with light exposure. In all of these preliminary experiments no attempt was made to control oxygen concentration or to measure light intensities. The stability of BP in the absence of light in both acetone and acetone-water solutions was a principal factor in choosing the acetone system for both sorption and extraction.

The energy spectrum of the cool white fluorescent light (33) used in the photodecomposition studies is shown in Figure 4, where it is compared to that of typical solar radiation normal to the earth's surface (34), as well as to the spectral sensitivity of the human eye (33). In an additional preliminary experiment, 20 percent acetone solutions of BP in sealed glass containers were mechanically shaken

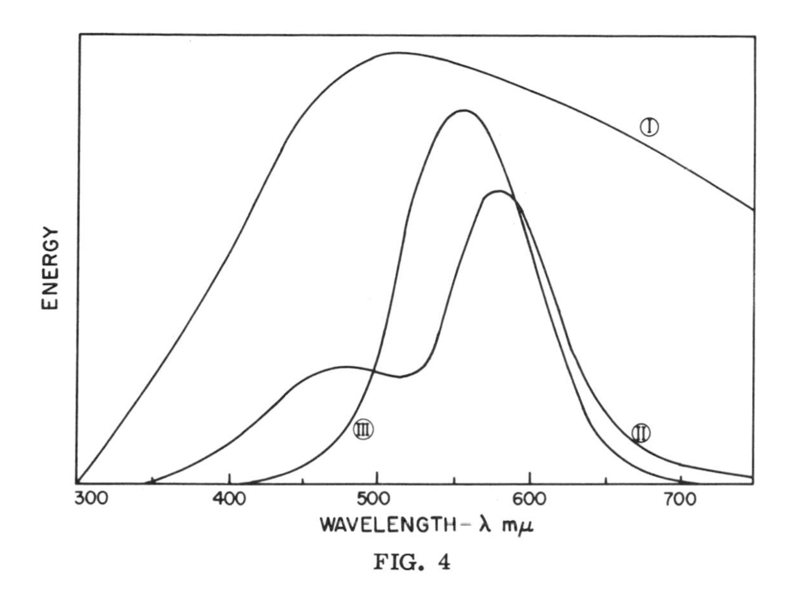

FIG. 4

Spectra for light sources and eye sensitivity with energy scale different for each. I. Typical solar radiation at earth's surface. II. Cool white fluorescent lamp. III. Sensitivity of human eye.

while exposed to two such 40 watt cool white linear fluorescent lamps. However, the subsequent comprehensive studies were performed in several specially designed illumination chambers, a cross-section of which is shown in Figure 5. The source of illumination was a cool white circular fluorescent lamp whose light output was adjusted by varying the applied voltage. The upper compartment containing the beaker and its calcium carbonate suspension was coated internally with flat black paint and was ventilated by air circulated by a fan in the lower compartment. This served to maintain the temperature in the upper chamber at 1 to 2°C. above room temperature. The suspension in the beaker was magnetically stirred by the fan motor. Illumination at the inner face of the beaker was measured at three vertical levels. The mean illumination was 0.31 ± 0.01 mw/cm^2 when the power supply was at 120 volts and the ambient temperature 21°C. The ambient temperature affected the illumination, which was also adjusted by varying the voltage supply to the lamp.

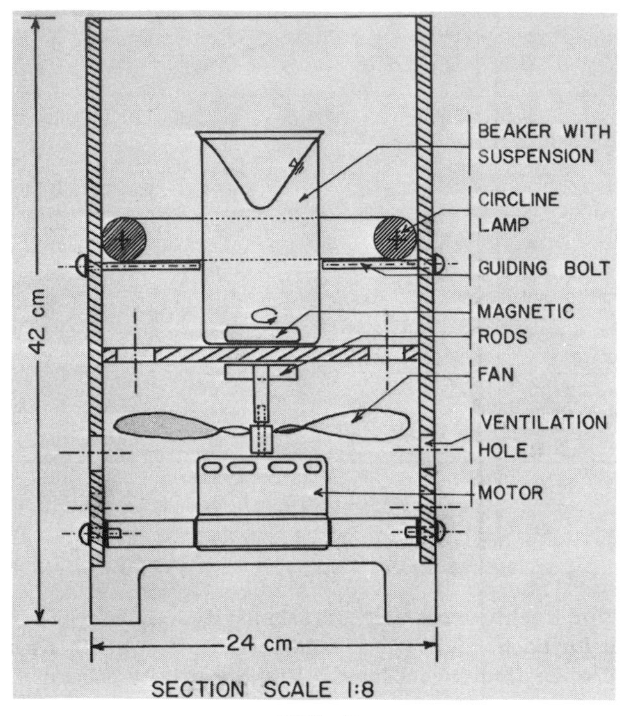

FIG. 5

Cross-section of illumination chamber.

The photodecomposition studies in the illumination chamber typically involved placing a 500 ml aqueous suspension of 5 g of calcium carbonate, with BP sorbed on its surface, into the upper compartment beaker, with stirring at a rate of about 800 rpm to maintain a uniform suspension. Unless otherwise noted the pH of the water was initially adjusted to 9 to 10 with sodium hydroxide to inhibit calcium carbonate dissolution. In order to achieve temperature equilibrium, the aqueous solution was usually placed in the beaker with the light on for 30 minutes prior to adding the calcium carbonate. Similarly, gases were bubbled into the solution for this period to assure equilibration in a gas bubbling experiment. At the termination of the exposure period the suspension was removed, filtered, and extracted with acetone and the BP analyzed spectrophotometrically, as described in the previous section on adsorption. For comparison, photodecomposition experiments were also performed in this chamber on BP in 20 percent acetone, as well as in acetone to which calcium carbonate was added to simulate the light distribution in the aqueous suspension experiments.

Before describing the results of the photodecomposition experiments it is useful to note the following conclusions from the preliminary experiments:

(a) Calcium carbonate suspensions in water showed no loss of sorbed BP when stirred or shaken in the dark for over 90 hours, whether bubbled with helium, air, or oxygen.

(b) Such suspensions did show a reduction in sorbed BP when exposed to cool white fluorescent light.

(c) The BP reduction was less with nitrogen bubbling than with air or oxygen.

The results of preliminary photodecomposition experiments in which aqueous suspensions of BP on calcium carbonate were shaken in flasks initially exposed to the atmosphere are shown in Figure 6. In these experiments, 4 g of calcium carbonate were immersed in 20 ml of water and the system was exposed to two 40 watt linear cool white fluorescent lamps. The results indicate that, for relatively long exposure times and over almost a ten fold range of BP surface concentration, the photodecomposition is first order with respect to BP, with the rate constant, k, ranging from 0.0012 to 0.0015 hr^{-1}. This corresponds to a reaction half-time of 250 to 200 hours under these experimental conditions.

The results of a series of photodecomposition experiments with aqueous calcium carbonate suspensions in the illumination chamber exposed to 0.31 mw/cm^2 at an ambient temperature of 21°C. are shown in Figure 7. In Experiments 4 through 7 and 9 through 11, the reaction vessel was simply exposed to air; in Experiment 8 air was bubbled through the suspension, while in Experiments 11 through 15 oxygen was bubbled. As was the case in the preliminary experiments shown in Figure 6, the photodecomposition rate is first order with respect to BP. For the suspensions simply exposed to air, k-values varied from 0.019 to 0.022 hr^{-1}. The k-value for the air bubbling experiment was 0.021 hr^{-1}. In contrast, k-values in the oxygen bubbling experiments varied from 0.027 to 0.028 hr^{-1}; significantly higher than with air.

The effect of oxygen was further studied at three concentrations by bubbling nitrogen or helium, air and oxygen, respectively, through the aqueous calcium carbonate suspensions with sorbed BP while exposed to light. Figure 8 shows the results of such experiments at an ambient temperature of 21°C. Experiments 29, 30, and 31 were of a preliminary nature and involved exposures to two 40 watt lamps

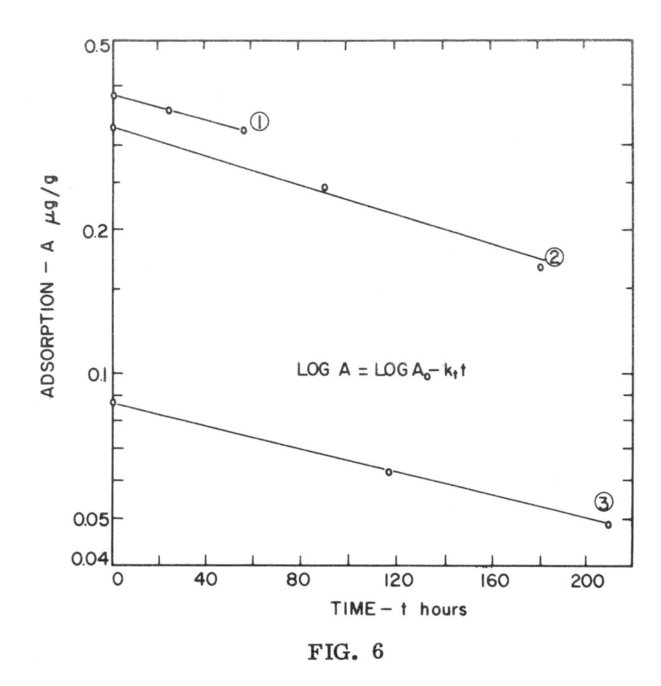

FIG. 6

BP degradation on aqueous calcium carbonate suspensions shaken under linear fluorescent lamps at room temperature.

with illumination not being measured. It is assumed that the decomposition was first order in BP concentration. Nevertheless oxygen and air bubbling is similar in their effects on decomposition, which was much greater than with nitrogen bubbling. As was noted previously, both the nitrogen and the helium used in these experiments had small but measureable concentrations of oxygen. A similar study was conducted in the illumination chamber; the results are shown in Figure 8 as Experiments 32, 33, and 34. These show relatively larger differences between the decomposition rates for air and oxygen exposures. The rate constants are 0.0081, 0.021, and 0.028 hr^{-1} for helium, air, and oxygen bubbling, respectively. Experiment 35 represents the photodecomposition of a BP solution in acetone in the illumination chamber at 0.31 mw/cm2 and with oxygen bubbling. In order to simulate the illumination conditions of the aqueous calcium carbonate suspensions, 5 g of the latter were added in this experiment. It is interesting to note that the first order decomposition

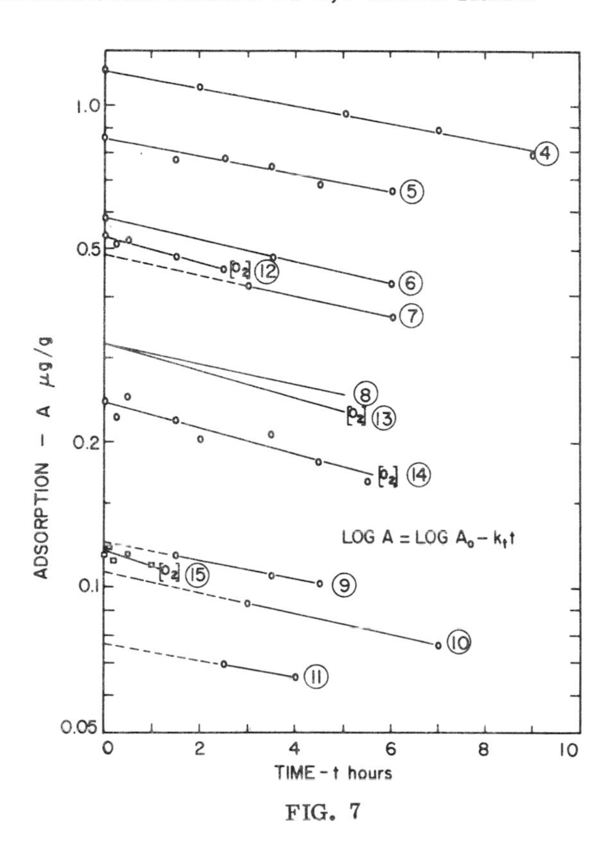

FIG. 7

BP degradation at 21°C. on aqueous calcium carbonate suspensions in illumination chamber at 0.31 mw/cm².

constant was 0.040 hr^{-1}, which is not too much larger than the value of 0.028 obtained in Experiment 34 for the oxygen bubbled aqueous suspension system.

Experiments on the effect of light intensity on the photodecomposition of BP sorbed on aqueous calcium carbonate were performed at an ambient air temperature of 21°C. in the illumination chamber exposed to air. Figure 9 shows the results for two such experiments, each with different initial surface concentrations of sorbed BP and different illumination times. It is apparent that the relationship between BP surface concentration and illumination is semi-logarithmic.

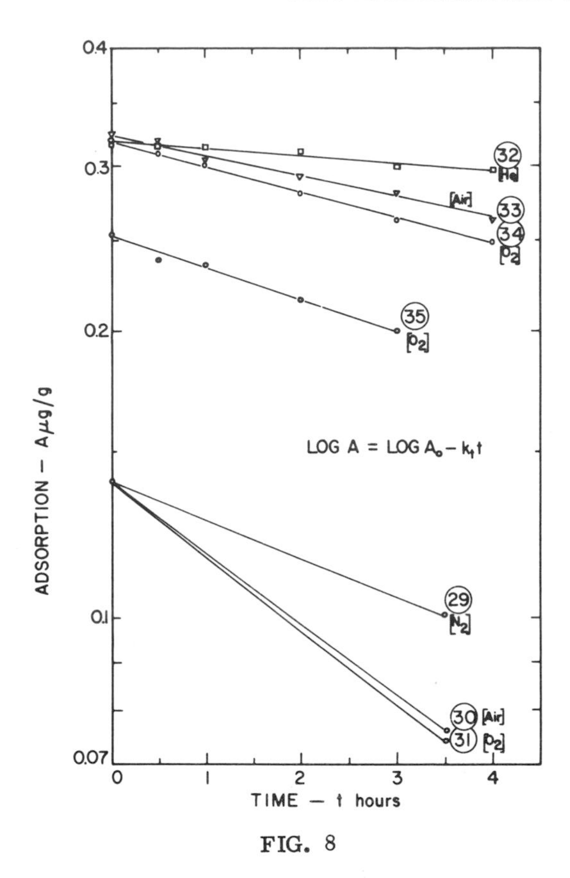

FIG. 8

Effect of oxygen concentration, controlled by gas bubbling, on BP degradation at room temperature. Expts. 29, 30, 31. Suspensions exposed to two 40 watt lamps. Expts. 32, 33, 34. Suspensions in illumination chamber at 0.31 mw/cm^2. Expt. 35. Acetone solution of BP in illumination chamber at 0.31 mw/cm^2. (Ordinate is μg/l BP for Expt. 35 only.)

The fact that k-values increase with illumination is shown in Figure 10, where, on a linear scale, the log of the relative BP surface concentration is plotted as a function of time. Although this plot is somewhat different than Figure 7, this indicates the same order decay of BP with time.

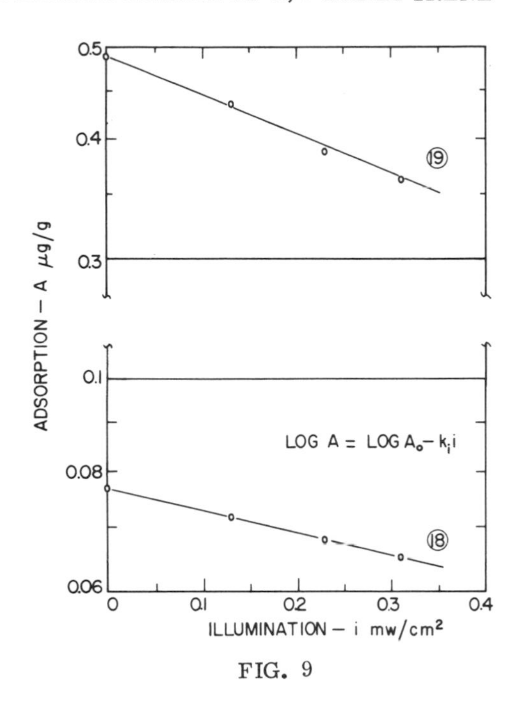

FIG. 9

Effect of illumination on BP degradation on aqueous calcium carbonate suspensions at 21°C. in illumination chamber. Expt. 18. At 4 hours. Expt. 19. At 6 hours.

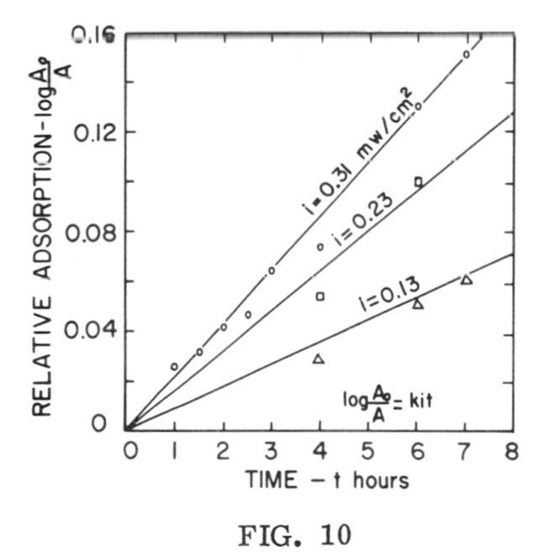

FIG. 10

Effect of illumination on BP degradation on aqueous calcium carbonate suspensions at 21°C. in illumination chamber.

The effect of pH value and, hence, calcium concentrations on photodecomposition was studied for the aqueous suspensions at an ambient temperature of 21°C. The systems were first exposed to air, and then sealed. The pH 9.9 solutions were obtained by first saturating the distilled water with BP free calcium carbonate, adding the calcium carbonate with the BP adsorbed, sealing, and exposing to fluorescent light at 0.31 mw/cm^2 in the illumination chamber. In this experiment the measured equilibrium calcium concentration was 5 mg/l. A low pH system was prepared by similarly saturating distilled water with calcium carbonate, lowering the pH with hydrochloric acid to 6.8, adding the BP sorbed calcium carbonate, and then sealing the system as previously prior to exposing to light. Here the measured calcium concentration was 100 mg/l. The exposures were for 6 hours and the rate constants for both pH 6.8 and 9.9 were 0.022 hr^{-1} which is in agreement with those obtained previously for the similar experiments of Figure 7.

The effect of ionic strength was also studied briefly. To the otherwise typical aqueous suspension at 21°C. was added 500 mg/l of sodium chloride, resulting in an increment of 0.0085M in ionic strength compared to that associated with the calcium carbonate solution in equilibrium with air. Photodecomposition experiments for 4 hours gave k-values of 0.018 and 0.019 hr^{-1} respectively, in the absence and presence of added salt, showing no significant effect of ionic strength in the range studied.

Finally the effect of temperature on the photodecomposition of the aqueous calcium carbonate suspensions of BP was studied in the illumination chamber. Duplicate experiments were performed with air bubbling at each ambient temperature of 5, 21, and 31°C. Because of the reduction in fluorescent lamp illumination at the reduced temperature, in order to obtain constant illumination over this temperature range, 0.13 mw/cm^2 was utilized, a value lower than the typical illumination of 0.31 mw/cm^2. At each temperature, the BP surface concentration was measured after 7 hours. Assuming first order decomposition of BP, the rate constants obtained were 0.0019, 0.0087, and 0.022 hr^{-1} at 5, 21, and 31°C., respectively. The results are also shown in Figure 11, where the log of the ratio of BP surface concentration, before and after the 7 hour exposure, is plotted on a log scale versus the reciprocal of temperature, a linear relationship being obtained.

III. DISCUSSION

The results of these kinetic studies of the photodecomposition of BP, both in acetone solution and in aqueous suspension sorbed onto

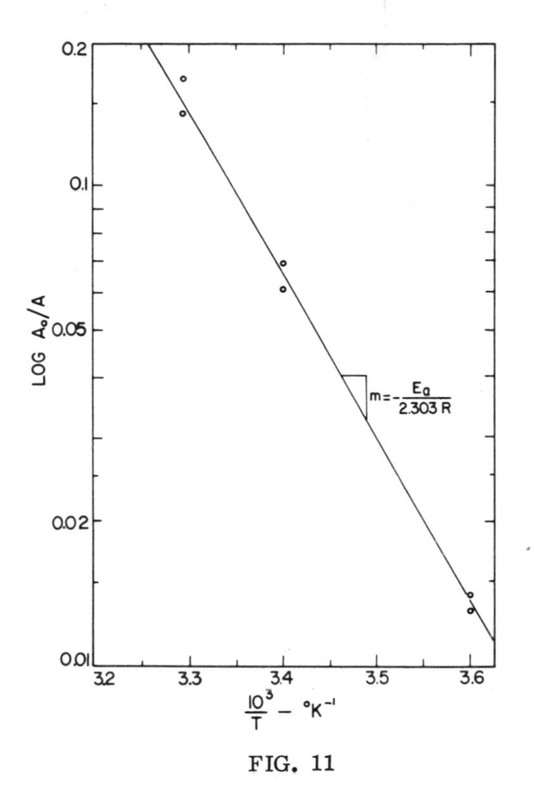

FIG. 11

Effect of temperature on BP degradation on aqueous calcium carbonate suspensions in an illumination chamber for 7 hours at 0.13 mw/cm^2.

calcium carbonate surfaces, indicate a first order dependency on BP concentration and light intensity, a more complicated dependency on oxygen concentration, and a temperature effect, all of which may be expressed in the rate equation as:

$$dA/dt = - k(A)\, I_0\, f(O_2) \exp(-E_a/RT) \qquad (19\text{-}1)$$

where (A) is the concentration of BP, I_o is the light intensity at the surface of the reaction vessel, E_a is an activation energy, R is the gas constant, and T is the absolute temperature. Generally the quantum yields of primary products in photochemical reactions are independent of the absorbed light intensity (35). With B as a primary product of a photochemical reaction, its quantum yield Φ_B would be defined as:

$$\Phi_B = (dB/dt)/I_a \tag{19-2}$$

where I_a is the intensity of absorbed light. If Φ_B was independent of light intensity and if Equation (19-2) was rearranged, one obtains:

$$dB/dt = \Phi_B I_a \tag{19-3}$$

where Φ_B becomes the rate constant, independent of I_a for the formation of B, a primary product in the photochemical reaction of some reactant, such as A. It must be emphasized that here the rate of formation of B is dependent on the absorbed light intensity, even though Φ_B is not. In fact there may be other secondary reactions in which the reactant or primary photochemical products are involved, so that either the disappearance of the reactant or the formation of products have rate expressions such that the quantum yields do depend on absorbed light intensity and, correspondingly, the reaction rate expression is not first order in I_a (35). In related studies it has been shown that, in the photooxidation of anthracene in benzene and carbon disulfide, the rate was directly proportional to I_a (36). Similarly it was established that, in the photooxidation of napthacene, rubrene, 9,10-dimethylanthracene, and 9,10-dimethyl-1,2-benzanthracene in benzene, the quantum yields were independent of I_a. It was noted that this eliminates processes involving more than one electronically excited species (37).

It should be emphasized at this point that the first order dependency of the rate of disappearance of BP on the light intensity noted in Equation (19-1) is not the absorbed light intensity, I_a, rather it is that at the surface of the reaction vessel, I_o. In dilute solutions where relatively little of the light is absorbed by the species A which is being photo-excited, the Beer-Lambert relationship takes the modified form:

$$I_a = I_o \epsilon_A b(A) \tag{19-4}$$

where ϵ_A is the molar absorptivity of A and b the path length through the absorbing solution. Inserting Equation (19-4) into Equation (19-3) one obtains:

$$dB/dt = \Phi_B I_o \epsilon_A b(A) \tag{19-5}$$

Thus, since the absorbed light intensity is proportional to (A), so is the rate of formation of B, without yet considering the possible functional relationship between Φ_B and (A). An indication of the fact that Equation (19-4) applies for the BP systems studied here is supplied by consideration of the appropriate ϵ_A values and concentrations for BP, as well as the likely light path lengths in the photodecomposition experiments. The effective light intensity from the fluorescent light source goes as low as approximately 350 mμ, as shown in Figure 4. Taking the maximum absorbance peak of approximately 384 mμ for BP as the basis of the calculation, one finds in acetone-water solutions an absorbance of 0.1 for 1 mg/1 solutions of BP when b is 1 cm, as shown in Figure 2. Thus, at this wave length, ϵ is 0.1 lit/mg/cm. For those photodecomposition experiments where the absorbed BP concentrations were the greatest, such as Expt. 4 in Figure 7, the maximum effective BP concentration was approximately 10 μg/l, assuming the BP to be "in solution," rather than adsorbed on a homogeneously dispersed suspension of calcium carbonate. Assuming a maximum light path length as the approximate diameter of the reaction vessel, namely 10 cm, one obtains a value for $\epsilon_A b(A)$ of approximately 0.01. For this value Equation (19-4) applies, as it would for the smaller concentrations and those excitation wave lengths with smaller values of ϵ_A. Thus one would expect the rate of formation of the decomposition product (or disappearance of BP) to be proportional to the light intensity at the surface of the reaction vessel, as observed in these experiments and predicted by Equation (19-5). It should also be noted that similar behavior would be expected in environmental waters where the concentrations of photodecomposing species are not large enough to significantly reduce the light intensity over path lengths of a few cm.

As was the case for BP in the present study, a first order dependency on BP concentration has been observed for its photodecomposition in aqueous solution under 366 mμ irradiation, in which either β-lactoglobulin or caffeine was added as a solubilizing agent (28). Other studies on the photooxidation or photodimerization of several PAH have, however, indicated a more complicated dependency of the rate on PAH concentration, usually taking the form KA/(K' + KA) where the constants K and K' have been related to specific proposed mechanisms (36, 37, 38, 39). However, at sufficiently low concentrations of A, such that K' $>>$ KA, it is apparent that the rate may then become essentially first order in A.

In a review of the photochemistry of PAH, it has been noted that oxygen can both quench excited singlet (fluorescent) states of PAH, as well as their triplet states, and may also cause photooxidation (32). The quenching of the fluorescence of several PAH by oxygen, both for vapor systems and in liquid solution has been measured, and it has been shown that the reaction in the liquid state is probably a diffusion limited one (40). The photooxidation of PAH by oxygen frequently results in the formation of cyclic peroxides, as has been shown for rubrene, anthracene, naphthacene, and pentacene (32). However, quinones may form from the decomposition of these peroxides (36) and are known to be photooxidation products for BP and pyrene (30, 31, 41).

The concentration of the oxygen dissolved in liquid systems may have a significant effect on the rate of photodecomposition of PAH, as shown in the present study. It has also been shown that this is the case for diphenylanthracene in benzene, where the rate increased with oxygen concentration. It was also demonstrated, however, that, except at a very low oxygen concentration, the rate of photodecomposition of anthracene in bromobenzene was essentially independent of oxygen concentration, and that the nature of the solvent, in particular its ability to quench the photo-excited state of the PAH, would influence the effect of oxygen concentration (38). Depending on the nature of the decomposition product, however, its rate of formation may also be reduced by an increase in oxygen concentration; this was shown to be the case in the photodecomposition of anthracene in benzene and chlorobezene, forming a dimer, dianthracene (36). However, a cyclic peroxide was also a product in these systems and its rate of formation increased with oxygen concentration. Other studies of the photodecomposition of 9,10-dimethylanthracene, 9,10-dimethyl-1,2-benzanthracene, naphthalene, and rubrene in benzene have also shown an increased rate with oxygen concentration; in these systems photooxidation was the principal chemical reaction (37).

In addition to anthracene forming a dimer photodecomposition product in benzene and chlorobenzene as noted above (36), such a dimer is also formed in bromobenzene (38). Similarly pyrene was found to form a dimer as one of its photodecomposition products when adsorbed onto soil (41). However, dimer formation does not necessarily occur, and was not found as one of the photodecomposition products of diphenylanthracene in benzene (38). Nevertheless, such a dimer may be considered as a possible product in the photodecomposition of BP reported here. Other photodecomposition products of BP, such as acids and alcohols, have also been reported (31, 41).

Several possible reaction schemes have been described for the photodecomposition of PAH in the condensed phase involving peroxides and dimers as the reaction products (36, 37, 38, 42). The likely processes are listed in Table 1 and will now be considered in relation to the rate equations that evolve. Processes 1 through 8, as well as the symbolism, are taken from the scheme of Stevens and Algar in which dimerization does not occur (42). In Table 1 all excited states are

TABLE 1

Possible Reactions in the Photodecomposition of PAH

Process	Number
$A + h\nu \longrightarrow {}^1A^*$	1
${}^1A^* \longrightarrow A + h\nu_f$	2
${}^1A^* \longrightarrow {}^3A^*$	3
${}^1A^* + {}^3O_2 \longrightarrow {}^3A^* + {}^3O_2$	4
${}^3A^* \longrightarrow A\ (+ h\nu_p)$	5
${}^3A^* + {}^3O_2 \longrightarrow A + {}^1O_2^*$	6
${}^1O_2^* \longrightarrow {}^3O_2$	7
${}^1O_2^* + A \longrightarrow AO_2$	8
$A + {}^1A^* \longrightarrow {}^3A^* + A$	9
$A + {}^1A^* \longrightarrow A_2$	10
${}^3A^* + {}^3O_2 \longrightarrow AOO$	11
${}^1A^* + {}^3O_2 \longrightarrow AOO$	12
$AOO + A \longrightarrow AO_2 + A$	13
${}^1A^* + {}^3O_2 \longrightarrow AO_2$	14

noted by an asterisk. Process 1 represents the primary excitation reaction involving the absorption of light to raise the PAH molecule to an excited singlet state. The superscripts 1 and 3 represent singlet and triplet states, respectively. Process 2 represents the fluorescence of ${}^1A^*$ where ν_f is the frequency of the emitted fluorescent light. Process 3 represents a radiationless intersystem crossing of the excited singlet state to a triplet state. Process 4 is a similar inter-system crossing of $^1A^*$ to $^3A^*$, in this case induced by collision with oxygen in its ground state $^3O_2$. It is, therefore, a quenching of the excited singlet state $^1A^*$ by an oxygen molecule. It should be emphasized that, unlike the typical PAH referred to here, the ground state of oxygen is a triplet one. Process 5 represents two possible steps: a radiationless relaxation of the triplet state in which the A molecules return to the ground state, and a similar change, but with the emission of radiation. This latter process is phosphorescence and the frequency of the emitted radiation, therefore, is designated ν_p.

Process 6 involves the quenching of the $^3A^*$ state by oxygen; but unlike Process 4 it involves an energy transfer in which A returns to the ground state and oxygen is raised to an excited singlet state $^1O_2^*$. The evidence for the likely participation of this excited oxygen species in these photochemical processes will be considered later. Process 7 is a radiationless relaxation of the excited oxygen species, returning to its ground triplet state. Finally, Process 8 is the interaction of the excited singlet oxygen species with a PAH molecule to give the photochemical oxidation product, the peroxide AO_2.

Before discussing dimer formation and other possible reaction mechanisms represented by Processes 9 through 14, it is of interest to consider the rate equation that evolves from the mechanism represented by Processes 1 through 8. The rate equation for the disappearance of A or the formation of AO_2 is developed by the usual steady-state treatment in which the excited state intermediates in the reaction are at small, but nevertheless finite concentrations, which are "constant" at a given time. The rate of formation of the oxidation product, the peroxide AO_2, is taken from Process 8 such that:

$$d(AO_2)/dt = k_8(A)\,(^1O_2^*) \qquad (19\text{-}6)$$

with k_8 as the rate constant for this bimolecular process. Since this rate expression involves an unknown quantity, $(^1O_2^*)$, it is necessary to use the steady-state treatment and express $(^1O_2^*)$ in terms of measurable variables. Such a treatment assumes that the rate of change of concentration of intermediate excited states is zero. Using this concept and all the processes in which $^1O_2^*$ appears, one obtains:

$$d(^1O_2^*)/dt = 0 = k_6\,(^3A^*)\,(^3O_2) - k_7(^1O_2^*) - k_8(^1O_2^*)\,(A) \qquad (19\text{-}7)$$

which may then be solved for $(^1O_2^*)$ to give:

$$(^1O_2^*) = k_6\,(^3A^*)\,(^3O_2)/\,\{k_7 + k_8\,(A)\} \qquad (19\text{-}8)$$

When Equation (19-8) is substituted into Equation (19-6), $(^1O_2^*)$ no longer appears; however, $(^3A^*)$ does and it is also unmeasurable. It and the other intermediate, $^1A^*$, are eliminated from the final rate expression by the same steady-state treatment:

$$d(AO_2)/dt = I_a \left[\frac{k_8\,(A)}{k_7 + k_8\,(A)}\right]\left[\frac{k_6\,(^3O_2)}{k_5 + k_6\,(^3O_2)}\right]\left[\frac{k_3 + k_4\,(^3O_2)}{k_2 + k_3 + k_4(^3O_2)}\right] \tag{19-9}$$

which is essentially the equation given by Stevens and Algar (42). The "I_a" term appears as a result of Process 1, using Equation (19-3) in which Φ is taken as unity, since this is the only process in this reaction scheme in which light is absorbed. Parenthetically it should be emphasized that "I_a" in this expression strictly refers to the light absorbed in Process 1, and is not I_o.

This scheme and Equation (19-9) successfully characterize the rates of photooxidation of several PAH in benzene (37,42). For example, it predicts that $1/\Phi_{AO_2} = I_a/\{d(AO_2)/dt\}$ is a linear function of $1/(A)$, which was found to apply in the systems studied. In one of the systems, 9, 10-dimethyl-1, 2-benzanthracene in benzene, at low oxygen concentrations where $k_4(^3O_2) << k_3$, Equation (19-9) successfully predicts that $1/\Phi_{AO_2}$ is a linear function of $1/(^3O_2)$. However, at higher oxygen concentrations this relationship becomes non-linear. This also was found experimentally. Similarly in the BP system reported in this paper there was an increase in the rate of loss of BP with increased oxygen concentration, and the reciprocal of this rate constant was a non-linear function of oxygen concentration. For Experiments 32, 33, and 34 shown in Figure 8, where helium, air, and oxygen, respectively, were bubbled into the calcium carbonate aqueous suspension, it was found that the rate constant for the photodecomposition of BP was proportional to $(P_{O_2})^{0.1}$.

There are other reaction mechanisms that result in rate equations for the formation of peroxides from PAH in the presence of oxygen that are quite similar in form to Equation (19-9). They involve the reaction of either an excited singlet or a triplet PAH, $^1A^*$ or $^3A^*$, with ground state oxygen to form a reaction intermediate, followed by reaction of this intermediate with a ground state PAH molecule to yield the final peroxide end product, AO_2. Such a scheme is represented by Processes 1 to 5, 11, 12, and 13 of Table 1 and has been utilized to develop PAH photooxidation kinetics (36,38,39). However, even though the form of the kinetic equation evolving from this scheme is satisfactory, it involves the formation and behavior of an oxygen complex with a PAH molecule, typically a "moloxide" AOO, which would have to have the unexpected property of not being able to collapse by itself to form a cyclic peroxide AO_2, but would transfer this oxygen to another PAH molecule (43). Others have expressed dissatisfaction with "moloxide" as a likely intermediate (32). It has been noted that, in systems where diffusion of such a large complex would be limited, the rate of photooxidation was high enough to indicate that the reactive intermediate was a more diffusible species (37, 44). However, the most convincing evidence in favor of the singlet excited molecule $^1O_2^*$ as the reactive intermediate in preference to the PAH oxygen complex, is the large number of experiments indicating that the photooxidation process follows a reaction scheme and forms reaction products identical with those arising from the interaction of organic species with $^1O_2^*$ produced nonphotochemically, either from the reaction of sodium hypochlorite and hydrogen peroxide, or a radiofrequency discharge in gaseous oxygen (44, 45). Also, it has been noted that when a radical capture agent was added in the photochemical oxidation of pyrene, the production of quinones was not inhibited, indicating that a singlet, rather than a triplet state reaction intermediate, was involved (41); this evidence is in agreement with $^1O_2^*$ as the reaction intermediate.

The principal difficulty in applying the $^1O_2^*$ mechanism to the BP photodecomposition results reported here is due to the experimental first order dependence of the rate of decomposition on BP concentration. If one were to properly use Equation (19-5), rather than Equation (19-3), to express the rate of loss of BP in the primary photochemical excitation Process 1, this would change the (A) in the numerator of Equation (19-9) to $(A)^2$. It then follows that the only way in which this modified Equation (19-9) could successfully describe the BP results reported here is if k_8 (A) >> k_7 which cancels one of the (A) terms in the numerator. However, considering the reported values for k_8/k_7 for several PAH systems, it is unlikely that this condition holds. Some typical values are 2.4 for anthracene in benzene, 28 in chloroform, and 17.6 for diphenylanthracene in benzene; an unusually high value is 600 for rubrene in benzene (36). Thus, even in the latter relatively favorable case, minimum concentrations of PAH of the order of 10^{-2}M are required to reduce the modified Equation (19-9) to an expression first order in (A). For the BP concentrations reported here, such as noted in Experiment 35 in Figure 8, it is highly unlikely that k_8 (A) >> k_7; in fact the reverse is likely to hold, which would lead to an expression for d (A)/dt that is second order in (A).

The possibility was examined that the rate of photodecomposition was second order in BP concentration. Although, in any given experiment, the data might be considered to reasonably approximate a second order rate, such a treatment resulted in a rate constant that changed systematically over the wide range of surface concentrations studied, unlike that for the first order expression, which was, therefore, considered to be the more appropriate one. The possibility that the BP on the calcium carbonate surfaces represented very high local concentrations such that k_8 (A) >> k_7 was also rejected, since the first order rate constants in Figure 8 for Experiment 35, involving BP in acetone solution, and Experiment 34, for BP sorbed on calcium carbonate, were comparable, implying similar mechanisms. It is also pertinent to note that, even though only ratios of k_8/k_7 have been experimentally determined for several PAH systems, rather than the individual k-values, with a maximum reported value as $10^4 M^{-1}$, theoretical considerations would seem to indicate that ratios significantly higher are not likely to be attained (46). It is, therefore, unlikely that the $^1O_2^*$ mechanism applies to the BP system reported here, at least in the scheme represented by Processes 1 through 8 in Table 1.

A simpler scheme which does lead to the first order decay rate observed here is represented by Processes 1, 2, 3, 4, 5, and 14. It involves the direct attack by ground state oxygen on the excited singlet state BP molecule, as shown in Process 14. Utilizing the steady-state treatment for this scheme in conjunction with Equation (19-5) leads to the rate equation:

$$d(AO_2)/dt = k_{14} I_o \epsilon_A (A) (O_2)/\{k_{14}(O_2) + k_4(O_2) + k_3 + k_2\} \tag{19-10}$$

Although this does not precisely describe the measured oxygen effect of $(P_{O_2})^{0.1}$ on the rate constant, it must be noted that this latter relationship is developed from the oxygen concentrations in pure oxygen, air, and helium. The latter concentration used to calculate this relationship could well have been in error; for example, at these very low oxygen concentrations it is likely to be difficult to purge oxygen from the mineral surface. The fact that Equation (19-10) describes a rate of formation of peroxide (or loss of BP) that is first order in BP concentration makes this an attractive scheme for describing the BP mechanism. The possibility of such a mechanism in the photooxidation of pyrene has been suggested (41); nevertheless, it should be emphasized that there is no independent evidence supporting it, in contrast to the numerous reports in favor of an $^1O_2^*$ intermediate in other PAH systems.

The kinetics of dimer formation in the photochemistry of PAH has also been considered and related to the photooxidation process (36, 38). The reaction schemes that were considered essentially involved Processes 1 to 5, 9, and 10 in Table 1. Because the final step in dimer formation involves the reaction of an excited state PAH intermediate with a ground state PAH molecule, as shown in Process 10, the kinetic equation that results has the same dependency on PAH concentration that arises in the $^1O_2^*$ scheme leading to Equation (19-9) and implies a second order dependency on PAH concentration when Equation (19-5) is utilized, as is required. Thus the same objections apply to this scheme for interpreting the BP results as in the case of the photooxidation with the $^1O_2^*$ intermediate. However, the final kinetic equation also predicts a reduction in the rate of dimer formation with increased oxygen concentration, as was observed for anthracene (36). Since the BP results indicate an increase in rate with oxygen concentration, this implies that if this proposed scheme were to apply, the rate of dimer formation in the BP system is relatively small compared to the photooxidation.

Finally the temperature effect on BP photodecomposition as shown in Figure 11 is of interest. Although frequently temperature changes will not significantly affect the rate of a photochemical reaction, particularly in unimolecular processes where the activation energy necessary to dissociate a molecule is readily supplied by the absorption of light, there are photochemical reactions, such as those involving diffusion processes, which are affected by temperature. However, in contrast to the temperature effect reported here, it was established that the photooxidation of anthracene in benzene was practically independent of temperature over an approximate 50°C. range (36). Nevertheless, there are several bi-molecular reactions shown in Table 1 which are likely to be diffusion limited and, hence, result in a temperature dependency when they are involved in a reaction mechanism. For example, Process 4, the quenching of the fluorescent state by oxygen is, with great likelihood, diffusion limited and its rate frequently increases with temperature (47). Also Process 8 is probably diffusion limited as well (42). If the same applies to Process 6, it is apparent that Equation (19-9), which incorporates these processes, would assume a complicated temperature dependency.

The rate constant for such a diffusion limited quenching reaction may take the approximate form:

$$k_D = 8RT/3000\eta \qquad (19\text{-}11)$$

where η is the viscosity of the solvent. It is assumed that Stokes Law applies (40). With an increase of temperature from 5 to 31°C., η in water decreases approximately by a factor of 2 (35); and since T increases by only about 10 percent over this range, if only one first order diffusion constant appeared in the final kinetic equation, one could not account for the large increase in photodecomposition rates for BP, from 0.0019 to 0.022 hr^{-1} over this temperature range. Also, the oxygen solubility significantly increases at the lower temperature, indicating an even larger difference in the rate constants than the decomposition rates imply. Therefore, it must be assumed that more than one diffusion process is involved in order to attribute the temperature effect to such a mechanism. This is not unreasonable in the reaction schemes discussed above.

It is of interest to note that a plot of the log of the first order decomposition rate constant versus 1/T is linear, as shown in Figure 11. Assuming an Arrhenius relationship as indicated in Equation (19-1), from this plot one can then calculate a value for the activation energy, E_a of 15.4 kcal/mole. However, this is misleading because no consideration was given to a change in oxygen solubility. If this was taken into account, an even larger value for E_a would be obtained. Such a large activation energy must be attributed to a process other than diffusion, if only a single temperature dependent rate constant was responsible.

Aside from considerations of mechanism, Equation (19-1), describing the rate of decomposition of BP sorbed on calcium carbonate, can be related to the fate of BP in the natural water environment. Since it explicitly involves the critical variables of temperature, oxygen concentration, and light intensity, conclusions may be drawn as to those regions in the aquatic environment where decomposition is likely to proceed at a high rate. For example, such photodegradation should be faster near the water surface where oxygen concentration and light intensity are higher. Such quantitative information is useful in assessing the movement and residence times of photodegradable species in natural waters.

ACKNOWLEDGMENT

This study was supported in part by Research Grant No. I-1 from the Health Research and Services Foundation, Pittsburgh, Pennsylvania for which the authors express their thanks.

REFERENCES

1. A portion of this work is abstracted from the Dissertation of Michael J. Suess, submitted in partial fulfillment of the requirement for the degree of Doctor of Science in Hygiene, Graduate School of Public Health, University of Pittsburgh, December 1967.

2. D. Hoffmann and E. L. Wynder, Cancer, 15, 93 (1962).

3. G. M. Badger, The Chemical Basis of Carcinogenic Activity, Charles C. Thomas, Publisher, Springfield, Illinois, 1962.

4. J. Borneff and R. Fischer, Arch. Hyg. Bakteriol., 146, 572 (1963).

5. J. Borneff, ibid., 148, 1 (1964).

6. W. Gräf, ibid., 148, 489 (1964).

7. W. Gräf, Med. Klin., 60, 561 (1965).

8. W. Gräf and H. Diehl, Arch. Hyg. Bakteriol., 150, 49 (1966).

9. L. Mallet and M. Heros, Compt. Rend., 250, 943 (1960).

10. J. Borneff, Der Landarzt, 40, 109 (1964).

11. J. Borneff and H. Kunte, Arch. Hyg. Bakteriol., 148, 585 (1964).

12. L. Mallet, Compt. Rend., 253, 168 (1961).

13. J. Bourcart and L. Mallet, ibid., 260, 3729 (1965).

14. J. Borneff and R. Knerr, Arch. Hyg. Bakteriol., 144, 81 (1960).

15. P. Ekwall and L. Sjoblom, Acta Chem. Scand., 6, 96 (1952).

16. H. Weil-Malherbe, Biochem. J. 40, 351 (1946).

17. W. Gräf and G. Nothafft, Arch. Hyg. Bakteriol., 147, 135 (1963).

18. N. Brock, H. Druckey, and H. Hamperl, Arch. Exp. Pathol. Pharmakol., 189, 709 (1938).

19. M. E. L. McBain and E. Hutchinson, Solubilization and Related Phenomena, Academic Press, New York, 1955 .

20. T. Bohm-Gossl and R. Kruger, Kolloid-Z. Z. Polymere, 206, 65 (1965).

21. J. Borneff and R. Knerr, Arch. Hyg. Bakteriol., 143, 390 (1959).

22. L. Mallet and C. Schneider, Compt. Rend., 259, 675 (1964).

23. J. Borneff and H. Kunte, Arch. Hyg. Bakteriol., 149, 226 (1965).

24. M. Kuratsune and T. Hirohata, Nat. Cancer Inst., Monogr. No. 9, p. 117 (1962).

25. J. W. Woenckhaus, C. W. Woenckhaus, and R. Koch, Z. Naturforsch., 17b, 295 (1962).

26. H. Tanimura, Tekko Rodo Eisei, 13, 174 (1964).

27. H. Kriegel and L. Herforth, Z. Naturforsch., 12b, 41 (1957).

28. G. Reske and J. Stauff, ibid., 18b, 774 (1963).

29. J. Borneff and R. Knerr, Arch. Hyg. Bakteriol., 143, 405 (1959).

30. Y. Masuda and M. Kuratsune, Intl. J. Air Water Pollut., 10, 805 (1966).

31. Anonymous, Sanitary Engineering and Water Resources News Quarterly, University of California, Berkeley, Vol. XIX, No. 1, January, 1969.

32. E. J. Bowen in Advances in Photochemistry, (W. A. Noyes, Jr., G. S. Hammond, and J. N. Pitts, Jr., eds.), Vol. 1, p. 23 Interscience Publishers, New York, 1963 .

33. General Electric, Fluorescent Lamps, Bul. TP-11; Light Measurement and Control, Bul. PT-118, General Electric, Large Lamp Dept., Nela Park, Cleveland, Ohio.

34. G. E. Hutchinson, The Treatise on Limnology, John Wiley and Sons, New York, 1957 .

35. J. G. Calvert and J. N. Pitts, Jr., Photochemistry, John Wiley and Sons, Inc., New York, 1966.

36. E. J. Bowen and D. W. Tanner, Trans. Faraday Soc., 51, 475 (1955).

37. B. Stevens and B. E. Algar, J. Phys. Chem., 72, 3468 (1968).

38. R. Livingston and V. S. Rao, J. Phys. Chem., 63, 794 (1959).

39. E. J. Bowen, Disc. Faraday Soc., 14, 143 (1953).

40. E. J. Bowen, Trans. Faraday Soc., 50, 97 (1954).

41. A. J. Fatiadi, Environ. Sci. Technol., 1, 570 (1967).

42. B. Stevens and B. E. Algar, J. Phys. Chem., 72, 3794 (1968).

43. C. S. Foote and S. Wexler, J. Amer. Chem. Soc., 86, 3880 (1964).

44. C. S. Foote, Acc. Chem. Res., 1, 104 (1968).

45. E. J. Corry and W. C. Taylor, J. Amer. Chem. Soc., 86, 3881 (1964).

46. C. S. Foote, S. Wexler, W. Ando, and R. Higgins, J. Amer. Chem. Soc., 90, 975 (1968).

47. E. J. Bowen and J. Sahu, J. Phys. Chem., 63, 4 (1959).

CHAPTER 20

PHOTODECOMPOSITION OF SOME CARBAMATE INSECTICIDES IN AQUATIC ENVIRONMENTS

Osman M. Aly and Mohammed A. El-Dib

National Research Center
Dokki, Cairo, U.A.R.

I. INTRODUCTION

The massive use of organic pesticides for agricultural, domestic, and industrial purposes has resulted in the increased occurrence of these compounds in natural waters (1, 2). The presence of such organicides in waters used as drinking water supplies constitutes a potential health hazard to the consumer. In addition, many aquatic organisms are capable of concentrating organic pesticides and consumption of these accumulator species or subsequent predators may expose human beings to high levels of these toxicants. The recent discovery of the N-alkyl and N, N-dialkyl carbamate esters as broad spectrum insecticides led to the extensive use of these compounds for the control of agricultural and other pests affecting human health. Very little information is available concerning the fate and persistence of these compounds in the aquatic environment.

Among the environmental factors that may influence the persistence of the carbamate insecticides in the aquatic environment would be the decomposition of these compounds under the influence of solar radiation. Many organicides have been shown to undergo drastic changes upon exposure to ultraviolet light and artificial or natural sunlight (3-10). Therefore, the effect of ultraviolet irradiation upon the stability of Sevin (A), Baygon (B), Pyrolan (C), and Dimetilan (D) in aqueous solutions was investigated.

A. SEVIN (1-Naphthyl-N-methylcarbamate)

B. BAYGON (o-Isopropoxyphenyl-N-methylcarbamate)

C. PYROLAN (1-Phenyl-3-methyl-5-pyrazolyl-dimethylcarbamate)

D. DIMETILAN (2-Dimethylcarbamoyl-3-methyl-5-pyrazolyl-dimethylcarbamate)

II. EXPERIMENTAL

A. Irradiation Procedure.

Aqueous solutions of the carbamate insecticides in different phosphate buffer solutions at pH values of 5.0, 7.0, and 9.0 were subjected to ultraviolet irradiation for varying periods of times. The irradiation source was two 15-Watt germicidal tubes which produce peak radiation at 254 mμ. The tubes were mounted, 10-cm apart, in an exposure cabinet lined with reflecting shields of aluminum sheets. Solutions of the insecticides, 20 mg/l, were prepared in 0.1 M phosphate buffer (pH 6.8), and were adjusted to the desired pH value by the addition of sodium hydroxide or phosphoric acid solutions. Irradiations were conducted in open 9-cm petri dishes placed at a distance of about 5-cm from the light source. Aliquots were withdrawn periodically and were analyzed for the parent insecticides and their hydrolysis products using the 4-aminoantipyrine colorimetric method (11).

The degradation was also followed by measuring the ultraviolet absorption spectra of solutions exposed to irradiation in 2-cm quartz cells at different time intervals. The temperature was maintained constant by circulating air through the exposure cabinet.

B. Chromatographic Procedure.

For qualitative analysis of the degradation products, solutions exposed to irradiation for 30 minutes were acidified to pH 3.0 and were extracted three times with an equal volume of a mixture (1:1) of chloroform-methylene chloride solvent. The extract was washed with distilled water, was dried over magnesium sulphate, and was evaporated to dryness under reduced pressure. The residue was dissolved in about 0.5-ml of chloroform and saved for chromatographic separation.

The chromatographic procedure was as follows: Thin-layer chromatoplates (20- x 20-cm) were prepared by slurrying 25 g of silica gel G (Fluka A.G., Buchs, Switzerland) with 65-ml of water by grinding for 60 seconds in a mortar. The resulting slurry was applied to five glass plates with a spreader set at a thickness of 250 microns. The plates were air dried for 30 minutes, were activated at 110°C. for 45 minutes, and then stored in a dessicator. Aluminium oxide coated plates were prepared in a similar manner.

Aliquots of the chloroform solutions of the irradiation products (5-10 μl) were applied with a micropipette, 1-cm apart and along a line 1.5-cm from the lower edge of the plates.

The chromatograms were developed by the ascending technique and the solvent was allowed to reach a distance of 10-cm from starting line. The developing solvents used were:

(a) Ether: hexane (6:4)
(b) Acetone: toluene: hexane (1:1:3)
(c) Benzene: methanol: acetic acid (8:1:1).

The plates were air-dried for five minutes after development and sprayed with a 0.02% solution of Rhodamine B in ethanol. The plates then were exposed to ultraviolet light for 5 minutes where the spots were visible as blue areas against a pink background.

III. RESULTS

A. Effect of pH on the Photodecomposition.

Results of the photodecomposition of Sevin, at pH values of 5.0, 7.0, and 8.0 at increasing time intervals, are shown in Figure 1. Generally, the concentration of Sevin decreased with an increase of the irradiation time, however, the photolysis process progressed at decreasing rates after prolonged exposure. The photodecomposition proceeded at increasing rates as the pH value of the solutions was increased. Thus, after an exposure time of 60 minutes, only 50% of Sevin was decomposed at pH 5.0 whereas 57% was decomposed at pH 7.0, and 78% at pH 8.0. 1-Naphthol appeared as a decomposition product after 5 minutes of exposure in all the irradiated solutions of Sevin (Table 1). However, its concentration decreased as the irradiation time was increased. At pH 5.0, 3 mg/l of 1-naphthol were produced after 15 minutes and decreased to 1.0 mg/l after 90 minutes; at pH 7.0, the concentration reached 2 mg/l after 30 minutes and was reduced to 1 mg/l after 120 minutes. At pH 8.0, on the other hand, 1.5 mg/l of 1-napthol were detected after 15 minutes and decreased to 1 mg/l after 45 minutes.

The fact that 1-napthol was not accumulated in the irradiated solutions suggested that this compound also underwent photolysis as soon as it appeared in the solution. Therefore, the photodecomposition of 1-naphthol was investigated at pH values of 5.0, 7.0, 8.0, and 9.0. The results are presented in Figure 2.

It can be seen that the concentration of 1-naphthol also decreased with increase of irradiation time. The photodecomposition, however, was affected by the pH value of the medium. At pH 5.0, 1-naphthol decomposed at a relatively slow rate, but, at higher pH values, the photodecomposition progressed at a faster rate.

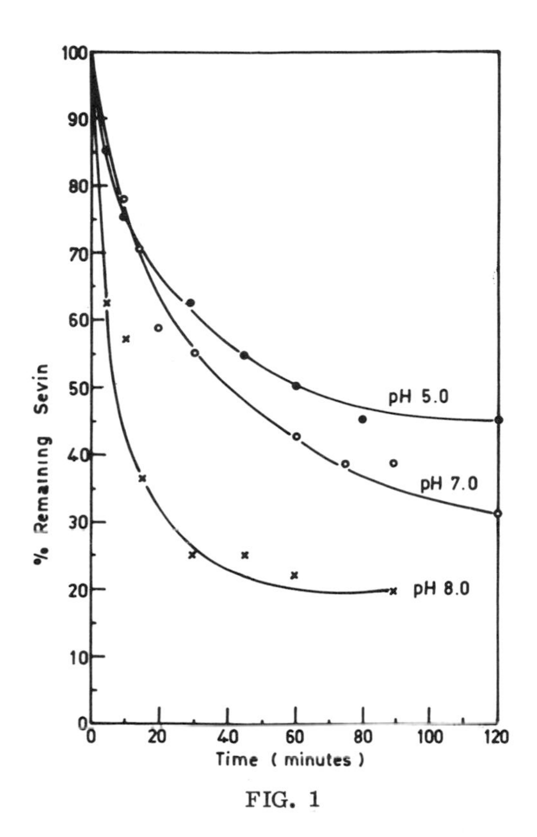

FIG. 1

Effect of pH on photodecomposition of Sevin.

The photodecomposition of Baygon at pH values of 5.0, 7.0, and 9.0 is shown in Figure 3. The decomposition progressed at decreasing rates at the three pH values similar to that observed with Sevin and 1-naphthol. Also, the pH value of the reaction medium affected the rate of photodecomposition being low at low pH values and increasing with the increase in pH.

The irradiation of Baygon resulted in the immediate appearance of o-isopropoxyphenol as a degradation product which was detected by the 4-aminoantipyrine colorimetric method. However, the concentration of the phenol was decreased with irradiation time similar to that observed with the 1-naphthol resulting from the photodecomposition of Sevin.

TABLE 1

Photodecomposition of Sevin at Different pH Values

pH	Time min.	Conc. mg/l	% Remaining	1-Naphthol mg/l
5.0	0	23.0	100.0	0
	5	19.7	85.6	2
	10	17.3	75.2	2
	15	16.3	70.8	3
	20	14.4	62.6	3
	45	12.4	53.9	2
	60	11.4	49.5	2
	90	10.4	45.2	1
	120	10.4	45.2	1
7.0	0	20.0	100.0	0
	10	15.3	76.5	1
	20	11.7	58.5	1
	30	11.0	55.0	2
	60	8.5	42.5	2
	75	7.7	38.5	2
	90	7.7	38.5	2
	120	6.2	31.0	2
8.0	0	20.0	100.0	0
	5	12.4	62.0	1
	10	11.4	57.0	1
	15	7.3	36.5	1.5
	30	5.1	25.5	1.5
	45	5.1	25.5	1.0
	60	4.4	22.0	1.0
	90	3.9	19.5	1.0

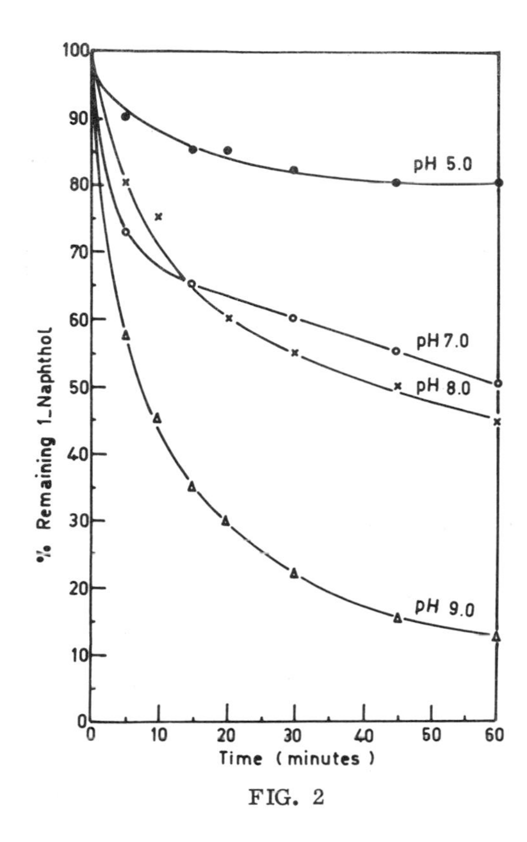

FIG. 2

Effect of pH on photodecomposition of 1-Naphthol.

The photodecompositions of Pyrolan and Dimetilan at pH values of 5.0, 7.0, and 9.0 are shown in Figures 4 and 5, respectively. Decomposition of the two carbamate esters progressed more rapidly than Sevin and Baygon. The pH of the irradiation medium appeared to have no effect on the photodecomposition of the two compounds. The times required for 50% decomposition of Pyrolan and Dimetilan at the three pH values were found to be about 6.5 minutes and 15 minutes, respectively. The photodecomposition of both compounds resulted in the appearance of the heterocyclic enol forms that were detected colorimetrically. However, after 60 minutes of irradiation time, the enols were completely decomposed. The amounts of phenols or heterocyclic enols produced during the irradiation of Sevin, Baygon, Pyrolan, and Dimetilan are shown in Table 2.

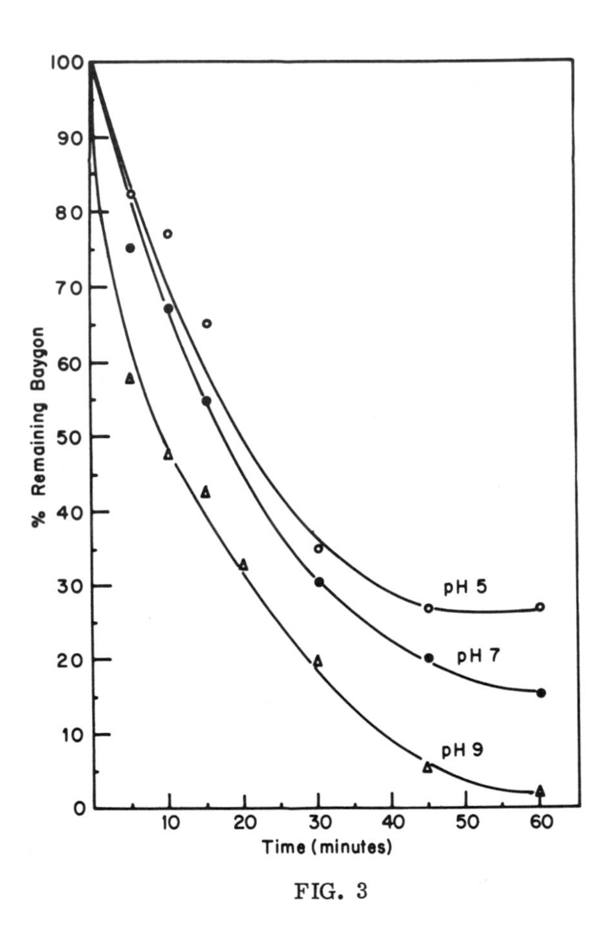

FIG. 3

Effect of pH on photodecomposition of Baygon.

TABLE 2

Concentrations of Phenols or Heterocyclic Enols Formed During the Irradiation of Sevin, Baygon, Pyrolan, and Dimetilan at Different pH Values[a]

Time min.	Compound Irradiated: Sevin	Baygon	Pyrolan	Dimetilan
	Concentration of phenol or heterocyclic enol produced mg/l			
pH 5.0				
0	0.0	0.0	0.0	0.0
10	2.0	1.5	1.0	3.0
15	3.0	3.0	1.0	2.0
20	3.0	--	1.0	1.5
30	--	4.0	2.0	0.8
45	2.0	9.0	2.0	0.0
60	2.0	8.0	1.0	0.0
90	1.0	8.0	--	0.0
pH 7.0				
0	0.0	0.0	0.0	0.0
10	1.0	4.0	1.0	2.5
15	1.0	5.0	2.0	3.0
30	2.0	7.0	2.0	1.0
45	2.0	9.0	1.5	0.0
60	2.0	8.0	0.8	0.0
90	2.0	7.0	--	0.0
pH 9.0				
0	--	0.0	0.0	0.0
10	--	7.0	0.0	2.5
15	--	7.5	1.0	1.5
30	--	8.0	1.0	0.0
45	--	7.0	1.0	0.0
60	--	7.0	0.0	0.0

[a]Initial concentration of each = 20 mg/l.

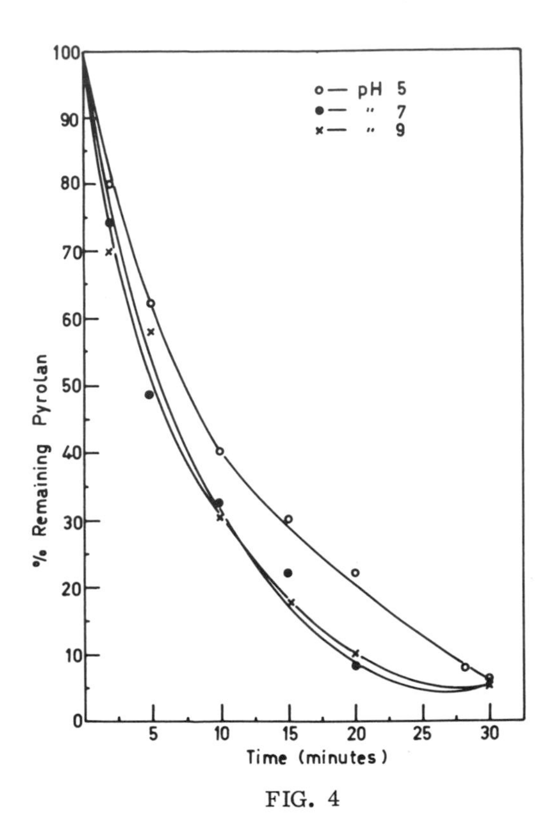

FIG. 4

Effect of pH on photodecomposition of Pyrolan.

Table 3 shows a summary of the times required for 50% decomposition of Sevin, 1-naphthol, Baygon, Pyrolan, and Dimetilan at the different pH values.

All the irradiated solutions of the carbamate esters and 1-naphthol acquired a yellowish-brown coloration after exposure to ultraviolet light. This color became darker as the period of exposure was lengthened.

B. Qualitative Changes During Photodecomposition.

The qualitative changes that occur during the ultraviolet irradiation of aqueous solutions of Sevin, 1-naphthol, Baygon, Pyrolan, and Dimetilan

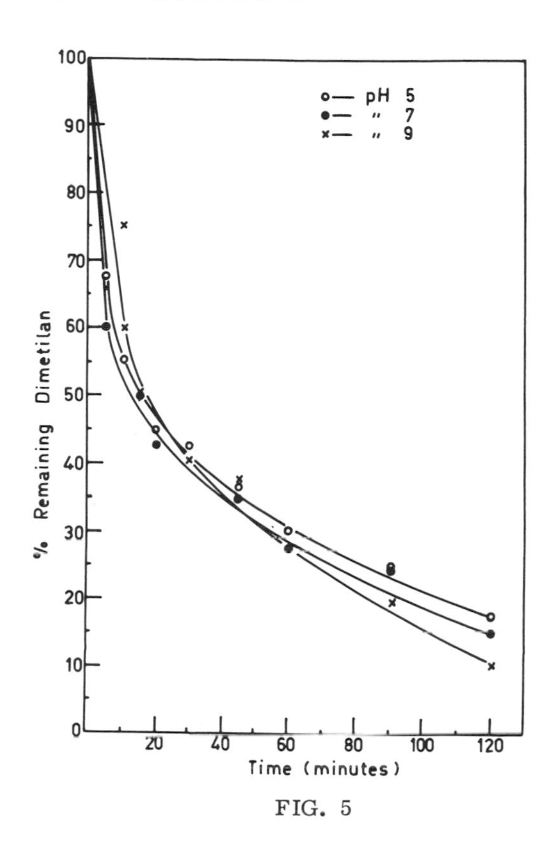

FIG. 5

Effect of pH on photodecomposition of Dimetilan.

were investigated by estimating the changes in the ultraviolet absorption spectra of these compounds during exposure.

Figure 6 shows the absorption spectra of aqueous solutions of Sevin (2 mg/l) at pH 7.0 and 8.0 before and after exposure to ultraviolet light irradiation. Sevin exhibited two absorption maxima; one at 280 mμ and a second at 220 mμ. After irradiation, a decrease in absorption at the wavelength of 220 mμ was observed indicating gradual decomposition of the original compound. The irradiated solutions acquired a yellowish-brown coloration and increased absorption occurred at 250 to 260 mμ. After 90 minutes of exposure, the absorption spectrum of Sevin at pH 8.0 was completely changed showing that the compound had

TABLE 3

Irradiation Time Required for 50% Decomposition of Sevin, 1-Naphthol, Baygon, Pyrolan, and Dimetilan[a]

Compound	Irradiation Time min.			
	pH 5	pH 7	pH 8	pH9
Sevin	60.0	39.0	8.0	---
1-Naphthol	120.0	60.0	43.0	7.5
Baygon	19.0	17.5	---	9.0
Pyrolan	7.5	6.0	---	6.0
Dimetilan	15.0	15.0	---	15.0

[a]Initial Concentration = 20 mg/l.

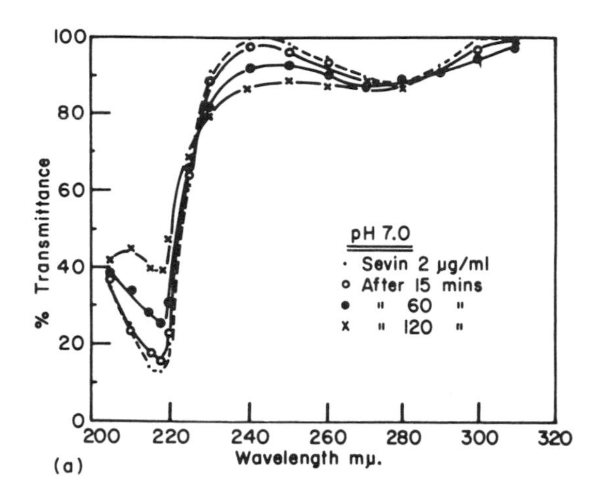

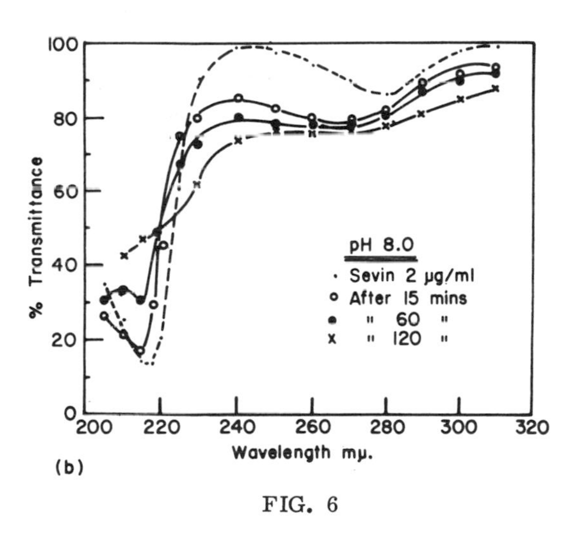

FIG. 6

Ultraviolet absorption spectra of Sevin before and after irradiation.

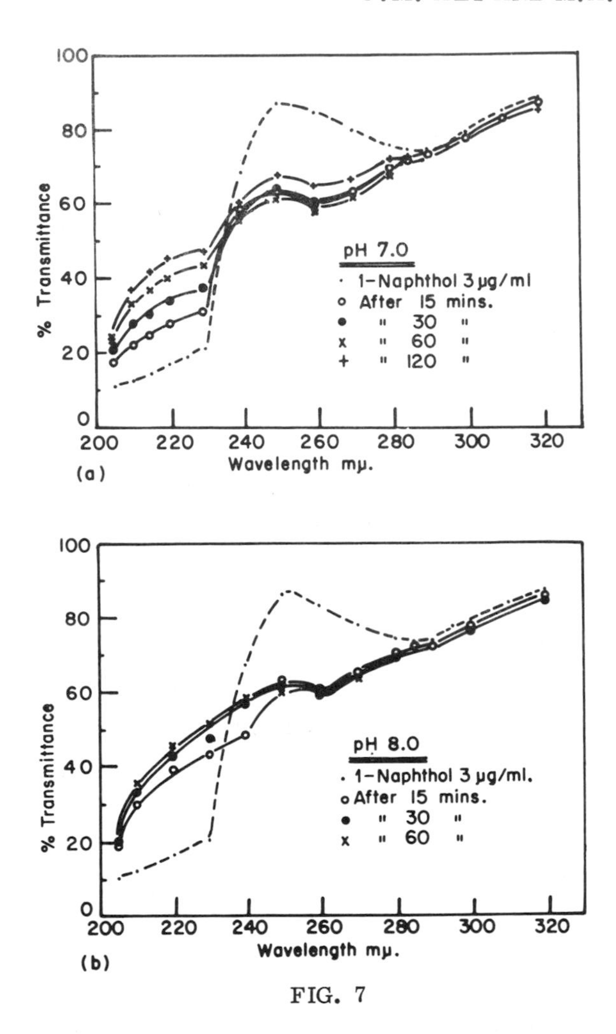

FIG. 7

Ultraviolet absorption spectra of 1-Naphthol before and after irradiation.

lost its identity. Similar changes in the ultraviolet absorption spectrum of 1-naphthol were observed during irradiation (Fig. 7).

Figures 8 and 9 show that the absorption spectrum of Baygon at pH values of 5.0 and 9.0 and that of Pyrolan at pH 7.0 changed completely after irradiation. The absorption generally increased with exposure times at wavelengths between 210 and 310 mμ. The characteristic absorption maximum of Baygon at 270 mμ and that of Pyrolan

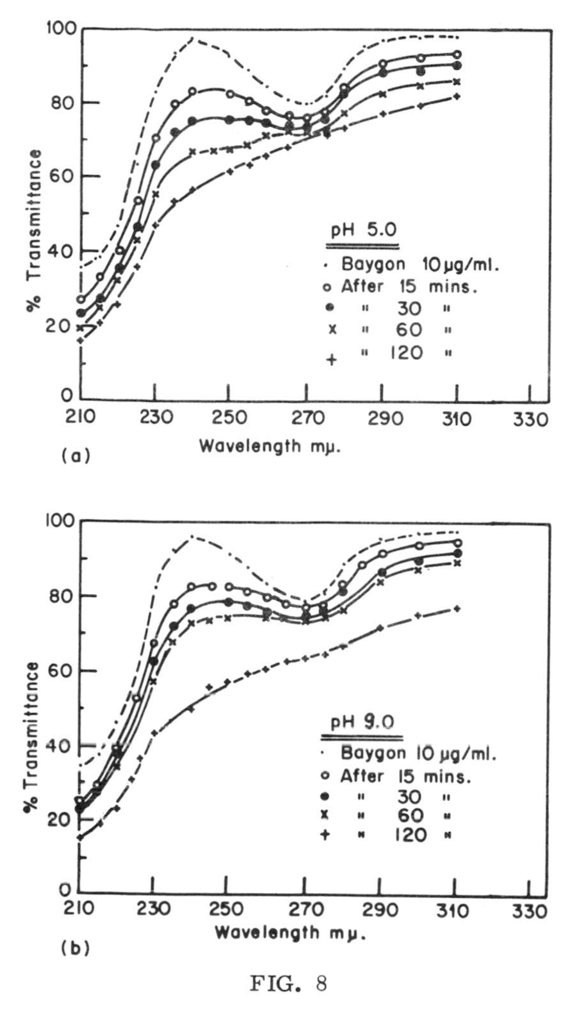

FIG. 8

Ultraviolet absorption spectra of Baygon before and after irradiation.

at 248 mμ disappeared completely after prolonged exposure and the solutions acquired dark brownish colorations.

The effect of the ultraviolet light on the absorption spectrum of Dimetilan is shown in Figure 10. The decrease of the absorption maximum at 238 mμ with exposure time is quite evident. However, after 15 minutes of irradiation, a new peak was rapidly formed at 278

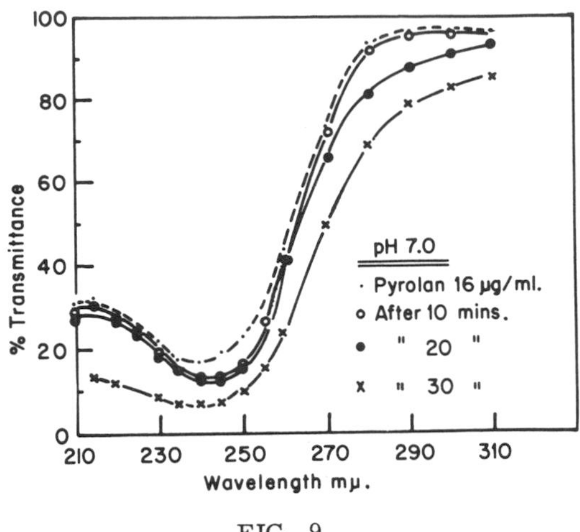

FIG. 9

Ultraviolet absorption spectra of Pyrolan before and after irradiation.

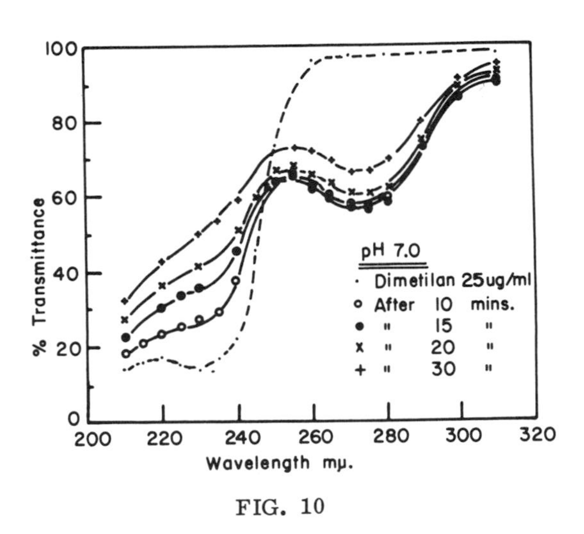

FIG. 10

Ultraviolet absorption spectra of Dimetilan before and after irradiation.

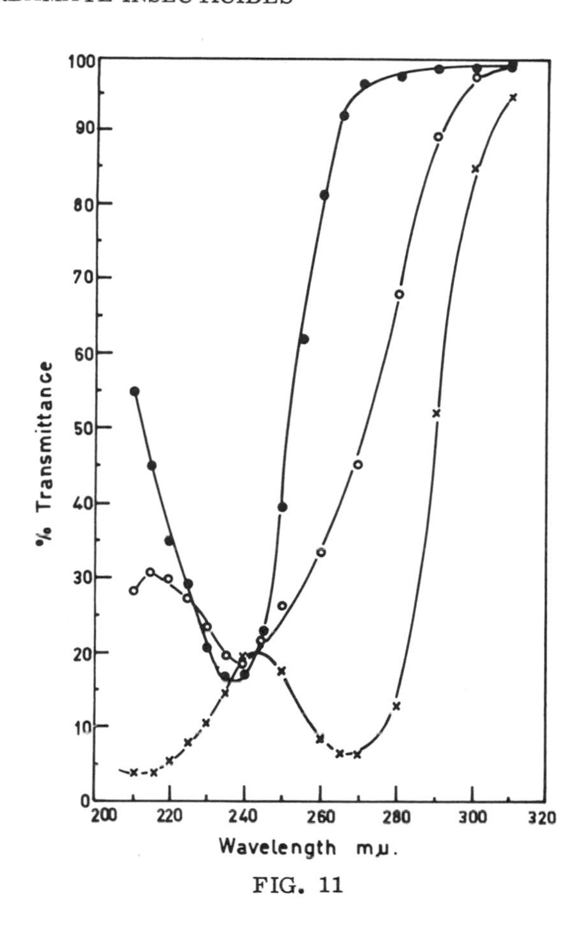

FIG. 11

Ultraviolet absorption spectra of 1-phenyl-3-methyl pyrazolone, 3-methyl pyrazolone and 2-dimethylcarbamoyl-3-methyl-5-pyrazolone.

o-1 -phenyl-3-methyl pyrazolone, (10 μg/ml.) •-3-methyl pyrazolone (25 μg/ml.)
x-2-dimethylcarbamoyl-3methyl-5-pyrazolone (15 μg/ml.)

mμ which is characteristic of 2-dimethylcarbamoyl-3-methyl-5-pyrazolone, the hydrolysis product of Dimetilan (Fig. 11). Increasing the irradiation time resulted in decreased absorption at this wavelength showing decomposition of the formed enol.

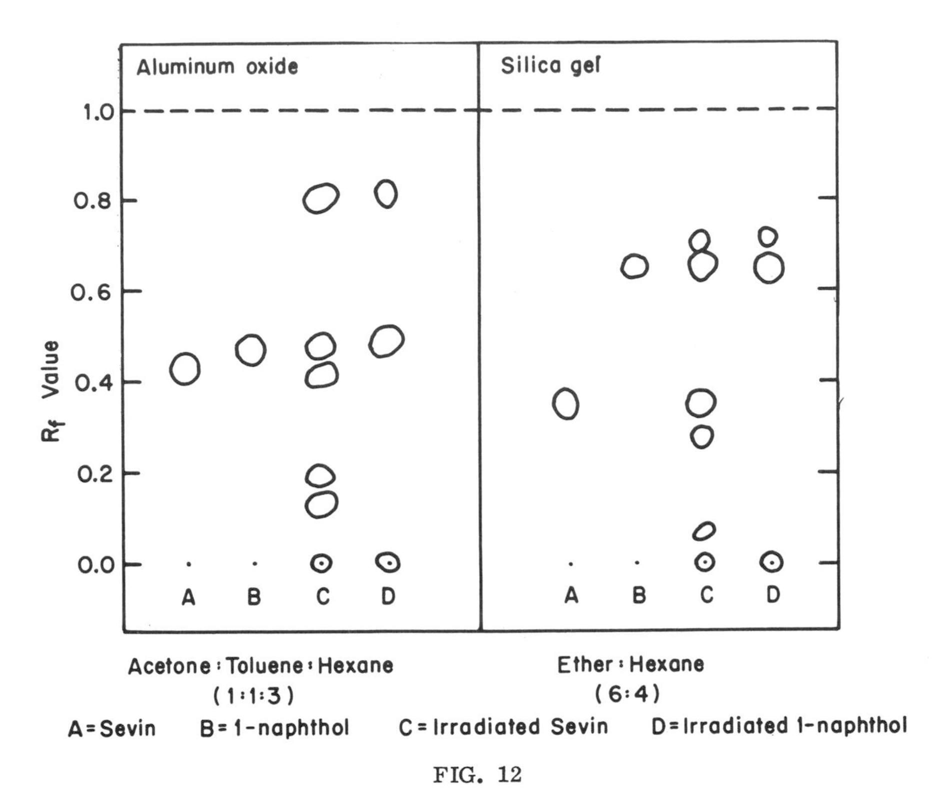

FIG. 12

Thin-layer chromatograms of irradiation products of Sevin and 1-Naphthol.

C. Chromatographic Separation of the Degradation Products.

Thin-layer chromatography was used for the separation and detection of the photodecomposition products of the four carbamate insecticides. R_f values of the unirradiated carbamate esters as well as their hydrolysis products on different adsorbents are shown in Table 4. The R_f values of the decomposition products are shown in Table 5 and the chromatograms are shown in Figures 12,13, and 14.

The chromatogram of irradiated Sevin solution showed, together with the unchanged compound, five degradation products. One of these showed the same chromatographic and color characteristics as 1-naphthol. The other four compounds had R_f values of 0.0, 0.14, 0.19, and 0.81. 1-Naphthol, on the other hand, produced two degradation

TABLE 4

R_f Values of Pure Carbamate Insecticides and Their Enol Forms

Compound	Adsorbent	Solvent	R_f Values
Sevin	Silca gel G	Ether:Hexane (6:4)	0.35
Sevin	Aluminium oxide	Acetone:Toluene:Hexane (1:1:3)	0.44
1-Naphthol	Silica gel G	Ether:Hexane (6:4)	0.64
1-Naphthol	Aluminium oxide	Acetone:Toluene:Hexane (1:1:3)	0.47
Baygon	Silica gel G	Benzene:Methanol:Acetic acid (8:1:1)	0.56
o-Isopropoxyphenol	Silica gel G	Benzene:Methanol:Acetic acid (8:1:1)	0.68
Pyrolan	Silica gel G	Acetone:Toluene:Hexane (1:1:3)	0.40
1-Phenyl-3-methyl-pyrazolone	Silica gel G	Acetone:Toluene:Hexane (1:1:3)	0.26
Dimetilan	Silica gel G	Acetone:Toluene (1:1:3)	0.22
2-Dimethylcarbamoyl-3-methyl-5-pyrazolone	Silica gel G	Acetone:Toluene:Hexane (1:1:3)	0.14
3-Methyl-5-pyrazolone	Silica gel G	Acetone:Toluene:Hexane (1:1:3)	0.00

TABLE 5

R_f Values of Irradiation Products of Carbamate Insecticides and 1-Naphthol

Compound	Adsorbent	Solvent	Number of Spots	R_f Values
Sevin	Aluminium oxide	Acetone:Toluene:Hexane (1:1:3)	5	0.0, 0.14, 0.19, 0.47, 0.81
Sevin	Silica gel G	Ether:Hexane (6:4)	5	0.0, 0.07, 0.28, 0.64, 0.7
1-Naphthol	Aluminium oxide	Acetone:Toluene:Hexane	2	0.0, 0.81
1-Naphthol	Silica gel G	Ether:Hexane (6:4)	2	0.0, 0.7
Baygon	Silica gel G	Benzene:Methanol:Acetic (8:1:1)	4	0.03, 0.39, 0.5, 0.69
Pyrolan	Silica gel G	Acetone:Toluene:Hexane (1:1:3)	4	0.07, 0.18, 0.26, 0.62
Dimetilan	Silica gel G	Acetone:Toluene:Hexane (1:1:3)	3	0.14, 0.29, 0.57

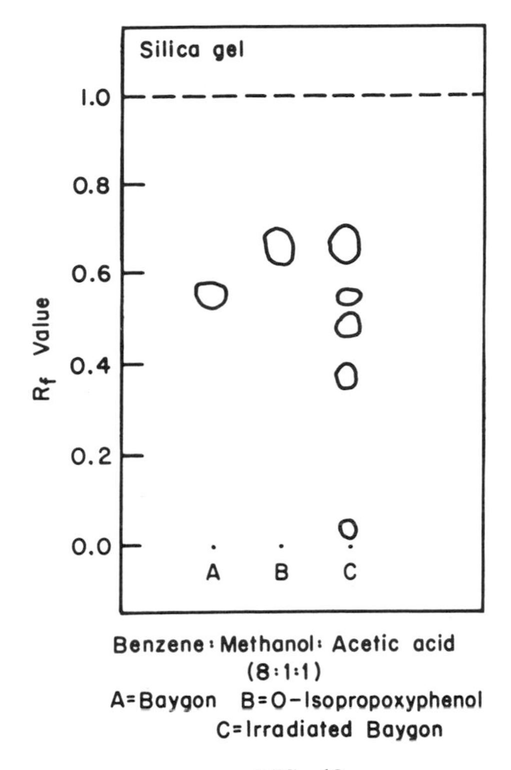

FIG. 13

Thin-layer chromatogram of irradiation products of Baygon.

products having R_f values of 0.0 and 0.81 and are chromatographically identical with two of the degradation products of Sevin. Similar results were obtained when silica gel was used as the adsorbent (Fig. 12). These results suggest that photodecomposition of Sevin resulted in the production of 1-naphthol and other degradation products; some of which are photolysis products of 1-naphthol.

Under the same irradiation conditions, Baygon produced four decomposition products, (R_f: 0.03, 0.39, 0.5, and 0.69). One of the spots was identified chromatographically as o-isopropoxyphenol.

Pyrolan showed the formation of 1-phenyl-3-methyl-pyrazolone and three other degradation products having R_f values of 0.07, 0.18, and 0.62. Similarly, Dimetilan produced three photolysis products. One of these showed the same R_f value as the hydrolysis product

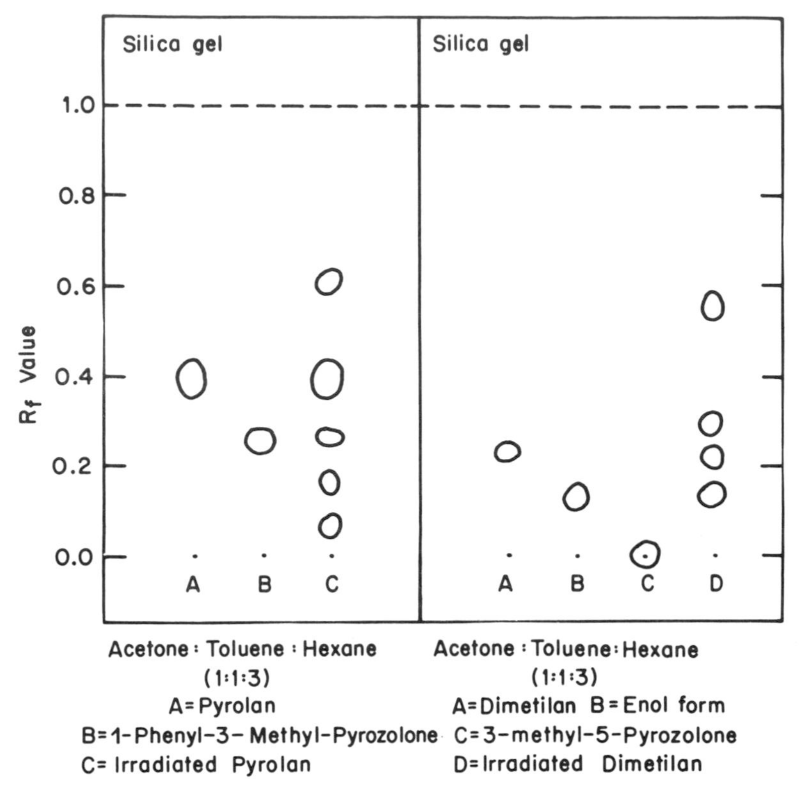

FIG. 14

Thin-layer chromatograms of irradiation products of Pyrolan and Dimetilan.

2-dimethylcarbamoyl-3-methyl-5-pyrazolone which showed that the decomposition occurred only at the carbamate ester linkage since no spots identical with 3-methyl-5-pyrazolone were produced.

IV. DISCUSSION

The results of this study indicate that carbamate insecticides in water are subject to photodecomposition under the effect of ultraviolet light. The pH value of the aqueous medium was found to be an important factor in determining the rate of the photolysis of Sevin and Baygon, being slow at low pH values and tending to increase with an increase of pH value. However, the decomposition of Pyrolan and Dimetilan was not affected by the pH of the irradiated medium. The primary effect of the ultraviolet light irradiation appears to be the cleavage of the ester bond resulting in the production of the phenol or heterocyclic enol of the four carbamate esters. This was confirmed both quantitatively by colorimetric analysis, and qualitatively by thin-layer chromatography. The hydrolysis products produced were further photodecomposed to other unidentified degradation products as shown by the quantitative analysis, the changes in the ultraviolet absorption spectra, and the chromatographic results. The effect of pH on the photolysis of Sevin and Baygon is analogous to that observed with 2, 4-dichlorophenoxyacetic acid (5) where the photodecomposition was much faster in alkaline than neutral or acidic media.

Sevin produced five degradation products, one of which was identified as 1-naphthol, two were shown to be degradation products of the latter, and two others are as yet unidentified. It is assumed that cleavage of the ester bond is not the only effect of the ultraviolet light but probably other modifications in the molecule tend to occur. Crosby et al. (7) studied the photodecomposition of some carbamate esters using thin-layer chromatography and the cholinesterase-inhibition method for the detection of these products. They showed that the photodecomposition of Sevin yielded, in addition to 1-naphthol, several cholinesterase inhibitory substances which indicate that these compounds retained the intact carbamate ester group, and that irradiation resulted in changes at other positions in the molecule. Similar results were reported for the irradiation products of other carbamate esters (7, 8, 10). Crosby et al. (7) also reported that ethanolic or hexane solutions of Baygon were only slightly affected by ultraviolet light. In the present investigation, the irradiation of aqueous solutions of Baygon resulted in the formation of several degradation products. This supports the view that photodecomposition of the carbamate esters is affected by the nature of the solvent.

By comparing the half-life times for the photodecomposition of the four carbamate insecticides under comparable conditions (Table 3), it can be seen that the order of decomposition was Pyrolan > Dimetilan > Baygon > Sevin.

Pyrolan, which showed the fastest rate of photodecomposition, exhibits a wavelength of maximum absorption at 248 mμ which lies very close to the peak radiation (254 mμ) of the ultraviolet light source used in this investigation. Dimetilan, Baygon, and Sevin exhibited relatively lower absorption values at this wavelength. These results suggest that the light absorption characteristics of the insecticide influences the extent of its photodecomposition by a specific light source. The ultraviolet spectrum of sunlight ranges between 292 and 400 mμ and, therefore, the rate of photodecomposition and the nature of degradation products induced by sunlight is expected to be different from those observed by this investigation. Eberle and Gunther (8) showed that Pyrolan was rapidly decomposed when ethanolic or hexane solutions were exposed to short-wave ultraviolet light (254 mμ), while natural sunlight did not induce any photodecomposition at all. On the other hand, both ultraviolet light and natural sunlight caused decomposition of Sevin (7) and Dimetilan (8). However, the extent of photodecomposition, particularly in the case of Sevin, was not the same under the different irradiation conditions (7,8). Intense ultraviolet irradiation generally resulted in the formation of a greater number of degradation products.

It seems reasonable, therefore, to suggest that photodecomposition may account for some loss of the carbamate insecticides in clear surface waters exposed to long periods of sunlight illumination. However, photolysis may be a minor factor in the decomposition of these compounds in highly turbid waters where the penetration of light will be greatly reduced.

ACKNOWLEDGMENTS

The authors wish to express their thanks to Professor F. M. Ramadan for sponsoring and encouragement of this investigation. Thanks are also due to the Rudolfs Research Conference for allowing us to present this paper.

REFERENCES

1. S. D. Faust and O. M. Aly, J. Amer. Water Works Assoc., 56, 267 (1964).

2. S. D. Faust and I. H. Suffet, Residue Reviews, 15, 44 (1966).

3. L. C. Mitchel, J. Assoc. Offic. Agr. Chemists, 44, 643 (1961).

4. L. W. Weldon and F. L. Timmons, Weeds, 9, 111 (1961).

5. O. M. Aly and S. D. Faust. J. Agr. Food Chem., 12, 541 (1964).

6. L. S. Jordan, C. W. Coggins, Jr., B. E. Day, and W. A. Clery, Weeds, 12, 1 (1964).

7. D. G. Crosby, E. Leitis, and W. L. Winterlin, J. Agr. Food Chem., 13, 204 (1965).

8. D. O. Eberle and F. A. Gunther. J. Assoc. Offic. Agr. Chemists, 48, 927 (1965).

9. D. G. Crosby and H. O. Tutass, J. Agr. Food Chem., 14, 596 (1966).

10. A. M. Abdel-Wahab and J. E. Casida, J. Agr. Food Chem., 15, 479 (1967).

11. O. M. Aly, J. Amer. Water Works Assoc. 59, 906 (1967).

CHAPTER 21

ENERGETICS AND BACTERIAL GROWTH

Perry L. McCarty

Stanford University

I. INTRODUCTION

Classical thermodynamics is concerned primarily with closed systems which are in equilibrium or which are undergoing reversible processes. Since bacterial cells are open systems in which irreversible processes are occurring, classical thermodynamics can only be used in a limited, but still useful way, for relating growth to substrate utilization. In a reversible process, all of the free energy released can be converted to useful work, while in an irreversible process only a portion of the free energy is available. Observations on the efficiency with which bacteria capture free energy together with classical thermodynamic concepts can lead to generalizations concerning expected cell yields from given energy reactions.

Some isolated cultures do not have the enzyme systems required to capture energy from a given reaction as efficiently as other cultures which tend to dominate in the natural environment. For this reason, observations on enrichment cultures which encourage the growth of more efficient organisms may be of more value than isolate studies in drawing conclusions about maximum efficiencies under natural conditions. A comparison of energy utilization by isolated and enrichment cultures is made in the following.

Classical thermodynamics is also limited in that it does not allow prediction of reaction rates except under certain specialized conditions. However, the rate of electron transport has been empirically observed to be nearly constant at a given temperature for a wide variety of autotrophic and heterotrophic bacteria. This observation together with anticipated efficiencies of energy transfer make possible reasonable estimates of maximum bacterial growth rates.

II. BACTERIAL GROWTH KINETICS

Monod (17) presented an empirical and now widely used equation for the growth rate of bacteria as a function of substrate concentration:

$$\mu = \mu_m \frac{S}{K_s + S} \qquad (21\text{-}1)$$

Equation (21-1) was modified by van Uden (18) to consider the influence of specific maintenance rate, maximum yield factor, and transport of substrate into the cell:

$$\mu = Y_m k_m \frac{S}{K_s + S} - b \qquad (21\text{-}2)$$

The specific maintenance rate, b, has also been termed more generally the organism decay rate (19) as a decrease in biomass may occur through death, lysis, endogenous metabolism, and predation, as well as through energy utilization for maintenance. Typical values for b have ranged from 0.01 day^{-1} to 0.05 day^{-1} in mixed cultures. In pure cultures, values as high as 0.43 and 0.67 day^{-1} have been noted (20). Organism decay rates under anaerobic conditions tend to be lower, in general, than under aerobic conditions (19). Values for K_s for aerobic decomposition are generally less than 20 mg/l (17,21), although in the methane fermentation of acetate and propionate, values of over 100 mg/l seem typical (19). When the value for S is large relative to K_s, the maximum growth rate occurs and equals the following:

$$\mu_m = Y_m k_m - b \qquad (21\text{-}3)$$

Frequently, the magnitude of b is sufficiently small so that it has little influence on the maximum specific growth rate, in which case:

$$\mu_m \simeq Y_m k_m \quad (b << Y_m k_m) \qquad (21\text{-}4)$$

The relationship between Y_m and the free energy of reaction will be considered for a number of heterotrophic and autotrophic bacterially

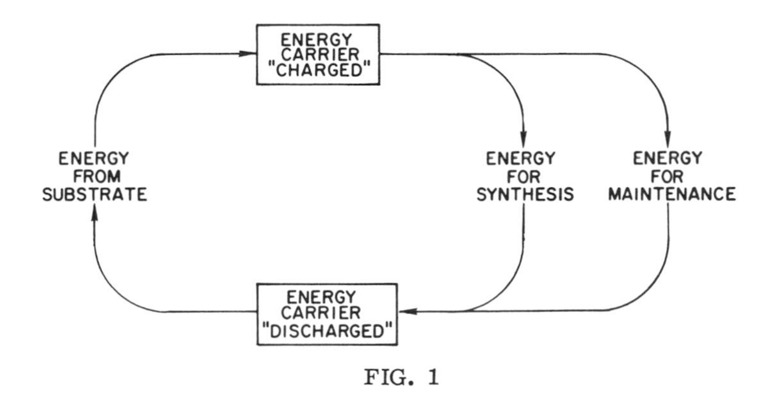

FIG. 1

Energy transport from substrate to biological synthesis and maintenance by way of an energy carrier.

mediated reactions. Then, typical values of k_m for similar reactions will be compared in order to determine what generalizations about this coefficient can be made. Material balances are essential in such comparisons, and, in the oxidation-reduction reactions typically used by bacteria for energy, this can perhaps best be done by use of electron transfer and electron equivalents.

III. ENERGETICS AND CELL YIELDS

Various formulations have been developed to relate the yield of bacteria to the energetics of bacterially mediated reactions. Baas-Becking and Parks (22) presented a formulation for the efficiency of energy utilization by chemosynthetic autotrophic bacteria. Servizi and Bogan (23) related growth yields of aerobic heterotrophs to free energy of substrate oxidation, while McKinney (24) proposed that the energy available to microorganisms was proportional to the change in heat energy liberated during metabolism. Laudelout et al. (27) estimated efficiency of energy utilization by autotrophs using a calorimetric procedure. While useful, these formulations and methods do not appear general in application to all groups of bacteria.

McCarty (13) presented a relationship between free energy of reaction and maximum cell yield which was felt to be applicable for both heterotrophic and chemosynthetic autotrophic bacteria. This model considered, as illustrated in Figure 1, that energy was transferred in two steps, first from the energy source to an energy carrier such as ADP, and then from the energy carrier to a synthesis or cell maintenance reaction. For each transfer a certain loss of energy was assumed to result and this was related to an energy transfer efficiency for the reaction. In evaluating the maximum cell yield according to Equation (21-4), the energy for maintenance is taken to equal zero indicating that all of the energy transferred is used for synthesis.

In heterotrophic growth an organic substrate is used, of which a portion is converted to end products for energy and another portion is synthesized into cellular material. For such growth a method must be devised to evaluate the proportion converted for each use. It was suggested (13) that the theoretical oxygen equivalent of the substrate (COD') should be used for this purpose. This has led to some confusion because of the units involved and it is now proposed that electron equivalents should be used. One electron equivalent is equal to 8 grams of COD' and the substitution makes no fundamental change in the relationship previously proposed. However, electron equivalents are more readily defined, are basic to the processes involved, and their use reduces the complexity of the chemical relationships which must be considered.

The number of electron equivalents in a unit mass of cells can be determined once its composition in terms of the major elements is known. For example, a typical empirical formulation for bacterial cells was given by Hoover and Porges (25) as $C_5H_7O_2N$. This is reasonably representative of the carbon to nitrogen ratio for a wide variety of bacteria (12, 13, 26) and of the electron equivalents per mole of carbon synthesized. The electron equivalents are given by the following half reaction:

$$C_5H_7O_2N + 9\ H_2O = 4\ CO_2 + HCO_3^- + NH_4^+ + 20\ H^+ + 20\ e^- \qquad (21\text{-}5)$$

Thus, one electron equivalent of cells equals 1/20 of an empirical mole, or 1/4 of a mole of carbon or 1/20 of a mole of ammonia synthesized. Since the molecular weight of the empirical cell formulation is 113, the cell dry weight (volatile solids basis) is 113/20 or 5.65 grams per electron equivalent. The cell carbon is 12(5)/20 or 3.0 grams per electron equivalent. Other desired conversions can readily be made.

Considering the energy for maintenance to be zero, the net energy loss from the synthesis of an equivalent of cells must equal the net energy change from the overall synthesis reaction:

$$A\ \Delta G_r + \Delta G_s = \text{Energy Lost} \qquad (21\text{-}6)$$

The energy lost is equal to the portion (1 - k) of the substrate energy which is not transferred to the energy carrier. Thus, energy loss equals (1 - k) A ΔG_r. With this substitution, Equation (21-6) can be solved for A:

$$A = -\frac{\Delta G_s}{k\ \Delta G_r} \qquad (21\text{-}7)$$

Here, k is the efficiency of transfer of substrate energy to the energy carrier expressed as a decimal fraction.

The quantity of energy required for synthesis (ΔG_s) is dependent, among other things, on the energy state of the carbon source used for synthesis. The energy required to convert a given carbon source into cells is postulated to consist of three major energy requirements as illustrated in Figure 2. First, carrier energy is required to raise the cell carbon source to some intermediate level, represented by pyruvate in the illustration. This conversion may require considerable energy as indicated for CO_2, or may release energy as indicated for glucose. Pyruvate appears to be a reasonable choice for the intermediate energy level as it represents about the same energy content on an electron equivalent basis as "active acetate," which is the usual level of entrance into the tricarboxylic acid cycle for synthesis of proteins and fats.

A second energy requirement for synthesis also may be important at times and was not considered in the original formulation. This is the energy for reduction of an oxidized nitrogen source to the ammonia level prior to synthesis into cellular material. Finally, a certain quantity of energy is needed to convert the intermediate carbon source and ammonia into cellular material. In summary, the energy for synthesis is given as follows:

$$\Delta G_s = \frac{\Delta G_p}{k^m} + \Delta G_c + \frac{\Delta G_n}{k} \qquad (21\text{-}8)$$

The ratio of electron equivalents of substrate used for energy to electron equivalent of cells formed, A, is finally obtained by combining Equations (21-7) and (21-8):

$$A = -\frac{\Delta G_p/k^m + \Delta G_c + \Delta G_n/k}{k\ \Delta G_r} \qquad (21\text{-}9)$$

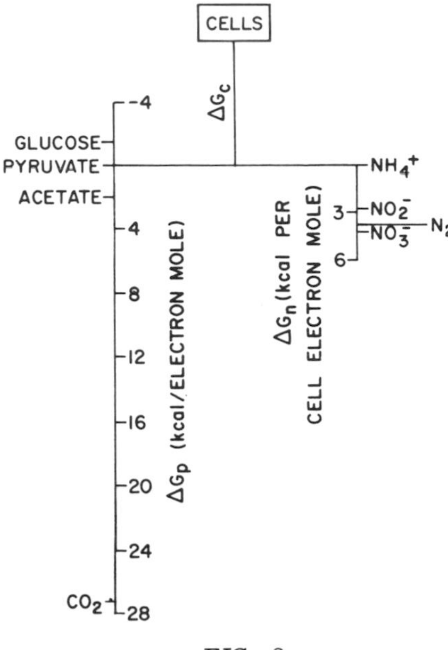

FIG. 2

Energy requirement for cell synthesis from various carbon and nitrogen sources.

The efficiencies for the different energy transfers have been assumed equal so that calculations from experimental measurements using this formulation will result in an overall average efficiency for the different processes involved.

A. Energy Calculations.

Table 1 contains typical half reactions of importance in many bacterially mediated reactions. These reactions are written to represent the quantity of substrate involved in the transfer of a single mole of electrons. The redox potential, $\Delta G^o(W)$, represents the free energy change per mole electron from a given energy reaction when all species are at unit activity except H^+ and OH^- which are taken at their activities in neutral water (44). The values for $\Delta G^o(W)$ were calculated from the free energies of formation listed in Table 2. The free

TABLE 1

Free Energies for Various Half Reactions
Reactants and products at unit activity except $[H^+] = 10^{-7}$

	Half Reaction		$\Delta G^o(W)$ kcal per mole electrons
1.	$1/6\ N_2 + 2/3\ H_2O$	$= 1/3\ NO_2^- + 4/3\ H^+ + e^-$	22.263
2.	$1/2\ H_2O$	$= 1/4 O_2 + H^+ + e^-$	18.675
3.	Fe^{++}	$= Fe^{3+} + e^-$	17.780
4.	$1/10\ N_2 + 3/5\ H_2O$	$= 1/5\ NO_3^- + 6/5\ H^+ + e^-$	17.128
5.	$1/2\ NO_2^- + 1/2\ H_2O$	$= 1/2\ NO_3^- + H^+ + e^-$	9.425
6.	$1/8\ NH_4^+ + 3/8\ H_2O$	$= 1/8\ NO_3^- + 5/4\ H^+ + e^-$	8.245
7.	$1/6\ NH_4^+ + 1/3\ H_2O$	$= 1/6\ NO_2^- + 4/3\ H^+ + e^-$	7.852
8.	$1/12\ H_2S + 1/12\ HS^- + 1/2\ H_2O$	$= 1/6\ SO_3^= + 5/4\ H^+ + e^-$	- 3.248
9.	$1/6\ S + 2/3\ H_2O$	$= 1/6\ SO_4^= + 4/3\ H^+ + e^-$	- 4.657
10.	$1/16\ H_2S + 1/16\ HS^- + 1/2\ H_2O$	$= 1/8\ SO_4^= + 19/16\ H^+ + e^-$	- 5.085
11.	$1/8\ S_2O_3^= + 5/8\ H_2O$	$= 1/4\ SO_4^= + 5/4\ H^+ + e^-$	- 5.091
12.	$1/8\ CH_4 + 1/4\ H_2O$	$= 1/8\ CO_2 + H^+ + e^-$	- 5.763
13.	$1/3\ NH_4^+$	$= 1/6\ N_2 + 4/3\ H^+ + e^-$	- 6.560

TABLE 1 (continued)

	Half Reaction		$\Delta G^o(W)$ kcal per mole electrons
14.	$1/8\ CH_3COO^- + 3/8\ H_2O$	$= 1/8\ CO_2 + 1/8\ HCO_3^- + H^+ + e^-$	- 6.609
15.	$1/92\ CH_3(CH_2)_{14}COO^- + 31/92\ H_2O$	$= 15/92\ CO_2 + 1/92\ HCO_3^- + H^+ + e^-$	- 6.657
16.	$1/44\ CH_3(CH_2)_6\ COO^- + 15/44\ H_2O$	$= 7/44\ CO_2 + 1/44\ HCO_3^- + H^+ + e^-$	- 6.657
17.	$1/14\ CH_3CH_2COO^- + 5/14\ H_2O$	$= 1/7\ CO_2 + 1/14\ HCO_3^- + H^+ + e^-$	- 6.664
18.	$1/30\ C_6H_5COO^- + 13/30\ H_2O$	$= 1/5\ CO_2 + 1/30\ HCO_3^- + H^+ + e^-$	- 6.892
19.	$1/14\ (CH_2)_2(COO^-)_2 + 3/7\ H_2O$	$= 1/7\ CO_2 + 1/7\ HCO_3^- + H^+ + e^-$	- 7.101
20.	$1/12\ CH_3CH_2OH + 1/4\ H_2O$	$= 1/6\ CO_2 + H^+ + e^-$	- 7.592
21.	$1/18\ COOHCH_2CH_2CHNH_2COO^- + 4/9\ H_2O$	$= 1/6\ CO_2 + 1/9\ HCO_3^- + 1/18\ NH_4^+ + H^+ + e^-$	- 7.602
22.	$1/12\ CH_3CHNH_2COOH + 5/12\ H_2O$	$= 1/6\ CO_2 + 1/12\ HCO_3^- + 1/12\ NH_4^+ + H^+ + e^-$	- 7.639
23.	$1/12\ CH_3CHOHCOO^- + 1/3\ H_2O$	$= 1/6\ CO_2 + 1/12\ HCO_3^- + H^+ + e^-$	- 7.873
24.	$1/6\ CH_2NH_2COOH + 1/2\ H_2O$	$= 1/6\ CO_2 + 1/6\ HCO_3^- + 1/6\ NH_4^+ + H^+ + e^-$	- 8.422
25.	$1/10\ CH_3COCOO^- + 2/5\ H_2O$	$= 1/5\ CO_2 + 1/10\ HCO_3^- + H^+ + e^-$	- 8.545
26.	$1/6\ CH_3OH + 1/6\ H_2O$	$= 1/6\ CO_2 + H^+ + e^-$	- 8.965
27.	$1/14\ CH_2OHCHOHCH_2OH + 3/14\ H_2O$	$= 3/14\ CO_2 + H^+ + e^-$	- 9.366

TABLE 1 (continued)

		Half Reaction	$\Delta G^{o}(W)$ kcal per mole electrons
28.	1/2 H_2	= $H^+ + e^-$	- 9.670
29.	1/24 $C_6H_{12}O_6$ + 1/4 H_2O	= 1/4 CO_2 + $H^+ + e^-$	-10.020
30.	1/2 $SO_3^=$ + 1/2 H_2O	= 1/2 $SO_4^=$ + $H^+ + e^-$	-10.595
31.	1/2 $HCOO^-$ + 1/2 H_2O	= 1/2 HCO_3^- + $H^+ + e^-$	-11.480
32.	Bacterial Cells:		
	1/20 $C_5H_7O_2N$ + 9/20 H_2O	= 1/5 CO_2 + 1/20 HCO_3^- + 1/20 NH_4^+ + $H^+ + e^-$	

TABLE 2

Free Energies of Formation for Various Compounds
(One Molar Concentration Unless Otherwise Noted)

Compound	Stand. State	ΔG_f^o kcal per mole	Ref.	Compound	Stand. State	ΔG_f^o kcal per mole	Ref.
Formate$^-$	aq	- 80.0	(1)	CH_4	g	- 12.14	(1)
Acetate$^-$	aq	- 88.99	(4)	CO_2	g	- 94.26	(1, 3)
Propionate$^-$	aq	- 87.47	(3)[a]	HCO_3^-	aq	-140.31	(1)
Butyrate$^-$	aq	-84.6	(4)	Fe^{+2}	aq	- 20.30	(1)
Palmitate$^-$	aq	- 74.0	(3)[a]	Fe^{+3}	aq	- 2.52	(1)
Glucose	aq	-217.02	(3)	H_2O	l	- 56.69	(1)
Glycerol	aq	-116.76	(4)	$H^+(10^{-7})$	aq	- 9.67	(1)[a]
Methanol	aq	- 41.80	(1)[a]	NH_4^+	aq	- 19.00	(1, 4)
Ethanol	aq	-43.39	(4)	NO_3^-	aq	- 26.41	(1)
Pyruvate$^-$	aq	-113.32	(4)	NO_2^-	aq	- 7.91	(1)[b]
Lactate$^-$	aq	-123.64	(4)	S	c	0.00	(1)
Benzoate$^-$	aq	- 52.25	(2)	$SO_3^=$	aq	-118.8	(1)
Glycine	aq	- 90.99	(3)[a]	$SO_4^=$	aq	-177.34	(1)
Alanine	aq	- 88.75	(4)	$S_2O_3^=$	aq	-127.2	(1)
Glutamate$^-$	aq	-166.11	(4)	HS^-	aq	+ 3.01	(1)
Succinate$^=$	aq	-164.97	(4)	H_2S	aq	- 6.54	(1)

[a] Data from original source were either not for aqueous solutions or for different concentrations. They were modified using the relation, $\Delta G = \Delta G^o + RT \ln \frac{\text{Final Conc.}}{\text{Init. Conc.}}$ to convert to molar aqueous concentration. $\Delta G = -RT \ln K_a$ was used to convert free energy of propionate and palmitate from acid to salt form. Vapor pressure data were used to determine aqueous free energy for methanol (5).

[b] Calculated from electrochemical reaction potentials.

energy of the electron at unit activity is equal to zero by convention (30) as are all elements when in their stable state under standard conditions and at unit activity.

Included in Table 1 is the half reaction for the typical empirical formulation for bacterial cells of $C_5H_7O_2N$ discussed previously. The free energy of formation of bacterial cells is not known and so a ΔG^o (W) value for this half reaction could not be computed.

In order to use Equation (21-9), values for ΔG_r, ΔG_p, ΔG_n, and ΔG_c are needed. A value for ΔG_r for a particular energy reaction is obtained by subtracting one half reaction from another as illustrated by the examples in Table 3. The molar concentrations of the various reactants and products involved in an electron equivalent of change are indicated. The free energy for the reaction is obtained as the difference in the ΔG^o(W) values from Table 1 for the two half reactions. Concentrations of reactants and products would be different from unit activity under natural conditions, however, the differences in free energy of reaction that would result from this correction are not too great and can usually be ignored with little effect on the final results. The major exceptions to this occur at low ΔG_r values such as occur with iron oxidation or with some methane fermentation reactions.

The relationship between various electron donors, electron acceptors, and ΔG_r for oxidation-reduction reactions are illustrated in Figure 3. When oxygen is used as the electron acceptor for the oxidation of organic matter, the difference between ΔG_r values is relatively small, varying from about -24 kcal for methane oxidation to about -30 kcal for formate oxidation. Cell yields from aerobic heterotrophic reactions are not greatly different for this reason. The same is true with nitrite and, to a lesser degree, with nitrate as the electron acceptor. However, with sulfate or carbon dioxide as an electron acceptor, the relative difference between ΔG_r for various organic compounds is great. Because of this, cell yields from such anaerobic heterotrophic reactions should vary widely.

Values for ΔG_r for autotrophic oxidations, by contrast to heterotrophic conditions, are widely different. Using oxygen as an electron acceptor, the value for ΔG_r varies from about -37 kcal with sulfite to -0.9 with Fe^{++} as electron donor. A wide range in growth yields from autotrophic reactions can be anticipated.

The value for ΔG_p is obtained by subtracting the redox potential for pyruvate (half reaction 25, Table 1) from that of the original carbon source. Values obtained for various carbon sources are illustrated in Figure 2. The energy requirement for nitrogen reduction per electron equivalent of cells formed (ΔG_n) is based upon the need for 1/20 mole of nitrogen per electron equivalent of cells formed as indicated by

TABLE 3

Examples of Calculations for ΔG_r

Electron Donor	Electron Acceptor	Half Reactions Combined (From Table 1)	Complete Reaction for One Equiv. of Change	ΔG_r kcal per Elect. Equiv.
HETEROTROPHIC				
Acetate	O_2	(14) - (2)	$1/8\ CH_3COO^- + 1/4\ O_2 = 1/8\ CO_2 + 1/8\ HCO_3^- + 1/8\ H_2O$	-25.28
Ethanol	NO_3^-	(20) - (4)	$1/12\ CH_3CH_2OH + 1/5\ NO_3^- + 1/5\ H^+ = 1/6\ CO_2 + 1/10\ N_2 + 7/20\ H_2O$	-24.72
Pyruvate	$SO_4^=$	(25)-0.8(14) -0.2(10)	$1/10\ CH_3COCOO^- + 1/40\ SO_4^= + 3/80\ H^+ = 1/10\ CH_3COO^- + 1/80\ H_2S + 1/80\ HS^- + 1/10\ CO_2$	- 2.24
Acetate	CO_2	(14) - (12)	$1/8\ CH_3COO^- + 1/8\ H_2O = 1/8\ CH_4 + 1/8\ HCO_3^-$	- 0.85
Glucose	Glucose[a]	(29) - (20)	$1/24\ C_6H_{12}O_6 = 1/12\ CH_3CH_2OH + 1/12\ CO_2$	- 2.43
AUTOTROPHIC				
Fe^{++}	O_2	(3) - (2)	$Fe^{++} + 1/4\ O_2 + H^+ = Fe^{3+} + 1/2\ H_2O$	- 0.90
S	NO_2^-	(9) - (1)	$1/6\ S + 1/3\ NO_2^- = 1/6\ SO_4^= + 1/6\ N_2$	-26.92
H_2	CO_2	(28) - (12)	$1/2\ H_2 + 1/8\ CO_2 = 1/8\ CH_4 + 1/4\ H_2O$	- 3.91

[a] Fermentation to Ethanol.

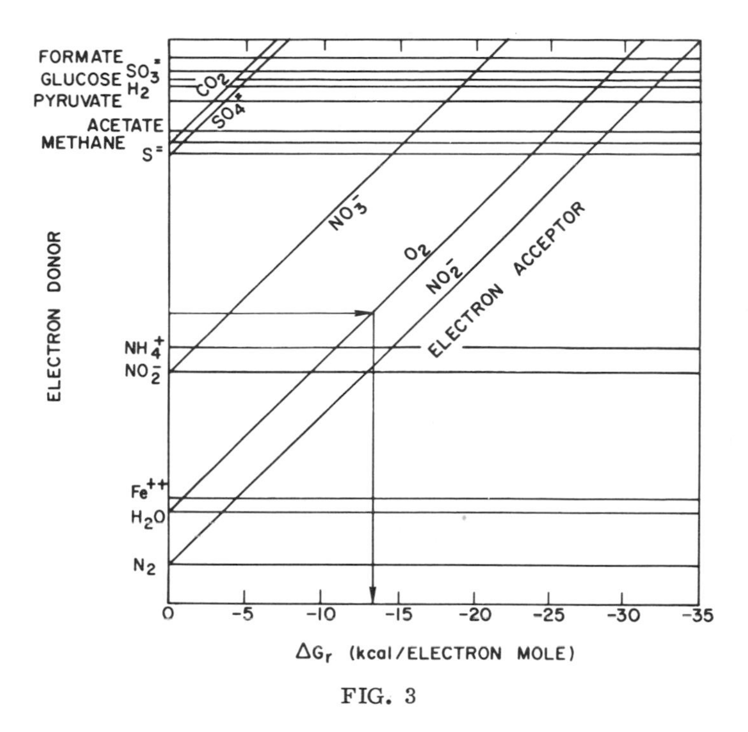

FIG. 3

Free energy from substrate energy conversion (ΔG_r) for various electron donors and electron acceptors.

Equation (21-5). Table 4 is a summary of the computed values for ΔG_n obtained by combining the appropriate half reactions from Table 1.

The value for ΔG_c was estimated previously from the yield value of 10.5 grams of bacterial dry weight per mole of ATP (Bauchop and Elsden(9, 28)), and from the free energy released by hydrolysis of ATP under physiological conditions of -12.5 kcal (Burton (29)). This resulted in a value of 932 cal per gram of cell oxygen equivalent (13) or 7.5 kcal per electron equivalent of cells as used in the present context (assuming cells to be 90 percent volatile organic material).

For a given average energy transfer efficiency, energy reaction, and carbon and nitrogen source for cell synthesis, an overall stoichiometric equation can be written for the products and reactants resulting from bacterial growth. Typical computations for both autotrophic and heterotrophic reactions are illustrated in Table 5. Similarly, if

TABLE 4

Values for ΔG_n Based upon 1/20 Mole Nitrogen Per Cell Electron Equivalent

Nitrogen Source	Equation No.	Reduction Equation	ΔG_n kcal per cell equiv.
NO_3^-	1.	$1/20\ NO_3^- + 1/10\ H^+ + 1/20\ H_2O = 1/20\ NH_4^+ + 1/10\ O_2$	4.17
NO_2^-	2.	$1/20\ NO_2^- + 1/10\ H^+ + 1/20\ H_2O = 1/20\ NH_4^+ + 3/40\ O_2$	3.25
N_2	3.	$1/40\ N_2 + 1/20\ H^+ + 3/40\ H_2O = 1/20\ NH_4^+ + 3/80\ O_2$	3.78
NH_4^+	4.	$1/20\ NH_4^+ = 1/20\ NH_4^+$	0.00

the reactants and products from bacterial growth are known, the average efficiency of energy transfer from Equation (21-9) can be determined. Computations of this type were made previously (13) and transfer efficiencies ranging from 12 to 100 percent were found. However, the efficiencies for autotrophic growth generally ranged from 30 to 60 percent, that for anaerobic heterotrophic growth from 40 to 70 percent, and that for aerobic heterotrophic growth from 20 to 50 percent. A more critical evaluation of the data available is given here and includes some more recently available growth information and excludes some previously used data which was not sufficiently documented or which was taken from old cells for which the energy of maintenance was likely to have been a large factor in reducing the reported growth yields.

B. Efficiencies of Energy Transfer.

Computed efficiencies of energy transfer using selected data from the literature are summarized in Tables 6 and 7 for heterotrophic

TABLE 5

Typical Computations for Overall Autotrophic and Heterotrophic Reactions Based Upon k = 0.60

AUTOTROPHIC GROWTH

Conditions:

Electron Donor - H_2, Electron Acceptor - O_2, Nitrogen Source - NO_3^-

Energy Calculations:

ΔG_r = (-9.670)-(18.675) = -28.345 (half react. 28 - half react. 2) (Table 1)

ΔG_p = (18.675)-(-8.545) = +27.220 (half react. 2 - half react. 25) (Table 1)

ΔG_n = 4.17 (Table 4)

$$A = -\frac{27.220/0.6 + 7.5 + 4.17/0.6}{0.6(-28.345)} = 3.51$$

Overall Reaction:

Synthesis Reaction (half react. 2 - half react. 32 + Eq. 1 (Table 4)) + Energy Reaction (A (half react. 28 - half react. 2)) normalized to one mole of H_2:

$$H_2 + 0.3O_2 + 0.11\ CO_2 + 0.03\ HCO_3^- + 0.03\ NO_3^- + 0.06\ H^+ = 0.03\ C_5H_7O_2N + 0.94\ H_2O$$

HETEROTROPHIC GROWTH

Conditions:

Electron Donor - Acetate, Elect. Accept. - CO_2, Nitrogen Source - NH_4^+

Energy Calculations:

ΔG_r = (-6.609)-(-5.763) = -0.846 (half react. 14 - half react. 12) (Table 1)

ΔG_p = (-6.609)-(-8.545) = 1.936 (half react. 14 - half react. 25) (Table 1)

ΔG_n = 0.0

$$A = -\frac{1.936/0.6 + 7.5}{0.6\ (-0.846)} = 21.2$$

Overall Reaction:

Synthesis Reaction (half react. 14 - half react. 32) + Energy Reaction [A(half react. 14 - half react. 12)] normalized to one mole of Acetate:

$$CH_3COO^- + 0.027CO_2 + 0.018NH_4^+ + 0.93H_2O = 0.018\ C_5H_7O_2N + 0.96\ CH_4 + 0.98\ HCO_3^-$$

TABLE 6

Efficiencies of Energy Transfer for Anaerobic Heterotrophic Bacterial Growth

Energy Reaction	Organism	ΔG_r kcal per elec. mole	A	% Eff. 100 k	Ref.
Glucose + 2 HCO_3^- = 2 Lactate + 2 H_2O + 2 CO_2	*S. faecalis*	-2.15	5.5	56	(9)
Glucose + 2 HCO_3^- = 2 Lactate + 2 H_2O + 2 CO_2	*S. faecalis*	-2.15	6.5	48	(10)
1.5 Glucose + 3 H_2O = 2 Propionate + 2 Acetate + 4 H_2O + 4 CO_2	*P. pentosaceum*	-3.37	3.0	64	(9)
Glucose = 2 Ethanol + 2 CO_2	*S. cerevisiae*	-2.43	6.1	45	(9)
8 Pyruvate + 2 $SO_4^=$ + 3 H^+ = 8 Acetate + H_2S + HS^- + 8 CO_2	*D. desulfuricans*	-2.24	5.7	60	(11)
6 Pyruvate + 2 $SO_3^=$ + 3 H_2O = 6 Acetate + H_2S + HS^- + 3 HCO_3^- + 3 CO_2	*D. desulfuricans*	-2.61	6.6	44	(11)
4 Lactate + 2 $SO_4^=$ + H^+ = 4 H_2O + 4 Acetate + H_2S + HS^- + 4 CO_2	*D. desulfuricans*	-1.77	5.5	85	(11)
Glucose = 1.4 Acetate + 1.4 H^+ + 1.6 CO_2 + 1.6 CH_4	Enrichment	-3.86	2.7	64	(12)
2 Octanoate + 8 H_2O = 11 CH_4 + 3 CO_2 + 2 HCO_3^-	Enrichment	-0.894	13.5	81	(12)
Acetate + H_2O = CH_4 + HCO_3^-	Enrichment	-0.846	15.7	75	(12)
4 Methanol = 3 CH_4 + CO_2 + H_2O	Enrichment	-3.20	5.54	41	(13)
4 Propionate + 6 H_2O = 7 CH_4 + CO_2 + 4 HCO_3^-	Enrichment	-0.90	13.5	80	(13)
4 Benzoate + 22 H_2O = 15 CH_4 + 9 CO_2 + 4 HCO_3^-	Enrichment	-1.13	8.1	101[a]	(13)

TABLE 6 (continued)

Energy Reaction	Organism	ΔG_r kcal per elec. mole	A	% Eff. 100 k	Ref.
Succinate + 1.7 H_2O = 0.84 CH_4 + 0.19 Acetate + 0.41 Propionate + 0.15 CO_2 + 1.4 HCO_3^-	Enrichment	-0.875	16.2	67	(13)
Lactate + 0.7 H_2O = 1.2 CH_4 + 0.07 Acetate + 0.14 Propionate + 0.5 CO_2 + 0.8 HCO_3^-	Enrichment	-1.92	11.7	40	(13)

[a]Theoretically impossible but may result from experimental error or poor thermodynamic data.

TABLE 7

Efficiencies of Energy Transfer for Aerobic Heterotrophic Bacterial Growth[a]

Energy Reaction	ΔG_r kcal per elec. mole	A	% Eff. 100 k
Glucose + 6 O_2 = 6 CO_2 + 6 H_2O	-28.7	0.27	81
Fructose + 6 O_2 = 6 CO_2 + 6 H_2O	-28.7	0.35	64
Lactose + 12 O_2 = 12 CO_2 + 11 H_2O	-28.7	0.35	64
Sucrose + 12 O_2 = 12 CO_2 + 11 H_2O	-28.7	0.34	65
2 Glycine + 3 O_2 = 2 CO_2 + 2 HCO_3^- + 2 NH_4^+	-27.1	0.92	32
Alanine + 3 O_2 = 2 CO_2 + HCO_3^- + NH_4^+ + H_2O	-26.3	0.92	40
2 Glutamate + 9 O_2 = 6 CO_2 + 4 HCO_3^- + 2 NH_4^+ + 2 H_2O	-26.3	0.81	45
Butanol + 6 O_2 = 4 CO_2 + 5 H_2O	-25.8	1.04	41
2 Benzoate + 15 O_2 = 12 CO_2 + 2 HCO_3^- + 4 H_2O	-25.6	1.19	39
Butyrate + 5 O_2 = 3 CO_2 + HCO_3^- + 3 H_2O	-25.5	0.50	76
2 Propionate + 7 O_2 = 4 CO_2 + 2 HCO_3^- + 4 H_2O	-25.3	0.74	57
Acetate + 2 O_2 = CO_2 + HCO_3^- + H_2O	-25.3	0.74	62

[a]Enrichment cultures as reported in Reference 26.

cultures and in Table 8 for autotrophic cultures. The values for heterotrophic aerobic organisms were taken from recent data by Burkhead and McKinney (26). They measured, simultaneously, the oxygen uptake, substrate utilization, and cell synthesis by actively growing enrichment cultures. Data of this type are needed to obtain a good estimate of A since, with aerobic heterotrophic growth, the value for this parameter is so small. Additional data of this type are scarce. Tables 6, 7, and 8 contain data from both pure and enrichment cultures. Data were included to indicate the efficiencies of energy transfer by the dominant cultures which would tend to develop in a natural environment. The efficiencies of energy transfer by many of the pure cultures included were in the same range as found for enrichment cultures. In general, efficiencies ranged from about 40 to 80 percent for both aerobic and anaerobic cultures and for both heterotrophic and autotrophic organisms. An average efficiency of about 60 percent is indicated.

Figure 4 illustrates a comparison between computed values of A using an energy transfer efficiency of 60 percent and observed values from Tables 6, 7, and 8. A scatter is apparent, but the trend is statistically significant. The data suggest that a reasonable correlation exists between free energy of reaction and cell yield for cultures dominating under natural conditions.

C. Efficiencies of Photosynthetic Bacteria.

Rabinowitch (31) and Larsen, et al. (32) reported the yields of various photosynthetic bacteria and concluded that the maximum quantum yield was one molecule of carbon dioxide converted to cells per 10± 2 quanta of light. They refuted earlier work by Warburg (33) which indicated the same amount of synthesis could be achieved with 4 or 5 quanta of light. Larsen et al., used various electron donors in their studies with the green sulfur bacterium, Chlorobium thiosulfatophilum, and obtained the values listed in Table 9. Since the quantum yield was found to be independent of the nature of the electron donor, they concluded that the primary photochemical reaction was the photolysis of H_2O, and the energy released during oxidation of the electron donor was not utilized for carbon dioxide assimilation.

A calculation of energy transfer efficiency for the photosynthetic bacteria can be made using Equation (21-9) and the above values. In this case, ΔG_p would be 27.22 kcal per electron equivalent of cells formed (difference between $\Delta G^o(W)$ for half reactions 2 and 25, Table 1) as for other autotrophic organisms using H_2O as the primary electron donor. Based on half reaction 32, the synthesis of one mole

TABLE 8

Efficiencies of Energy Transfer for Autotrophic Bacterial Growth

Energy Reaction	Organism	ΔG_r kcal per elec. mole	A	% Eff. 100 k	Ref.
Aerobic:					
$2\ S + 2\ H_2O + 3\ O_2 = 2\ SO_4^= + 4\ H^+$	*T. thiooxidans*	-23.3	3.50	58	(7)
$S_2O_3^= + H_2O + 2\ O_2 = 2\ SO_4^= + 2\ H^+$	*T. thiooxidans*	-23.8	8.23	39	(6)
$2\ NH_4^+ + 3\ O_2 = 2\ NO_2^- + 2\ H_2O + 4H^+$	*Nitrosomonas*	-10.8	9.40	56	(8)
$2\ H_2 + O_2 = 2\ H_2O$	*H. facilis*	-28.3	4.00	52	(15)
$2\ H_2 + O_2 = 2\ H_2O$	*H. ruhlandii*	-28.3	2.60	66	(16)
$4\ Fe^{++} + O_2 + 4H^+ = 4\ Fe^{3+} + 2\ H_2O$	unnamed	- 6.90[a]	25	46	(17)
Anaerobic:					
$5\ S_2O_3^= + H_2O + 8\ NO_3^- = 10\ SO_4^= + 4N_2 + 2\ H^+$	*T. denitrificans*	-22.2	3.6	63	(6)
$4\ \text{Formate} + CO_2 + 2\ H_2O = 4\ HCO_3^- + CH_4$	Enrichment	- 5.72	10.1	75	(13)
$4\ H_2 + CO_2 = CH_4 + 2\ H_2O$	Enrichment	- 3.91	12.5	83	(14)

[a] Based on $[H^+] = 10^{-2.7}$

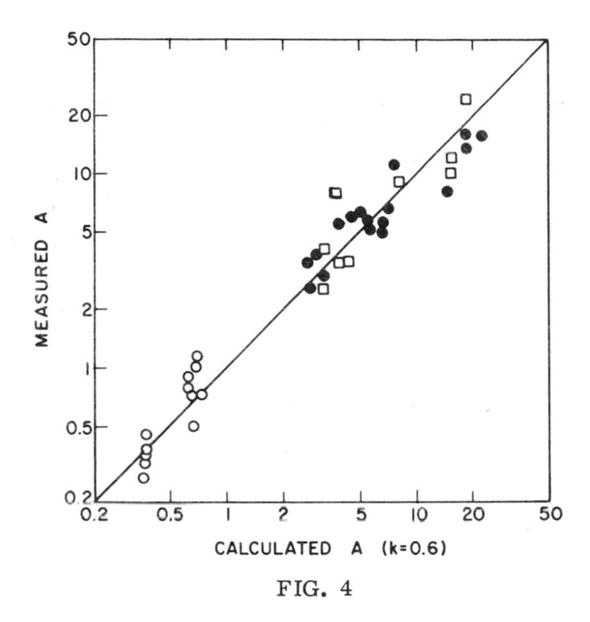

FIG. 4

Comparison between calculated and measured ratio of electron equivalents of substrate converted for energy to that synthesized (A). Open circles represent aerobic heterotrophic growth, shaded circles represent anaerobic heterotrophic growth, and squares represent autotrophic growth.

of carbon dioxide represents the formation of 4 electron equivalents of cells, so that a yield of 10 quanta per molecule of carbon dioxide would equal 2.5 quanta moles per electron equivalent synthesized. Since the energy per mole quanta of infrared light (747mμ) is 39,000 cal, the term A ΔG_r would equal 2.5 (39) or 97.5 kcal per electron equivalent of carbon dioxide synthesized. Substitution of these values into Equation (21-9) gives an energy transfer efficiency of 57 percent. Thus, by assuming that the free energy released from the oxidation of the electron donors listed in Table 9 is not captured by the photosynthetic bacteria as indicated by Larsen et al., energy transfer efficiencies for the photochemical and synthesis reactions are similar to those for nonphotosynthetic bacteria.

TABLE 9

Photosynthetic Yield of Chlorobium thiosulfatophilum

Electron Donor	Simplified Synthesis Reaction		Quanta Required for Assimilation of One Molecule CO_2
H_2	$CO_2 + 2\ H_2$	$= (CH_2O) + H_2O$	9
$S_2O_3^=$	$CO_2 + 1/2\ S_2O_3^= +$		
	$3/2\ H_2O$	$= (CH_2O) + SO_4^= + H^+$	9
$S_4O_6^=$	$CO_2 + 2/7\ S_4O_6^= +$		
	$13/7\ H_2O$	$= (CH_2O) + 8/7\ SO_4^= +$	
		$12/7\ H^+$	9

D. Limitations of Energy-Yield Relationship.

Efficiency of energy transfer of 60 to 80 percent appears to represent the maximum which would be obtained under natural conditions. Less efficient energy utilization can perhaps result from what Gunsalus and Shuster (34) termed "nutrient limitation," a condition which might occur in high energy aerobic growth where a high percentage of substrate carbon is converted to cells. Such diversion may result in: (a) the accumulation of polymeric products either in storage form or as unusable waste, (b) dissipation as heat by "ATPase mechanisms," and (c) activation of shunt mechanisms, by passing energy-yielding reactions. In addition, organisms living under adverse conditions such as in the presence of inhibiting materials, unbalanced ionic concentrations, or other than optimum pH values, may have smaller cell yields because of higher energy expenditure for maintenance of favorable balances within the cell.

TABLE 10

Comparison of Calculated Efficiencies of Energy Transfer for Formate Oxidation by Autotrophic and Heterotrophic Pathways

Value	Autotrophic Pathway	Heterotrophic Pathway
A (Ref. 13)	2.1	2.1
ΔG_r	-30.2	-30.2
ΔG_p	+27.2	- 2.94
ΔG_n	0.0	0.0
m	+1	- 1
k	0.71	0.11

Pure culture studies have indicated that some bacteria do not have enzyme systems required to obtain energy efficiently and so exhibit low growth yields. Bauchop and Elsden (9) indicated the growth yield with P. lindneri during the fermentation of glucose to ethanol was less than half that by S. cerevisiae because P. lindneri ferments glucose by an inefficient pathway yielding only half the ATP as obtained by S. cerevisiae. The energy transfer efficiency computed from Equation (21-9) using the data furnished for P. lindneri is only 17 percent, while that for S. cerevisiae is 45 percent. This suggests that, in a competitive anaerobic environment where glucose is the major available substrate for growth, S. cerevisiae would probably dominate over P. lindneri.

Another example of an inefficient pathway for energy transfer is given by organisms which use formate. Limited radiotracer studies by Quayle (35) indicated that metabolism of formate as a sole carbon and energy source by Pseudomonas oxalatucus was autotrophic, the bulk of the cell carbon being fixed from carbon dioxide. The low values for cell yield from this compound by methane fermentation and aerobic oxidation (13) support this observation. The calculations in Table 10 indicate the efficiencies of energy transfer, assuming heterotrophic growth, are extremely low, while when assuming autotrophic growth, they are in the normal range. Thus, although conversion of formate directly into cells requires much less energy theoretically than the conversion of formate first into carbon dioxide and then synthesis of the carbon dioxide into cells, the organisms developed in mixed cultures on this single carbon compound apparently do not have the required enzyme system for the direct conversion, and so must grow by the less efficient autotrophic pathway.

These examples indicate a major limitation of Equation (21-9) for the calculation of efficiencies of energy transfer. Pathways for the energy and synthesis reactions must be assumed. If the assumptions are incorrect, then the calculated energy transfer efficiencies will be in error. This may be viewed in another way, if the calculated energy transfer efficiencies are either very low or very high, then there is a reasonable chance that the assumed pathways are incorrect.

An example where both the assumed pathway for energy transfer and for carbon synthesis is open to question is in the autotrophic reaction by which hydrogen and carbon dioxide are converted to methane gas. Three possible schemes for this are listed in Table 11 along with the computed efficiencies of energy transfer using yield values by Shea et al. (14). In the first case, carbon dioxide is reduced directly with molecular hydrogen, and energy for synthesis (ΔG_p) is relatively low, which is, in fact, lower than for most heterotrophic synthesis reactions. This is not a known pathway for autotrophic synthesis and probably does not occur as suggested by the low energy transfer efficiency calculated. The inability to carry out this direct reduction of carbon dioxide is similar to the observation made by Larsen et al. (32) with the anaerobic photosynthetic bacteria.

In the second and third cases, the normal autotrophic energy requirement for synthesis from carbon dioxide is assumed. In this reaction, oxygen is produced. Since methane fermentation is carried out by obligate anaerobes, the oxygen must be eliminated by some scheme, and hydrogen is the only available reducing material for this purpose. The reduction of oxygen with hydrogen yields a significant quantity of energy. In case 2 this energy was assumed to be available to the organisms while in case 3 it was not. The energy transfer efficiency in case 2 is a more typical value and so perhaps indicates the energy from hydrogen oxidation is available, although the transfer efficiency in case 3 is not sufficiently different to be ruled out. The anaerobic autotrophic values listed in Table 8 were computed using the assumptions of case 3.

TABLE 11

Possible Pathways and Calculated Energy Transfer Efficiencies for Hydrogen Fermentation to Methane Using A = 12.5

	Case	Half React. Used - Table 1	ΔG_p kcal	Half React. Used - Table 1	$A\ \Delta G_r$ kcal	k
1.	CO_2 reduced directly by H_2 in synthesis reaction.	(28)-(25)	-1.13	A ((28)-(12))	-48.8	0.15
2.	Normal autotrophic energy requirement, energy from O_2 reduction available.	(2) -(25)	+27.2	A ((28)-(12)) + ((28)-(2))	-77.1	0.64
3.	Normal autotrophic energy requirement, energy from O_2 reduction not available.	(2) -(25)	+27.2	A ((28)-(12))	-48.8	0.83

IV. ELECTRON TRANSFER RATES

Equations (21-2), (21-3), and (21-4) indicate the relationship between the transport rate of substrate and the growth rate of bacteria. If, in Equation (21-4), the yield coefficient, Y_m, is defined as grams of bacteria synthesized (dry volatile weight) per mole electrons transferred for energy, then the transport rate, k_m, can be defined in electron moles transferred for energy per gram of bacteria per day. This allows the transport rate to be expressed in equivalent terms which can be readily compared. Y_m can be estimated as follows:

$$Y_m = \frac{c}{dA} \tag{21-10}$$

Here, A is as previously defined; c is the grams of cells formed per electron mole of carbon synthesized, and from Equation (21-5) equals 113/20 or 5.65 grams per electron mole. The value, d, equals the electrons actually transferred from a donor molecule divided by the electron equivalents per molecule as given by the half reactions in Table 1. Normally, d equals 1.0, but it may be less than unity in certain fermentations as will be illustrated in the next section.

In heterotrophic reactions, one fraction of the organic substrate is synthesized into cells (f_s) while the other portion is utilized for energy (f_e), where $f_s + f_e = 1$. The value of A for heterotrophic reactions is by definition equal to f_e/f_s so that the fraction of organic substrate utilized for energy is $f_e = A/(A+1)$. Values for k_m can be determined from conventional units as follows:

Heterotrophic reactions:

$$k_m = n\left(\frac{A}{A+1}\right)\left(\frac{\text{moles substrate utilized/day}}{\text{grams of bacteria present}}\right) \tag{21-11}$$

Autotrophic reactions:

$$k_m = \frac{n\,(\text{moles substrate utilized for energy/day})}{(\text{grams of bacteria present})} \tag{21-12}$$

The value, n, is the number of electron moles transferred per mole of substrate utilized for energy.

Table 12 is a summary of values for k_m based upon available data. In some cases, a value for the maximum rate of reaction per unit of time and per unit weight of organisms was given so that values for k_m could be computed directly. In a few cases, μ_m values only were available and so k_m was estimated by using Equations (21-4) and (21-10). Values for A in this case were computed using Equation (21-9) and an efficiency of energy transfer of 60 percent. Whether k_m or μ_m values were available for the determination of the k_m values is also indicated in Table 12.

The rate of substrate utilization is expected to be a function of temperature and should obey the Arrhenius equation:

$$k_m = Be^{-Ea/RT} \tag{21-13}$$

Values for Ea were available in some cases and are listed in Table 12. The values compare favorably with others reported in the past (17,45), and are valid only for temperatures between about 10° and 40° C.

While there is some spread in the computed k_m values, the differences are not great when the highly different reactions and the uncertainty of the experimental methods used by the different investigators is considered. Values for k_m appear to vary between about 1 and 2 electron moles/gm-day at 25° C. and increase with increasing temperature as suggested by the Arrhenius equation. The close agreement between such widely varying energy reactions suggests that the rate of electron transport is the rate determining step in most bacterially mediated reactions. More closely controlled experiments are needed to determine the general validity of this observation.

V. GROWTH RATES

Maximum growth rates of bacteria can be calculated by using a typical value for the energy transfer efficiency and by assuming that k_m is a temperature dependent constant. Y_m can then be found from Equations (21-9) and (21-10), and the growth rate can be determined from Equation (21-4). The maximum growth rates computed for a variety of bacterially mediated reactions determined in this way are listed in Table 13. Also listed are generation times computed as $\ln(2)/\mu_m$ and multiplied by 24 to convert from days to hours.

The generation times for organisms using glucose under aerobic and anaerobic conditions are not too different. This similarity results since, when glucose is anaerobically fermented, only four moles of

TABLE 12

Electron Transfer Rates

Electron Donor Half Reaction	Electron Acceptor	d	Data Available	k_m Electron moles/gm-day 6–15°C.	16–25°C.	26–35°C.	Ea kcal per mole	Ref.
Glucose + $6H_2O = 6\ CO_2 + 24H^+ + 24e^-$	O_2	1.0	k_m	0.12	0.35	1.0	18.2	(36)
Glucose + $6H_2O = 6CO_2 + 24H^+ + 24e^-$	O_2	1.0	μ_m	--	1.1	2.8	13.5	(17)
Maltose + $13H_2O = 12CO_2 + 48H^+ + 48e^-$	O_2	1.0	μ_m	--	0.9	1.9	11.7	(17)
Acetate + $3H_2O = CO_2 + HCO_3^- + 8H^+ + 8e^-$	CO_2	1.0	k_m	--	0.6	1.1	9.9	(19)
Propionate + $2H_2O$ = Acetate + $CO_2 + 6H^+ + 6e^-$	CO_2	6/14	k_m	--	1.0	1.0	--	(19)
Butyrate + $H_2O + HCO_3^-$ = Acetate + $CO_2 + 4H^+ + 4e^-$	CO_2	4/20	k_m	--	--	1.0	--	(19)
$NH_4^+ + 2H_2O = NO_2^- + 8H^+ + 6e^-$	O_2	1.0	k_m	--	--	2.9	--	(27)
$NH_4^+ + 2H_2O = NO_2^- + 8H^+ + 6e^-$	O_2	1.0	μ_m	0.42	0.6	1.3	12.9	(37)
$NH_4^+ + 2H_2O = NO_2^- + 8H^+ + 6e^-$	O_2	1.0	μ_m	0.7	1.1	2.8	16.8	(38)
$NO_2^- + H_2O = NO_3^- + 2H^+ + 2e^-$	O_2	1.0	k_m	--	--	2.0	--	(27)
$NO_2^- + H_2O = NO_3^- + 2H^+ + 2e^-$	O_2	1.0	μ_m	0.53	0.7	1.2	9.8	(37)
$NO_2^- + H_2O = NO_3^- + 2H^+ + 2e^-$	O_2	1.0	μ_m	1.3	1.8	3.2	10.3	(38)

TABLE 12 (continued)

Electron Donor Half Reaction	Electron Acceptor	d	Data Available	k_m Electron moles/gm-day 6-15°C.	16-25°C.	26-35°C.	Ea kcal per mole	Ref.
$Fe^{++} = Fe^{3+} + e^-$	O_2	1.0	k_m	--	--	0.6	--	(39)
$Fe^{++} = Fe^{3+} + e^-$	O_2	1.0	k_m	--	--	2.0	--	(40)
$H_2 = 2H^+ + 2e^-$	CO_2	1.0	k_m	--	--	3.1	--	(14)
$H_2 = 2H^+ + 2e^-$	$SO_4^=$	1.0	k_m	--	--	2.4	--	(41)

TABLE 13

Computed Growth Rates and Generation Times for Typical Bacterially Mediated Energy Reactions (for k = 0.6, k_m = 2 electron moles/gm-day, and $\Delta G_n = 0$)

Electron Donor	Electron Acceptor	n	ΔG_r kcal	ΔG_p kcal	μ_m day^{-1}	Generation Time Hours
Glucose	O_2	24	-28.7	-1.48	29	0.6
Glucose	Pyruvate[a]	4	- 2.15	-1.48	13	1.3
Acetate	NO_3^-	8	-23.7	1.94	15	1.1
Acetate	$SO_4^=$	8	- 1.52	1.94	1.0	17
Acetate	CO_2	8	- 0.85	1.94	0.27	31
Methane	O_2	8	-24.4	27.2	3.1	5.3
H_2	O_2	2	-28.3	27.2	3.6	4.6
H_2	CO_2	2	- 3.91	27.2	0.5	33
Fe^{++}	O_2[b]	1	- 6.90	33.2	0.7	22
NH_4^+	O_2	6	-10.8	27.2	1.4	12
NO_2^-	O_2	2	- 9.25	27.2	1.2	14

[a]For lactic acid fermentation; electron donation reaction is Glucose + $2HCO_3^-$ = 2 Pyruvate + $2CO_2$ + 2 H_2O + $4H^+$ + $4e^-$, electron acceptance reaction is 2 Pyruvate + $4H^+$ + $4e^-$ = 2 Lactate.

[b]Computed for pH = 2.7, $\Delta G_f^o = -3.70$ kcal for $(H^+) = 10^{-2.7}$.

electrons are transferred per mole of glucose, while under aerobic conditions 24 moles are transferred. By assuming the maximum electron transport rate, k_m, is constant, the rate of utilization of glucose is much greater on a molar basis (24/4 or six times in this example) under anaerobic conditions than under aerobic conditions, a commonly observed phenomena termed the "Pasteur effect" (42).

The generation time with acetate, however, is markedly different depending upon the electron acceptor available since the number of electrons transferred is generally the same for both aerobic and anaerobic conditions. An exception might be in the methane fermentation of acetate since, in this case, carbon dioxide does not really act as an electron acceptor. The acetate molecule is simply split into two fragments, with the methyl carbon converted to methane and the carboxyl carbon to carbon dioxide (43). Thus, the actual number of electrons transferred is probably less than for other acetate conversions, although the computed generation time for methane fermentation, assuming eight mole electrons per mole of acetate, agrees well with actual observations.

Computed generation times for autotrophic growth generally are much longer than for heterotrophic growth. This results largely because of the high value for ΔG_p under autotrophic conditions. In most cases when oxygen is the electron acceptor, values for ΔG_r are similar for autotrophic and heterotrophic reactions. The most apparent exception is the oxidation of Fe^{++}. The value for ΔG_r shown in Table 13 for this reaction is for a pH of 2.7, an optimum value for the growth of iron oxidizing organisms. At a more neutral pH value, the value for ΔG_r would be only -0.90 kcal assuming equal soluble concentration for the reduced and oxidized iron species. The significant reduction in ΔG_r at a higher pH value may be one explanation for the more optimum growth at lower pH values.

The reasonable agreement between the computed generation times based upon the assumptions made and those frequently measured suggests that the growth rate of organisms is largely a function of the energy released by the energy reaction, the energy required for synthesis, and temperature. Lower growth rates than calculated can result either from reduced efficiencies of energy transfer or from reduced rates of electron transport compared with values normally obtained by the dominant organisms under natural conditions.

VI. SUMMARY

Classical thermodynamics can be used to indicate the equilibrium direction and state for a given closed system or to indicate the quantity of energy released from a given change from one state to another. Bacteria which can capture more of the energy released by a chemical change will have a higher cell yield and so will have a competitive advantage in a natural environment.

Classical thermodynamics was used in order to estimate the maximum efficiency by which energy is transferred from substrate to cellular synthesis by cultures which tend to dominate in the natural environment. Energy was postulated to be transferred in two steps, first from the energy substrate to an energy carrier, and then from the energy carrier to cell synthesis.

The energy available per equivalent of substrate used for energy varies widely with the available electron donor and electron acceptor. Also the energy requirement for cell synthesis differs to a marked degree with the nature of the carbon and nitrogen source used. By considering these differences, the maximum efficiency of energy transfer was calculated for a variety of autotrophic and heterotrophic bacteria living under both aerobic and anaerobic conditions. The efficiency of energy transfer by dominant organisms growing under conditions to encourage maximum cellular yield varied within a reasonably narrow range of 40 to 80 percent, with a typical value of about 60 percent. Considering the differing energy requirements of various bacteria, the differences in efficiency between autotrophic and heterotrophic bacteria does not seem as great as generally considered.

The correlation between energy availability and maximum cellular yield is illustrated by the comparison in Figure 4. The correlation is good, but not perfect, and this is a point of frequent debate. The relative differences in cellular yield corresponding to different energy transfer efficiencies of 40 to 80 percent is about one to four. This may seem a large difference from some points of view, but when the range of energy available from various bacterially mediated reactions is considered to vary over about two orders of magnitude, the difference does not appear so great.

Another observation of interest is that the maximum rate of electron transport during energy metabolism is relatively constant at a given temperature, varying from 1 to 3 electron moles per gram of bacteria per day at a temperature of about 30°C. This constancy suggests that electron transport is a major rate limiting step in energy metabolism by bacteria. By considering the relative constancy of the maximum electron transport rate, and typical efficiencies of energy transfer, the growth rates of bacteria utilizing different substrates were estimated. Such computations suggest reasons for differences in the molar rate of glucose utilization under aerobic and anaerobic conditions (the Pasteur effect) and help explain the different generation times between heterotrophic and autotrophic bacteria and between growth under aerobic and under anaerobic conditions.

The above generalizations, of course, have several limitations and cautions expressed by others must be recognized (9, 34, 45, 46, 47). Numerous examples of individual organisms which are relatively inefficient in capturing available free energy can be presented, and those which have rates of electron transport well below the optimum can certainly be found. Individual organisms have limited temperature ranges for optimum growth, have enzyme systems which are tailored best for certain reactions, and have specific environmental requirements of salinity, pH, redox potential, and medium composition under which both their growth rate and cellular yield will be a maximum. Nevertheless, in a natural environment those organisms which are most efficient in energy capture and utilization under a specified set of conditions should tend to dominate. The maximum efficiencies of energy transfer and maximum rates of electron transfer by the dominating organisms appear in general to be similar.

APPENDIX

The following symbols have been adopted for use in this paper:

A = electron equivalents of substrate converted for energy per electron equivalent of cells synthesized.

b = specific maintenance rate, units equivalent biomass per unit biomass present per unit time.

B = constant, $time^{-1}$.

c = grams of cells formed per electron mole of carbon synthesized.

d = electrons transferred in energy reaction per molecule of substrate used divided by the electron equivalents per molecule as given in Table 1.

Ea = energy of activation, kcal per mole.

f_e = fraction of organic substrate used for energy.

f_s = fraction of organic substrate used for synthesis.

ΔG_c = free energy of conversion of one electron equivalent of intermediate to one electron equivalent of cells.

ΔG_n = free energy per electron equivalent of cells for reduction of nitrogen source to ammonia.

ΔG_p = free energy of conversion of one electron equivalent of cell carbon source to intermediate.

ΔG_r = free energy per electron equivalent of substrate converted for energy.

ΔG_s = carrier free energy required for synthesis of one electron equivalent of cells.

$\Delta G^o(W)$ = redox potential, free energy change per mole electron when all species are at unit activity except H^+ and OH^- which are at activity of neutral water.

k = efficiency of energy transfer.

k_m = electron transport rate, electron moles transferred for energy per gram of bacteria per day.

K_s = saturation coefficient, or growth limiting substrate concentration at which specific growth rate is one half μ_m.

m = constant, equal to +1 when ΔG_p is positive, and -1 when ΔG_p is negative.

n = number of electron moles transferred per mole of substrate utilized for energy.

R = universal gas constant, 1.99 cal $mole^{-1}$ deg^{-1}.

S = concentration of growth limiting substrate.

T = temperature, degrees Kelvin.

μ = specific growth rate, biomass formed per unit biomass present per unit time.

μ_m = maximum specific growth rate occurring at high substrate concentration.

Y_m = maximum yield factor, units biomass formed per unit mass of energy source consumed, if no energy required for maintenance; expressed in grams of bacteria synthesized (dry volatile weight) per mole electrons transferred for energy.

REFERENCES

1. Handbook of Chemistry and Physics, 44, Chem. Rubber Pub. Co., 1963.

2. G. S. Parks and H. M. Huffman, Free Energies of Some Organic Compounds, Chem. Catalog Co., New York, 1932.

3. Lange Handbook of Chemistry, 10, McGraw-Hill, New York, 1961.

4. K. Burton and H. A. Krebs, Biochem. J., 54, 94 (1953).

5. J. H. Perry, Chemical Engineers Handbook, 3, McGraw-Hill, New York, 1950.

6. K. Baalsrud and K. S. Baalsrud, "The Role of Phosphate in CO_2 Assimilation," Phosphorus Metabolism, 2, 544, John Hopkins Press, Baltimore, Md., 1952.

7. K. G. Vogler, J. Gen. Physiol., 26, 103 (1942).

8. T. Hofman and H. Lees, Biochem. J., 52, 140 (1952).

9. T. Bauchop and S. R. Elsden, J. Gen. Microbiol., 23, 457 (1960).

10. J. T. Sokatch and I. C. Gunsalus, J. Bacteriol., 73, 452 (1957).

11. J. C. Senez, Bacteriol. Rev., 26, 95 (1962).

12. R. E. Speece and P. L. McCarty, "Nutrient Requirements and Biological Solids Accumulation in Anaerobic Digestion," First International Conference on Water Pollution Research, Pergamon Press, London, 1964, p. 305.

13. P. L. McCarty, "Thermodynamics of Biological Synthesis and Growth," Second International Conference on Water Pollution Research, Pergamon Press, New York, 1965, p. 169.

14. T. G. Shea, W. A. Pretorius, R. D. Cole, and E. A. Pearson, Water Res., 2, 833 (1968).

15. A. Schatz, J. Gen. Microbiol., 6, 329 (1952).

16. L. Parker and W. Vishniac, J. Bacteriol., 70, 216 (1955).

17. J. Monod, Recherches sur la Croissance des Cultures Bacteriennes, Hermann and Cie, Editors, Rue de la Sorbonne, 1942.

18. N. van Uden, Arch. für Mikrobiol., 58, 145 (1967).

19. A. W. Lawrence and P. L. McCarty, J. Water Pollut. Contr. Fed., 41, R1 (1969).

20. A. G. Marr, E. H. Nilson, and D. J. Clark, Ann. New York Acad. Sci., 102, 536 (1963).

21. E. Stumm-Zollinger, Appl. Microbiol., 14, 654 (1966).

22. L. Baas-Becking and G. S. Parks, Physiol. Rev., 7, 85 (1927).

23. J. A. Servizi and R. H. Bogan, Proc. Amer. Soc. Civil Eng., 89, SA3, 17 (1963).

24. R. E. McKinney, Microbiology for Sanitary Engineers, McGraw-Hill Book Company, Inc., New York, 1962.

25. S. R. Hoover and N. Porges, Sewage Ind. Wastes, 24, 306 (1952).

26. C. E. Burkhead and R. E. McKinney, Proc. Amer. Soc. Civil Eng., 95, SA2, 253 (1969).

27. H. Laudelout, P. C. Simonart, and R. van Droogenbroeck, Arch. für Mikrobiol., 63, 256 (1968).

28. T. Bauchop, J. Gen. Microbiol., 18, vii (1958).

29. K. Burton, Nature, 181, 1594 (1958).

30. L. G. Sillen and A. E. Martell, Stability Constants of Metal-Ion Complexes, Spec. Pub. No. 17, The Chemical Society, London, 1964.

31. E. I. Rabinowitch, "The Light Factor II. Maximum Quantum Yield of Photosynthesis," Photosynthesis and Related Processes, II, Part 1, p. 1083, Interscience Publishers, New York, 1951.

32. H. Larsen, C. S. Yocum, and C. B. van Niel, J. Gen. Physiol., 36, 161 (1952).

33. O. Warburg, Amer. J. Bot., 35, 194 (1948).

34. I. C. Gunsalus and C. W. Shuster, "Energy-Yielding Metabolism in Bacteria," The Bacteria, 2, 1, Academic Press, New York, 1961.

35. J. R. Quayle, Ann. Rev. Microbiol., 15, 119 (1961).

36. H. Ng, J. Bacteriol., 98, 232 (1969).

37. F. E. Stratton and P. L. McCarty, Environ. Sci. Technol., 1, 405 (1967).

38. G. Knowles, A. L. Downing, and M. J. Barrett, J. Gen. Microbiol., 38, 263 (1965).

39. G. A. Din, I. Suzuki, and H. Lees, Can. J. Biochem., 45, 1523 (1967).

40. J. V. Beck, J. Bacteriol., 79, 502 (1960).

41. K. R. Butlin and J. R. Postgate, "Microbiological Formation of Sulphide and Sulphur," Symposium Microbial Metabolism, VIth International Congress of Microbiology, Fondazione Emanuele Paterno, Rome, 1953, p. 126.

42. K. C. Dixon, Biol. Rev., 12, 431 (1937).

43. H. A. Barker, Bacterial Fermentations, John Wiley and Sons, Inc., New York, 1956.

44. J. C. Morris and W. Stumm, "Redox Equilibria and Measurements of Potentials in the Aquatic Environment," Equilibria Concepts in Natural Water Systems, Amer. Chem. Soc., Washington, D. C., 1967, p. 270.

45. C. Lamanna and M. F. Mallette, Basic Bacteriology, 3rd ed., The Williams and Wilkins Co., Baltimore, 1965.

46. L. J. Hetling, D. R. Washington, and V. S. Rao, "Kinetics of the Steady-State Bacterial Culture. II. Variations in Synthesis," Proc. 19th Industrial Waste Conference, Purdue Univ. Eng. Extension Series No. 117, 687 (1964).

47. E. D. Schroeder and A. W. Busch, Proc. Amer. Soc. Civil Eng., 94, SA2, 193 (1968).

CHAPTER 22

KINETICS OF BACTERIAL GROWTH DURING AEROBIC OXIDATION OF ORGANICS

M. A. Caglar, A. R. Thompson, C. W. Houston, and V. C. Rose

University of Rhode Island

I. INTRODUCTION

A type of pollution commonly encountered in industrially developed areas is the presence of petroleum fractions in aqueous systems. Instances of such contamination frequently make headlines as "oil slicks" or "oil pollution." Due to their physico-chemical properties, which differ markedly from other organic compounds, hydrocarbons are insoluble and are not diluted when they enter a water source. They remain as a separate phase, usually in stagnant areas, until they are degraded. The rate of biodegradation of hydrocarbons varies considerably depending upon the suitability of environmental conditions for rapid proliferation of microorganisms involved and for the bringing of hydrocarbons into contact with the microorganisms.

Following a discussion of mechanisms of microbial oxidation of straight chain paraffins and the physico-chemical factors affecting the process, some aspects will be reported of a study concerned with the microbial oxidation of n-heptane by pure cultures of a pseudomonad in batch fermentors with pH control. Studies with Na-heptanoate and glucose were also made in an attempt to explain the anomalous behavior of the hydrocarbon systems.

A. Mechanisms of Microbial Oxidation of Paraffinic Hydrocarbons.

The discovery of the utilization of hydrocarbons as a sole source of carbon and energy dates from Sohngen's demonstration in 1906 of the bacterial consumption of methane (1). Research since the 1940's has provided sufficient evidence that oxidation of hydrocarbons is a widespread property of microorganisms. Senez and associates (2) were the first to suggest that normal alkanes were attacked at the C-1 position, after they identified n-heptanoic acid in cultures of *Pseudomonas aeruginosa* growing on n-heptane.

The requirement of molecular oxygen for the enzymatic breakdown of hydrocarbons is well established since the process does not occur in its absence (3). Direct evidence for the involvement of oxygenases in the process has been obtained by oxygen-18 incorporation studies (4). Studies on the mechanism suggest that the initial attack is on the C-1 position and that molecular oxygen is incorporated into the oxidative product of the hydrocarbon molecule. Details of the mechanism of incorporation are not clear. Of the various hypotheses, those of Kallio in the United States and Senez in France are most widely upheld (1). The equations relative to these hypotheses are shown in Figure 1.

Kallio suggests that a hydroperoxide is formed as an intermediate. Bacterial oxidation of 1-alkyl hydroperoxides occurs at a rate which suggests their possible role as intermediates. In addition, oxygen-18 incorporation data support the hypothesis.

Senez's hypothesis involves the production of an olefinic intermediate. 1-Heptene has been isolated and identified by infrared spectroscopy in cultures of *P. aeruginosa* growing at the expense of n-heptane (5). No alkane dehydrogenase, however, was detected for alkanes having ten or more carbon atoms (1).

There is agreement in the subsequent steps. Once the corresponding acid is formed, reactions are similar to those in the breakdown of fatty acids. However, unusual compounds are the dicarboxylic acids formed by diterminal oxidation and the methyl ketones. Experimental evidence for diterminal oxidation has been presented (4). After the formation of the dicarboxylic acid, breakdown is believed to be by β-oxidation.

Methyl ketones have been observed in cultures grown on hexane and lower alkanes. Foster (4) has proposed a pathway for this which involves a free radical equilibrium, which is analogous to the chemical oxidation of alkanes involving multiple sites of free radical attack. Either two independent mechanisms exist (one for attack at the terminal carbon and another for attack at the 2-carbon) or both products are the result of a common mechanism.

INITIAL STEPS :

KALLIO'S HYPOTHESIS:

$$R\text{-}CH_2\text{-}CH_3 \xrightarrow{O_2} R\text{-}CH_2\text{-}\overset{OOH}{\overset{|}{C}}H_2 \xrightarrow[-HOH]{+2H} R\text{-}CH_2\text{-}CH_2OH$$

SENEZ'S HYPOTHESIS:

$$R\text{-}CH_2\text{-}CH_3 + DPN \longrightarrow R\text{-}CH{=}CH_2 + DPNH + H^+$$

$$R\text{-}CH{=}CH_2 \xrightarrow[+HOH]{} R\text{-}CH_2\text{-}CH_2\text{-}OH$$

SUBSEQUENT STEPS :

$$\underset{\text{alcohol}}{R\text{-}CH_2\text{-}OH} \longrightarrow \underset{\text{aldehyde}}{R\text{-}CHO} \longrightarrow \underset{\text{acid}}{R\text{-}COOH} \longrightarrow \beta\text{-Oxidation}$$

acid $\downarrow$ Diterminal Oxidation

$$HOOC\text{-}R'\text{-}COOH \longrightarrow \beta\text{-Oxidation}$$

FREE-RADICAL EQUILIBRIUM :

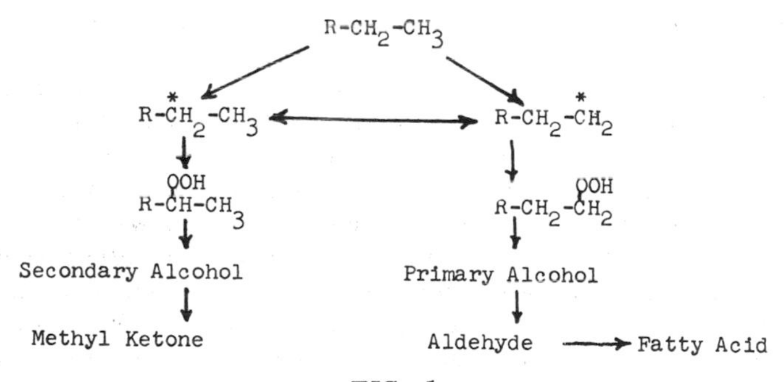

FIG. 1

Mechanisms of paraffinic hydrocarbon oxidation.

B. Physico-Chemical Factors Affecting Hydrocarbon Oxidation.

An important factor in the resistance of hydrocarbons to degradation is their insolubility in the aqueous phase and the resulting separation of phases. In general, hydrocarbon solubility in water increases with temperature and decreases with increasing chain length. Ring compounds are more soluble than straight chain compounds (6). The low solubility of hydrocarbons in water is due to the fact that the partial molal volume of a gas is smaller in water than in organic solvents (7). The presence of electrolytes in water has an adverse effect in that the solubility of hydrocarbons decreases as the salt concentration increases (6). Data on hydrocarbon solubility in water are scarce and are somewhat inconsistent. Perhaps the most extensive list is the one compiled by Irman (8), who was interested in the correlation between the structure of a hydrocarbon and its solubility in water. Table 1 lists the solubility of a few representative hydrocarbons (8, 9). Branching of aliphatics increases solubility due to the higher vapor pressure of the compound rather than to the branched structure (9).

Some of the methods available to increase the solubility of hydrocarbons involve the use of salts of fatty acids which act as surface active agents (10). Formation of such solutions is accompanied by the lowering of free energy (11). There are reports that the use of such surface active agents interferes with oxygen supply by increasing the total resistance to oxygen transfer (12). There are, however, reports to the contrary (13). Methods to check the two opposing hypotheses have been suggested (14).

The supply of oxygen to microbial systems oxidizing hydrocarbons is significant because of the absolute necessity of molecular oxygen for the process (3). The solubility of oxygen in water is much lower than in hydrocarbons, as seen in Table 2 (15, 16, 17), and, as stated above, is due to the smaller partial molal volume of a gas in water than in a hydrocarbon (7). Thus, at the same percentage of dissolved oxygen saturation and at the same temperature, a hydrocarbon, such as heptane, will contain a concentration of dissolved oxygen which is about 10 times that in water. When heptane, containing a low percentage of dissolved oxygen saturation, is shaken with water of a higher percentage saturation, the oxygen will be "extracted" from water into the hydrocarbon phase. When equilibrium is reached, the percentage of dissolved oxygen saturation will be equal in both phases; however, the water phase will have been considerably impoverished. Depending upon the initial levels of oxygen saturation, oxygen may be "extracted" by either phase from the other, and different equilibrium values may be obtained. However, the ratio of dissolved oxygen concentration in

TABLE 1

Solubility of Hydrocarbons in Water
(T = 25° C.)

Hydrocarbon	Solubility ppm
Pentane	39
Hexane	9.5
2-Methylpentane	14
2,2-Dimethylbutane	18
Heptane	2.9
2,4-Dimethylpentane	3.6
Octane	1.4
2,2,4-Trimethylpentane	2.4
Nonane	0.22
Decane	0.052
Cyclopentane	156
Cyclohexane	55
Benzene	1780
Toluene	538
Ethylbenzene	159
Isopropylbenzene	53

one phase to that in the other will be reasonably constant over a small temperature range. This, then, can be considered a case of the distribution of a solute (oxygen) between two solvents (hydrocarbon and water) according to definite distribution ratio.

Hydrocarbons form films on the surface of stagnant waters and it has been reported (12, 18) that they decrease the dissolved oxygen content of the water. This decrease becomes significant as the thickness of the hydrocarbon film increases. Long term effects were not reported. This observed decrease could be transient. Once the equilibrium is established, oxygen could then be transferred from the hydrocarbon layer to the water and increase its dissolved oxygen content.

In agitated systems, the equilibrium would be achieved sooner due to the effects of turbulent diffusion and dispersion of the hydrocarbon as fine droplets. If some dissolved oxygen is removed from the aqueous

TABLE 2

Solubility of Oxygen in Various Solvents
(Air-saturated Liquids)

Liquid	Temperature °C.	Pressure mmHg	Solubility mg/l	Reference
Water	20	750	8.9	(15)
Cyclohexane	20	750	73	(15)
Octanol	20	750	89	(15)
n-Heptane	20	750	97	(15)
n-Heptane	24	752	88	(16)
Isooctane	25	745	91.5	(16)
n-Dodecane	25	748	54.5	(16)
Gasoline	26	744	60	(16)
Benzene	26	748	43.2	(16)
Toluene	26	751	57.9	(16)
t-Butylbenzene	25	760	43.2	(17)
1-Octene	25	760	65.4	(17)

phase (e.g., by microorganisms), the equilibrium would be upset and, to re-establish the distribution ratio, some oxygen would be transferred from the hydrocarbon phase to the aqueous phase. Thus, the hydrocarbon phase would act as an "oxygen reservoir."

It has been reported (19) that, in agitated and air sparged systems, hydrocarbons may contribute to the total dissolved oxygen content and that this contribution is proportional to the hydrocarbon to aqueous system ratio. The contribution increases as the hydrocarbon dispersion approaches perfect homogeneity (19).

In one experiment, using an aerated and agitated vessel containing an aqueous medium but no microorganisms, heptane was fed as vapor into the air which was being sparged. After the establishment of a given ratio of hydrocarbon to aqueous medium, hydrocarbon feed was stopped, but the air sparging was continued. As the heptane was stripped by the air and removed from the system, the ratio of hydrocarbon to aqueous medium decreased. The effect of this decrease on the total dissolved oxygen concentration in the system is seen in

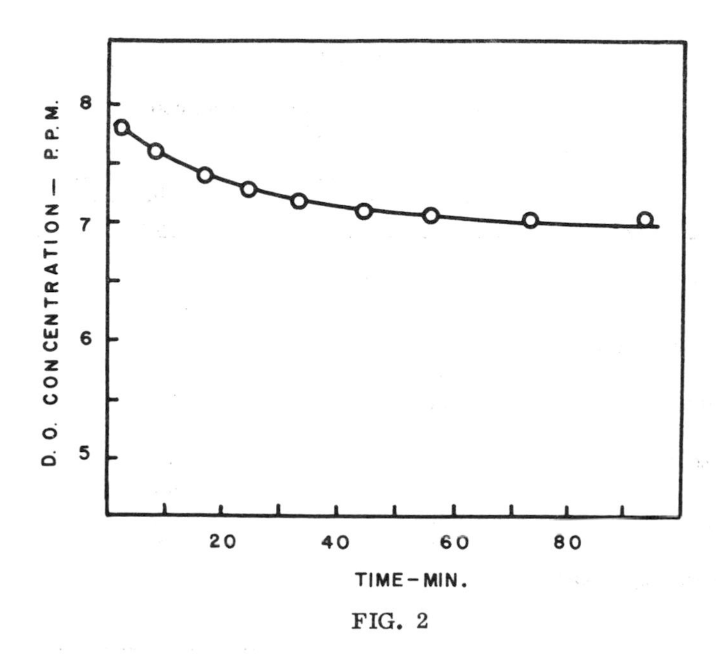

FIG. 2

Change in the dissolved oxygen concentration of the medium as the hydrocarbon to water ratio decreases by the votalization of heptane.

Figure 2. It is obvious that as the heptane content of the system decreases, its contribution to dissolved oxygen concentration also decreases. When, finally, all heptane is removed from the system, the dissolved oxygen concentration does not decrease further but levels off at a new value, which is the saturation value for the aqueous sytem.

A less important factor is the solubility of water in hydrocarbons (Table 3) (20). Contrary to early ideas, the amount of water is enough to sustain some microorganisms in various hydrocarbon products (21).

II. METHODS AND EQUIPMENT

A. Organism.

The organism used in these studies was isolated from a soil sample taken from a local fuel oil depot. It was identified as a strain of *Pseudomonas aeruginosa*.

TABLE 3

Solubility of Water in Paraffinic Hydrocarbons
(P = 760 mm Hg)

Hydrocarbon	Temperature °C.	Solubility mg/l
Hexane	20	101
	30	179
2, 3-Dimethylbutane	20	110
	30	192
Heptane	20	96
	30	172
2-Methylhexane	20	103
	30	182
Octane	20	95
	30	168
2, 4-Dimethylhexane	20	98
	30	180

B. Medium and Culture Conditions.

The medium used in enrichment cultures and for the isolation of the organism is shown in Table 4. n-Heptane (1%) was added as the carbon and energy source. To determine optimum values of pH, temperature, and concentration of nutrients, a sequence of factorial experiments was designed. The optimum pH value and temperature were found to be 6.9 and 30° C., respectively. A medium suitable for growth of dense cultures, as indicated by a study of depletion of various nutrients, was developed. The composition is given in Table 4.

C. Fermentor and Related Equipment.

Shake-flask studies were conducted using a "Psychrotherm" incubator-shaker equipped with temperature and speed control, manufactured by New Brunswick Scientific Company. Two fermentors, also made by New Brunswick Scientific Company, were used. One of these had a 14-liter capacity and a maximum agitation speed of 700 r.p.m.;

TABLE 4

Compositions of Media Used

Enrichment Medium		Developed Medium	
KH_2PO_4	= 0.5 g/l	KH_2PO_4	= 2.0 g/l
K_2HPO_4	= 1.0 g/l	Na_2HPO_4	= 0.4 g/l
$(NH_4)_2SO_4$	= 1.0 g/l	$(NH_4)_2HPO_4$	= 3.0 g/l
$MgSO_4$	= 0.2 g/l	$(NH_4)_2SO_4$	= 0.8 g/l
$MnCl_2 \cdot 4H_2O$	= 0.1 g/l	$MgCl_2$	= 0.3 g/l
$CaCl_2 \cdot 2H_2O$	= 0.01 g/l	$MnCl_2 \cdot 4H_2O$	= 0.03 g/l
$FeSO_4 \cdot 7H_2O$	= 0.005 g/l	$CaCl_2 \cdot 2H_2O$	= 0.01 g/l
		$FeSO_4 \cdot 7H_2O$	= 0.03 g/l
		pH value adjusted to 6.9	

the other had a five-liter capacity and a maximum speed of 1500 r.p.m. pH value was controlled by an on-off type dual mode controller and recorded on a Leeds and Northrup Speedomax recorder.

The oxygen analyzer was a Beckman Model 777 Laboratory Analyzer, equipped with a dual-speed Speedomax recorder. The oxygen probe was of the polarographic type and was sterilized with 1% β-propiolactone solution.

Because of its high volatility, heptane was condensed from the effluent air by refrigeration.

A diagram of the equipment set-up is shown in Figure 3.

D. Analytical Methods.

1. Culturing methods.

Two methods of culturing were used in the experiments: cells were grown in batch shake flasks without external pH control

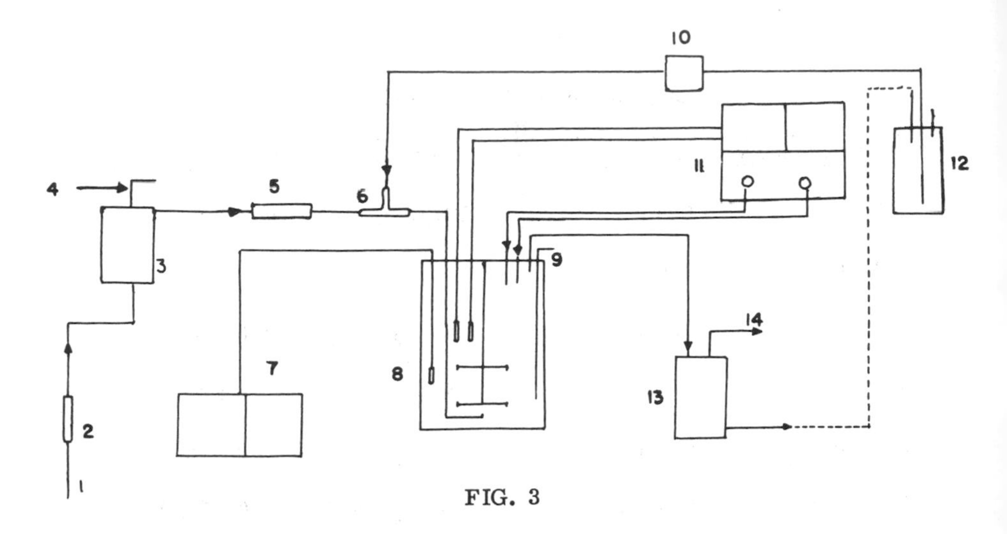

FIG. 3

Equipment diagram: 1. air inlet, 2. rotameter, 3. air humidifier, 4. water, 5. air filter, 6. heptane-air mixing, 7. oxygen analyzer and recorder, 8. fermentor, 9. sampling line, 10. metering pump, 11. pH controller and recorder, 12. heptane reservoir, 13. heptane recovery system, 14. heptane-free effluent gas.

and in fermentors with or without external pH control. In the fermentor studies, for the most part, a semi-batch technique was used. When the cells were in log phase, part of the culture was pumped out and fresh medium was pumped in to dilute the cultures to desired cell concentrations. When it appears, the expression "dilute culture" refers to a cell concentration of about 1 g/l of dry cells and "dense culture" refers to a cell concentration of about 6 g/l of dry cells.

2. Growth monitoring.

Microbial propagation was followed by dry cell weight per unit volume, viable cell counts, and turbidity. The latter was determined with a "Spectronic 20" Spectrophotomer (Bausch and Lomb).

3. Uptake studies.

An Industrial Instruments Conductivity Bridge and a Barnstead Purity Meter were used in the study of total mineral uptake.

The latter is equipped with a temperature compensator. The cell used for measurements had a constant equal to 10.7. A calibration line was plotted for KC1 solutions (mg/l) versus electrolytic conductivity. Samples were diluted twenty times and conductivity was measured. In the absence of a temperature compensator, a constant temperature bath was used, since electrolytic conductivity varies considerably with temperature. The microorganisms need not be removed by centrifugation because of the negligible contribution of cells to total conductivity.

4. Oxygen analysis.

Dissolved oxygen measurements were made with a Beckman Oxygen Analyzer. In cases in which the built-in calibration was not adequate, the Modified Winkler Method (22) was used.

5. Heptane degradation products.

The methods were reported previously (19).

III. RESULTS AND DISCUSSION

In this study, microbial degradation and utilization of n-heptane were studied in a controlled and reproducible environment in a batch formentor. Emphasis was on the process kinetics of the utilization under conditions that are nearly optimum rather than on the mechanisms.

In another report (19), it was indicated that heptanoic (or heptoic) acid was identified in the culture broth, in concentrations up to 60 mg/l only in the stationary phase. That the acid is a product of heptane's oxidation by the organism is suggested by the fact that the organism is adapted to rapid growth on heptanoic acid. During growth, the acid is further metabolized. When growth ceases, heptanoic acid accumulates in the medium.

In batch cultures, with heptane or glucose as the carbon source, the pH value decreases to about 4.6 (if it is not buffered) and growth ceases. With readjustment of pH, growth resumes. In studies with heptanoic acid as the sole carbon source, sodium heptanoate was used. The pH value of the culture was increased due to the sodium ions left in the medium when the heptanoate ion was taken up by the cells. By balancing heptanoate, alkali, phosphate, and ammonium ions, in the initial formulation, the pH stayed constant within 0.1 pH unit (around 7.0). This effect was used to advantage in various studies related to mineral uptake. Thus, the magnitude and direction of the pH changes were found to be a function of the electrolyte imbalance as well as of metabolic activity.

With heptane, hydrocarbon emulsions which form, interfere with optical density measurements of growth in young cultures. Similarly, erroneous results are obtained with very dilute cultures. It was reported (19) that electrolytic conductivity of the mineral medium decreases as the culture grows. This decrease parallels growth and can be related to the number of viable cells. When cells begin to die, some of the ions are released into the medium, and conductivity shows some increase.

Since the initial electrolytic conductivity of the medium may vary slightly depending upon the substrate used, autoclaving, and the inoculum added, calculations were made as the percent decrease in the total mineral content at the time of measurement (relative to its initial value). "Percent Mineral Uptake" at any time 't' is defined as:

$$\frac{\text{initial conductivity} - \text{conductivity at time t}}{\text{initial conductivity}} \times 100 \qquad (22\text{-}1)$$

If there is a decrease in conductivity, percent mineral uptake increases; and, conversely, if conductivity increases, percent mineral uptake decreases.

In Figure 4, for cultures grown on heptane at 20 °C., optical density and percent mineral uptake are plotted against time. It can be seen that cultures exhibit the usual lag and that they show no decline in optical density at the stationary phase. When the culture is studied by electrolytic conductivity measurements, however, no lag in mineral uptake is observed. There is, however, a decrease in percent uptake late in the culture. This decline is attributed to the release of ions by dying cells.

In Figure 5, variation in mineral uptake with temperature is shown for cultures grown on heptane at 20 and 25 °C. Rate of mineral uptake, as indicated by the slopes of the curves, increases with increase in temperature. The rate of mineral uptake was observed to parallel growth rate of the culture.

The curves for mineral uptake by cultures grown at 30° C. on glucose, heptane, and heptanoate are shown in Figure 6. In heptane and glucose cultures, the sudden rise in mineral uptake follows the decrease in pH. In contrast, however, in heptanoate cultures which have a constant pH, mineral uptake is much less and does not show the sudden increase. The sharp decrease in conductivity (indicating increase in mineral uptake) in glucose and heptane cultures is attributed to the uptake of alkali ions to maintain proper intracellular conditions, when the pH of the medium is continually decreasing.

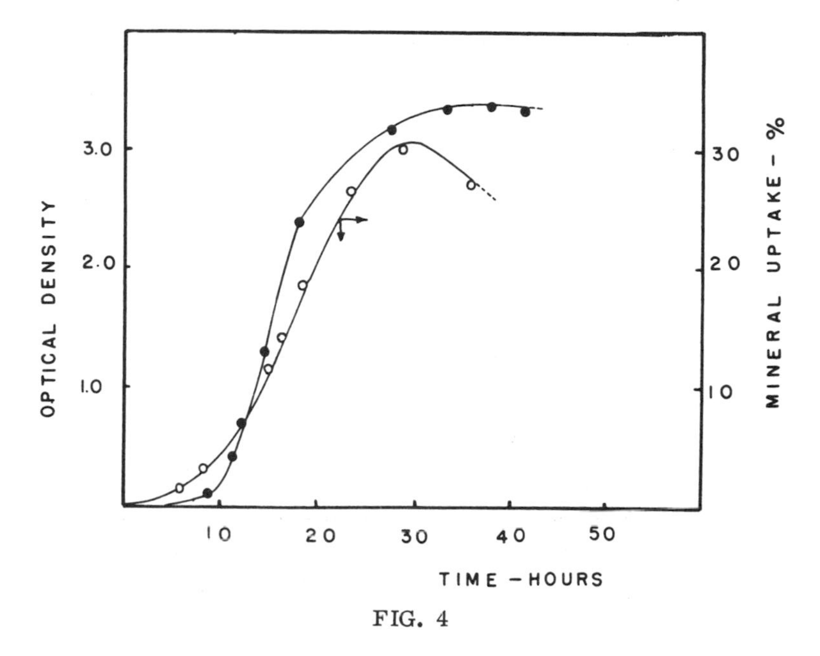

FIG. 4

Growth curves based on optical density and mineral uptake at 20°C., with heptane as the substrate. • optical density. o mineral uptake.

With heptanoate, uptake of minerals is mainly for metabolic purposes; the need for ions to balance pH changes does not exist, since the pH of the culture remains constant. This is evident from the flattened curve for hepanoate.

In the decline phase, dying cells in heptane and glucose cultures release large quantities of unbound ions; the conductivity increases markedly. Therefore, the percent mineral uptake (calculated from the conductivity measurements) decreases. In contrast, with heptanoate cultures, the release is much less, and the effect upon conductivity measurement is gradual.

When the pH value is adjusted by the addition of acid or alkali, the electrolytic conductivity varies considerably due to the addition of the electrolytes. Compensation for this external addition must be made if conductivity is to be used for monitoring growth. Such compensation, however, is somewhat complicated. One method, used in our laboratory, is to determine uptake rate at discrete intervals. If one

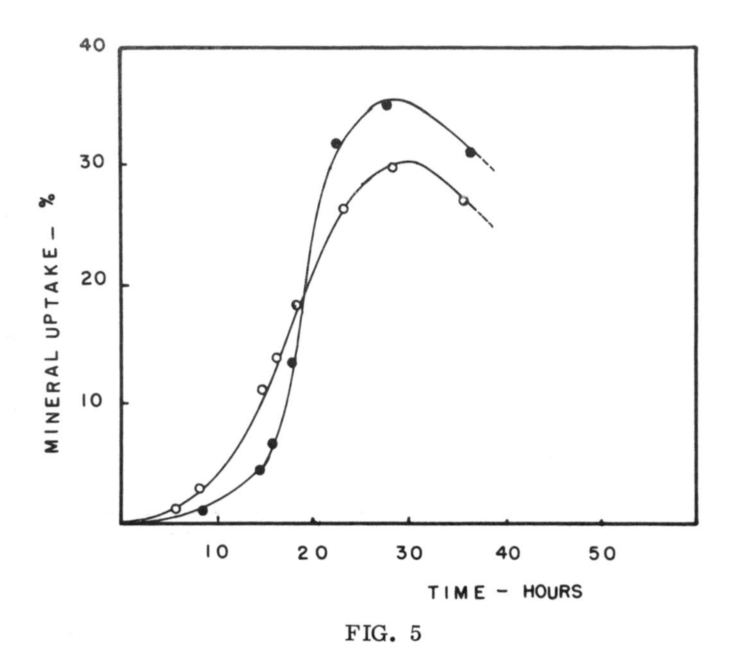

FIG. 5

Mineral uptake curves at 20°C. and 25°C. with heptane as the substrate. o growth at 20°C. ● growth at 25°C.

assumes that a change of 0.1 pH unit does not have a significant effect on cell metabolism, the pH controller can be turned off, and conductivity measurements taken at intervals. Cell concentrations may be measured simultaneously. When the pH change is about 0.1 pH unit, the controller is turned on to adjust the pH to its set value. Uptake rates were calculated from such data. When the procedure was repeated at intervals, the calculated rates gave a good indication of mineral uptake rates at constant pH values.

To operate the fermentors with sufficient aeration, the critical dissolved oxygen concentration was determined using the method outlined by Bennet (23). A modified form of the recorder tracing is shown in Figure 7. At point 1, aeration was stopped, but agitation was continued. It can be seen that the oxygen uptake was independent of the dissolved oxygen concentration until it reached a low value

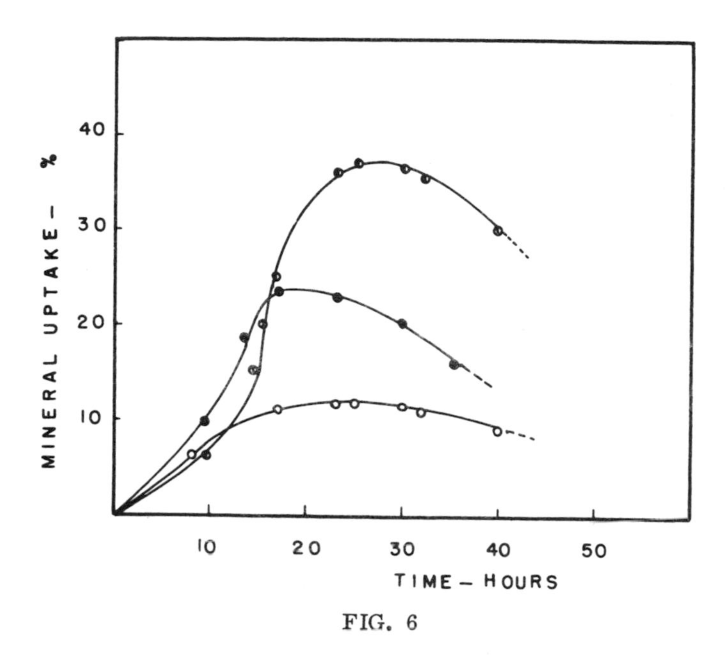

FIG. 6

Mineral uptake at 30°C. for different substrates. ◐ heptane. o heptanoate-(Na salt). • glucose.

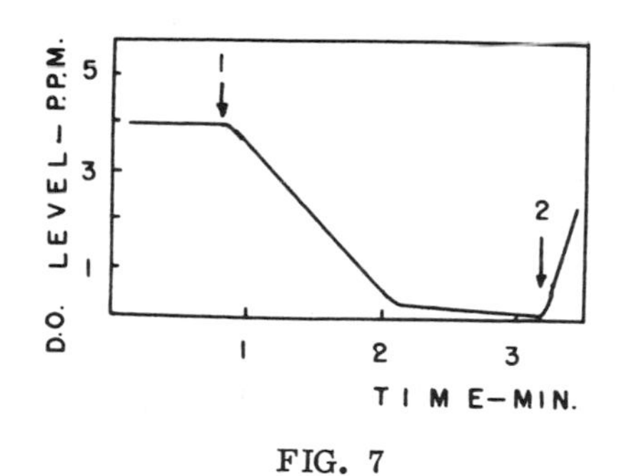

FIG. 7

Determination of critical dissolved oxygen concentration.

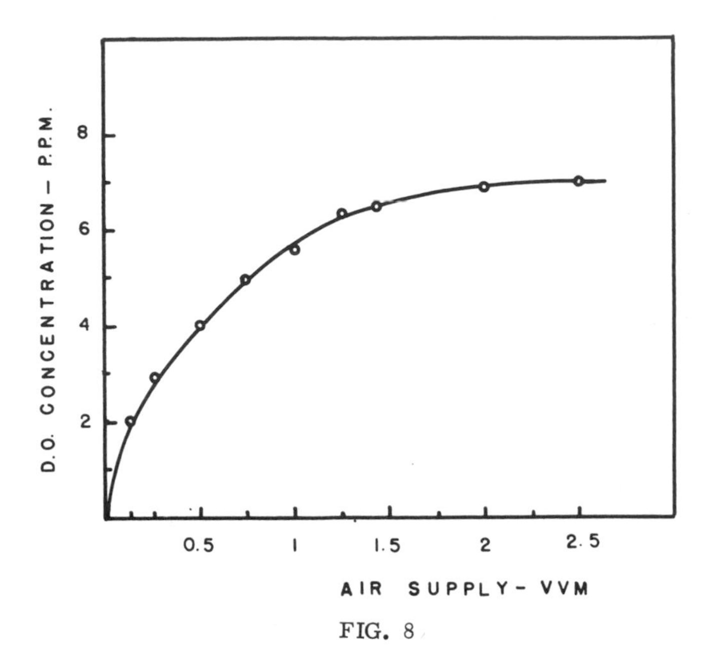

FIG. 8

Variation of dissolved oxygen concentration with changes in air supply (heptane to aqueous medium ratio kept constant).

NOTATION: VVM = volumes of air per unit working volume of fermentor per minute.

(about 0.3 mg/l). At this point it became dependent upon concentration. The value of the dissolved oxygen concentration at this inflection point is the critical concentration. Dissolved oxygen concentrations were always kept well above this value.

The effect of air supply, given as volumes of air per volume of fermentor per minute, on dissolved oxygen concentration is shown in Figure 8. The data for this figure were taken from a continuous culture study, the details of which will be reported elsewhere. The dilution rate was 0.2 hr^{-1}. The ratio of heptane to aqueous medium was kept constant by changing the heptane feed rate. The amount of heptane needed to compensate for losses, due to volatilization by higher flow rates of air, was calculated. Changes in cell concentration were not significant, and the steady state was reached within a short time after each disturbance in air flow rate. An air supply of 0.5 VVM at 30 °C. corresponds to approximately 4 ppm. Part of the data reported, including that in Figure 7, was taken under these conditions.

To study the kinetics of growth on various substrates under various conditions, batch fermentors were used. A semi-batch technique was used so that nearly equal cell concentrations were used in the various runs. A culture in the log phase was pumped out, and fresh medium was added until similar turbidities were achieved. As a basis for comparison, the specific growth rate was used. For unicellular growth, the rate of growth (24) can be expressed by:

$$\frac{dN}{dt} = f(N, S, ---) \qquad (22\text{-}2)$$

where N is cell concentration, S is substrate concentration, and t is time. The specific growth rate of the cells, μ, is:

$$\mu = \frac{1}{N}\frac{dN}{dt} \qquad (22\text{-}3)$$

The values of μ for the organism under different conditions and on various substrates are shown in Tables 5 and 6.

In dense cultures, growth on heptane exhibits some peculiarities. As reported previously (19), heptane was pumped into the air stream, volatilized, and fed into the culture. In this way, better dispersion and higher growth rates were obtained. The rate controlling step was found to be the creation of sufficient interfacial area for rapid mass transfer of heptane to the cells. It can be seen in Table 5 that the μ values increase significantly with temperature in dense cultures using heptane. Changes in μ for hepanoate and glucose are not so great. In dilute cultures, however, changes in μ values are of about the same order of magnitude for all three substrates. This can be explained on the basis of heptane supply which seems to be the controlling factor in growth. Since the solubility of heptane changes significantly with temperature, more heptane can be transferred in an equal time at the higher temperatures. A temperature change from 20°C. to 30°C. does not affect the growth rate of the organism being used significantly, as shown by glucose grown cultures. The dispersion of heptane is about the same at the three temperatures reported in Table 5. As previously mentioned, the amount supplied was adjusted to compensate for losses due to volatilization. Based on vapor pressure data for n-heptane (25), at one atmosphere at different temperatures, the amount of heptane stripped from the fermentor was calculated for various air flow rates.

TABLE 5

μ Values on Different Substrates (hr^{-1}).

Temperature °C.	Heptanoate	Heptane	Glucose
Dense Cultures:			
20	0.316	0.156	0.389
25	0.359	0.212	0.404
30	0.375	0.260	0.420
Dilute Cultures:			
20	0.349	0.294	0.421
25	0.392	0.327	0.446
30	0.429	0.362	0.473

TABLE 6

μ Values for Growth on Heptane (hr^{-1}).

Dense Cultures:	Temperature °C.		
Air Flow VVM	20	25	30
0.25	0.136	0.179	0.229
0.50	0.145	0.196	0.247
0.75	0.152	0.205	0.254
1.00	0.156	0.212	0.260

Since the mean temperature of effluent gases was 18°C., the assumption was made that they were saturated with heptane vapor at that temperature as they left the recovery unit.

Since the hydrocarbon solubility in aqueous systems increases with temperature (6) and a similar effect is seen on the mass transfer coefficient for heptane, the results might suggest that heptane is supplied to cells primarily through dissolution rather than contact with the droplets of dispersed heptane. As shown in Table 6, the μ values increase with increasing aeration. This is due not only to an increased oxygen supply but also to better dispersion of heptane and perhaps to a high rate of removal of volatile metabolites. Dissolved oxygen concentrations in these cultures were far above the critical concentration, and the agitation rate was high enough to insure good distribution. This observation also suggests the importance of heptane supply and its dispersion.

It has been observed that the effect of increasing aeration is more pronounced for heptane grown cultures than for those with either glucose or heptanoate. This effect cannot be explained solely on the basis of the amount of oxygen needed for optimum growth on the respective compounds. If that were the case, the effect of aeration on heptanoate cultures would be very high in comparison to that on glucose cultures, but the observed difference is not significant. A comparison of the effect on heptane and heptanoate, however, reveals a marked difference. Heptane has limited solubility, and the uniformity of its distribution in the culture is controlled by the degree of dispersion, whereas glucose and heptanoate are highly soluble. The above effects, therefore, may be attributed to the availability of the substrates to the cells rather than to structure.

IV. SUMMARY

Pure cultures of a pseudomonad were grown on n-heptane in batch and semi-batch fermentors. n-Heptanoate and glucose were used as substrates for comparative purposes. In batch cultures with no external pH control, the electrolytic conductivity was found to decrease with growth. Percent mineral uptake (percent decrease in conductivity relative to the initial conductivity) was found to correlate with population and the rate of mineral uptake to parallel the growth rate of the culture. The magnitude and the rate of mineral uptake were higher in cultures in which the pH changed than in those in which it was steady.

To study the kinetics of growth on various substrates, both dilute and dense cultures, having dry cell concentrations of about 1 g/l and 6 g/l, respectively, were used. The basis of comparison was specific growth rate, μ . Heptane was volatilized in the air stream and fed into the system in the vapor phase. Changes in μ with temperature and aeration rate were more significant for dense cultures growing on heptane than those on other substrates; with dilute cultures the changes in μ values were about the same order of magnitude for all three substrates. The rate controlling step was found to be the creation of sufficient interfacial area for rapid mass transfer of heptane to the cells and for the uniform distribution of heptane which is sparingly soluble.

ACKNOWLEDGMENT

The work upon which this publication is based was supported in part by funds provided by the United States Department of the Interior, Office of Water Resources Research through the R. I. Water Resources Center as authorized under the Water Resources Research Act of 1964 (P. L. 88-379).

REFERENCES

1. E. J. McKenna and R. E. Kallio, Ann. Rev. Microbiol., 19, 183 (1965).

2. J. C. Senez and M. Konovalschikoff, Compt. Rend. Acad. Sci., 24, 2873 (1956).

3. R. W. Hansen and R. E. Kallio, Science, 125, 1198 (1957).

4. J. W. Foster, in Oxygenases (O. Hayalshi, ed.) Academic Press, New York, 1962, p. 241.

5. J. Chouteau, E. Azoulay, and J. C. Senez, Nature, 194, 576 (1962).

6. A. N. Guseva and E. I. Parnov, Neftekhimiya, 5, 786 (1965).

7. D. D. Eley, Trans. Faraday Soc., 35, 1421 (1939).

8. F. Irman, Chem. Ingr. Tech., 37, 789 (1965).

9. C. McAuliffe, Science, 163, 478 (1969).

10. R. Durand, Compt. Rend., 226, 409 (1948).

11. J. W. McBain, J. Amer. Chem. Soc., 62, 2855 (1940).

12. F. Edeline, Trib. CEBEDEAU, 20 (279), 95 (1967). (in French) through Chem. Abstr., 67, 47012d (1968).

13. S. A. Zieminski, C. C. Goodwin, and R. L. Hill, Tappi, 43, 1029 (1960).

14. K. H. Mancy and D. A. Okun, J. Water Pollut. Contr. Fed., 37 (2) 212 (1965).

15. K. Novak, Chem. Prumsyl, 12, 658 (1967).

16. A. B. McKeown and R. R. Hibbard, Anal. Chem., 28, 1490 (1956).

17. J. A. Petrocelli and D. H. Lichtenfels, Anal. Chem., 31, 2017 (1959).

18. H. Mann and E. Stehr, Arb. Deut. Fisherei-Verbandes, No. 11, 7 (1959).

19. M. A. Caglar, A. R. Thompson, C. W. Houston, and V. C. Rose, Biotechnol. Bioeng., XI, 417 (1969).

20. B. A. Englin, Khim. I. Technol. Topliv. I. Masel, 10 (9), 42 (1965), (in Russian); through Chem. Abstr., 63, 146089 (1965).

21. P. Edmonds and J. J. Cooney, Appl. Microbiol., 15, 411 (1967).

22. A.P.H.A., Standard Methods for the Examination of Water and Waste Water, 12th Edition, A.P.H.A. Publication Office, New York, 1965.

23. G. F. Bennet and L. L. Kempe, Biotechnol. Bioeng., VI, 347 (1964).

24. S. Aiba, A. E. Humphrey, and N. F. Mills, Biochemical Engineering, Academic Press, New York, 1965, p. 100.

25. J. H. Perry (Editor), Chemical Engineers Handbook, 4th Edition, McGraw-Hill Co., New York, 1963.

CHAPTER 23

KINETICS OF BIOLOGICALLY MEDIATED AEROBIC OXIDATION OF ORGANIC COMPOUNDS IN RECEIVING WATERS AND IN WASTE TREATMENT

Elisabeth Stumm-Zollinger
and
Robert H. Harris

Harvard University

I. INTRODUCTION

The general understanding of microbial activities has made rapid advances in the past 15 years. In part, this has resulted from the important role that microbiology has played in the field of molecular biology, the study of life on the molecular level. Molecular biology has attracted talented researchers who introduced physical and chemical concepts to the study of microorganisms and developed a large variety of new ideas and technical means (1). In contrast, the understanding of specific microbial activities and chemical transformations in natural waters has made little progress despite an increasingly pressing concern for the quality of water resources. In an excellent review titled "Current Concepts of Aquatic Microbiology" Jannasch (1) discussed the difficulties and technical limitations inherent to the study of microbial activities in natural waters and elaborated on a few novel and promising developments. Because of their pertinence, some of the viewpoints developed by Jannasch will be briefly presented as they seem relevant to the discussion.

Most natural waters provide a nutritionally complex habitat for a multiplicity of metabolic types of microorganisms of large physiological and biological diversity. The cellular functions of each organism are susceptible to stimulation and retardation by various mechanisms. Futhermore, rather than being in steady state, a natural water represents a habitat usually in a transient state of continual change to which the organisms quickly respond physiologically and in number. A bewildering array of physiological phenomena results. In addition, studies in aquatic microbiology are frequently handicapped by methodological difficulties. Historic ties with medical microbiology have to be cut and an entirely new methodology has to be developed. Extremely low concentrations of energy sources, essential nutrients, and microorganisms in natural waters present problems since it is often not permissible to extrapolate from experimental data collected at higher concentrations (2).

A serious dilemma arises from the fact that, on the one hand, the scientist concerned with the quality of water resources is primarily interested in the overall elimination rate of substrates by microorganisms. Individual elimination patterns and metabolic mechanisms that underlie these phenomena are of concern only in so far as they are of predictive value. On the other hand, it is recognized that the fate of organic compounds in natural waters cannot be studied in a gross manner. It is impractical to treat as one group the large variety of unrelated microorganisms with their immense metabolic diversity (1). However, experiments with heterogeneous communities easily lead to erroneous conclusions (3) and to confusion concerning their interpretation (see examples given by Gaudy (4)); collective parameters are often misleading (5). Recent descriptions of new metabolic types of organisms, some of which are extremely abundant in natural waters (6), illustrate the importance of detailed studies despite the fact that these studies may not readily yield answers to problems of water quality. Although it is important to measure the kinetics of dissimilative activities of individual bacterial groups, it is also necessary to assign to these groups a position of relevance in the natural system. Insight into ecological interactions cannot be gained solely by detailed studies in well-defined systems. It often requires imaginativeness and insight to find a balance between definedness and undefinedness of the experimental system (1).

Since this presentation obviously cannot be a comprehensive review, emphasis is placed on recent developments. This chapter is organized along the following lines:

(a) The significance of the continuous growth system as a tool in aquatic microbiology is demonstrated by illustrating its usefulness in various study areas. A short discussion follows of a biological method suitable for assay of organic compounds at the relatively small concentrations typically encountered in natural waters.

(b) The kinetics of microbially mediated aerobic oxidation of organic compounds is initially considered on the cellular level.

Growth characteristics of organisms isolated from natural waters and of laboratory stock cultures are compared and discussed in relation to the typically low substrate concentrations in natural waters. Evidence will be presented that deviations from Monod's growth-rate substrate-concentration relationship may be encountered at very low microbial population densities. A discussion of mutual substrate interference follows where it is shown that sequential substrate dissimilation is observed in complex media at low substrate concentrations.

(c) The discussion is then extended to systems containing heterogeneous communities of organisms, where it is shown that mixed microbial populations often behave like pure cultures. Experimental difficulties with mixed microbial cultures are illustrated by considering substrate elimination from multisubstrate media by heterogeneous microbial communities. In conclusion, it is demonstrated that water quality and structure of the microbial community are important determinants of the kinetics of microbial dissimilatory activities.

Natural waters and waste treatment systems do not receive separate treatment in this chapter since they may be assumed to represent complementary portions of an entire study area (7) despite the ecological differences. Although many findings reported here have been observed on marine organisms, there is no reason, in principle, to distinguish between fresh water and marine microbiology.

II. ADVANCES IN METHODOLOGY

A. The Chemostat.

The chemostat was introduced to quantitative biological science by Monod (8) and by Novick and Szilard (9). Their theoretical treatment was expanded and experimentally verified by Herbert et al. (10) in 1956. The mathematical relationships between growth, metabolic activity, and the concentrations of organisms and limiting substrate in a chemostat have frequently been applied to studies of water quality and are so well known (10) that it is not necessary to derive them here. The principal characteristics of a simple continuous growth system are summarized in Figures 1 and 2 and in Table 1. Assuming a completely mixed system, constant hydraulic flow rate, and constant influent substrate properties, the biological system in a chemostat will pass through a transient state in which the growth rate will adjust to the concentration of the limiting substrate. Eventually the system will reach a steady state of constant bacterial concentration, constant suboptimal exponential growth rate, and constant concentration of the limiting nutrient. The growth rate and the concentration of the limiting

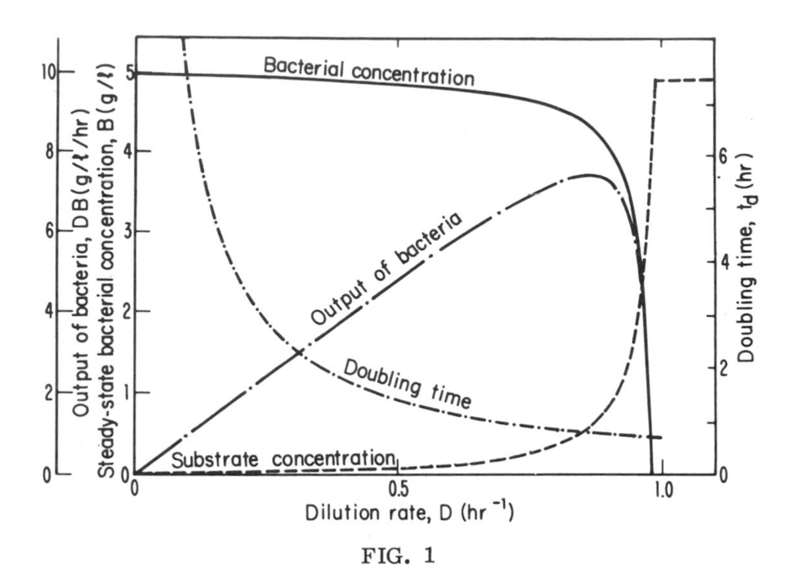

FIG. 1

Steady-state relationships in a continuous culture (theoretical). The steady-state values of substrate concentration (S), bacterial concentration (B), and output of bacteria (DB) at different dilution rates are calculated from Equations (23-6, 23-7) for an organism with the following growth constants: $\mu_m = 1.0$ hr^{-1}; y = 0.5; K_S = 0.2g/l; and a substrate concentration (S_R) in the inflowing medium of 10 g/l. The symbols are defined in Table 1. Reprinted from Ref. (10), p. 608, by courtesy of Cambridge University Press.

nutrient are determined exclusively by the experimentally chosen dilution rate (reciprocal of detention time). The concentration of the organisms depends on the concentration of the limiting nutrient in the feed solution as well as on the dilution rate. The bacterial concentration is maximal when growth rate and steady state nutrient concentration are minimal. While the concentration of the growth limiting nutrient is independent of its concentration in the feed solution, this is not true for the other nutrients whether they are measured individually or collectively. For example, Eckhoff and Jenkins (11) observed an almost linear relationship between influent COD (mg/1) and effluent COD (mg/l) in a system containing a large variety of organic nutrients other than the limiting substrate.

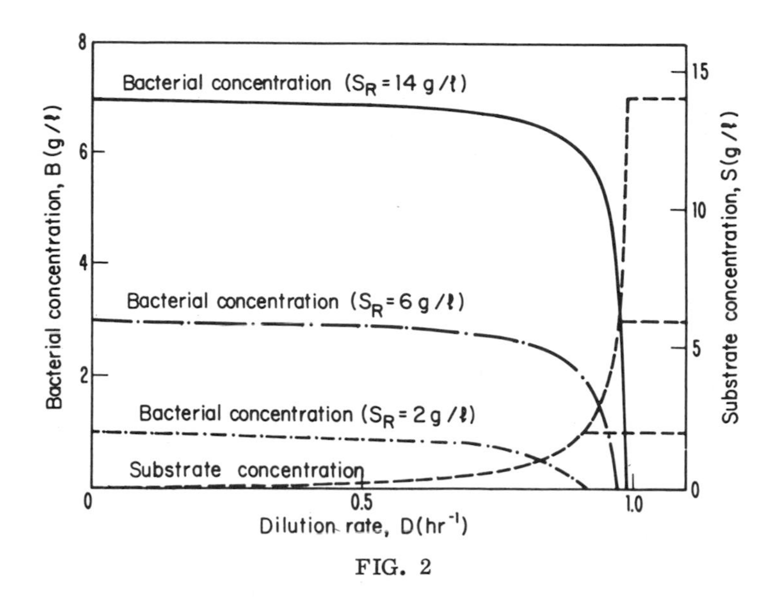

FIG. 2

Effect of varying the concentration of substrate in the inflowing medium (S_R) on the steady-state relationships in a continuous culture (theoretical). The values are calculated from Equations (23-6, 23-8) at three different influent substrate concentrations for an organism with the following growth constants: μ_m = 1.0 hr^{-1}; y = 0.5; and K_S = 0.1 g/l. The symbols are defined in Table 1. Reprinted from Ref. (10), p. 608 by courtesy of Cambridge University Press.

The continuous culture apparatus was initially employed in kinetic studies as a means of creating a constant environment so that microorganisms would remain in a time invarient state of constant exponential growth during the period of experimental observation. With an increasing awareness of the potential for its use as an experimental tool, the chemostat is now applied to a wide range of research problems in aquatic microbiology.

1. Model of a natural aquatic habitat.

The chemostat has certain characteristics in common with a natural aquatic habitat. Both represent systems open to the flow of matter and energy. Both provide the organism with an environment of

TABLE 1

Systemic Equations on Kinetics of Bacterial Growth and Substrate Utilization in Batch and Continuous Culture Systems

Growth Rate:	$\frac{dB}{dtB} = \mu$	(23-1)
Yield:	$y = -\frac{\Delta B}{\Delta S} \approx -\frac{dB}{dS}$	(23-2)
Substrate Utilization:	$-\frac{dS}{dt} = \frac{B}{y} \mu_m \frac{S}{K_S + S}$	(23-3)
Relation between μ and S:	$\mu = \mu_m \frac{S}{K_S + S}$	(23-4)
<u>Continuous Cultures</u>:		
Growth Rate:	$\mu B - DB = 0; \quad \mu = D$	(23-5)
Substrate Concentration:	$S = K_S \left(\frac{D}{\mu_m - D} \right)$	(23-6)
Bacterial Concentration:	$B = y \left(S_R - S \right)$	(23-7)
	$B = y \left(S_R - K_S \frac{D}{\mu_m - D} \right)$	(23-8)

B	Concentration of bacteria (g/l)
μ	specific growth constant (hr^{-1})
S	substrate concentration (g/l)
S_R	substrate concentration in feed solution (g/l)
K_S	saturation constant; numerically equal to the substrate concentration (S) at which $\mu = 1/2\ \mu_m$ (g/l)
μ_m	maximum value of μ at saturation levels of substrate (hr^{-1})
y	yield constant
D	dilution rate (hr^{-1})

competition for a limiting substrate (energy source or essential nutrient). Both operate at extremely low stationary concentrations of nutrient and function at similarly low concentrations of organisms (12).

In other characteristics, the chemostat differs from the natural aquatic habitat and from waste water systems. The chemostat is strongly selective (9, 13) for the species attaining the highest growth rate under the given conditions, this selectivity leading to the eventual displacement of all other species originally present. At steady state the population in a chemostat will consist of a homogeneous group of organisms in which all the individuals grow at identical growth rates in a chemically homogeneous environment.

In contrast to the chemostat, the natural aquatic habitat can sustain a diversified community of organisms. Such habitats are partitioned into a large number of ecological niches and can be visualized as consisting of as many individual open systems as there are species present, where the population of each is controlled by its individual limiting factor (1). The large majority of microorganisms do not float freely but settle on the surface of mineral particles and organic debris (14, 15) or on living organisms which frequently excrete organic nutrients (16, 17). Furthermore, the biological properties of a natural habitat oscillate, frequently changing from one steady state to another with periodical changes in the environment.

We may conclude from the above, that the chemostat does not simulate a natural aquatic habitat, although it does provide a valuable tool for simulating individual environmental factors while simultaneously establishing the conditions of steady-state growth desirable for kinetic studies. For example, the chemostat is of value in the study of bacterial growth in dilute systems where concentration conditions are similar to those found in natural waters (18). The results of growth measurements in similarly dilute batch cultures may be misleading. It has been shown that, in a batch system of low nutrient concentration, autolysis may give rise to growth which can not be attributed to the substrate under study (19). Furthermore, it is often not permissible to extrapolate from experimental growth data which were observed in batch systems of higher concentrations of nutrients and organisms (20, 21).

2. Enrichment.

The majority of the information on bacterial growth is based on experimental work with enteric bacteria. Little is known regarding the physiology and nutrition of microorganisms indigenous to natural waters and to waste water treatment systems. To some extent this has resulted from the nature of conventional enrichment techniques where growth conditions preferentially meet the growth requirements of one type of organism present in the mixed community, and where, as a routine, nutrients are offered at high concentrations (several g/l). These procedures lead to the preferential isolation of organisms of high nutrient specificity which are at a selective advantage at high nutrient concentrations.

The chemostat enrichment technique was developed for the separation of organisms which exhibit similar growth characteristics and nutrient requirements and which demonstrate a selective advantage at low nutrient concentrations. Although this technique was used in 1950 by Novick and Szilard in studies of mutation rates (9), it was Jannasch who first recognized and demonstrated the value of the chemostat for the isolation of the nutritionally inconspicuous heterotrophic organisms which are primarily responsible for the degradation of the large variety of organic compounds in natural waters (13, 22).

By the nature of a chemostat operating at a constant flow of medium with the undiscriminating removal of cells of all species, the species which grows fastest under the specific conditions will become predominant. This can be illustrated with results obtained by Jannasch (13) who separated two species (*Achromobacter* sp. and *Aerobacter* sp.; both isolates from seawater) by operating chemostats at two different dilution rates. The growth characteristics of the two isolates and the relationships of substrate concentration and growth rate (dilution rate) are given in Figure 3. At a dilution rate of 0.4 hr^{-1} *Achromobacter* outgrew *Aerobacter*, while *Aerobacter* became predominant at a dilution rate of 0.6 hr^{-1}, when the maximum growth rate of *Achromobacter* was exceeded.

Similar experiments were conducted with sea water inocula. The chemostat units were fed with fresh sea water supplemented with nutrients. The systems were operated at various dilution rates for periods corresponding to 5 to 10 retention times. The organism predominating at each dilution rate was then isolated. The results, presented in Table 2, show the efficiency of the chemostat enrichment technique for the separation and isolation of organisms from natural waters. The growth characteristics of the microbial strains (Table 3) can be used in predicting which strain will predominate in a mixed system under a given set of conditions. They also indicate the rapidity of separation to be expected. The shift in the relative frequency of *Methanobacterium soehngenii* and *Sarcina methanica* in the methane fermentation cultures as described by Lawrence and McCarty (23) may be cited as another

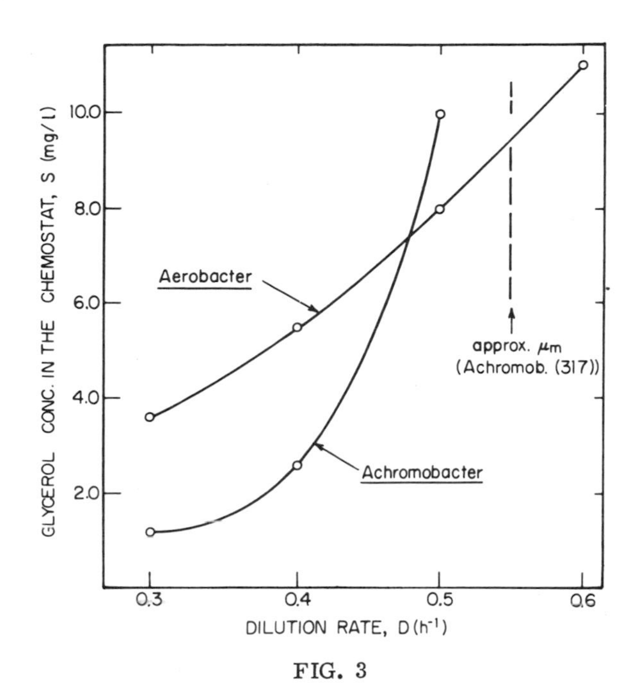

FIG. 3

Plot of substrate concentration (S) vs. dilution rate (D) in two glycerol limited continuous cultures. (Aerobacter (412); Achromobacter (317)). The values of S are calculated from Equation (23-6) using the approximate growth constants of the two organisms. (Aerobacter: K_S = 11.0mg glycerol/l; μ_m = 1.2 hr^{-1}. Achromobacter: K_S = 1.0mg glycerol/l; μ_m = 0.55 hr^{-1}). Reprinted from Ref. (13), p. 171, by courtesy of H. W. Jannasch, Woods Hole, Mass., and Springer-Verlag.

example of an enrichment process effected by differential dilution rates in a continuous growth system.

During his work with aerobic isolates from natural waters, Jannasch observed a positive correlation between μ_m values and K_S values (24). Organisms that could grow at high growth rates (high μ_m) demonstrated poorer growth at low substrate concentrations (high K_S) than organisms with low maximal growth rates (μ_m). This correlation may explain the observation that isolates from agar plates (where high growth rate at

TABLE 2

Continuous Culture Enrichments of Various Bacteria From Inshore Seawater in 16 Experiments That Varied the Dilution Rate and the Concentration of the Limiting Substrate in the Reservoir[a,b]

Conc. of lactate in the reservoir mg/l	Dilution rate, hr^{-1}			
	0.05	0.10	0.25	0.50
100.0	A, C	Pseudomonas, C	E. coli, C	E. coli, C
10.0	A, C	Pseudomonas, C	Aerobacter	Aerobacter
1.0	Achromobacter	Vibrio	Pseudomonas, B	Aerobacter
0.1	A	Achromobacter	A, B	A, B

[a]Reprinted from Ref. (32), p. 1617, by courtesy of American Society for Microbiology.

[b]A, no appreciable enrichment within eight retention times; B, heavy wall growth; C, visible turbidity.

TABLE 3

Growth Parameters as Obtained from Continuous or Batch Culture in Filter-sterilized Offshore Seawater at 20°C. for Two Limiting Substrates [a]

Culture	Lactate		Glucose	
	K_S mg/l	μ_m hr^{-1}	K_S mg/l	μ_m hr^{-1}
Vibrio spp. (strain 204)	0.8	0.15	5.5	0.40
Achromobacter spp. (strain 208)	1.0	0.15	3.0	0.35
Spirillum spp. (strain 101)	3.0	0.45		
Aerobacter spp. (strain 417)	6.0	0.50	5.0	0.45
E. coli (strain 415)	9.0	0.80	8.0	0.65
Serratia marinorubra	15.0	1.10		

[a]Reprinted from Ref. (32), p. 1617, by courtesy of American Society for Microbiology.

high substrate concentration is of selective value) often show no metabolic activity at a low substrate concentration usually found in natural waters. In addition, since isolates with low maximal growth rates (μ_m) but relatively high growth rates at low substrate concentrations (low K_S) are easily outgrown on common enrichment media, the μ_m - K_S correlation observed by Jannasch may explain the difficulty in culturing naturally-occurring organisms in the laboratory.

3. Non-steady state growth.

An analysis of growth and metabolic activity of microorganisms in response to shifts in environmental conditions is of importance in studies of natural waters and of waste treatment systems. Using the chemostat technique, Krishnan and Gaudy (25) demonstrated that the excretion of metabolites or end products during the transition to a

TABLE 4

Changes in the Concentration of Cellular Nitrogen (N_e) and Medium Nitrogen (N) in a Continuous Culture After the Microorganism's Growth Rate Suddenly Falls to Zero [a]

Time min.	N mg/l	N_e mg/l
0	1	99
1	2.6	97.4
2	4.2	95.8
3	5.8	94.2

Calculations made for a continuous culture with: $D = 1.0\ hr^{-1}$; $N_o = 100$ mg/l.

[a]Reprinted from Ref. (26), p. 263, by courtesy of Nature, Little Essex St., London.

higher influent substrate concentration can deviate considerably from the steady state excretion rates before and after the shift. Mateles et al. (26) have proposed an interesting application of the chemostat to studies of transient growth rates. In an attempt to understand the growth response of organisms to a sudden change in dilution rate, a continuous growth culture was operated with an ammonium salt as the sole nitrogen source and limiting nutrient. Rather than variations in the bacterial concentration, changes in ammonium salt concentration were measured during the shift from one steady state to another. Small relative changes in bacterial concentration were reflected by large, easily observable relative changes in ammonium salt concentration (Table 4). Using this technique, Mateles et al. demonstrated that there may be a significant lag in the adjustment of growth rate to fluctuations in the dilution rate (Fig. 4); the transient growth behavior of microorganisms may thus significantly depart from Monod's model.

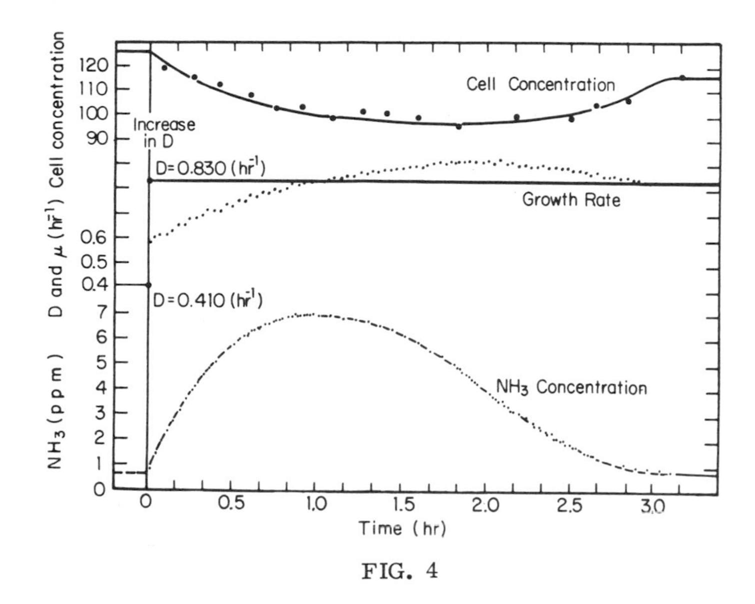

FIG. 4

Response of a continuous growth system to a sudden increase in the steady-state dilution rate (at time zero). Ammonium nitrogen was the limiting nutrient. Reprinted from Ref. (26), p. 264, by courtesy of Nature, Little Essex St., London.

4. Mixed cultures in a chemostat.

As previously discussed, a heterogeneous system is subject to strong selective forces in a continuous growth culture. Therefore, a chemostat is in principle not well suited to maintaining mixed bacterial communities in steady state. However, two or more types of organisms have been shown to coexist at steady state in a chemostat as a result of a physiological interdependence (23, 27, 28, 29). For example, an essential niacin-like growth substance elaborated by the yeast *Saccharomyces cerevisia* and required by the bacterium *Proteus vulgaris* led to a dependence of the bacterium on the yeast and a coexistence of the two organisms in a continuous culture (28). A detailed analysis of the coexistence of *Bacillus polymyxa* and *Proteus vulgaris* revealed a more complex interaction (27). It was observed that a defined medium deficient in both niacin and biotin supported neither *Proteus vulgaris* nor *Bacillus polymyxa* in pure culture. A

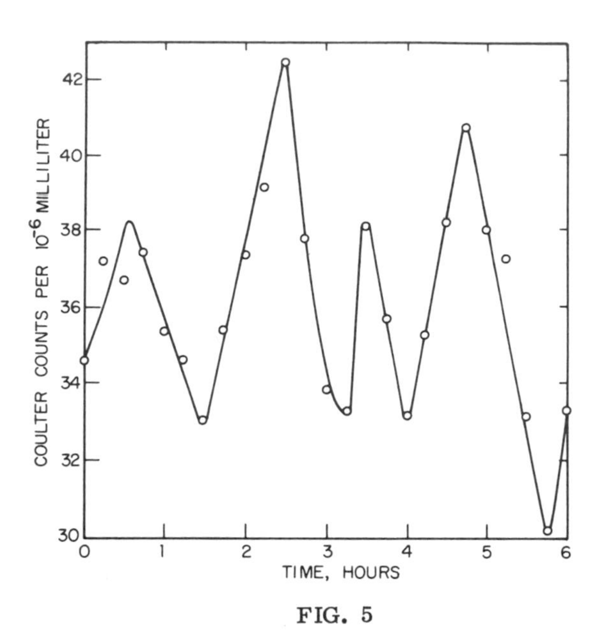

FIG. 5

Total counts of organisms (P. vulgaris + B. polymyxa) vs. time in a chemostat operating at a steady-state dilution rate of 0.25 hr^{-1} (37°C.). There existed mutualism on a minimal defined medium where each strain required a substance elaborated by the other. Accumulation of an inhibitor led to growth inhibition until continued pumping diluted the inhibitor. Reprinted from Ref. (27), p. 491, by courtesy of the National Research Council of Canada.

mixed culture grew well, since each species supplied the vitamin required by the other. A continuous growth culture of the two organisms exhibited oscillatory population levels (Fig. 5). P. vulgaris was shown to produce an inhibitor which accumulated to inhibit growth of B. polymyxa. This caused, eventually, a decrease in growth of P. vulgaris due to the vitamin interdependency. After dilution of the inhibitor by continued pumping, growth of both species resumed.

The community of microorganisms involved in methane fermentation is an excellent example of a system of several types of organisms coexisting through interdependency (23). The transformation of fatty acids to methane and carbon dioxide is a multistep process requiring several species of methanogenic bacteria. Most longer chain fatty acids are converted by a given bacterial species to methane, CO_2, and a fatty acid of lower molecular weight which is fermented further in a similar manner by a different bacterial species.

B. Uptake Kinetics and Bioassay.

The study of the fate of individual dissolved organic nutrients in natural waters is handicapped by low nutrient concentrations. Conventional bioassay methods cannot be applied since the individual organic compounds almost always occur far below the concentration necessary to support growth sufficient to be measured as a turbidity increase. Hobbie and Wright (30) have developed a suitable bioassay method which utilizes the kinetics of substrate uptake as the response reaction. The bioassay organism used by these authors was a motile, rod-shaped bacterium isolated from Lake Erken, Sweden. Small amounts of labelled and unlabelled glucose were added to natural waters and the uptake of labelled glucose by the test organism was followed for 2-3 hours. The concentration of glucose present in the original water sample was then calculated from:

$$\frac{C\alpha t}{c} = \frac{K_T + S}{V} + \frac{A}{V} \qquad (23\text{-}9)$$

where C is counts per min. of one microcurie of ^{14}C in the counting assembly; α, number of microcuries added per sample; t, incubation time (hr); c, radioactivity (counts per min.) of the organisms retained on a filter; K_T, uptake constant; S, concentration (mg/l) of glucose originally present in the natural water; A, concentration (mg/l) of added glucose (labelled and unlabelled); and V, maximum uptake velocity. Concentrations as small as 0.01 mg/l glucose (30) and 0.001 mg/l amino acid (31) were so determined in natural waters (Table 5). Amino acids, succinate, and citrate as well as a large number of carbohydrates did not interfere with the glucose assay. Hobbie and Wright (30) and Jannasch (1) suggested that glucose concentrations obtained by this method be considered as probable rather than absolute concentrations, since the predetermined uptake constants (K_T, V) of the test organism may vary depending on the biological properties of the unknown sample.

TABLE 5

Application of Bioassay to Synthetic Solutions and to Natural Waters [a]

Water	Date	Notes	Glucose mg/l Recovered	Standard Deviation mg/l
Inorganic medium		0.010 mg/l glucose added	0.011	0.003
Inorganic medium		0.010 mg/l glucose added	0.013	0.003
Inorganic medium		0.025 mg/l glucose added	0.025	0.006
Inorganic medium		0.025 mg/l glucose added	0.026	0.004
Inorganic medium		0.100 mg/l glucose added	0.100	0.008
Lake Erken	30 July 1964	Naturally eutrophic lake east of Uppsala, 1 m.	0.010	0.002
Lake Erken	30 July 1964	19 m sample	0.020	0.002
Pond in Botanical Garden	23 July 1964	Small eutrophic pond in Uppsala	0.011	0.005
River Fyris	6 Aug. 1964	Mildly polluted river in Uppsala, above outfall from sewage treatment plant	0.024	0.003
River Fyris	6 Aug. 1964	Immediately below outfall	0.059	0.007
Lake Ekoln	23 July 1964	Mildly polluted lake south of Uppsala, surface sample	0.009	0.002

[a] Reprinted from Ref. (30), p. 473 by courtesy of Limnology and Oceanography.

III. KINETICS OF GROWTH AND METABOLIC ACTIVITY AT THE CELLULAR LEVEL

A. Direct Determination of Metabolic Rates in Natural Waters.

There is little information available on in-situ rates of microbial transformations of dissolved organic materials in natural waters. This lack of information is due to the fact that performing direct measurements of such rates in the field is impeded by the minute concentrations of organic substances and of microorganisms; by the complexity of natural waters as the medium for analytic investigations; and by the difficulty in avoiding physiological and structural shifts in the natural microbial community as the result of the experimentally imposed environmental disturbances.

In an attempt to assess natural metabolic reaction rates, Jannasch (32) fed nonsupplemented filter-sterilized sea waters from various areas to chemostats which were inoculated with raw sea water or with pure culture marine isolates. In all cases, even at very low dilution rates, bacterial numbers decreased with time, nearing complete washout. However, in experiments with inshore waters, the rate of decrease of bacterial numbers in the chemostat was distinctly slower than the dilution rate, indicating that growth did occur. From this difference between the experimentally imposed dilution rate and the bacterial washout rate, Jannasch (33) was able to calculate specific growth rates for various marine organisms in sea water. The values obtained ranged from 0.05 hr^{-1} to 0.005 hr^{-1}.

B. Indirect Determination of Metabolic Rates in Natural Waters.

Since there is a scarcity of information that was obtained by direct measurements of microbial activities in natural waters, metabolic rates in the aquatic habitat must be estimated indirectly. For example, if the population density and nutrient concentration can be measured, Equation (23-3) (Table 1) can be used to assess microbial metabolic rates providing it is possible to determine three constants: the yield coefficient (y), the maximal specific growth rate (μ_m), and the saturation constant (K_S). In addition, the aquatic microbiologist must verify that metabolic relationships experimentally obtained under laboratory conditions also apply to the natural aquatic habitat.

1. Growth constants.

Jannasch (34) performed growth studies in a continuous culture on natural water isolates growing in supplemented natural water.

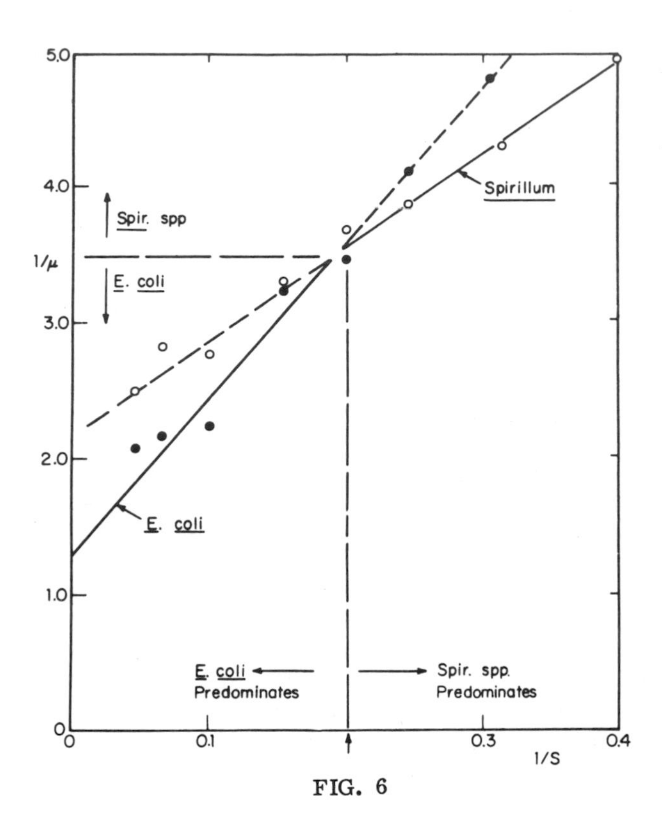

FIG. 6

Reciprocal plot of concentration of limiting substrate (S, mg lactate/l) vs. growth rate (μ, hr^{-1}), demonstrating the ranges of successful competition between Spirillum spp. strain 101 (O) and E. coli, strain 415 (●). Reprinted from Ref. (32), p. 1617, by courtesy of American Society of Microbiology.

Examples of the growth constants determined for these isolates are presented in Table 6. A comparison of the constants of typical laboratory stock cultures with those of isolates from natural waters show that the latter are generally better adapted to low nutrient concentrations. Therefore, natural water isolates would be expected to be successful competitors in the natural aquatic habitat with the organisms more typically studied in the laboratory. For example, Figure 6 shows that Spirillum spp. would be able to compete successfully with E. coli

TABLE 6

Comparison of Growth Parameters of Various Types of Microorganisms

Organisms	Substrate	μ_m hr^{-1}	K_S mg/l	Ref.
Unidentified, belonging to	glucose	0.5	<1	(46)
coliform group, laboratory	galactose	--	<1	..
stock culture	benzoate	--	approx. 20	..
E. coli, laboratory	glucose	--	4	(20)
stock culture	mannitol	--	2	..
	lactose	--	20	..
Aerobacter cloakae, laboratory stock culture	glycerol	0.85	12.3	(10)
Serratia marinorubra, laboratory stock culture	lactate	1.1	15.0	(32)
Aerobacter spp.[a]	glucose	0.45	5.0	(32)
	lactate	0.5	6.0	..
E. coli [a]	glucose	0.65	8.0	(32)
	lactate	0.15	9.0	..
Vibrio spp., seawater	glucose	0.4	5.5	(32)
isolate	lactate	0.15	0.8	..
Achromobacter spp.,	glucose	0.35	3.0	(32)
seawater isolate	lactate	0.15	1.0	..
Spirillum spp. seawater isolate	lactate	0.45	3.0	(32)
Unidentified bacterium isolated from fresh water	glucose 8°C.	--	0.005[b]	(30)

[a]Isolated from seawater but thought to be a contaminant.
[b]Uptake rate constant.

TABLE 7

Growth Constants of Several Marine Isolates for Two Limiting Substances, Measured at 20°C. before (A) and after (B) Growth on 10^{-2}M of the Corresponding Substrate for at Least Five Transfers [a]

Isolate	Constant[b,c]	A Lactate	A Glucose	B Lactate	B Glucose
Achromobacter sp. (strain 208)	K_S	1.0	3.0	5.0	12.0
	μ_m	0.15	0.35	0.40	0.90
Vibrio sp. (strain 204)	K_S	0.8	5.5	9.5	---
	μ_m	0.15	0.40	0.80	---
Spirillum sp. (strain 101)	K_S	3.0	---	12.0	---
	μ_m	0.45	---	1.10	---
Pseudomonas sp. (strain 201)	K_S	9.0	10.5	15.0	18.0
	μ_m	0.80	0.95	1.40	1.60

[a] Reprinted from Ref. (34), p. 722, by courtesy of American Society for Microbiology.
[b] Constants: K_S in 10^{-5} M; μ_m in hr^{-1}.
[c] Maximum error for K_S values, 12.5%; for μ_m values, 6.4%.

under conditions of undiscriminating elimination of microbial cells at lactate concentrations below 5 mg/l.

The growth constants of microorganisms may vary with changes in the environmental conditions. Jannasch (34) found that the constants given in Table 6 increased considerably as a result of repeated transfer of the cultures on agar plates containing 10^{-2}M of the limiting substrate (Table 7, A, B). No appreciable changes were observed in cultures that remained undisturbed in seawater for similar periods of time. Attempts to reduce the growth parameters by transferring the cultures back into medium containing little or no carbon source (10^{-6}M glucose) were unsuccessful.

2. Substrate concentration.

The growth constants given in Table 6 must be evaluated on the basis of actual nutrient concentrations occurring in natural waters and in waste treatment systems. Examples of nutrient concentrations in natural systems are given in Table 8. Most concentrations in Table 8 are collective values comprising a group of chemical compounds. As emphasized by Degens (35), only a small number of organic molecules, accounting for less than 10% of the total organic matter in the oceans, have thus far been identified. Furthermore, most dissolved organic nutrients present in natural waters are of higher molecular weight than the organic nutrients usually fed to experimental cultures in laboratory investigations. For example, only 10% of the dissolved organic compounds in the ocean has a molecular weight less than 400 (35,36). The molecular weight of the majority (90%) of the organic compounds ranges from 3,000 to 5,000. Similarly, in secondary sewage effluents, 40% of the dissolved organic material resists dialysis and is probably composed of compounds of high molecular weight (37). These materials are biologically degraded, but the mechanisms and the kinetics of the degradation processes are poorly understood.

Although there are difficulties in quantitatively and qualitatively identifying organic compounds in natural waters, the concentrations of individual monomeric molecules have been determined in a few instances. Growth activities of heterotrophic bacteria in the natural habitat, therefore, can be assessed on the basis of Equation (23-4) (Table 1), using the data of Tables 6 and 8. The growth rates so estimated are in agreement with the general observation that growth rates and metabolic rates in natural waters are extremely low. Jannasch (34) suggests that some organisms of the natural bacterial community may occur in a dormant stage while others are associated with particles which provide an environment of higher nutrient concentrations than measured in the bulk of the water.

When estimating the concentrations of substrate available to microorganisms in a natural water the following factors must be taken into consideration:

(a) Floc formation. In an experimental system organisms are usually well dispersed. Microorganisms in waste treatment systems, in contrast, are generally agglomerated into flocs. Typical floc sizes, determined by various methods, fall into the range of 20 - 200 μ . Microorganisms in natural waters also have been reported to occur in clumps (38). In such flocculated systems the physical transport of the nutrient to the center of the agglomerate may become the rate limiting

TABLE 8

Concentrations of Various Nutrients in Natural Waters and in Waste Water Treatment Systems

System	Nutrient	Concentration mg/l	Ref.
Sewage	Total hexoses	9.05	(57)
Soluble fraction	Total amino acids	9.77	
	Total pentose	0.77	
Sewage	Total amino acids	10.0[a]	(58)
Soluble fraction	Total carbohydrates	90.0[a]	
	Total soluble acids	34.0[a]	
Sewage treatment	Total amino acids	0.12	(57)
Plant effluent	Total carbohydrates	0.6	
	Total soluble acids	3.3	
Eutrophic lake	Glucose	.01 -.02	(30)
Eutrophic pond	Glucose	.01	
Mildly polluted river	Glucose	.024	
Polluted river	Glucose	.059	
Mildly polluted lakes	Glucose	.009	
Eutrophic coastal ponds	Dissolved carbohydrates	1.57 -3.54	(59)
Coastal Pacific Ocean	Dissolved carbohydrates	0.1 -0.4	(60)
Coastal lagoon		8.0	
Pacific Ocean	Total dissolved amino acids	.218	(35)
	Total free amino acids	.06	

[a]Approximate values calculated from given mg/l C values.

step. The nutrient concentration at the medium-cell membrane interface may differ considerably from the nutrient concentration in the bulk of the growth medium.

(b) The effect of surfaces. It is generally observed that in dilute media bacterial growth occurs preferentially on solid surfaces. This observation has given rise to the speculative concept that the proliferation of bacteria in dilute nutrient solutions is enhanced by the adsorption of organic matter at solid-liquid interfaces (39). Corroborative evidence for this concept has been presented by Heukelekian and Heller (40) based on their studies of the relation of bacterial growth and nutrient concentration in the presence or absence of solid surfaces. In these experiments E. coli did not multiply at substrate concentrations below 0.5 mg/l unless glass beads were added to provide additional surface. The growth enhancing effect of glass beads was observed in solutions containing nutrient concentrations below 25 mg/l.

Few investigations of the effect of solid surface on bacterial growth have been performed in continuous growth systems. Button (41) investigated the effect of clay particles (10^{12} particles/l) on the growth behavior of E. coli, Cryptococcus albus, and of an isolate from nature, (probably Rhodotorula sp.) in continuous culture. The clay particles were added either continuously or in a single injection. In the latter case, the adsorption of a mono-molecular layer of glucose (the limiting nutrient) on the clay surface would have reduced the glucose concentration level in the bulk medium from 4 mg/l to 2 mg/l at the instant clay was injected. However, Figure 7 shows that the presence of clay had no effect on the growth rate of E. coli. With Cryptococcus albus a temporary reduction in the number of microorganisms was observed, although the population eventually returned to the initial steady state level and subsequent clay additions had no effect. This temporary reduction in population numbers was thought to result from a perturbation of the cell membrane of the organisms. According to Button's interpretation, the extent of adsorption of glucose onto the suspended materials was very small.

(c) Association with phytoplankton. Allen (17) was able to show for several algal cultures isolated from oxidation ponds (Chlamydomonas pseudogloea, C. reinhardii, C. moewusii, C. eugametos, and two Chlamydomonas sp.) that 10-45% of the organic material synthesized by young growing algal cells was excreted into the surrounding medium in a soluble form. Glycolic and oxalic acid could be identified among the excretion products. Similarly, according to Hellebust (42), excretion by marine organisms amounts to as much as 35-40% of the photoassimilated carbon. The annual excretion of dissolved organic

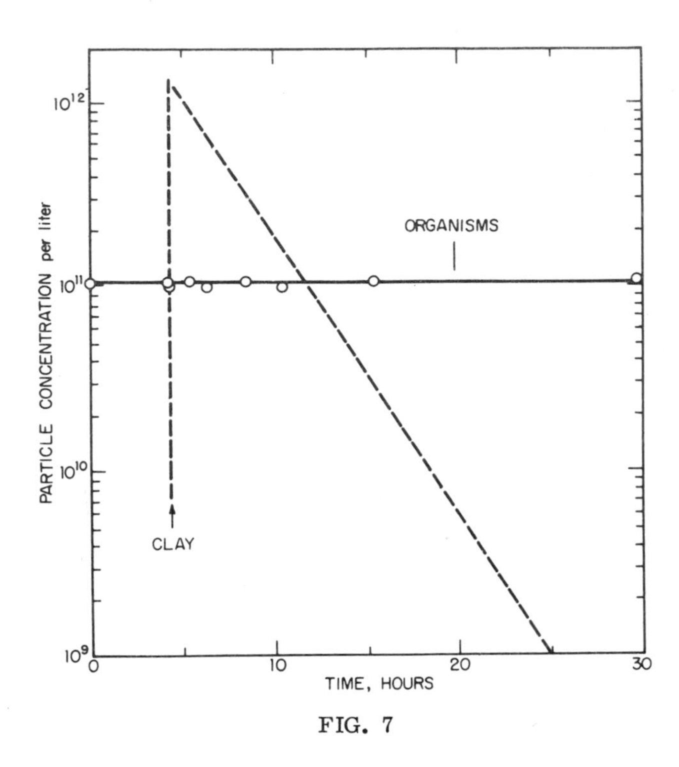

FIG. 7

Steady-state population of E. coli growing at 0.35 hr^{-1} before and after the sudden introduction of 10^{12} clay particles/l into the reactor. The growth limiting nutrient was glucose. Reprinted from Ref. (41), p. 98, by courtesy of Limnology and Oceanography.

substances into the sea by healthy marine phytoplankton equals approximately 10% of the dissolved carbon present in the open oceans. The rate of excretion of organic compounds appears to remain nearly constant whether the organisms are actively growing or in a stationary state. Simple monomeric compounds that were identified as excretion products include mannitol, aspartic acid, glutamic acid, arginine, proline, and arabinose. Only small quantities of polysaccharides and proteins were excreted by actively growing organisms.

In light of these investigations it appears significant that microorganisms have a strong tendency to associate with phytoplankton. Wuhrmann (14) enumerated the heterotrophic microorganisms present in the various biotopes of two experimental rivers which were fed waters of different quality (one extremely clean, the other slightly polluted). The carrier plant density in the rivers was relatively large, as is generally observed in shallow waters. In both rivers the microbes which attached to carrier plants comprised more than 99% of the total bacterial community. As a result of such associations the nutrient concentration in the immediate vicinity of a bacterium may frequently differ considerably from the nutrient concentration in the bulk medium.

3. The bacterial concentration.

According to Equation (23-3) (Table 1) the rate of metabolic transformation of nutrients is linearly proportional to the bacterial concentration. The bacterial concentration is thus an important parameter in the assessment of the rate of transformation of organic compounds in any bacterial culture. Jannasch (24) was able to show by using continuous growth systems that the growth constants can vary with the bacterial concentration in cultures of low population density. In dilute experimental cultures this effect was reflected by a deviation of the steady state population densities from calculated values (Fig. 8). Equation (23-10) defines the dependence of bacterial concentration on the concentration of the limiting nutrient in the feed solution and on the dilution rate in a continuous growth system:

$$S_R = \frac{B}{y} + K_S \frac{D}{\mu_m - D} \qquad (23\text{-}10)$$

The broken line in Figure 8 represents the calculated function given by Equation (23-10) using conventionally determined growth constants for $D/\mu_m = 0.5$ and also, to a first approximation the calculated function for the other D/μ_m values at an S_R greater than approximately 10 mg/l. As can be seen, when a marine strain (strain 101) is grown at various dilution rates and lactate concentrations, the measured steady state population densities deviate from the calculated densities, the deviation becoming larger with decreasing growth rates. Furthermore, when the lactate concentration in the feed solution (S_R) was lowered below a certain minimum (threshold concentration) a steady state population could no longer be maintained at the given dilution rate and washout occurred. These threshold substrate concentrations are illustrated by the solid points in Figure 8. Theoretically, in a continuous culture washout occurs ($B \rightarrow 0$) as S_R approaches S (Equation (23-7)). Since at steady state the chemostat substrate concentration (S) is a function of only one variable, the dilution rate, Equation (23-6), washout would

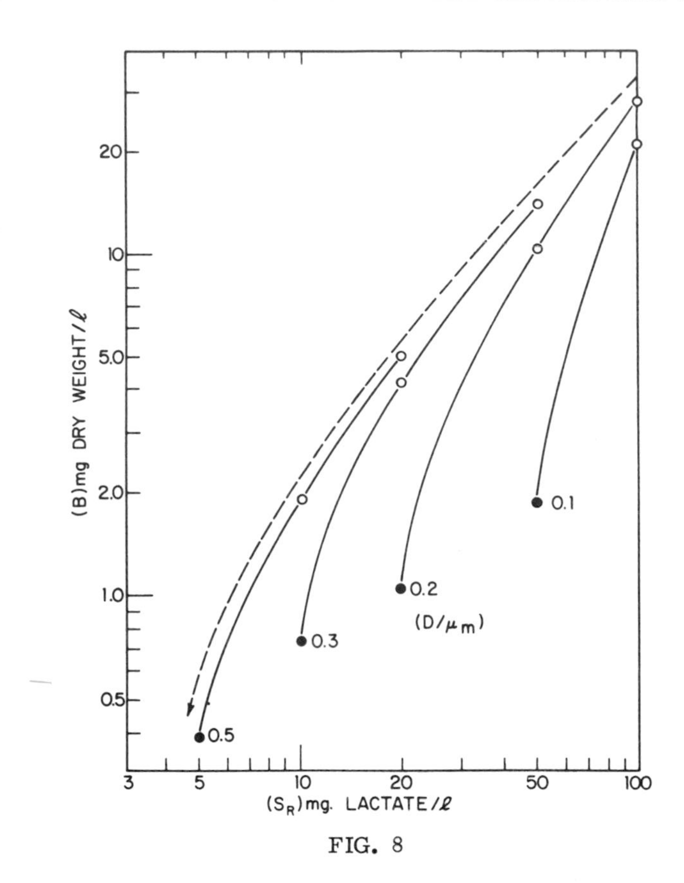

FIG. 8

Steady-state population density vs. concentration of the limiting substrate in the feed reservoir (S_R) at four different growth rates (strain 101). The broken line represents the calculated curve at $D/\mu_m = 0.5$. Growth constants were: $K_S = 0.5$; $\mu_m = 0.45$. Reprinted from Ref. (24), p. 268, by courtesy of Limnology and Oceanography.

be expected when the feed substrate concentration (S_R) is chosen equal to, or less than, the S value determined by the given dilution rate. For example, from Equation (23-6) and the growth constants of strain 101 (Tables 9 and 10), washout would have been expected at a D/μ_m of 0.3 when the feed lactate concentration was equal to, or less than 1.3 mg/l. However, the observed threshold concentration was 10 mg/l (Fig. 8).

TABLE 9

Threshold Concentrations of Three Growth Limiting Substrates (in mg/l) in Sea Water at Several Relative Growth Rates of Six Strains of Marine Bacteria and the Corresponding Maximum Growth Rates (in hr^{-1}) [a]

Strain	D/μ_m	Lactate	Glycerol	Glucose
208	0.5	0.5	1.0	0.5
Achromobacter	0.1	0.5	1.0	0.5
aquamarinus	0.05	1.0	5.0	1.0
μ_m		0.15	0.20	0.34
102	0.3	0.5	no	0.5
Spirillum	0.1	1.0	growth	5.0
lunatum	0.05	1.0		10.0
μ_m		0.45		0.25
101	0.5	5	5	
Spirillum sp.	0.3	10	10	no
	0.2	20	100	growth
	0.1	50		
μ_m		0.45	0.60	
204	0.3	1	5	1
Vibrio sp.	0.1	5	5	5
	0.05	10	10	20
μ_m		0.15	0.35	0.40
317	0.5	20	50	
Achromobacter	0.3	50	50	
sp.	0.1	100	>100	
	0.05	>100		no data
μ_m		0.85	0.70	
201	0.5	20	50	20
Pseudomonas sp.	0.3	50	50	50
	0.2	100	>100	>100
	0.1	>100		
μ_m		0.80	0.65	0.80

[a]Reprinted from Ref. (24), p. 268, by courtesy of Limnology and Oceanography.

TABLE 10

Growth Constants (determined at high population densities) of Two Strains of Marine Bacteria Compared with Threshold Concentrations of Lactate in Sea Water at Two Relative Growth Rates [a]

Strain	K mg/l	μ_m hr^{-1}	Threshold Concentration mg/l	
			$D/\mu_m = 0.5$	$D/\mu_m = 0.1$
101	3.0	0.45	5	50
201	8.0	0.80	20	100

[a]Reprinted from Ref. (24), p. 268, by courtesy of Limnology and Oceanography.

These results suggest that at low microorganism densities and low growth rates, growth constants determined by conventional techniques are no longer applicable. These constants are critical for describing the relationship between growth rate and substrate concentration (Equation, 23-4). Jannasch suggests that the observed threshold concentrations (Tables 9 and 10) are at least one to two orders of magnitude too large to be explained by maintenance metabolism or endogenous respiration. His argument is substantiated by the observation that the washout substrate concentration increased with decreasing growth rate (Fig. 9).

As an explanation of his results, Jannasch suggests the implication of certain metabolic products as growth stimulants. When S_R and, consequently, the bacterial concentration reached their minimum values, production of a growth stimulating metabolite would not meet the demand for maintenance of the growth rate concomitant with the dilution rate. This hypothesis is in accord with the experimental observation that at lower dilution rates (lower metabolic rates) washout occurs at higher population densities. It should be emphasized that these studies were performed with pure cultures and, therefore, exemplify a principle rather than reproduce data of a more complex system. The smallest threshold concentrations found in Jannasch's study are still larger than the concentrations of similar compounds reported for natural waters.

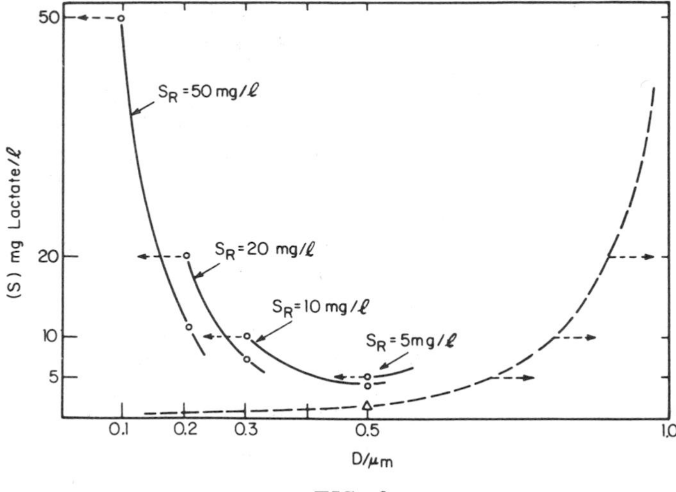

FIG. 9

Steady-state concentrations of limiting substrate (S) vs. relative growth-rate at four different substrate concentrations in the feed reservoir (S_R). Horizontal portions of the curves (dashed) indicate washout of the culture ($S = S_R$). The broken line indicates the theoretical relationship for all four influent substrate concentrations ($\Delta = K_S$). The organism and growth constants are the same as for Figure 8. (For symbols see Table 1). Reprinted from Ref. (24), p. 269, by courtesy of Limnology and Oceanography.

The relatively high and constant concentrations of dissolved organic matter in the sea has been explained by assuming an unavailability of the material to microbial breakdown. However, in the light of Jannasch's studies, standing concentrations of organic substrates not oxidized by bacteria may well be characteristic for sea water as a suboptimal environment for microbial life.

4. Effect of organic compounds other than the carbon and energy source on the kinetics of bacterial growth.

It is well known that microorganisms synthesize and excrete growth regulating organic substances such as vitamins and amino acids into the surrounding medium. These organic compounds, the so-called growth factors, are required in minute amounts by certain microorganisms for growth in addition to the carbon and energy source. One generally distinguishes between essential growth factors (in the absence of which no growth occurs) and stimulatory growth factors (which merely serve to enhance growth). The latter affect organisms in various ways: they facilitate the enzymatic adjustment (adaptation) to a change in the environment, increase the logarithmic growth rate, or lead to a variation in the yield coefficient. There is not sufficient information available for a quantitative evaluation of the effects of growth factors on the kinetics of dissimilation of organic substances in natural waters. Nevertheless, a brief consideration of the significance of growth factors in natural microbial communities seems indispensable to the context of this discussion.

Burkholder (43) determined the production of several vitamins in pure cultures of 1054 isolates of marine bacteria. About one-half of the cultures studied produced physiologically significant quantities of biotin and thiamin, while nicotinic acid and vitamin B_{12} were synthesized and excreted by somewhat less than one-half of the strains. An analysis of the vitamin production and vitamin requirements of some of these strains is given in Table 11. Similar vitamin requirements and vitamin excretions were observed by other investigators with microorganisms from natural waters and from soil (Table 12).

In a mixed microbial community, excretion of and the requirement for growth factors will lead to interactions among microorganisms. These interactions are thought to be among the major biological determinants of the species structure of natural microbial communities. From laboratory investigations, it is suggested that growth factors are of significance in the microbial metabolic activity of natural waters both directly on the level of the physiological behavior of the individual organism and indirectly by their effect on the community structure.

5. Multisubstrate media.

Natural waters and waste waters represent complex media which contain a large variety of organic nutrients. Since enzymes exhibit, in most cases, a strong substrate specificity, the complexity of the medium will lead to increased complexity of the enzyme mediated processes in the microbial cell.

In 1942, Monod (20) observed diphasic growth of heterotrophic organisms incubated in a medium containing two carbon sources. This observation was interpreted as reflecting sequential utilization of the two substrates. This phenomenon of sequential and selective nutrient utilization is not restricted to carbon sources. For example, it has been reported that some microorganisms deplete the culture medium of ammonium before utilizing nitrate as a nitrogen source (44).

TABLE 11

Vitamin Production and Vitamin Requirements in Marine Microbial Isolates [a]

No. of cultures	Patterns of requirements				No. of vitamin producers			
	Biotin	Thiamine	Nicotinic acid	B_{12}	Biotin	Thiamine	Nicotinic acid	B_{12}
29	+					3	29	1
7		+			6		7	
2			+		1			
3				+	3	1	3	
3	+			+			3	
5		+		+	3		5	
1		+	+		1			
1	+	+	+					
1	+	+		+			1	

[a] Reprinted from Ref. (43), p. 135, by courtesy of C. C. Thomas.

TABLE 12

Growth Factors and the Development of Soil Bacteria[a]

Vitamin Required or Excreted	Percentage of Total Bacteria Which Require the Vitamin	Percentage of Total Bacteria Which Excrete the Vitamin
Thiamine	19.4	35.5
Biotin	16.4	19.6
Vitamin B_{12}	7.2	29.9
Pantothenic acid	4.6	---
Folic acid	3.0	---
Nicotinic acid	2.0	---
Riboflavin	0.6	39.2
Terregens factor	---	22.4
One or more vitamins	27.1	50.5

[a]Reprinted from Ref. (61), p. 428, by courtesy of John Wiley and Sons, Inc.

The problem of sequential substrate utilization has been studied in great detail by many researchers as part of the general problem of integration and coordination of physiological processes in a living cell. These studies have revealed two mechanisms which specifically regulate the dissimilatory activities of heterotrophic microorganisms growing in a multisubstrate environment - enzyme repression and enzyme inhibition (45). If a substrate (or more specifically, a metabolic intermediate related to the substrate) represses further synthesis of an unrelated dissimilatory enzyme, the phenomenon is called catabolite repression. The term catabolite inhibition applies if a metabolic intermediate inhibits an existing unrelated dissimilatory enzyme. The two mechanisms are thought to economize substrate utilization by coordinating rates of catabolic (dissimilatory) and anabolic (synthetic) processes in the cell (45). Catabolite repression has been observed more frequently than catabolite inhibition. Both mechanisms lead to sequential substrate utilization although inhibition manifests itself immediately in the dissimilation rates while catabolite repression alters substrate utilization rates only after a delay. Figures 10, 11, and 12 illustrate growth and substrate utilization patterns under conditions of concurrent substrate utilization (Fig. 10), catabolite repression (Fig. 11), and catabolite inhibition (Fig. 12). In any given culture the substrate

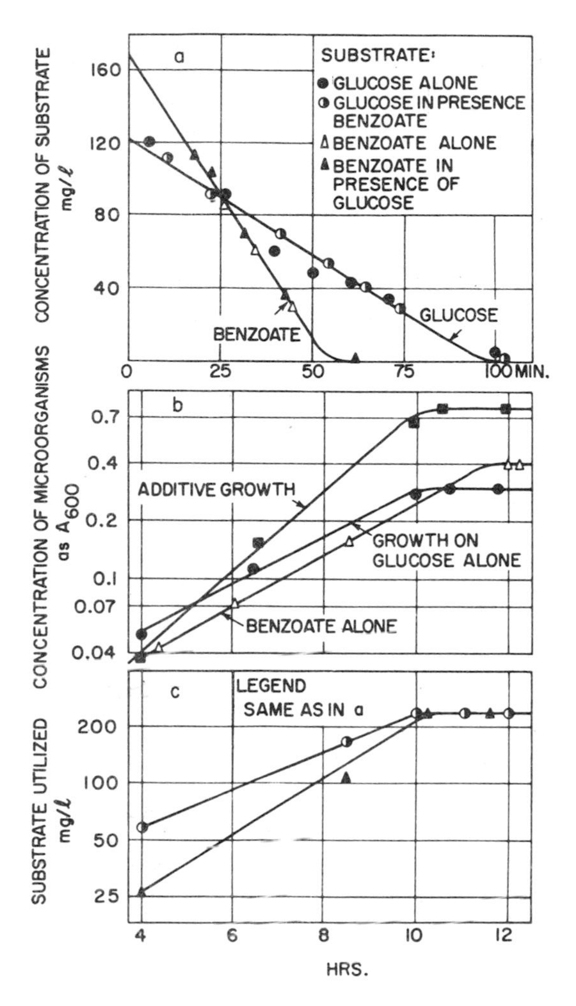

FIG. 10

Additive growth and concurrent utilization of substrates (glucose, benzoate). Neither substrate interfers with the activity or the formation of the enzymes related to the second substrate. (a) Simultaneous activity of glucose and benzoate catabolizing enzymes is shown by concurrent utilization of the substrates at constant bacterial concentration. (b) The growth rate constant (μ) observed in a medium containing both substrates is larger than either of the growth rate constants in the single substrate media. (c) The additive rate of growth is a manifestation of the similarly additive and concurrent substrate elimination patterns. Reprinted from Ref. (46), p. 659, by courtesy of American Society for Microbiology.

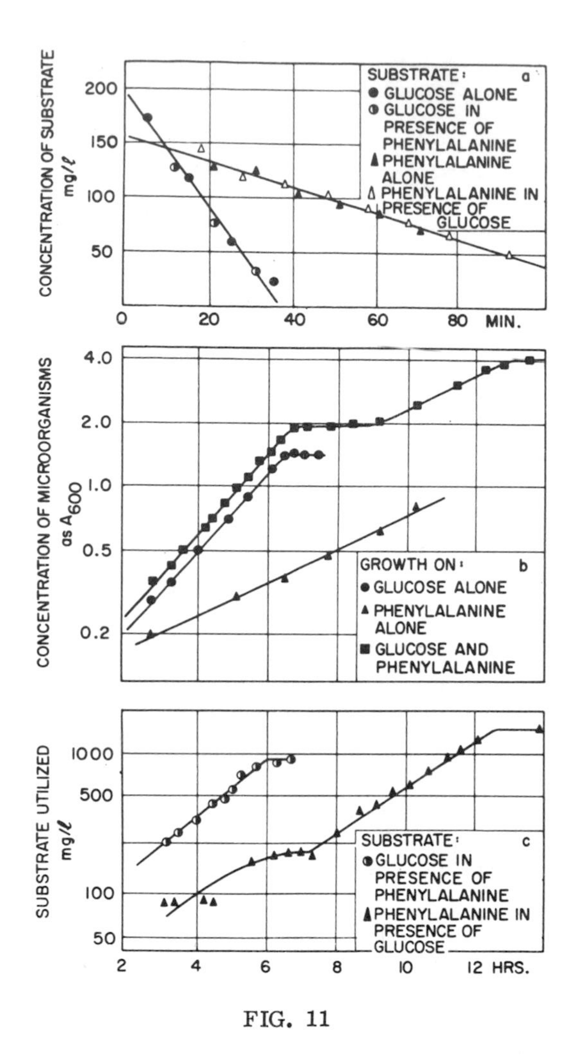

FIG. 11

Catabolite repression (glucose, L-phenylalanine). (a) Concurrent utilization of L-phenylalanine and glucose in a short-term experiment at constant bacterial concentration shows concurrent activity of the enzyme systems necessary for the utilization of glucose and L-phenylalanine. (b) Diauxic growth. (c) Sequential utilization of substrates is caused by repression of the formation of new L-phenylalanine-catabolizing enzymes in the presence of glucose. The elimination of L-phenylalanine in (a) is mediated by the L-phenylalanine enzyme present in the organisms at the time of glucose addition. Since no new enzyme is formed, the elimination rate remains constant with time and in a growing culture (c) soon becomes negligible. Reprinted from Ref. (46), p. 661, by courtesy of American Society for Microbiology.

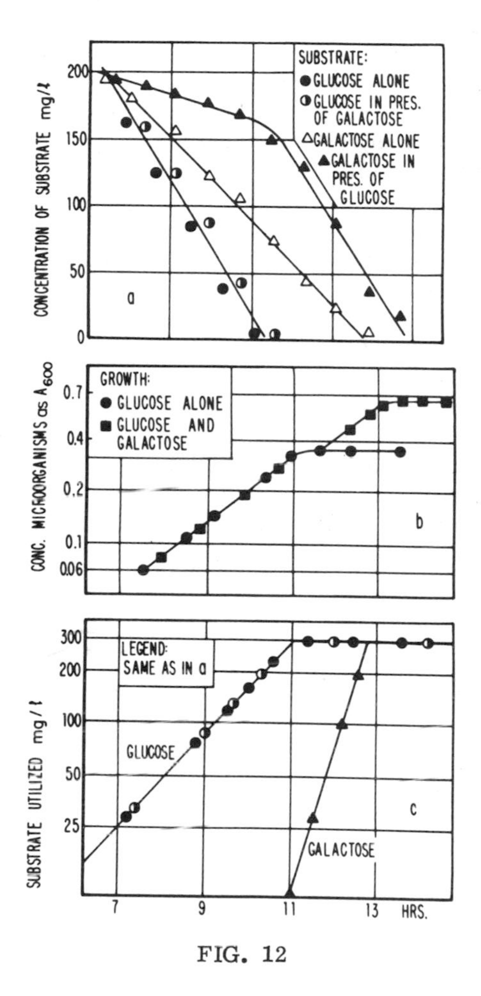

FIG. 12

Catabolite inhibition (glucose, galactose). (a) Glucose inhibits the activity of enzymes related to galactose catabolism at constant bacterial concentration. This leads to diphasic (diauxic) growth (b) and to sequential elimination of the two substrates (c). Reprinted from Ref. (46), p. 660, by courtesy of American Society for Microbiology.

elimination pattern will depend on the strain of organism, the types of substrates, external physical parameters (46), and intracellular physiological factors (47).

Most studies on regulation of enzyme activity and enzyme formation in a multisubstrate medium have been performed with laboratory cultures under laboratory conditions (batch cultures, high substrate concentrations). The complexity of the medium was usually maintained within narrow limits to facilitate interpretation of the experimental data. Nevertheless, the following observations indicate that the results obtained in the laboratory may bear relevance with respect to microbial substrate elimination in natural waters and in waste waters: (a) Catabolite repression was observed in highly complex culture media containing many amino acids and other growth factors besides the carbon and energy sources (46). (b) It was observed in very dilute culture media where growth activities were limited by nutrient concentrations (48). Figure 13 relates the rate of protein synthesis (growth) to the differential rate of enzyme synthesis (β - galactosidase) under conditions of rate limiting substrate concentration. It is seen that below a threshold growth rate the enzyme synthesis is not repressed (the threshold growth rate is one-sixth of the maximal growth rate in the particular experimental system). Above the threshold growth rate, the degree of repression increases with increasing growth rate and thus with increasing substrate concentration (Equation, 23-3).

IV. KINETICS OF AEROBIC OXIDATION OF ORGANIC COMPOUNDS BY HETEROGENEOUS MICROBIAL COMMUNITIES

A. Quantitative and Qualitative Characteristics of Heterogeneous Microbial Communities.

The metabolic activities of heterogeneous microbial communities depend not only on the physiological response of the individual cells to the environment, but also on the species structure of the community, on shifts in this structure and on interactions between the various microbial species.

Wuhrmann (14) attempted to establish a quantitative relationship between the rate of elimination of specific organic substances in a natural water and the quantity and quality of the biomass in continuous contact with the water. Artificial channels were used for the study where most environmental factors could be controlled. The channels were fed raw or purified sewage diluted with various quantities of natural river water. When each channel was populated by a biocenosis characteristic for the quality of the feed water, ground water supplemented with glucose (10 - 15 mg/l) was fed to the channels and the

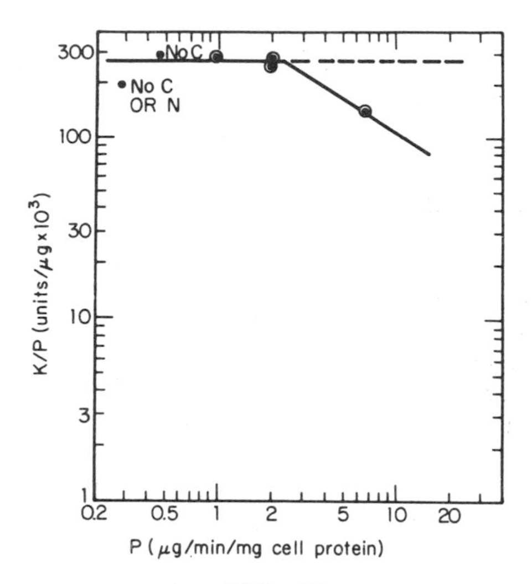

FIG. 13

Repression of enzyme formation at growth limiting nutrient concentrations. Relationship between rate of protein synthesis (P) and differential rate of β-galactosidase synthesis (K/P, or the ratio of rate of enzyme synthesis to rate of total protein synthesis) in glycerol limited chemostats. P is controlled by the experimentally determined dilution rate. "No C" refers to cells maintained without a source of carbon, "No C or N" refers to cells maintained without a source of either carbon or nitrogen (batch culture). Reprinted from Ref. (47), p. 157, by courtesy of the American Society for Microbiology.

decrease in glucose concentration was followed over the length of the channels. The specific substrate elimination rates measured at various points along the three channels which had been fed with sewage - river water varied considerably with the character of the biocenosis. The biota characteristic of unpolluted river water did not eliminate glucose at all from solution during the time of observation. The following conclusions were drawn by Wuhrmann from the experimental results: (a) the rate of dissimilation of an organic substance by the microphytes in a natural water is highly dependent on the absolute quantity of biomass in intimate contact with the water; (b) densely packed bacterial masses show smaller specific dissimilation rates than microphytic association with large specific surface areas in contact with the surrounding water; (c) an important determinant of the rates of microbial dissimilatory activities is the species structure of the microphytic community; and (d) qualitative and quantitative characteristics of the community are determined to a large extent by the nature and concentration of organic substances present in the aquatic habitat.

B. Substrate Elimination From Multisubstrate Media.

Substrate elimination from multisubstrate media by mixed microbial cultures has been studied in our own laboratory (3, 46) as well as by several other investigators (48, 49, 50, 51). In our laboratory, the experimental cultures were subcultures of a batch culture fed daily with synthetic sewage (nutrient broth 0.6g/l; glucose 0.2g/l, and oleate 0.03g/l). The batch culture had originally been inoculated with activated sludge from a sewage treatment plant. The subcultures were adapted to the substrates under study for approximately 16 hours prior to the experiment. Similar procedures were employed by the other investigators. In these studies the elimination of some substrate pairs (e.g., glucose - benzoate) was found to occur simultaneously, while other substrate pairs (e.g., glucose - sorbitol; glucose - galactose; glucose - L-phenylalanine; benzoate - L-phenylalanine) were utilized sequentially. The observed sequential substrate elimination was interpreted in terms of catabolite inhibition or repression since ecological succession could be excluded as cause for the diphasic elimination and growth patterns.

An analysis of the experimental mixed cultures employed in our study revealed that the heterogeneity of the cultures had been diminished considerably during the prolonged propagation in the laboratory. The majority (75%) of the organisms in the batch culture fed with synthetic sewage belonged to the microbial group of coliforms. During the adaption period to an individual substrate the coliform bacteria were further enriched (95%). Similar shifts in the species structure of heterogeneous cultures during the adaption period were reported by Prakasam and Dondero (52).

When fresh water isolates (in pure culture or combined to a mixed culture) or field activated sludge were incubated with the substrate pairs, glucose - galactose or glucose-L-phenylalanine, the two substrates were eliminated simultaneously rather than sequentially. For activated sludge, Wuhrmann (53) also observed concurrent elimination of glucose, acetate, and aspartate, while Prakasam and Dondero (51) reported simultaneous utilization of glucose and sorbitol in similar inocula.

Current information does not permit a generalized interpretation of the observed differences between laboratory cultures and isolates from natural waters or activated sludge since the investigations were performed with a small number of substrate combinations and with a limited number of microbial communities. It should be pointed out, however, that the organisms present in fresh water and in the activated sludge generally exhibited slower metabolic rates than did the laboratory organisms. For example, the elimination rate of glucose by a glucose adapted laboratory culture was three times higher than the rate of elimination under identical experimental conditions by a culture composed of fresh-water isolates. These differences in metabolic rates may be relevant in the light of the observation that, in E. coli and in A. aerogenes, the degree of catabolite repression was correlated with the rate of metabolism of the repressing substrate (47, 54).

From the above results, it appears that sequential substrate elimination and diauxic growth may be relevant in habitats where a small number of substrates predominate and where, therefore, microorganisms capable of fast growth on the particular substrates are enriched. Such conditions are often encountered in the biological treatment of industrial wastes and, perhaps, in those polluted environments that are characterized by high organism densities but a scarcity in the number of species.

C. Instability of Heterogeneous Cultures.

Experimental investigations with natural heterogeneous cultures often lead to ambiguous results because of the quick response of such cultures to the experimental environment. This is illustrated in the study by Vaccaro and Jannasch (55), on the relationship between the rate of glucose uptake and glucose concentration in sea water. ^{14}C-labelled glucose was added in various concentrations to off-shore waters of the Pacific Ocean and the rate of glucose uptake was measured. In undisturbed samples the uptake rate as a function of glucose concentration was sigmoid (Fig. 14a). In preaerated sea water samples or in samples preincubated with 50μg/l of glucose for 12 hours prior to the uptake measurements, the uptake rate - glucose concentration relationship was a hyperbolic function (Fig. 14b), typical of pure culture systems. Similar relationships were observed in mixed microbial communities in the Atlantic Ocean by the same investigators and in fresh water samples by Parsons and Strickland (56).

Vaccaro and Jannasch (55) concluded from their results that the preincubation of the seawater samples may have effected either the induction of an enzyme system in the organisms or a shift in the species structure of the natural community towards predominance of a single species. The observed dependency would thus be an experimental artifact. This study suggests that before experimental kinetic data obtained on heterogeneous communities can be extrapolated to natural

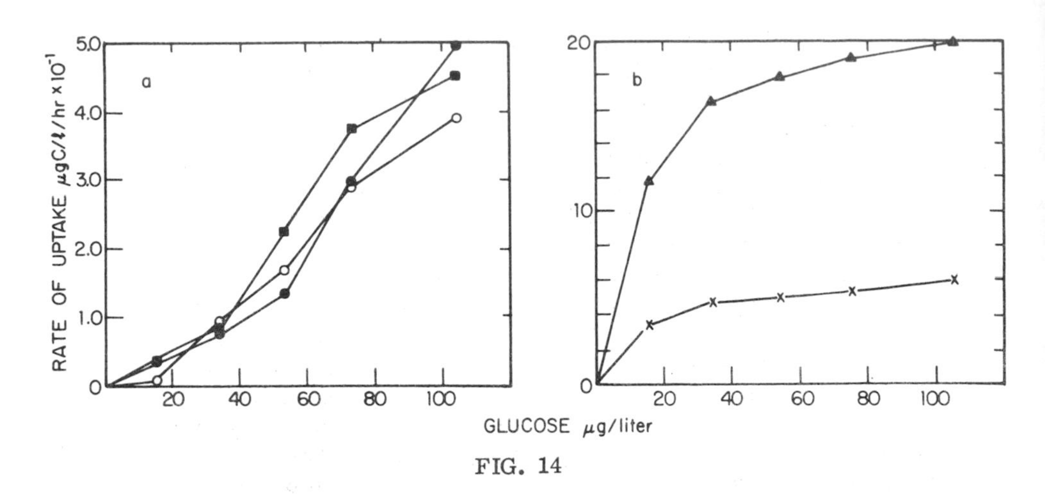

FIG. 14

Glucose uptake by mixed bacterial communities of the Pacific Ocean off Peru during a 3-hr exposure to uniformly labeled ^{14}C glucose. (a) Glucose uptake by three sea water samples from various depths (closed circles, 10m; open circles, 50m; squares, 100m) immediately after collection. (b) Glucose uptake by sea water sample (50m) after aeration for 12 hr in absence (triangles) and in presence (crosses) of glucose. Reprinted from Ref. (55), p. 540 by courtesy of Limnology and Oceanography.

ecological systems it must be established that the species structure of the community was preserved throughout the period of the experimental observations. Vaccaro and Jannasch emphasize that, since reaction kinetics of various species differ, the exhibition of single reaction kinetics by a heterogeneous culture must be interpreted with caution. Single reaction kinetics may be found in natural microbial communities where one species is predominant in the community or where the differences between the kinetics of the various species are too small to be resolved.

REFERENCES

1. H. W. Jannasch, Verh. Int. Ver. Limnol., 17, 1 (1969).

2. O. Maaloe and N. O. Kjeldgard, Control of Macromolecular Synthesis, W. A. Benjamin, Inc., New York, 1966.

3. E. Stumm-Zollinger, J. Water Pollut. Contr. Fed., 40, R213 (1968).

4. A. F. Gaudy, Jr., and E. T. Gaudy, Annu. Rev. Microbiol. 20, 319 (1966).

5. K. Wuhrmann, Sewage Ind. Wastes, 26, 212 (1954).

6. H. G. Schlegel, ed., Anreicherungskultur und Mutantenauslese, Zbl. Bakt. Abt. I, Suppl. Vol. 1, Gustav Fischer Verlag, Stuttgart, Germany, 1965.

7. H. Heukelekian and N. C. Dondero, in Principles and Applications in Aquatic Microbiology, (H. Heukelekian and N. C. Dondero, eds.), John Wiley, New York, 1963, p. 441.

8. J. Monod, Ann. Inst. Pasteur, 79, 390 (1950).

9. A. Novick and L. Szilard, Proc. Nat. Acad. Sci., 36, 708 (1950).

10. D. Herbert, R. Elsworth, and R. C. Telling, J. Gen. Microbiol., 14, 601 (1956).

11. D. W. Eckhoff and D. Jenkins, in Proc. Int. Water Pollut. Res. Conf., 3rd, Munich, 1966.

12. H. W. Jannasch, Biotech. Bioeng., 7, 279 (1965).

13. H. W. Jannasch, Arch. Mikrobiol., 59, 165 (1967).

14. K. Wuhrmann, in Principles and Applications in Aquatic Microbiology, (H. Heukelekian and N. C. Dondero, eds.) John Wiley, New York, 1963, p. 167.

15. H. W. Jannasch, Ber. Limnol. Flusstation Freudenthal, 7, 21 (1956).

16. G. E. Fogg, in Physiology and Biochemistry of Algae, (R. A. Lewin, ed.) Academic Press, New York, 1962, p. 475.

17. M. B. Allen, Arch. Mikrobiol., 24, 163 (1956).

18. H. W. Jannasch, Arch. Mikrobiol., 45, 323 (1963).

19. J. R. Postgate and J. R. Hunter, J. Appl. Bacteriol., 26, 295 (1963).

20. J. Monod, Recherches sur la croissance des cultures bacteriennes, Hermann, Paris, 1942.

21. H. W. Jannasch, Symp. on Organic Matter in Natural Waters, University of Alaska, 1968.

22. H. W. Jannasch, in Zbl. Bakt, Abt. I, Suppl. Vol. I. (H. G. Schlegel, ed.), 1965, p. 498.

23. A. W. Lawrence and P. L. McCarty, J. Water Pollut. Contr. Fed., 41, R1 (1969).

24. H. W. Jannasch, Limnol. Ocean., 12, 264 (1967).

25. P. Krishnan and A. F. Gaudy, Jr., Biotech. Bioeng., 7, 455 (1965).

26. R. I. Mateles, D. Y. Ryu, and T. Yasuda, Nature, 208, 263 (1965).

27. H. T. Yeoh, H. R. Bungay, and N. R. Krieg, Can. J. Microbiol., 14, 491 (1968).

28. A. Shindala, H. R. Bungay, N. R. Krieg, and K. Culbert, J. Bacteriol., 89, 693 (1965).

29. R. K. Guthrie, B. H. Cooper, J. K. Ferguson, and H. E. Allen, Can. J. Microbiol., 11, 947 (1965).

30. J. E. Hobbie and R. T. Wright, Limnol. Ocean., 10, 471 (1965).

31. J. E. Hobbie and R. T. Wright, Science, 159, 1463 (1968).

32. H. W. Jannasch, Appl. Microbiol., 16, 1616 (1968).

33. H. W. Jannasch, J. Bacteriol., 99, 156 (1969).

34. H. W. Jannasch, J. Bacteriol., 95, 722 (1968).

35. E. T. Degens, Symp. on Organic Matter in Natural Waters, University of Alaska, 1968.

36. I. A. Breger, Symp. on Organic Matter in Natural Waters, University of Alaska, 1968.

37. R. J. Bunch, E. F. Barth, and M. B. Ettinger in "Advances in Biological Waste Treatment," (W. W. Eckenfelder, Jr. and J. McCabe, eds.), MacMillan, New York, 1963, p. 77.

38. G. E. Jones and H. W. Jannasch, Limnol. Ocean, 4, 269 (1956).

39. C. E. Zobell and D. Q. Anderson, Biol. Bull., 71, 324 (1936).

40. H. Heukelekian and A. Heller, J. Bacteriol., 40, 547 (1940).

41. D. K. Button, Limnol. Ocean., 14, 95 (1969).

42. J. A. Hellebust, in Estuaries (G. H. Lauff, ed.), Publ. 83, AAAS, Washington, DC, 1967, p. 76.

43. P. R. Burkholder, in Symp. Marine Biology (C. H. Oppenheimer, ed.), C. C. Thomas, Springfield, Ill., 1963, p. 133.

44. A. C. R. Dean and Sir Cyril Hinshelwood, Growth, Function and Regulation in Bacterial Cells, Clarendon Press, Oxford, 1966.

45. B. Magasanik, Cold Spring Harbor Symp. Quant. Biol., 26, 249 (1961).

46. E. Stumm-Zollinger, Appl. Microbiol., 14, 654 (1966).

47. A. F. Gaudy, Jr., E. T. Gaudy, and K. Komolrit, Appl. Microbiol., 11, 157 (1963).

48. D. Kennel and B. Magasanik, Biochem. Biophys. Acta, 81, 418 (1964).

49. M. N. Bhatla and A. F. Gaudy, Jr., Appl. Microbiol., 13, 345 (1965).

50. K. Wuhrmann, personal communication.

51. T.B.S. Prakasam and N. C. Dondero, Bacteriol. Proc., 10, A55 (1967).

52. T.B.S. Prakasam and N. C. Dondero, Proc. Ind. Waste Conf., 19th Eng. Bull. of Purdue University, (1964).

53. K. Wuhrmann, F. von Beust, and T. K. Ghose, Schweiz. Zeits. Hydrol., 20, 284 (1958).

54. F. C. Neidhardt, J. Bacteriol., 80, 536 (1960).

55. R. F. Vaccaro and H. W. Jannasch, Limnol. Ocean, 12, 540 (1967).

56. T. R. Parsons and J. D. H. Strickland, Deep Sea Res., 8, 211 (1961).

57. J. V. Hunter and H. Heukelekian, J. Water Pollut. Contr. Fed., 37, 1142 (1965).

58. H. A. Painter, M. Viney, and A. Bywaters, J. Inst. Sewage Pur., 5, 3 (1961).

59. G. E. Walsh, Limnol. Ocean., 10, 570 (1965).

60. G. J. Lewis and N. W. Rakestraw, J. Marine Res., 14, 253 (1955).

61. M. Alexander, Introduction to Soil Microbiology, John Wiley, New York, 1961, p. 428.

CHAPTER 24

DECOMPOSITION OF NATURALLY-OCCURRING ORGANIC POLYMERS

Walter J. Nickerson

Rutgers - The State University

I. INTRODUCTION

A complex array of polymeric substances is produced in nature in vast quantities. However, most of the naturally produced polymeric substances do not accumulate since they are degraded microbiologically and their components are returned eventually to the carbon, nitrogen, or other elemental cycles. It is obvious that if there was no natural agency for the degradation of such substances as chitin (found in insect cuticle, butterfly wings, lobster shells, etc.) or keratin (the sclero-protein comprising the bulk of wool, hair, feathers, and hooves) which have been deposited over the world for eons, we would, today, be knee deep in feathers and butterfly wings!

There is a parallel between the geological era in which production of a polymer evolved and the number of microbial agencies capable of degrading the polymer. In general, the earlier a polymer appeared in nature, the more numerous are the agencies for its destruction. Cellulose, one of nature's oldest polymers, appeared in primitive algae and is attacked by several bacteria, both aerobic and anaerobic, and by a great many fungi (Table 1). Lignin, on the other hand, may be nature's newest polymer since it probably did not appear until the Ordovician-Silurian period of the Paleozoic era with the development of pteridophytes. The lignin of tree ferns, characteristic of the later Paleozoic era, accumulated and eventually became coal. Possibly, the higher basidiomycetous fungi which degrade lignin aerobically had not evolved yet. Today, lignin that escapes degradation aerobically accumulates under anaerobic conditions.

TABLE 1

Naturally-Occurring Polymers Degraded by Microorganisms

Insoluble polymer	Source	Chemical nature	Degraded by
Cellulose	Plants	β-1, 4-poly-glucosan	Cellulomonas, Cellvibrio, Clostridia, many fungi
Chitin	Insects, crustacea, many fungi	β-1, 4-poly N-acetylglucos-amine	Streptomyces sp.
Keratin	Wool, hair, feathers, hooves	Scleroprotein	Streptomyces fradiae
Lignin	Woody plants	Methoxylated aromatic propane network	Higher basidio-mycetous fungi
Rubber	Plant latex	Polyisoprene	Streptomyces sp., a few filamentous fungi

II. MICROBIAL ATTACK ON POLYMERIC SUBSTANCES

Although the naturally produced polymeric substances are quite different in their chemical composition, there are many points in common in the manner in which they are degraded by microorganisms. At first glance, it seems improbable that such insoluble polymeric substances as cellulose, chitin, keratin, lignin, and rubber could be attacked at all. These things cannot enter a microbial cell. The question then arises: how does the cell "detect" the presence of the insoluble substance and secrete the adaptive enzyme that can hydrolyze the substrate? A microorganism that can attack an insoluble substance first establishes intimate contact with the surface of the polymer. Then, activity at the cell surface causes an initial attack on the polymer whereby some oligomeric material is solubilized. This soluble oligomer can enter the cell and serves as an inducer for the synthesis of

TABLE 2

Stages in Microbial Attack on an Insoluble Substrate

a. Intimate contact established between microbial surface and polymer surface.
b. Activity at microbial surface - initial attack on polymer.
c. Oligomeric substances released by initial attack.
d. Oligomeric substances enter cell - induce elaboration of specific hydrolytic enzyme.
e. Hydrolytic enzyme binds to insoluble substrate.
f. Substrate is hydrolyzed and solubilized.
g. Solubilized products enter cell and support growth of cell.

a specific hydrolytic enzyme that is secreted into the medium. The secreted enzyme binds to the insoluble substrate and catalyzes hydrolytic cleavage of specific linkages. The substrate is solubilized thereby and provides a nutrient source for growth of the microorganisms (Table 2).

This analysis of how a microorganism attacks an insoluble substrate is based on studies conducted in the author's laboratory on the degradation of keratin (1,2), rubber (3), and citrus peels (4), and on an analysis of publications from other laboratories on degradation of cellulose (5,6,7), and of chitin (8,9). Our studies on keratin degradation may serve to illustrate the steps listed in Table 2. A strain of _Streptomyces fradiae_ was isolated from soil in an area where keratin degradation has been proceeding for many years (a chicken yard). This organism grew readily on wool as sole source of carbon and nitrogen. Filaments of the organism developed in intimate contact with fibers of wool. Cytochemical tests demonstrated that reactive sulfhydryl (-SH) groups were present in cell-wool interfaces, and low molecular weight -SH peptides were found in the culture broth. The N-ethyl maleimide derivatives of these -SH peptides were chromatographed and found to be different from the derivatives of cysteine and glutathione (10). Thus, the activity at the cell surface of _S. fradiae_ in contact with wool represents a type of protein disulfide reductase activity which originally was described from yeast (11,12), and which has been found since to be widely distributed in plant tissues (13) and microorganisms (14,15).

The -SH peptides appear in the culture broth in the early stages of incubation. It is likely, therefore, that they enter the cell and signal the induction of the hydrolytic enzyme that has been termed keratinase. The enzyme is elaborated by S. fradiae as a conjugate in which a basic protein is bound to a highly acidic polymer. Keratinase-conjugate binds to wool according to the Freundlich adsorption equation, and peptides are rapidly released from wool by the hydrolytic action of the enzyme (16,17).

In general, other polymer-degrading systems follow the pattern outlined in Table 2. However, the nature of the activity at the microbial surface has not been elucidated in every case. An early event in fungal attack on cellulose (cotton) is the hydration and swelling of the crystalline portion of the fiber. Both cellobiose and sophorose are effective inducers of cellulase secretion by fungi (18). For elaboration of chitinase by Streptomyces, N, N′-diacetyl chitobiosamine and N-acetylglucosamine have been reported to induce secretion of this enzyme (19). The binding of hydrolytic enzymes to their insoluble substrates merits attention. In 1963, reports from three different laboratories showed that binding of enzyme to insoluble substrate followed the Freundlich adsorption equation: Walker and Hope (20) for the action of salivary amylase on starch granules; McLaren (21) for three different systems including cellulase-cellulose; and the report from the author's laboratory for the action of keratinase on wool (16). In all of these cases, the rate of hydrolysis of the insoluble substrate is proportional to the amount of enzyme adsorbed onto the surface of the polymer.

Thus we see that an ecological relationship has evolved whereby polymer production in nature is balanced by polymer degradation brought about largely by microbial agencies. It is only through man's activities that this balance is upset. But now it is apparent that man must stop dumping huge amounts of waste into rivers, must stop burning discarded rubber tires, and, in short, must stop poisoning the environment in which all things live.

Let us look at some examples of how microbial processes can be employed to achieve this goal (Table 3). As a first principle it may be stated that if the waste material to be converted, transformed, or utilized is concentrated, then the microbial population employed to conduct the process must be concentrated. This condition can be achieved by carrying out the process in a fermentor in which conditions for microbial growth can be optimized.

III. MICROBIAL UTILIZATION OF WASTE MATERIALS

One of the first successful efforts to combine the solution of a waste disposal problem with the production of a worthwhile product was brought into operation about twenty years ago at Rhinelander,

TABLE 3

Microbial Utilization of Waste Materials

Waste material	Organism(s)	Products	References
Sulfite pulp liquor	Candida utilis	Torula yeast	(22, 23)
Whey	Saccharomyces fragilis	Yeast-lactalbumin ("wheast")	(25)
Potato culls and peelings	Endomycopsis fibuliger and C. utilis	Torula yeast	(27)
Feathers	Streptomyces fradiae	Water-soluble peptides	(29)
Citrus pulp and peels	Sclerotium rolfsii	Oxalic acid and water-soluble polysaccharides	(4)
Cellulosic wastes	Trichoderma viride	Glucose	(6)

Wisconsin. Torula yeast was produced in a continuous fermentation of spent liquor from the sulfite-pulping process. Of the sulfite liquor solids, about 20% consists of sugars, principally pentoses, and organic acids readily utilizable by Candida utilis. The process has been described in detail by Inskeep et al. (22) and Wiley (23).

More recently, a solution to the problem of disposal of whey from the manufacture of cottage cheese has been found. The lactose-utilizing yeast Saccharomyces fragilis grown on sterilized whey is recovered centrifugally and most of the lactalbumin of whey is removed with the yeast, thus providing a high protein feed supplement (24, 25). Several hundred thousand tons of whey solids are obtained annually from cheese manufacturing in the United States which poses a serious waste disposal problem.

In the two instances of microbial utilization of waste materials just discussed, the carbon sources were water-soluble sugars. Let us look at some developments in microbial utilization of insoluble wastes. In 1956, a process was developed whereby the amylase-secreting yeasts, Endomycopsis fibuliger or E. chodati, grew on starch waste material together with Candida utilis which grew at the expense of sugars solubilized by the amylase-secreting species (26). More recently, Tveit (27) in Sweden has adapted this system to deal with the wastes obtained from peeling potatoes. In practice, urea is supplied as a nitrogen source, and the fermentor is inoculated with equal numbers of each species. After incubation for 24 hours, the cell population consists typically of 1425 x 10^6 C. utilis cells/ml and 175 x 10^6 E. fibuliger cells/ml. It has been estimated that by next year (1970), 50% of all potatoes consumed as food will be in some processed form (28). Since the process of peeling potatoes entails 20-30% loss by weight, it is obviously necessary that an efficient and constructive solution to the waste disposal problem must be employed.

As a by-product of the poultry industry, vast quantities of chicken feathers are produced annually in the United States and in several other countries. Feathers consist largely of keratinaceous protein. As the feathers are largely wasted - perhaps the largest waste of protein from any single source - they present a waste-disposal problem. As an outgrowth of our studies on keratin degradation, a large batch of chopped feathers was added to a mineral salts medium in a 1000-liter fermentor. After autoclaving, the fermentor was inoculated with a strain of Streptomyces fradiae. After incubation for 36 hours, the feathers were solubilized completely. The broth then was filtered and spray dried; the product was a light colored powder (ca. 60% yield) consisting principally of water-soluble peptides. It proved satisfactory in chick diets (supplied at ca. 25% of the total protein in the diet). The powder also proved suitable for chemical use as a cheap source of peptides. It could be reacted with fatty acid chlorides to form non-irritating, non-toxic surfactants (29).

Another solid waste disposal problem created by modern food technology results from the fondness of many people for orange juice. At present, the only use for the vast tonnages of orange peels is cattle fodder. For sometime we have been interested in a plant pathogen, Sclerotium rolfsii, that is able to solubilize a wide variety of plant materials. When inoculated into a suspension of orange peels in tap water, the fungus grew rapidly and the peels were solubilized within 36 hours (4). This fungus produces a number of worthwhile products and we are actively pursuing this line of investigation.

From the foregoing examples of microbial utilization of organic wastes, and from the principles in Table 2 on the stages involved in microbial attack on an insoluble substrate, the message emerges clearly -- keep waste organic matter out of the aquatic environment.

Let us consider what happens when certain organic wastes are not subject to microbial utilization, but are dumped into rivers. Saccharomyces fragilis is not commonly distributed in nature and, in any event, could not compete with enteric bacteria for utilization of lactose. Therefore, if large quantities of whey are dumped into a river, the growth of enteric bacteria present through fecal contamination is encouraged greatly. It is also worth noting that lactose positive enteric pathogens are being isolated with increasing frequency. If sulfite pulp liquors are dumped into streams, a variety of bacteria including Escherichia coli, Aerobacter aerogenes, and species of Pseudomonas utilize the pentose components of the waste. The two examples just given illustrate the fact that dumping concentrated, metabolizable organic wastes into an aquatic environment serves to increase the frequently undesirable microbial population.

The fate of naturally-occurring, insoluble polymeric substances deposited in an aquatic environment is varied. Lignin is decomposed slowly by aerobic higher basidiomycetous fungi, normally not found in water, and accumulates under anaerobic conditions. In general, the insoluble polymeric materials, aside from cellulose, are decomposed slowly in an aquatic environment. The microorganisms responsible for their degradation are generally soil borne and are not present in any large number in water. Secondly, the soluble product(s) arising from the initial attack brought about at the microbial-polymer interface (Table 2) are largely washed away and the requisite hydrolytic enzymes are not induced.

IV. DISPOSABLE SYNTHETIC POLYMERS

At the Third Rudolfs Research Conference, Kallio and McKenna (30) discussed the effect of hydrocarbon structure on bacterial utilization of alkanes. Highly branched hydrocarbons were especially resistant to microbial attack, and substances containing terminal neopentyl groups were almost completely resistant. Such studies (30, 31) indicated that resistance of alkyl benzene sulfonate detergents to microbial degradation resided in the alkyl chain. Detergents with straight, unsubstituted alkyl groups proved to be biodegradable.

Although the total output of high polymer synthetics by the chemical industry of the world may be small in comparison with the total synthesis of polymers by plants on land and in the sea, nevertheless vast tonnages of synthetic polymers are intended for disposal each year. Nylon, polyethylene, polypropylene, and most other synthetic polymers are not subject to microbial degradation and accumulate when disposed as waste. It is possible that a strain of microorganism may evolve

NYLON 66

$$\left[-NH-(CH_2)_6-NH-CO-(CH_2)_4-CO- \right]_n$$

FIG. 1

Unit structure of Nylon 66.

PENICILLIN

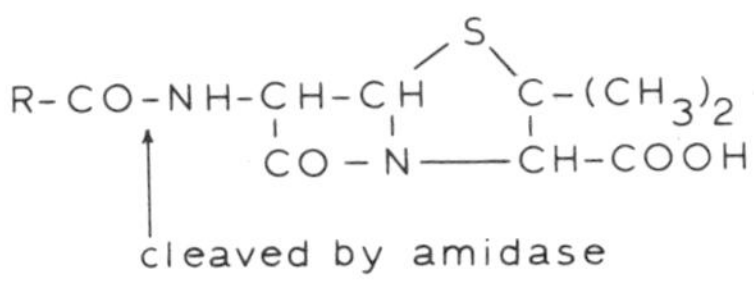

R-CO-NH-CH-CH(-S-)C-$(CH_3)_2$
CO - N —— CH-COOH

cleaved by amidase

FIG. 2

Cleavage of the penicillin molecule by amidase.

that can handle such materials. But we have only to consider the millions of years that have elapsed after the appearance of lignin before a group of fungi evolved capable of attacking this polymer - years during which our fossil reserves of coal developed. Several years ago this author drew attention to "the prospect of the world being covered at some future date with a Saran wrapping" (32). The use of disposable plastics has increased greatly in recent years and is a growing cause of concern. To make matters worse, we don't have the least notion of "the ground rules" (to use a favorite phrase of Kallio, 30) that govern susceptibility or resistance of a synthetic polymer to microbial attack. For example, the resistance of nylon to microbial attack is inexplicable at present. It is a linear polyamide without substituents (Fig. 1). Although many bacteria can elaborate amidases, apparently none can attack nylon. At the opposite extreme, the amide linkage in penicillin - a most improbable molecule that is a creation of fermentation technology and that probably is never produced under natural conditions - is hydrolyzed readily by amidases elaborated by a variety of microorganisms (Fig. 2). Perhaps, if nylon had some substituents, it might prove susceptible to microbial attack.

In the same vein, a tremendous number of microorganisms can utilize hydrocarbons as a source of carbon for growth. Why, then, is polyethylene resistant? The author is sure that we don't want to encourage the development of an extraterrestrial "Andromeda Strain" (33) to solve these waste disposal problems, but the time has surely arrived when the polymer chemist and microbiologist must cooperate to achieve disposable polymers that have the desired physical properties and yet are biodegradable. This end was achieved with synthetic detergents, and may be feasible with synthetic polymers.

REFERENCES

1. J. J. Noval and W. J. Nickerson, J. Bacteriol., 77, 251 (1959).

2. W. J. Nickerson and J. J. Noval, U. S. Pat. 2,988, 487 (1961).

3. W. J. Nickerson and G. Dozsa, Appl. Microbiol., in press.

4. M. Faber, G. W. Elmer, and W. J. Nickerson, Appl. Microbiol., in press.

5. K. Selby and C. C. Maitland, Biochem. J., 104, 716 (1967).

6. M. Mandels and E. T. Reese, Dev. Indust. Microbiol., 5, 5 (1964).

7. B. Norkrans, Adv. Appl. Microbiol., 9, 91 (1967).

8. L. R. Berger and D. M. Reynolds, Biochim. Biophys. Acta, 29, 522 (1959).

9. C. Jeuniaux, Arch. Intern. Physiol. Biochim., 65, 135 (1957).

10. J. J. Noval and W. J. Nickerson, in Symposium on Sulfur in Proteins, (R. Benesch, ed.), Academic Press, New York, 1959, p. 55.

11. W. J. Nickerson and G. Falcone, Science, 124, 318 (1956).

12. W. J. Nickerson and G. Falcone, in Symposium on Sulfur in Proteins, (R. Benesch, ed.), Academic Press, New York, 1959, p. 409.

13. M. D. Hatch and J. F. Turner, Biochem. J., 76, 556 (1960).

14. C. M. Brown and J. S. Hough, Nature, 211, 201 (1966).

15. P. Reichard, The Biosynthesis of Deoxyribose, John Wiley, New York, 1968, p. 27.

16. W. J. Nickerson, J. J. Noval, and R. S. Robison, Biochim. Biophys. Acta, 77, 73 (1963).

17. W. J. Nickerson and S. C. Durand, Biochim. Biophys. Acta, 77, 87 (1963).

18. M. Mandels, F. W. Parrish, and E. T. Reese, J. Bacteriol., 83, 400 (1962).

19. M. Mandels and E. T. Reese, J. Bacteriol., 73, 269 (1957).

20. G. J. Walker and P. M. Hope, Biochem. J., 86, 452 (1963).

21. A. D. McLaren, Enzymologia, 26, 237 (1963).

22. G. C. Inskeep, A. J. Wiley, J. M. Holderby, and L. P. Highes, Ind. Eng. Chem., 43, 1702 (1951).

23. A. J. Wiley, in Industrial Fermentations, (L. A. Underkofler and R. J. Hickey, eds.), Chemical Publishing Co., New York, 1954, Vol. 1, p. 307.

24. A. E. Wasserman, J. Dairy Sci., 44, 379 (1961).

25. M. E. Powell and K. Robe, Food Proc., 25 (2), 80, 95 (1964).

26. L. J. Wickerham and C. C. Kuehner, U. S. Pat. 2, 764, 487 (1956).

27. M. T. Tveit, U. S. Pat. 3, 105, 799 (1963).

28. I. B. Douglass, Purdue Univ. Eng. Bull. Ext. Ser., 106, 99 (1960).

29. W. J. Nickerson, in Fermentation Advances, (D. Perlman, ed.), Academic Press, New York, 1969, p. 64.

30. E. J. McKenna and R. E. Kallio, in Principles and Applications in Aquatic Microbiology, (H. Heukelekian and N. C. Dondero, eds.), John Wiley, New York, 1964, p. 1.

31. E. J. McKenna, Effect of Hydrocarbon Structure on Mechanisms of Microbial Alkane Metabolism, Ph. D. Thesis, University of Iowa, 1966.

32. W. J. Nickerson, Ind. Eng. Chem., 48, 1411 (1956).

33. M. Crichton, The Andromeda Strain, Alfred A. Kropf, New York, 1969.

AUTHOR INDEX

Numbers in parentheses are reference numbers and indicate that an author's work is referred to although his name is not cited in the text. Underlined numbers give the page on which the complete reference is listed.

D

Q

R

S

Y

Z

SUBJECT INDEX

A

B

C

U

W

Z